DNA Replication Controls: Volume 1

Special Issue Editor
Eishi Noguchi

MDPI • Basel • Beijing • Wuhan • Barcelona • Belgrade

MDPI

Special Issue Editor
Eishi Noguchi
Drexel University College of Medicine
USA

Editorial Office
MDPI AG
St. Alban-Anlage 66
Basel, Switzerland

This edition is a reprint of the Special Issue published online in the open access journal *Genes* (ISSN 2073-4425) from 2016–2017 (available at: http://www.mdpi.com/journal/genes/special_issues/dna_replication_controls).

For citation purposes, cite each article independently as indicated on the article page online and as indicated below:

Author 1; Author 2. Article title. *Journal Name* **Year**, *Article number*, page range.

First Edition 2017

Volume 1
ISBN 978-3-03842-568-7 (Pbk)
ISBN 978-3-03842-569-0 (PDF)

Volume 2
ISBN 978-3-03842-572-4 (Pbk)
ISBN 978-3-03842-573-1 (PDF)

Volume I – Volume II
ISBN 978-3-03842-574-8 (Pbk)
ISBN 978-3-03842-613-4 (PDF)

Table of Contents

About the Special Issue Editor

Eishi Noguchi was born in Osaka, Japan. After graduating from Shudo High school in Hiroshima, he studied at Kyushu University in Fukuoka, where he earned his Ph.D. in 1997 in Molecular Biology, under the guidance of Professor Takeharu Nishimoto. His graduate work involved understanding cell cycle control mechanisms, and he obtained training in mammalian cell biology and budding yeast genetics. Dr. Noguchi moved to the U.S. in 2000 to perform his postdoctoral studies with Professor Paul Russell at The Scripps Research Institute in La Jolla, California, where he initiated research projects concerning Cell Cycle Checkpoints and DNA replication in a fission yeast model. During this time, he identified the replication fork protection complex (FPC) that is responsible for stabilizing the replication fork in a configuration that is recognized by checkpoint sensors. In 2004, he took the position of Assistant Professor and started his research group within the Department of Biochemistry and Molecular Biology at Drexel University College of Medicine, Philadelphia, Pennsylvania. The goal of the research in the Noguchi laboratory is to understand the molecular mechanisms that ensure accurate duplication of chromosomal DNA. Dr. Noguchi is currently an Associate Professor and Graduate Program Director at Drexel University College of Medicine. The major directions of his research group include mechanisms of DNA replication at difficult-to-replicate genomic regions, alcohol-mediated genomic instability, and lifespan regulation via genome maintenance mechanisms.

Preface to "DNA Replication Controls"

The conditions for DNA replication are not ideal owing to endogenous and exogenous replication stresses that lead to arrest of the replication fork. Arrested forks are among the most serious threats to genomic integrity because they can break or rearrange, leading to genomic instability which is a hallmark of cancers and aging-related disorders. Thus, it is important to understand the cellular programs that preserve genomic integrity during DNA replication. Indeed, the most common cancer therapies use agents that block DNA replication, or cause DNA damage, during replication. Therefore, without a precise understanding of the DNA replication program, development of anticancer therapeutics is limited.

This volume, DNA Replication Controls, consists of a series of new reviews and original research articles, and provides a comprehensive guide to theoretical advancements in the field of DNA replication research in both prokaryotic and eukaryotic systems. The topics include DNA polymerases and helicases; replication initiation; replication timing; replication-associated DNA repair; and replication of difficult-to-replicate genomic regions, including telomeres, centromeres and highly-transcribed regions. We will also provide recent advancements in studies of cellular processes that are coordinated with DNA replication and how defects in the DNA replication program can result in genetic disorders, including cancer. We believe that this volume will be an important resource for a wide variety of audiences, including junior graduate students and established investigators who are interested in DNA replication and genome maintenance mechanisms.

Eishi Noguchi
Special Issue Editor

Review

Regulation and Function of Cdt1; A Key Factor in Cell Proliferation and Genome Stability

Pedro N. Pozo [1] and Jeanette Gowen Cook [1,2,*]

[1] Curriculum in Genetics and Molecular Biology, The University of North Carolina at Chapel Hill, Chapel Hill, NC 27599, USA; ppozo@email.unc.edu

[2] Department of Biochemistry and Biophysics, The University of North Carolina at Chapel Hill, Chapel Hill, NC 27599, USA

* Correspondence: jean_cook@med.unc.edu; Tel.: +1-919-843-3867

Academic Editor: Eishi Noguchi

Received: 6 November 2016; Accepted: 14 December 2016; Published: 22 December 2016

Abstract: Successful cell proliferation requires efficient and precise genome duplication followed by accurate chromosome segregation. The Cdc10-dependent transcript 1 protein (Cdt1) is required for the first step in DNA replication, and in human cells Cdt1 is also required during mitosis. Tight cell cycle controls over Cdt1 abundance and activity are critical to normal development and genome stability. We review here recent advances in elucidating Cdt1 molecular functions in both origin licensing and kinetochore–microtubule attachment, and we describe the current understanding of human Cdt1 regulation.

Keywords: cell cycle; DNA replication; genome instability; pre-RC; re-replication; ubiquitylation; cyclin-dependent kinase; geminin; Origin Recognition Complex (ORC); Minichromosome Maintenance (MCM)

1. Introduction

Origin licensing, the loading of replicative DNA helicases onto origin DNA, is the first committed step of DNA replication and is essential for cell proliferation. Numerous control mechanisms in eukaryotic cells regulate both origin licensing and subsequent replication initiation to ensure complete and precise genome duplication [1–6]. Perturbations to origin licensing and replication initiation can result in cell death or in genome instability leading to oncogenesis [1,7,8]. For these reasons, origin licensing control is intimately coordinated with mechanisms that govern cell cycle progression. In this review, we focus specifically on current understanding of the regulation and function of the Cdt1 protein (Cdc10-dependent transcript 1). Unlike other essential licensing proteins, Cdt1 lacks enzymatic activity and shares little resemblance to any other protein of known molecular function, yet it is essential for origin licensing in all eukaryotes tested. In mammalian cells, small changes in Cdt1 control can lead to catastrophic consequences for genome stability, suggesting that Cdt1 regulation is unusually important. Moreover, the recent finding that Cdt1 has a second essential role in the cell cycle during mitosis underscores the importance of fully understanding its function [9]. These features make Cdt1 unique among the core licensing factors and warrant a thorough up-to-date synthesis of the current knowledge about Cdt1 function, structure, regulation, and how its dysregulation contributes to disease. In this review, we focus on understanding mammalian Cdt1, and we are informed by key mechanistic insights gleaned from model experimental systems including *Saccharomyces cerevisiae*, *Schizosaccharomyces pombe*, *Xenopus laevis*, *Drosophila melanogaster*, and cultured mammalian cells.

2. Cdt1 Function

Mammalian cells replicate billions of DNA base pairs with high fidelity and then accurately segregate duplicated genomes to daughter cells each cell cycle. These incredible feats are under strict regulation and are tightly linked to cell cycle progression. Cdt1 is required for both DNA replication and chromosome segregation, and although these functions are not yet fully elucidated, recent advances inspire increasingly detailed models of Cdt1's role.

2.1. Origin Licensing

The first step in eukaryotic DNA replication occurs in G1 and is the sequential loading of replication factors at numerous sites in the genome, known as origins of replication. Origins are sites where DNA replication initiates during S phase. A typical eukaryotic cell contains between 400 (yeasts) and as many as >350,000 (human) potential origins [10–12]. Broad distribution of origins on chromosomes ensures complete genome duplication within the time allotted for S phase. Replication factor loading at origins, known as origin licensing, was first described nearly three decades ago using *X. laevis* egg extracts to determine what factors can induce unscheduled DNA re-replication in vitro [13]. The study concluded that DNA replication requires the recruitment of a "Licensing Factor" to DNA during mitosis, thereby setting the stage for DNA synthesis in the subsequent S phase. Furthermore, DNA that was replicated cannot replicate again until the following cell cycle because of the inability of the factor(s) to access chromatin. These results provided the first model for the control of DNA replication where a Licensing Factor binds DNA, is required for the initiation of DNA replication, and becomes deactivated until the following mitosis [13]. Since then, numerous studies have provided experimental support for the now-established "replication licensing system" to control precise genome duplication once-and-only-once per cell cycle [2,14]. The core licensing factors have since been identified, and they assemble into a chromatin-bound macromolecular complex, known as the pre-replication complex (pre-RC). Pre-RC assembly is a highly cell cycle-regulated process governed in part by the cyclical fluctuation of cyclins and the activity of the Cyclin-Dependent Kinases (CDKs) they activate.

The assembly of pre-RCs occurs during a period of low CDK activity in late mitosis and G1 phase. Biochemical and genetic studies in yeast, *Xenopus*, and mammalian cells identified the minimal licensing factors essential for pre-RC assembly [15–19]. These factors are Origin Recognition Complex (ORC), Cell Division Cycle 6 (Cdc6), Minichromosome Maintenance (MCM), and Cdt1. Eukaryotic ORC is a heterohexamer composed of six distinct subunits, Orc1 through Orc6. ORC is the only licensing component that directly binds origin DNA, and it is required for the nucleation of the pre-RC. Cdc6 is a monomeric protein that is recruited to DNA by protein–protein interactions with ORC [16,20,21]. Cdc6 and the Orc1–Orc5 subunits are members of the AAA+ family of ATPases which are prevalent in many DNA metabolic processes [22–24]. The MCM complex is the core component of the replicative DNA helicase, and its successful loading onto origin DNA is synonymous with origin licensing. Like ORC, MCM is a heterohexamer composed of six distinct subunits, Mcm2 through Mcm7, which are also AAA+ proteins. In this review, we will specifically discuss Cdt1 regulation and function; for in-depth reviews of ORC, Cdc6, and MCM, the reader is referred to excellent contributions by others in the field [14,23–25].

Our understanding of the molecular events in origin licensing (illustrated in Figure 1) comes primarily from pioneering work using both *X. laevis* egg extracts and purified budding yeast licensing proteins [2,5]. Importantly, the strong conservation of origin licensing proteins throughout eukaryotic evolution, combined with many corroborating studies in mammalian cells, gives confidence that licensing functions elucidated in model systems are applicable to human cells; though aspects of their regulation vary by species. Pre-RC assembly begins with ORC loading onto presumptive origin DNA. Interestingly, ORC DNA binding—particularly in metazoan genomes—is largely independent of DNA sequence, but is highly influenced by local chromatin characteristics [26–28]. ORC recruits the Cdc6 protein to chromatin to await the arrival of Cdt1 bound to the MCM complex to form a pre-RC [2,5].

In a process not yet fully understood [29,30], the concerted action of ORC, Cdc6 and Cdt1 results in topological loading of an MCM heterohexamer onto DNA with double-stranded DNA passing through the MCM central channel [18,19]. Cdc6 and then Cdt1 are released, followed by a second round of Cdc6 and Cdt1-MCM recruitment [31]. The second MCM complex is loaded such that the MCM N-termini face one another to create double hexameric rings. This arrangement sets each MCM complex in the correct orientation to establish bidirectional forks upon origin firing [32,33]. Only the correct loading of MCM double hexamers renders a locus competent for subsequent replication initiation, or "firing", during S phase. MCM loaded in G1 is not active as a helicase, and origin DNA is thought to remain double-stranded until origin firing. Origin firing requires phosphorylation events from CDKs and a replication-specific kinase, Dbf4-dependent kinase (DDK). These kinases promote the recruitment of additional essential helicase components, Cdc45 and GINS, to activate DNA unwinding [34–36].

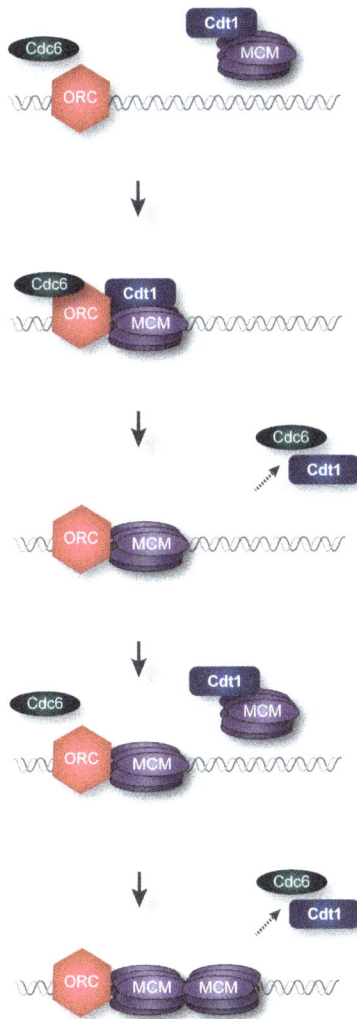

Figure 1. Origin Licensing. Minichromosome Maintenance (MCM) hexamers are loaded by Cdt1, Cdc6, and Origin Recognition Complex (ORC) at presumptive chromosomal origins during G1 phase.

Origin licensing can begin as early as telophase, as soon as nuclear envelopes have formed around the segregated mitotic chromosomes, though it is not clear if licensing begins this early in all species or cell types [37–39]. Licensing continues throughout G1 and ceases at the G1/S phase transition. Somewhat surprisingly, eukaryotic cells load many more MCM double hexamers than the number of DNA-bound ORCs [40]. At least 10-fold more origins can be licensed than are strictly required for complete replication under normal circumstances, though the degree of origin licensing likely varies among cells, tissues, and species [41–43]. In vitro, loaded MCM double hexamers can slide along DNA away from ORC, leaving space near ORC for another round of MCM loading [18,19], and recent results suggest that MCMs may also slide in vivo [42,44]. In a typical S phase, some MCM complexes that had been loaded in G1 are activated as part of the regular replication program, whereas others initiate replication in response to nearby stalled or damaged replication forks to ensure replication completion. Origins that are only utilized under the latter conditions of replication stress are termed "dormant" origins, and they safeguard the genome against under-replication. [45–47].

Notably, Cdt1 is essential for MCM loading in all eukaryotes in which it has been tested, but its precise molecular function in origin licensing is not fully clear [48–50]. Cdt1 interacts directly with the MCM complex in solution and with both ORC and Cdc6 [51–55]. In the absence of Cdt1, MCM complexes are never recruited to DNA [48,56,57]. In that regard, one likely role for Cdt1 in licensing is as a molecular bridge or "courier" to deliver soluble MCM complexes to DNA-bound ORC/Cdc6. In support of that model, recent single molecule studies using purified yeast licensing proteins discovered that Cdt1 is rapidly released upon successful loading of each MCM complex [31]. By following individual labeled proteins, Ticau et al. showed Cdt1 and Cdc6 release between the two rounds of MCM loading. This rapid shuttling between the bound and soluble states for both Cdt1 and Cdc6 suggests that each molecule could participate in many origin licensing events. Perhaps for this reason, the levels of both Cdc6 and Cdt1 are highly regulated during the cell cycle to prevent inappropriate origin licensing.

The MCM complex is a hexameric ring even in solution before it is loaded [18,36,58]. MCM loading is therefore not a process of assembling the heterohexamers on DNA from their component subunits, but rather, loading pre-assembled hexamers onto DNA. DNA passes through a side "gate" between the Mcm2 and Mcm5 subunits, and much speculation currently swirls around the mechanism and dynamics of MCM gate opening and closing [59,60]. Moreover, the MCM double hexamer central channels contain double-stranded DNA in G1 but the active MCM helicase at replication forks encircles single-stranded DNA and displaces the second strand [35,61,62]. At least in vitro, yeast Cdt1 is not released from the complex until MCM is successfully loaded [31]. Its persistence during the actual loading reaction suggests that Cdt1 does more than simply hand MCM off to ORC and Cdc6. Cdt1 may be required to maintain MCM in the proper orientation or conformation for successful DNA loading. If so, then how Cdt1—or ORC and Cdc6 for that matter—load the two MCM complexes in opposite orientations remains to be discovered [14,30].

2.2. De-Regulated Origin Licensing

The requirement that normal DNA replication produce exactly one copy of each chromosome puts important constraints on origin utilization. Specifically, each origin that fires, must fire no more than once per cell cycle. Origin re-firing results from re-licensing DNA that has already been duplicated. A second round of initiation from the re-licensed origins leads to duplicating sequences more than once, a phenomenon known as re-replication. Interestingly, re-replication is induced in the final cell cycles of some tissues to increase DNA copy number, most notably in *D. melanogaster*, but such cells are not normally destined to divide again. Re-replication is distinct from scheduled genome re-duplication which results from skipping cytokinesis; re-duplication typically produces quantile increases in ploidy whereas developmentally programmed re-replication targets only some loci [63–65]. In contrast to developmentally programmed re-replication, unscheduled re-replication is an aberrant phenomenon associated with genome instability [3,6]. Indeed, re-replication can be the initiating event for gene amplification [66], a frequent observation in cancer cells. Partial re-replication can be experimentally induced by deregulating MCM loading factors, and in human cells, re-replicated sequences are detectable essentially randomly throughout the genome [67]. Re-replication is typically associated with markers of replication stress and evidence of DNA damage response pathway activation [68–71].

To avoid re-replication, all origin licensing activity ends once S phase begins. There is no known means to directly reverse inappropriate origin licensing, so a network of overlapping inhibitory mechanisms is needed to prevent all origin licensing outside of G1 phase. These licensing controls target each member of the pre-RC from the onset of S phase through mitosis. Mammalian Cdt1 is inhibited by at least four distinct pathways, suggesting that it is among the most important to inhibit; we discuss each of these mechanisms in more detail in Section 4. Many licensing factors are inactivated by phosphorylation via the same CDK activity that triggers origin firing (in human cells primarily cyclin A/Cdk2). Interestingly, the outcomes of these phosphorylations may vary depending on the licensing factor being targeted and in which organism, though the end result is always to inhibit origin relicensing. For example, in *S. cerevisiae*, Cdc6 phosphorylation by CDK targets it for ubiquitin-mediated proteolysis, whereas phosphorylation of human and *Xenopus* Cdc6 induces nuclear export [20,72–74]. On the other hand, in *S. cerevisiae*, MCM and Cdt1 are subjected to CDK-mediated nuclear export [56,75]. In *S. cerevisiae*, ORC subunits are inhibited by CDK-dependent phosphorylation by disrupting their ATPase activity [76] and blocking interaction with Cdt1 [77], whereas in human and *Xenopus*, CDK-dependent ORC phosphorylation induces release from origins and/or degradation of the Orc1 subunit [78–80]. Regardless of the species-specific details, the aggregate result is inhibiting pre-RC assembly by neutralizing interactions or triggering licensing factor degradation.

Incomplete origin licensing in G1 can also be a source of genome instability. In untransformed human cells, significantly slowing origin licensing induces a delay in S phase onset by delaying the activation of Cdk2 [81–83]. This "origin licensing checkpoint" requires p53, meaning that p53-deficient cells can enter S phase with severely underlicensed chromosomes which renders them susceptible to S phase failure [81–83]. Despite extensive documentation of the licensing checkpoint phenomenon in several labs and in multiple cell lines, precisely how licensed or unlicensed DNA is detected to affect Cdk2 activity is still unclear. Moreover, "sufficient" origin licensing is not simply a matter of the total number of loaded MCM hexamers per genome since their distribution is also critical. A recent study by Moreno et al. found that moderate licensing inhibition that does not cause a cell cycle delay, nonetheless increases the likelihood that regions of unreplicated DNA persist through mitosis [84]. Thus, Cdt1 activity and origin licensing must be efficiently blocked in S phase and G2 to prevent re-replication but must be fully induced in G1 to ensure sufficient origin licensing and complete genome duplication.

2.3. Cdt1-Associated Chromatin Modifiers

Licensing factors must have local access to origin DNA to assemble and load MCM helicases. The chromatin environment at origins thus has a large impact on origin licensing. Post-translational histone modifications, such as methylation and acetylation, can greatly affect DNA accessibility which may facilitate ORC binding, MCM loading, and/or origin firing. In addition, at *S. cerevisiae* origins which have been mapped with high resolution, nucleosome positioning also plays a role in determining ORC localization and activity (reviewed in [10,27,85]). In the majority of eukaryotic genomes, DNA sequence is a minor determinant of origin location. The model that has emerged is that ORC is recruited to DNA not by a specific nucleotide sequence, but rather by aspects of local chromatin structure and DNA accessibility. Some evidence supporting this model is the presence of a BAH (Bromo Adjacent Homology) domain in Orc1, the largest subunit of ORC. The BAH domain specifically recognizes histone post-translational modifications (PTMs) enriched at replication origins, and is required for proper ORC DNA loading [86,87].

Once ORC has bound, the local chromatin environment may require additional modifications to permit efficient origin licensing. Several histone-modifying enzymes associate with licensing components and are predicted to modify nucleosomes to promote DNA accessibility; some of these enzymes have been identified as Cdt1 partners. One such chromatin modifier is histone acetyltransferase bound to Orc1 (Hbo1), which as its name implies, was first discovered as an Orc1-binding protein and later shown to bind the Mcm2 subunit of MCM, and Cdt1 [88–90]. Hbo1 is highly conserved, and orthologs in *D. melanogaster* and *S. cerevisiae* have also been linked to DNA replication [91,92]. In human cells, Hbo1 is responsible for the bulk of histone H4 acetylation genome-wide [93]. Since histone H4 acetylation generally correlates with active chromatin and accessible DNA, increased local histone acetylation could promote origin licensing. In addition, Hbo1 was specifically detected at several known human replication origins during G1 coincident with Cdt1 origin association [90]. Further studies found that Cdt1 promoted chromatin openness in association with Hbo1 during G1, likely increasing local chromatin accessibility and facilitating MCM loading [94]. In addition to Hbo1, early proteomic screens for Cdt1-interacting proteins discovered the GRWD1 protein (glutamate-rich WD40 repeat containing 1), a histone binding protein [95]. Follow up studies suggested that GRWD1 regulates chromatin openness during MCM loading at replication origins [95] and may cooperate with a chromatin remodeler, SNF2H [96]. On the other hand, during S phase and G2 Cdt1 may contribute to inhibiting origin licensing by recruiting the HDAC11 histone deacetylase. Local histone deacetylation would presumably reduce chromatin accessibility and inhibit origin relicensing [94,97]. Interestingly, association of the inhibitor protein geminin with Cdt1 during S phase enhanced the recruitment of HDAC11 to origins to further inhibit origin licensing [94].

2.4. Cdt1 in Chromosome Segregation

Surprisingly, human Cdt1 is required not only for origin licensing but also for mitosis. As a consequence, asynchronously-growing cells, depleted of Cdt1, accumulate in both G1 phase and G2 phase because they can neither license origins, nor progress through the metaphase-to-anaphase transition. This essential mitotic function was first discovered in a screen for Cdt1-interacting proteins that identified human Hec1 (Highly Expressed in Cancer 1), a component of the NDC80 kinetochore–microtubule attachment complex [9]. Hec1 is conserved from yeast to mammals, but the mitotic Cdt1 function is not evident in either budding or fission yeast [57,98]; more work is required to determine if Cdt1 has mitotic functions in invertebrates such as *D. melanogaster* or *Caenorhabditis. elegans* or in non-mammalian vertebrates such as *X. laevis*.

A fraction of human Cdt1 molecules localize to kinetochores in mitosis, and this localization requires Hec1; Hec1 localization is unaffected by Cdt1 depletion. Cdt1 interacts with and is recruited to kinetochores via a unique "loop" domain in Hec1 that interrupts an otherwise long coiled-coil central span. Both depletion of Cdt1 prior to mitosis or mutationally altering the Hec1 loop domain to block Cdt1 binding and recruitment resulted in prometaphase arrest with an unsatisfied spindle assembly checkpoint [9]. Importantly, the mitotic defect in Cdt1-depleted cells can be separated from potential indirect effects of incomplete DNA replication by depleting Cdt1 after origin licensing is complete and S phase has already begun [9].

It is not yet clear precisely how Cdt1 promotes stable kinetochore–microtubule attachments since it is not required for the localization of any other kinetochore proteins tested thus far. One clue to its function came from analysis of the conformation of the NDC80 complex in vivo using super-resolution microscopy. The structure of the NDC80 complex (Hec1/Nuf2/Spc24/Spc25) indicates that the loop region of Hec1 where Cdt1 binds is a point of flexibility in an otherwise long and rigid coiled-coil domain. Prior work by Wang et al. supported the notion that the loop region corresponds to a hinge or joint in the complex [99]. The N-terminal domains of Hec1 and Nuf2 directly contact kinetochore microtubules, whereas the Spc24 and Spc25 subunits connect the complex to other kinetochore proteins [100,101]. In prometaphase, prior to attachment, the two ends of the NDC80 complex are relatively close together, whereas at stably-attached kinetochores in metaphase, the two ends of the complex are considerably further apart [101]. Mutation of the loop domain or depletion of Cdt1 prevented this extended NDC80 conformation [9]. Thus, Cdt1 supports a microtubule-dependent conformational extension in its partner, the NDC80 complex, by interaction with the major point of flexibility conferred by the loop region of Hec1.

Many important questions about Cdt1 mitotic function remain: what other (if any) microtubule-associated or kinetochore partners bind Cdt1? The Hec1-interacting domain on Cdt1 is not yet known, but identifying this region is a first step towards generating separation-of-function alleles that are impaired for only origin licensing or only kinetochore–microtubule attachment. How, precisely, does Cdt1 affect the conformation of NDC80? Moreover, as described below (see Section 4.4), Cdt1 is heavily phosphorylated during G2 phase and mitosis. What role does Cdt1 phosphorylation play in its intermolecular interactions and function at kinetochores? Clearly, much remains to be learned about this novel role for Cdt1 and how it relates to the more famous origin licensing function.

3. Cdt1 Structure

In most species, Cdt1 is a ~60–70 kDa protein; *S. pombe* Cdt1 is somewhat smaller at ~50 kDa whereas the *D. melanogaster* Cdt1 is ~82 kDa. (*D. melanogaster* Cdt1 is named "double-parked", abbreviated Dup, but nearly all other species use "Cdt1" as the protein and gene name). Although each subunit of ORC and MCM, Cdc6 and Cdt1 are conserved in all eukaryotic genomes examined, the degree of sequence conservation is lowest for Cdt1 compared to the other licensing proteins. Indeed, the low sequence similarity between human and *S. cerevisiae* Cdt1 coupled with the unusual history of metazoan Cdt1 being identified first, led to a brief period in the field when it was not clear if budding yeast had a Cdt1 ortholog. Focused sequence searches coupled with functional tests ultimately identified the yeast Cdt1 ortholog [57]. Unlike nearly all other licensing components which are homologous to AAA+ ATPases, Cdt1 is not an enzyme, and the Cdt1 protein sequence bears little similarity to other proteins of known molecular activity. Although the Cdt1 sequence gives little insight into its function, some information about interacting regions, post-translational modifications, and domain structures is available which we describe here.

3.1. Functional Domains

Multispecies Cdt1 protein sequence alignments reveal regions that are relatively well-conserved and regions which share considerably less conservation. Not surprisingly, the regions of low conservation are particularly prominent in comparisons of mammalian and fungal Cdt1 species. Using human Cdt1 as a reference, Figure 2 includes pairwise sequence comparisons between human Cdt1 and Cdt1 sequences from several model organisms in four Cdt1 domains, the N-terminus (amino acids [aa] 1–166), the central domain (aa 167–374), a short "linker" region (aa 375–406), and the MCM binding domain (aa 407–546). Sites of protein–protein interactions and phosphorylations are also marked. The N-terminal sequences of both model yeast Cdt1 sequences are generally quite short and they bear little resemblance relative to their metazoan counterparts. On the other hand, sections of higher relative homology suggest regions important for functions that are conserved in all species, such as interaction with other origin licensing components.

Traditional truncation and mutagenesis approaches identified Cdt1 domains required for protein interactions and for specific aspects of origin licensing function [54,102,103]. The most comprehensive of these studies by Ferenbach et al. validated and/or delineated the MCM binding domain, geminin binding domain, and minimal licensing activity domain using recombinant fragments of *X. laevis* Cdt1 added to oocyte lysates. The shortest fragment that complemented Cdt1-depleted lysates for licensing corresponds to human Cdt1 aa 243–546 [54]. The finding that the N-terminal 242 amino acids (corresponding to human aa 1–170) are dispensable for licensing activity, plus the fact that this region is the least-well-conserved is consistent with the notion that the N-terminal region is the target of species-specific regulation rather than essential for Cdt1 function.

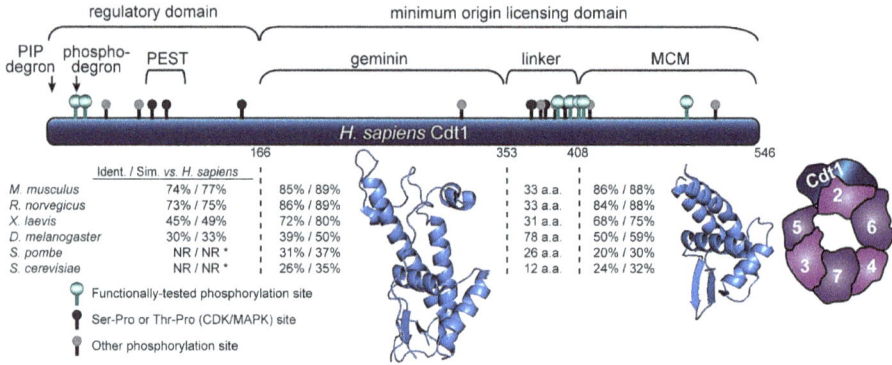

Figure 2. Human Cdt1 structure. Diagram of Cdt1 divided into four segments based on alignments and structural studies. Pairwise comparisons to the human sequence for representative eukaryotic Cdt1 orthologs within each segment are reported as % identity/% similarity; NR indicates regions in fungal sequences too short or dissimilar for comparison. Regions responsible for recognition by E3 ubiquitin ligases (degrons), a region enriched in proline, glutamic acid, serine, and threonines (PEST domain), geminin binding, MCM) binding, and a putative linker domain (enriched in phosphorylation sites) are marked. Phosphorylation sites in human Cdt1 that are conserved in at least one other vertebrate sequence are marked as ball-and-stick icons: green = Cyclin-Dependent Kinases (CDK)/Mitogen-Activated Protein Kinases (MAPK) sites validated by mutagenesis and functional studies, dark gray = putative CDK/MAPK sites (serine-proline or threonine-proline) identified by mass spectrometry [104], light gray = conserved sites detected by mass spectrometry distinct from the CDK/MAPK substrate consensus. Ribbon diagrams of the two segments for which structures have been determined are shown; central domain PDB 2WVR (human) and C-terminal domain PDB 3A4C (mouse) [105,106]. A diagram of the yeast MCM2-7 complex bound to full-length Cdt1 derived from tracing the single-particle analysis results from Sun et al. 2013 is also shown.

3.2. Crystal Structures/Cryo-EM Structures

Currently, no atomic structure for full-length Cdt1 from any species is available. One challenge for structure studies of Cdt1 is that both the N-terminal domain and part of the linker domain are predicted to be intrinsically disordered. Using two different prediction tools, the N-terminal 166 amino acids of human Cdt1 has a probability of disorder at each position greater than 65% [107,108]. The linker is relatively short, but it also contains a region of high predicted disorder. Trimming these regions to isolate the central domain or the C-terminal domain yielded fragments that were compatible with crystallography, and their exclusion from the structural studies is consistent with the notion that they are flexible. The atomic structure of the central domain was first solved for mouse Cdt1 (aa 172–368) in complex with the geminin inhibitor protein [105], and the corresponding human Cdt1 protein fragment (aa 166–353) was later crystallized [106]. A recent search of a database of protein structures for nearest neighbors to this central domain identified some similarity to winged-helix domains [109]. Otherwise, the central domain structure is relatively unique.

The C-terminal domain (human 408–546) interacts with the MCM complex. This isolated fragment can directly bind a C-terminal fragment from the Mcm6 subunit suggesting that this interaction is one of the direct contacts between Cdt1 and the MCM complex in vivo [110]. This protein fragment was characterized by both X-ray crystallography and Nuclear Magnetic Resonance as a winged helix domain [102,103,111]. Interestingly, this winged-helix shows some structural similarity to the central domain of Cdt1 itself [102]. Although winged helices are most well-known for roles in nucleic acid binding, the C-terminal Cdt1 winged-helix is unlikely to form stable interactions with DNA. Positions of key alpha helices are incompatible with DNA binding compared to winged-helix domains in canonical DNA binding proteins, and Cdt1 lacks charged patches that would stabilize DNA binding [102]. Moreover, Cdt1 chromatin association in cells requires ORC [48,77], and purified yeast Cdt1 does not bind origin DNA in the absence of ORC [19]. It is most likely therefore that the C-terminal Cdt1 winged-helix is of the type that mediates protein–protein rather than protein-nucleic acid interactions. In support of that model, mutational alteration of a subset of charged surface residues of the C-terminal domain impaired MCM binding in vitro [102,111], and several of the corresponding mutations to budding yeast Cdt1 impaired cell growth [102]. These biochemical data corroborated findings from a separate co-crystallographic study which demonstrated a direct Cdt1–Mcm6 interaction conferred by the Cdt1 C-terminal domain [103].

Although these studies provide important structural information, key aspects of Cdt1 structure are still not known. Yeast Cdt1 can directly bind the Orc6 subunit of ORC [55], but the Cdt1 domain responsible is not known nor are potential Cdt1 regions that bind other subunits of MCM. As-yet uncharacterized Cdt1 interactions with ORC and MCM are likely required for origin licensing and/or regulating Cdt1 function. In that regard, a recent paper described a novel and still uncharacterized "PEST" (rich in proline [P], glutamic acid [E], serine [S], and threonine [T]) domain in mouse Cdt1 (Figure 2) [112]. Cdt1 is abundant during G2 phase but is poorly associated with chromatin [38,112,113]. Truncating the PEST domain caused premature re-association of Cdt1 with chromatin during G2 and increased the likelihood of re-replication [112]. Given that Cdt1 chromatin binding requires ORC interaction [48], this PEST domain may indicate a region required for ORC binding.

Several studies using single particle electron microscopy coupled with labeling strategies have suggested how full-length budding yeast Cdt1 interacts with the MCM complex and in a licensing intermediate containing ORC, Cdc6, Cdt1 and MCM [29,32]. These models are consistent with the biochemical studies detecting Cdt1 in direct contact with Mcm6 [77]. In addition to this contact with Mcm6, Cdt1 appears to contact additional MCM subunits, especially extensive interaction with Mcm2 (Figure 2). This location is relatively close to the interface of Mcm2 with Mcm5 through which DNA passes during the loading reaction [59,60]. In this position, Cdt1 is well-placed to affect the conformation of the MCM complex during loading in ways that may stabilize either the open or closed MCM conformation.

4. Cdt1 Regulation

To properly license origins for DNA replication in G1 and block origin licensing from the onset of S phase through mitosis, multiple independent mechanisms control human Cdt1 abundance and function (illustrated in Figure 3). Although other licensing proteins are also under cell cycle control, Cdt1 is subject to the most extensive regulation in human cells, suggesting that it is perhaps the most important licensing factor to regulate in mammalian cells. The ultimate outcome is a collection of cell cycle-dependent regulatory mechanisms that allow Cdt1 to function efficiently in its origin licensing role during G1, prevent origin relicensing in S and G2 phase, and permit Cdt1 participation in kinetochore microtubule attachment during mitosis.

4.1. Transcriptional Control

The first Cdt1 ortholog was cloned 20 years ago in a screen for fission yeast transcripts that are upregulated at the G1 to S transition [98]. In fission yeast, the transcription factor driving Cdt1 expression is Cell Division Cycle 10 (Cdc10), which is responsible for transcriptional induction of many genes important for the G1 to S phase transition [114]. The analogous function in metazoans is the responsibility of the E2F family of transcription factors, though the protein sequences of Cdc10 and E2F themselves are unrelated [115,116]. The human *CDT1* gene has three putative E2F responsive elements in its promoter region, is activated by E2F with peak expression in late G1, and is inhibited by the Rb tumor suppressor [117]. Other studies have suggested that Cdt1 is also under the transcriptional control of the c-*Myc* proto-oncoprotein and the Gli1 component of the hedgehog signaling pathway [118,119]. Of note, Cdt1 protein abundance in proliferating cells peaks in G1 and G2 rather than S phase (Figure 3), but the transcriptional upregulation generally supports Cdt1 expression during proliferation. Aside from the documented regulation by E2F and possibly c-myc and Gli1, little else is known about how the production of Cdt1 is regulated. For instance, no evidence for alternative splicing, regulation by microRNAs or translational control has yet emerged, though such possibilities should be explored.

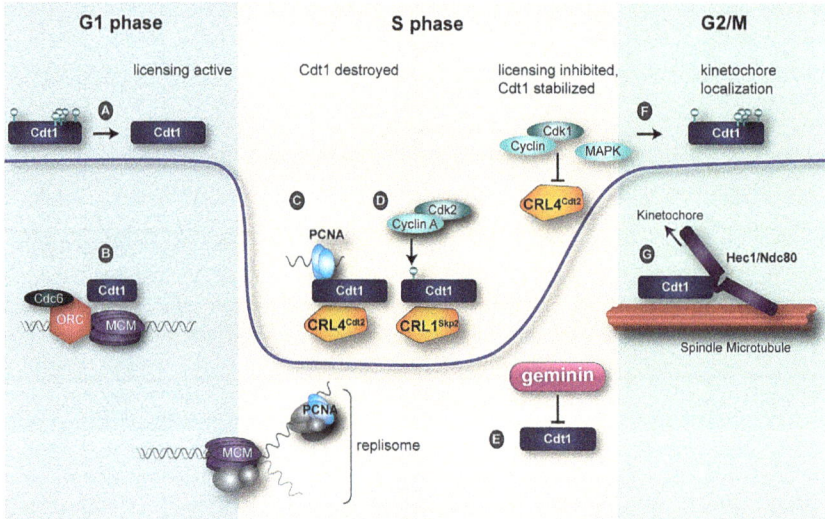

Figure 3. Human Cdt1 regulation during a single cell cycle. The blue line indicates relative Cdt1 protein abundance. (**A**). Cdt1 is dephosphorylated in early G1 by an unknown phosphatase; (**B**) Cdt1 participates with ORC and Cdc6 to load MCM hexamers onto DNA; (**C**) Proliferating Cell Nuclear Antigen (PCNA) loaded at DNA replication forks is bound by the Cdt1 PCNA-Interacting Protein (PIP) degron, and the complex is recognized for ubiquitylation and subsequent proteasome-mediated destruction by $CRL4^{Cdt2}$; (**D**) Cdt1 is phosphorylated at Thr29 by cyclin A/Cdk2 to create a phosphodegron recognized for ubiquitylation by $CRL1^{Skp2}$. The combined action of two E3 ubiquitin ligases drives Cdt1 degradation in S phase; (**E**) The geminin protein begins to accumulate in early S phase, and peaks in late S phase and G2. Geminin binding blocks Cdt1 origin licensing function; (**F**) During late S phase and G2, mitotic kinases—especially cyclin A/Cdk1 and the stress-activated MAP kinases p38 and c-Jun N-terminal Kinase (JNK)—phosphorylate Cdt1; Cdk1 also inactivates $CRL4^{Cdt2}$; (**G**) A subset of Cdt1 molecules is recruited to kinetochores in mitosis through interaction with the loop domain of Hec1. Cdt1 is required for stable kinetochore–microtubule attachment.

4.2. Ubiquitin Mediated Proteolysis in S Phase

A key aspect of re-replication control in metazoans is ubiquitin-mediated Cdt1 degradation during S phase. This regulation occurs in all eukaryotes except for budding yeast in which a Cdt1-MCM complex is exported from the nucleus to the cytoplasm during S phase [56,75]. In mammalian cells, Cdt1 degradation in S phase is mediated by two independent E3 ubiquitin ligase complexes, $CRL1^{Skp2}$ (also known as SCF; reviewed in [120]) and $CRL4^{Cdt2}$ [121,122]. Like many substrates of $CRL1^{Skp2}$, Cdt1 is only bound for productive ubiquitylation once it is phosphorylated by CDK. In human Cdt1, this "phosphodegron" is created by Cdk2-mediated phosphorylation at threonine 29. Thr29 phosphorylation is then recognized by Skp2, a substrate adaptor, to trigger ubiquitylation [123–125]. A nearby serine at position 31 that matches the minimal CDK substrate consensus is also phosphorylated in cells, but it plays a minor role in recruiting Cdt1 to $CRL1^{Skp2}$.

Although manipulations that block Cdt1 Thr29 phosphorylation prevent ubiquitylation by $CRL1^{Skp2}$, such manipulations do not substantially stabilize Cdt1 during S phase. Even in the absence of

CRL1^{Skp2} targeting, Cdt1 is ubiquitylated and degraded by a second E3 ubiquitin ligase, CRL4^{Cdt2} [126] (Cdt2 was identified in the same screen that discovered Cdt1, but the Cdt1 and Cdt2 sequences are unrelated [98]). Unlike targeting by CRL1^{Skp2}, Cdt1 ubiquitylation by CRL4^{Cdt2} is not stimulated by Cdt1 phosphorylation, but instead requires a ternary interaction among Cdt1, the substrate adapter Cdt2, and DNA-loaded Proliferating Cell Nuclear Antigen (PCNA). PCNA is a homotrimer that is loaded by Replication Factor C at replication forks and serves as the processivity factor for DNA polymerase during DNA replication. DNA-bound PCNA is also a platform for a host of proteins that bind PCNA through short linear motifs known as PCNA-Interacting Protein (PIP) boxes [127]. The Cdt1 PIP box is special in that it not only binds PCNA but also triggers degradation and is thus termed a "PIP degron." PCNA is only loaded during DNA synthesis, and this loading event is required for Cdt1 recognition by CRL4^{Cdt2}; thus this mode of Cdt1 degradation has been termed "replication-coupled destruction" [127]. Since the trigger for CRL4^{Cdt2}-mediated degradation is PCNA DNA loading, PIP degron-containing Cdt1 proteins are also degraded after DNA damage because PCNA is loaded during DNA repair [128].

Mutations to the human Cdt1 PIP degron alone have only modest effects on S phase degradation in otherwise unperturbed cells. On the other hand, a combination of PIP degron mutations with mutations that block Cdt1 phosphorylation at Thr29 stabilizes Cdt1 during S phase and induces substantial re-replication [129]. Near the end of S phase, human Cdt1 re-accumulates, but this re-accumulation is not strictly because PCNA is no longer DNA loaded. CRL4^{Cdt2} is globally inhibited as cells approach G2, leading to re-accumulation of all of its substrates [130]. Cdt1 is clearly not targeted by CRL1^{Skp2} in G2 phase either, although cyclin A-dependent activity is still high. The mechanism preventing CRL1^{Skp2}-mediated Cdt1 degradation in G2 is still unknown. One potential addition to Cdt1 stability control is the recent report that Cdt1 abundance is sensitive to the deubiquitylating enzyme, Usp37 [131]. Thus, Cdt1 re-accumulation could be as much a consequence of increased deubiquitylation as it is a result of decreased ubiquitylation.

Somewhat surprisingly and despite being a particularly potent inducer of S phase destruction [132], the PIP degron is not conserved in all Cdt1 proteins—not even among all mammalian Cdt1 sequences. PIP degron sequences are not evident in the cow, pig, sheep, or rabbit Cdt1 sequences, though Cdt1 PIP degrons are found in nematode, fruit fly, zebrafish, chicken, rat, mouse, baboon and many other species (J.G.C. unpublished observation and [121]). Moreover, CRL1^{Skp2}-mediated degradation to reinforce CRL4^{Cdt2}-mediated degradation may not be universal among metazoans (e.g., *X. laevis* Cdt1). In species where it appears that only one E3 ligase targets Cdt1 during S phase, the presence of stronger licensing inhibitory mechanisms that target other pre-RC components may have allowed the second E3 pathway to be lost.

4.3. Inhibition by Geminin

Unlike nearly all components of the replication licensing system, human Cdt1 was not cloned strictly on the basis of sequence homology to a yeast ortholog. In fact, the fission yeast Cdt1 was not directly investigated as a licensing protein until after the metazoan Cdt1 proteins were functionally characterized. Human Cdt1 was isolated both by sequence similarity to *D. melanogaster* and *X. laevis* orthologs and as the target of a re-replication inhibitor protein, geminin [48,133,134]. Geminin itself was cloned from biochemical screens for *X. laevis* proteins that are degraded in mitosis [135]. Of note, neither budding nor fission yeast harbor a geminin ortholog. Human geminin is abundant during S phase and G2, is degraded at anaphase, and is least abundant during G1 phase. Geminin is a substrate of the APC/C (Anaphase-Promoting Complex/Cyclosome) [135], an E3 ubiquitin ligase which promotes geminin degradation from late mitosis and throughout G1 phase [136,137].

Artificially elevating geminin concentration in G1 blocks MCM loading, but the mechanism of that inhibition was not known at the time geminin was first characterized [135]. An effort to gain insight into how geminin inhibits licensing by identifying partners yielded a tight-binding partner in human lysates, human Cdt1. Moreover, supplementing geminin-inhibited *X. laevis* lysates with additional human Cdt1 reversed the inhibitory effects of geminin on origin licensing and DNA replication [134]. Mutations in Cdt1 that alter geminin binding also have higher licensing activity in vitro compared to wild-type Cdt1 [138]. Like Cdt1, geminin has at least one alternative function outside the licensing system; geminin regulates gene expression during development [139–141].

Geminin is a dimer that forms a stable 2:1 complex with monomeric Cdt1 both on chromatin and in the nucleoplasm. Geminin binds the central region of Cdt1 (Figure 2), and indeed both high-resolution structures of this Cdt1 domain are in complex with geminin; possibly because the tight binding facilitated crystallization [105,106]. Interestingly, the human crystal structure consists of a Cdt1 and geminin heterohexamer composed of two Cdt1 and four geminin polypeptides; this structure is essentially a dimer of the trimer observed in the mouse structure. Based on this and other observations, De Marco et al. suggested that the trimer is permissive for licensing, whereas the hexamer corresponds to the inhibited form [106]. If true, then only high concentrations of geminin, such as those found in mid-to-late S phase and G2, would be effective for forming hexamers and preventing re-replication. In this scenario, re-replication control in early S phase should rely more heavily on Cdt1 degradation and mechanisms that target ORC, Cdc6, and MCM than on geminin because geminin is less abundant in early S phase.

How does geminin inhibit Cdt1 activity? In vitro, geminin prevents the association of Cdt1 with MCM complexes [51,142] and also blocks Cdt1 binding to Cdc6 [51]. A simple stearic occlusion model seems unlikely however, if the primary binding site for MCM is the Cdt1 C-terminal domain but geminin binds the central domain. Nonetheless, geminin dimers bound to the Cdt1 central domain could conceivably project far enough towards the C-terminal domain to interfere with stable MCM binding. Alternatively, geminin binding may induce a conformational change in the Cdt1 central domain that propagates to the C-terminal domain. It may also be that Cdt1 forms multiple contacts with the MCM ring, and that geminin interferes with MCM binding sites in Cdt1 that are separate from those at the C-terminal Cdt1–Mcm6 interface (e.g., the diagram based on Sun et al. 2013 in Figure 2). Testing these ideas directly will ultimately require a structure including both the Cdt1 central and C-terminal domains with and without geminin.

4.4. Cdt1 Phosphorylation

Cdt1 is phosphorylated at many serine and threonine (but not tyrosine) sites at different times during the cell cycle and in response to different cues. Phosphoproteomic analyses have identified dozens of phosphorylation sites detectable in proliferating human cells. Figure 2 marks only those human Cdt1 phosphorylation sites from the PhosphoSite Plus database [104] that are conserved in at least one other mammalian Cdt1. Those marked in dark grey are Ser-Pro or Thr-Pro sites which conform to the minimal recognition sequence for both CDKs and MAP kinases (mitogen-activated protein kinases). The green symbols mark Ser-Pro and Thr-Pro sites that have been identified in proteomics screens and also tested for functional consequences.

As discussed above (Section 4.2), phosphorylation at Thr29 targets Cdt1 for ubiquitylation by the CRL1[Skp2] E3 ubiquitin ligase. Which kinase (or kinases) is most responsible for Thr29 phosphorylation? In vitro, Cdt1 can be phosphorylated by Cdk4 (activated in G1 by cyclin D), Cdk2 (activated in S by cyclin E and cyclin A) and Cdk1 (activated by cyclin A and cyclin B) [124,125]. Moreover, Cdt1 isolated from cell lysates co-precipitates cyclin A, but not cyclin E or cyclin B. CDK interaction depends on a cyclin binding motif known as a Cy motif in Cdt1 (aa 68–70) [125]. It is most likely that Thr29 phosphorylation to induce Cdt1 ubiquitylation by CRL1[Skp2] is carried out by cyclin A/Cdk2 in early S phase, and Cdt1 phosphorylation in late S and G2 is carried out by cyclin A/Cdk1.

Interestingly, Thr29 can also be phosphorylated in vitro by the MAP kinase, Jnk1 (Jun kinase) [143]. Miotto and Struhl noted that treating cells to activate the stress kinase, Jnk1, also blocks Hbo1 recruitment to several selected origins [143]. Mutating Thr29 enhanced Hbo1 residence at origins, suggesting that Jnk1-mediated Cdt1 phosphorylation at Thr29 inhibits Hbo1 recruitment. Despite the ability of Jnk1 to phosphorylate Thr29, conditions that activate Jnk1 in cells (without causing DNA damage) had no effect on Cdt1 stability; the stability of Thr29-phosphorylated Cdt1 could indicate Jnk1-mediated inhibition of CRL1^{Skp2} or some other mechanism to prevent Cdt1 degradation [90]. In that study, the authors mapped multiple Cdt1 phosphorylation sites and determined which sites are sensitive to Jnk1 inhibitors. In addition to Thr29, at least two other phosphorylation sites showed the same Jnk1-sensitive pattern as Thr29: Ser93 and Ser318, but the outcomes of these phosphorylations have yet to be determined [143].

A concurrent study of Cdt1 phosphorylation by stress-activated MAP kinases focused on distinct sites in the linker domain and C-terminus. Chandrasekaran et al. mapped a collection of five sites that can be phosphorylated by either JNK or p38 MAPK isoforms: amino acids 372, 402, 406, 411 and 491 (Figure 2) [144]. Functional tests of phosphomimetic substitutions at these five positions led to the inference that Cdt1 phosphorylation not only inhibits its licensing activity in cells, but surprisingly, also blocks binding to the CRL4^{Cdt2} E3 ubiquitin ligase. As a result, this phospho-isoform of Cdt1 is resistant to degradation by CRL4^{Cdt2}, though it is still sensitive to CRL1^{Skp2} [144,145]. The molecular mechanism of licensing inhibition is not yet known, but the concentration of phosphorylation sites in the linker domain hints that phosphorylation could affect the positioning of the central and C-terminal domains relative to one another. Another possibility is that the N- and C-termini are in close proximity to each other to allow phosphorylation in the linker to disrupt CRL4^{Cdt2} binding to the PIP degron.

Unlike Thr29 (and Ser31), these more C-terminally located phosphorylation positions are responsible for the commonly-observed mitotic Cdt1 gel mobility shift by sodium dodecyl sulphate-polyacrylamide gel electrophoresis (SDS-PAGE). This gel shift is evident not only in response to cellular stresses that activate p38 and JNK, but also during G2 and mitosis in unperturbed cell cycles and in quiescent cells [9,112,129,144]. Cdt1 is robustly phosphorylated in mitotic cells and dephosphorylated in early G1, though the phosphatase responsible is not known. Both p38 and JNK are active during mitosis alongside Cdk1 [146]. Since the phosphorylation sites match the consensus sequence for both classes of proline-directed kinases, it is currently impossible to know which kinase(s) are truly responsible for Cdt1 mitotic phosphorylation. Regardless of how many kinases target these sites, the result is that beginning in late S phase, Cdt1 re-accumulates in a form that is not active for origin licensing and is no longer subject to rapid degradation (Figure 3). A potential role for geminin in Cdt1 protection from CRL1^{Skp2} during G2 has also been reported [113,147], but attempts to definitively confirm this relationship have utilized geminin and Cdt1 manipulation which frequently induces DNA damage. Separating potential geminin-mediated direct effects on Cdt1 stability from established indirect effects related to re-replication dependent DNA damage (and subsequent CRL4^{Cdt2}-mediated Cdt1 degradation) requires careful interpretation [148]. Nonetheless, the Cdt1 stabilization in G2 may serve the dual purposes of allowing Cdt1 to function with the NDC80 complex at kinetochores and providing a large pool of Cdt1 in the subsequent G1 to license origins in the next cell cycle.

Finally, the majority of detectable phosphorylation sites in human Cdt1 remain unstudied. Several of these match the consensus for CDK-mediated (or MAPK-mediated) phosphorylation. Are there additional CDK sites, and if so, are they dependent on the same Cy motif that directs phosphorylation at the N-terminus? Is there a MAPK docking site in Cdt1 that facilitates phosphorylation by p38 or JNK? The N-terminal PEST domain is in close proximity to several candidate sites which may function to inhibit Cdt1 chromatin binding in G2, though this idea has not been explicitly tested [112]. Of further note, all of the consequences of phosphorylation thus far lead to Cdt1 inhibition or changes in stability. It remains equally possible that phosphorylation at one or more novel sites promotes Cdt1 function in either licensing or mitosis.

5. Cdt1 in Disease

Although numerous mechanisms restrain Cdt1 function, pathological dysregulation of Cdt1 can still occur, particularly in cells whose control mechanisms have been compromised. Moreover, mutations in Cdt1 itself can cause pathological consequences.

5.1. Overexpression and Oncogenesis

Cdt1 overexpression can result in genotoxic stress leading to aberrant cell proliferation and predisposition to oncogenic transformation. Experimentally increased Cdt1 abundance outside of G1 phase promotes re-replication and genome instability [68,149,150]. Therefore, it is quite possible that more moderate overexpression from spontaneously deregulated transcriptional controls has the same effect on genome stability in vivo [7,151]. Over time, higher-than-normal levels of Cdt1 protein may not be fully restrained during S phase and G2 by the ubiquitin ligases, kinases, and geminin inhibition described in Section 4. This means that at some low frequency, Cdt1 may promote origin re-licensing and re-replication. Interestingly, cells expressing higher than normal Cdt1 exhibit a more aggressive and chemoresistant phenotype [152]. In addition, the genome instability from Cdt1 likely includes not only gene amplification and chromosome damage from re-replication [66], but also changes in chromosome number which may reflect Cdt1's role in chromosome segregation [7,9].

Cdt1 transcription is driven by the E2F family of transcription factors (see Section 4.1), and one of the most frequently-mutated regulatory pathways in cancers is the Rb-E2F pathway [153,154]. In fact, many cancer-derived cell lines exhibit higher-than-normal expression of Cdt1 [117]. Cdt1 overexpression could also lead to rapid origin licensing and shorter G1, thus proliferate more rapidly but with less fidelity. To our knowledge, a direct and quantitative correlation between Cdt1 abundance and the extent of genome instability in cancers has not yet been investigated. It may be that Cdt1 expression levels will identify particular cancers that are most likely to progress or are more or less susceptible to specific therapeutic interventions [155].

5.2. Meier-Gorlin Syndrome

Avoiding either over- or under-licensing origins is critical for successful cell proliferation during development. The need for not only effective licensing inhibition after S phase but also efficient origin licensing in G1 is most apparent in the phenotypes associated with a rare primordial dwarfism syndrome, Meier-Gorlin Syndrome (MGS). MGS patients harbor hypomorphic mutations in genes encoding pre-RC components, including Cdt1, some ORC subunits, and Cdc6. These patients have extremely short stature, small external ears, and focal hypoplasias, likely due to slow cell proliferation [156–158]. Furthermore, *de novo* mutations in the gene encoding geminin, have also been described in MGS patients [159]. In these instances, the mutations disrupt protein motifs required for normal geminin degradation in G1 phase [159]. As a result, geminin could inappropriately inhibit Cdt1 and result in slow origin licensing and G1 delay. The Cdt1 MGS patient genotypes are compound heterozygous missense mutations combined with nonsense mutations (presumed null alleles) [156,158]. In addition, the *CDT1* mutations are present across most of the *CDT1* gene and translate to amino acid substitutions located in regions that are presumably important for Cdt1 regulation and activity. Cdt1 is an essential gene, so these alleles are likely hypomorphic rather than null.

Although origin licensing defects appear to be one molecular underpinning of MGS, not all MGS patients have mutations in origin licensing components. Recently, hypomorphic mutations in the *CDC45* gene that encodes one of the helicase activating subunits were identified in an additional cohort of MGS patients [160]. Cdc45 is not required for origin licensing in G1, but rather it is required for origin firing and replication fork progression during S-phase as part of the fully-active helicase (Cdc45-MCM2-7-GINS). These *CDC45* mutations result in splicing defects leading to a significant reduction in Cdc45 protein [160]. The change in Cdc45 protein abundance likely impairs DNA synthesis, thus hindering cell proliferation and genome stability in early development.

Interestingly, the pre-RC proteins affected in MGS include Cdt1, Cdc6, and subunits of ORC, but not MCM subunits. Do mutations in MCM lead to MGS phenotypes? Analyses of a spontaneous mutation in the mouse Mcm4 subunit suggests that dwarfism is not a universal outcome of licensing disruption. In these studies, hypomorphic Mcm4 mutations resulted in mice with increased micronuclei (a sign of chromosome instability) and increased tumor incidence, but otherwise grew to normal size [161]. Mouse embryonic fibroblasts from crosses between the hypomorphic and null alleles

Genes **2017**, *8*, 2

proliferate normally but are sensitive to replication stress [161]. These findings are in contrast to cells bearing MGS alleles in other licensing proteins that proliferate slowly [157,162]. These differences could reflect the degree of impairment by the different mutations, or they could reflect qualitative differences in the roles of the altered proteins. Such complexity certainly creates challenges for predicting the precise phenotypes of any newly-identified Cdt1 mutations, but the general expectation is impaired cell proliferation and/or genome instability.

6. Summary

Metazoan Cdt1 is regulated by a large number of independent mechanisms including inhibitor binding, phosphorylation, and ubiquitylation. It is likely that even more regulatory mechanisms will continue to be discovered, perhaps from follow-up studies to the proteomic detection of Cdt1 acetylation or sumoylation [104]. The need for such extensive regulation may be because Cdt1 is an integral player in both DNA replication and chromosome segregation, meaning Cdt1 deregulation has potent effects on genome stability and cell proliferation. The mitotic Cdt1 function clearly arose in eukaryotic evolution after the split between unicellular and multicellular species, so the presumed ancestral function was origin licensing. Why would Cdt1 evolve to have a role in kinetochore–microtubule attachment, a function that does not involve loading proteins onto DNA? It is becoming increasingly common to discover second cell cycle functions for origin licensing proteins, such as non-licensing roles for individual ORC subunits or geminin [139–141,163–165]. Perhaps it is generally useful to re-purpose proteins that are already under cell cycle-dependent control for a second cell cycle function. Alternatively, it may be that Cdt1 has biophysical properties that are uniquely suited to its molecular roles in both origin licensing and kinetochore–microtubule attachment. Based on our limited current knowledge, we can attempt to draw parallels between Cdt1's two targets: the MCM and NDC80 complexes. In both cases, a multisubunit complex undergoes important conformational changes. For NDC80, the change manifests as a molecular extension in vivo, and for MCM it is the presumed opening and closing of the Mcm2–Mcm5 gate. It is thus tempting to speculate that Cdt1 stabilizes a particular (extended?) conformation in the MCM complex, and that its two cell cycle roles are in fact related. We look forward to future developments in the field that will continue to shed light on the regulation and function of this unique protein.

Acknowledgments: We are grateful to members of the Cook lab and D. Varma and colleagues for helpful comments. The illustration of the yeast Cdt1-MCM complex from Sun et al. 2013 in Figure 2 was generated with permission from the authors and publisher. P.N.P. is supported by the National Institutes of Health F31GM121073; J.G.C. is supported by the National Institutes of Health R01GM102413 and a grant from the W.M. Keck Foundation.

Author Contributions: P.N.P. and J.G.C. conceived and wrote the paper; J.G.C. produced the illustrations.

Conflicts of Interest: The authors declare no conflict of interest.

References

1. Yekezare, M.; Gomez-Gonzalez, B.; Diffley, J.F. Controlling DNA replication origins in response to DNA damage—Inhibit globally, activate locally. *J. Cell Sci.* **2013**, *126*, 1297–1306. [CrossRef] [PubMed]
2. Siddiqui, K.; On, K.F.; Diffley, J.F. Regulating DNA replication in eukarya. *Cold Spring Harb. Perspect. Biol.* **2013**, *5*, 1288–1302. [CrossRef] [PubMed]
3. Truong, L.N.; Wu, X. Prevention of DNA re-replication in eukaryotic cells. *J. Mol. Cell Biol.* **2011**, *3*, 13–22. [CrossRef] [PubMed]
4. Blow, J.J.; Ge, X.Q.; Jackson, D.A. How dormant origins promote complete genome replication. *Trends Biochem. Sci.* **2011**, *36*, 405–414. [CrossRef] [PubMed]
5. Masai, H.; Matsumoto, S.; You, Z.; Yoshizawa-Sugata, N.; Oda, M. Eukaryotic chromosome DNA replication: Where, when, and how? *Annu. Rev. Biochem.* **2010**, *79*, 89–130. [CrossRef] [PubMed]
6. Li, C.; Jin, J. DNA replication licensing control and rereplication prevention. *Protein Cell* **2010**, *1*, 227–236. [CrossRef] [PubMed]

7. Arentson, E.; Faloon, P.; Seo, J.; Moon, E.; Studts, J.M.; Fremont, D.H.; Choi, K. Oncogenic potential of the DNA replication licensing protein CDT1. *Oncogene* **2002**, *21*, 1150–1158. [CrossRef] [PubMed]
8. Blow, J.J.; Gillespie, P.J. Replication licensing and cancer—A fatal entanglement? *Nat. Rev. Cancer* **2008**, *8*, 799–806. [CrossRef] [PubMed]
9. Varma, D.; Chandrasekaran, S.; Sundin, L.J.; Reidy, K.T.; Wan, X.; Chasse, D.A.; Nevis, K.R.; DeLuca, J.G.; Salmon, E.D.; Cook, J.G. Recruitment of the human Cdt1 replication licensing protein by the loop domain of Hec1 is required for stable kinetochore-microtubule attachment. *Nat. Cell Biol.* **2012**, *14*, 593–603. [CrossRef] [PubMed]
10. Prioleau, M.N.; MacAlpine, D.M. DNA replication origins-where do we begin? *Genes Dev.* **2016**, *30*, 1683–1697. [CrossRef] [PubMed]
11. Mechali, M. Eukaryotic DNA replication origins: Many choices for appropriate answers. *Nat. Rev. Mol. Cell Biol.* **2010**, *11*, 728–738. [CrossRef] [PubMed]
12. Besnard, E.; Babled, A.; Lapasset, L.; Milhavet, O.; Parrinello, H.; Dantec, C.; Marin, J.M.; Lemaitre, J.M. Unraveling cell type-specific and reprogrammable human replication origin signatures associated with G-quadruplex consensus motifs. *Nat. Struct. Mol. Biol.* **2012**, *19*, 837–844. [CrossRef] [PubMed]
13. Blow, J.J.; Laskey, R.A. A role for the nuclear envelope in controlling DNA replication within the cell cycle. *Nature* **1988**, *332*, 546–548. [CrossRef] [PubMed]
14. Deegan, T.D.; Diffley, J.F. MCM: One ring to rule them all. *Curr. Opin. Struct. Biol.* **2016**, *37*, 145–151. [CrossRef] [PubMed]
15. Maine, G.T.; Sinha, P.; Tye, B.K. Mutants of *S. cerevisiae* defective in the maintenance of minichromosomes. *Genetics* **1984**, *106*, 365–385. [PubMed]
16. Cocker, J.H.; Piatti, S.; Santocanale, C.; Nasmyth, K.; Diffley, J.F. An essential role for the Cdc6 protein in forming the pre-replicative complexes of budding yeast. *Nature* **1996**, *379*, 180–182. [CrossRef] [PubMed]
17. Gillespie, P.J.; Li, A.; Blow, J.J. Reconstitution of licensed replication origins on Xenopus sperm nuclei using purified proteins. *BMC Biochem.* **2001**, *2*, 15. [CrossRef] [PubMed]
18. Evrin, C.; Clarke, P.; Zech, J.; Lurz, R.; Sun, J.; Uhle, S.; Li, H.; Stillman, B.; Speck, C. A double-hexameric MCM2–7 complex is loaded onto origin DNA during licensing of eukaryotic DNA replication. *Proc. Natl. Acad. Sci. USA* **2009**, *106*, 20240–20245. [CrossRef] [PubMed]
19. Remus, D.; Beuron, F.; Tolun, G.; Griffith, J.D.; Morris, E.P.; Diffley, J.F. Concerted loading of Mcm2–7 double hexamers around DNA during DNA replication origin licensing. *Cell* **2009**, *139*, 719–730. [CrossRef] [PubMed]
20. Saha, P.; Chen, J.; Thome, K.C.; Lawlis, S.J.; Hou, Z.H.; Hendricks, M.; Parvin, J.D.; Dutta, A. Human CDC6/Cdc18 associates with Orc1 and cyclin-cdk and is selectively eliminated from the nucleus at the onset of S phase. *Mol. Cell. Biol.* **1998**, *18*, 2758–2767. [CrossRef] [PubMed]
21. Coleman, T.R.; Carpenter, P.B.; Dunphy, W.G. The Xenopus Cdc6 protein is essential for the initiation of a single round of DNA replication in cell-free extracts. *Cell* **1996**, *87*, 53–63. [CrossRef]
22. Neuwald, A.F.; Aravind, L.; Spouge, J.L.; Koonin, E.V. AAA+: A class of chaperone-like ATPases associated with the assembly, operation, and disassembly of protein complexes. *Genome Res.* **1999**, *9*, 27–43. [PubMed]
23. Li, H.; Stillman, B. The origin recognition complex: A biochemical and structural view. *Subcell. Biochem.* **2012**, *62*, 37–58. [PubMed]
24. Borlado, L.R.; Mendez, J. CDC6: From DNA replication to cell cycle checkpoints and oncogenesis. *Carcinogenesis* **2008**, *29*, 237–243. [CrossRef] [PubMed]
25. Bell, S.D.; Botchan, M.R. The minichromosome maintenance replicative helicase. *Cold Spring Harb. Perspect. Biol.* **2013**, *5*, a012807. [CrossRef] [PubMed]
26. Leonard, A.C.; Mechali, M. DNA replication origins. *Cold Spring Harb. Perspect. Biol.* **2013**, *5*, a010116. [CrossRef] [PubMed]
27. Dorn, E.S.; Cook, J.G. Nucleosomes in the neighborhood: New roles for chromatin modifications in replication origin control. *Epigenetics* **2011**, *6*, 552–559. [CrossRef] [PubMed]
28. MacAlpine, D.M.; Almouzni, G. Chromatin and DNA replication. *Cold Spring Harb. Perspect. Biol.* **2013**, *5*, a010207. [CrossRef] [PubMed]
29. Riera, A.; Li, H.; Speck, C. Seeing is believing: The MCM2–7 helicase trapped in complex with its DNA loader. *Cell Cycle* **2013**, *12*, 2917–2918. [CrossRef] [PubMed]

30. Yardimci, H.; Walter, J.C. Prereplication-complex formation: A molecular double take? *Nat. Struct. Mol. Biol.* **2014**, *21*, 20–25. [CrossRef] [PubMed]

31. Ticau, S.; Friedman, L.J.; Ivica, N.A.; Gelles, J.; Bell, S.P. Single-molecule studies of origin licensing reveal mechanisms ensuring bidirectional helicase loading. *Cell* **2015**, *161*, 513–525. [CrossRef] [PubMed]

32. Sun, J.; Evrin, C.; Samel, S.A.; Fernandez-Cid, A.; Riera, A.; Kawakami, H.; Stillman, B.; Speck, C.; Li, H. Cryo-EM structure of a helicase loading intermediate containing ORC-Cdc6-Cdt1-MCM2–7 bound to DNA. *Nat. Struct. Mol. Biol.* **2013**, *20*, 944–951. [CrossRef] [PubMed]

33. Li, N.; Zhai, Y.; Zhang, Y.; Li, W.; Yang, M.; Lei, J.; Tye, B.K.; Gao, N. Structure of the eukaryotic MCM complex at 3.8 A. *Nature* **2015**, *524*, 186–191. [CrossRef] [PubMed]

34. Labib, K. How do Cdc7 and cyclin-dependent kinases trigger the initiation of chromosome replication in eukaryotic cells? *Genes Dev.* **2010**, *24*, 1208–1219. [CrossRef] [PubMed]

35. Tanaka, S.; Araki, H. Helicase activation and establishment of replication forks at chromosomal origins of replication. *Cold Spring Harb. Perspect. Biol.* **2013**, *5*, a010371. [CrossRef] [PubMed]

36. Ilves, I.; Petojevic, T.; Pesavento, J.J.; Botchan, M.R. Activation of the MCM2–7 helicase by association with Cdc45 and GINS proteins. *Mol. Cell* **2010**, *37*, 247–258. [CrossRef] [PubMed]

37. Symeonidou, I.E.; Kotsantis, P.; Roukos, V.; Rapsomaniki, M.A.; Grecco, H.E.; Bastiaens, P.; Taraviras, S.; Lygerou, Z. Multi-step loading of human minichromosome maintenance proteins in live human cells. *J. Biol. Chem.* **2013**, *288*, 35852–35867. [CrossRef] [PubMed]

38. Xouri, G.; Squire, A.; Dimaki, M.; Geverts, B.; Verveer, P.J.; Taraviras, S.; Nishitani, H.; Houtsmuller, A.B.; Bastiaens, P.I.; Lygerou, Z. Cdt1 associates dynamically with chromatin throughout G1 and recruits Geminin onto chromatin. *EMBO J.* **2007**, *26*, 1303–1314. [CrossRef] [PubMed]

39. Dimitrova, D.S.; Prokhorova, T.A.; Blow, J.J.; Todorov, I.T.; Gilbert, D.M. Mammalian nuclei become licensed for DNA replication during late telophase. *J. Cell Sci.* **2002**, *115*, 51–59. [PubMed]

40. Edwards, M.C.; Tutter, A.V.; Cvetic, C.; Gilbert, C.H.; Prokhorova, T.A.; Walter, J.C. MCM2–7 complexes bind chromatin in a distributed pattern surrounding the origin recognition complex in Xenopus egg extracts. *J. Biol. Chem.* **2002**, *277*, 33049–33057. [CrossRef] [PubMed]

41. Woodward, A.M.; Gohler, T.; Luciani, M.G.; Oehlmann, M.; Ge, X.; Gartner, A.; Jackson, D.A.; Blow, J.J. Excess Mcm2–7 license dormant origins of replication that can be used under conditions of replicative stress. *J. Cell Biol.* **2006**, *173*, 673–683. [CrossRef] [PubMed]

42. Powell, S.K.; MacAlpine, H.K.; Prinz, J.A.; Li, Y.; Belsky, J.A.; MacAlpine, D.M. Dynamic loading and redistribution of the Mcm2–7 helicase complex through the cell cycle. *EMBO J.* **2015**, *34*, 531–543. [CrossRef] [PubMed]

43. Hyrien, O. How MCM loading and spreading specify eukaryotic DNA replication initiation sites. *F1000Res* **2016**, *5*. [CrossRef] [PubMed]

44. Gros, J.; Kumar, C.; Lynch, G.; Yadav, T.; Whitehouse, I.; Remus, D. Post-licensing Specification of Eukaryotic Replication Origins by Facilitated Mcm2–7 Sliding along DNA. *Mol. Cell* **2015**, *60*, 797–807. [CrossRef] [PubMed]

45. McIntosh, D.; Blow, J.J. Dormant origins, the licensing checkpoint, and the response to replicative stresses. *Cold Spring Harb. Perspect. Biol.* **2012**, *4*, 235–242. [CrossRef] [PubMed]

46. Ge, X.Q.; Jackson, D.A.; Blow, J.J. Dormant origins licensed by excess Mcm2–7 are required for human cells to survive replicative stress. *Genes Dev.* **2007**, *21*, 3331–3341. [CrossRef] [PubMed]

47. Ibarra, A.; Schwob, E.; Mendez, J. Excess MCM proteins protect human cells from replicative stress by licensing backup origins of replication. *Proc. Natl. Acad. Sci. USA* **2008**, *105*, 8956–8961. [CrossRef] [PubMed]

48. Maiorano, D.; Moreau, J.; Mechali, M. XCDT1 is required for the assembly of pre-replicative complexes in *Xenopus laevis*. *Nature* **2000**, *404*, 622–625. [PubMed]

49. Nishitani, H.; Lygerou, Z.; Nishimoto, T.; Nurse, P. The Cdt1 protein is required to license DNA for replication in fission yeast. *Nature* **2000**, *404*, 625–628. [PubMed]

50. Rialland, M.; Sola, F.; Santocanale, C. Essential role of human CDT1 in DNA replication and chromatin licensing. *J. Cell Sci.* **2002**, *115*, 1435–1440. [PubMed]

51. Cook, J.G.; Chasse, D.A.; Nevins, J.R. The regulated association of Cdt1 with minichromosome maintenance proteins and Cdc6 in mammalian cells. *J. Biol. Chem.* **2004**, *279*, 9625–9633. [CrossRef] [PubMed]

52. Fernandez-Cid, A.; Riera, A.; Tognetti, S.; Herrera, M.C.; Samel, S.; Evrin, C.; Winkler, C.; Gardenal, E.; Uhle, S.; Speck, C. An ORC/Cdc6/MCM2–7 complex is formed in a multistep reaction to serve as a platform for MCM double-hexamer assembly. *Mol. Cell* **2013**, *50*, 577–588. [CrossRef] [PubMed]

53. Zhang, J.; Yu, L.; Wu, X.; Zou, L.; Sou, K.K.; Wei, Z.; Cheng, X.; Zhu, G.; Liang, C. The interacting domains of hCdt1 and hMcm6 involved in the chromatin loading of the MCM complex in human cells. *Cell Cycle* **2010**, *9*, 4848–4857. [CrossRef] [PubMed]

54. Ferenbach, A.; Li, A.; Brito-Martins, M.; Blow, J.J. Functional domains of the Xenopus replication licensing factor Cdt1. *Nucleic Acids Res.* **2005**, *33*, 316–324. [CrossRef] [PubMed]

55. Chen, S.; de Vries, M.A.; Bell, S.P. Orc6 is required for dynamic recruitment of Cdt1 during repeated Mcm2–7 loading. *Genes Dev.* **2007**, *21*, 2897–2907. [CrossRef] [PubMed]

56. Tanaka, S.; Diffley, J.F. Interdependent nuclear accumulation of budding yeast Cdt1 and Mcm2–7 during G1 phase. *Nat. Cell Biol.* **2002**, *4*, 198–207. [CrossRef] [PubMed]

57. Devault, A.; Vallen, E.A.; Yuan, T.; Green, S.; Bensimon, A.; Schwob, E. Identification of Tah11/Sid2 as the ortholog of the replication licensing factor Cdt1 in Saccharomyces cerevisiae. *Curr. Biol.* **2002**, *12*, 689–694. [CrossRef]

58. Boskovic, J.; Bragado-Nilsson, E.; Saligram Prabhakar, B.; Yefimenko, I.; Martinez-Gago, J.; Munoz, S.; Mendez, J.; Montoya, G. Molecular architecture of the recombinant human MCM2–7 helicase in complex with nucleotides and DNA. *Cell Cycle* **2016**, *15*, 2431–2440. [CrossRef] [PubMed]

59. Bochman, M.L.; Schwacha, A. The *Saccharomyces cerevisiae* Mcm6/2 and Mcm5/3 ATPase active sites contribute to the function of the putative Mcm2–7 'gate'. *Nucleic Acids Res.* **2010**, *38*, 6078–6088. [CrossRef] [PubMed]

60. Samel, S.A.; Fernandez-Cid, A.; Sun, J.; Riera, A.; Tognetti, S.; Herrera, M.C.; Li, H.; Speck, C. A unique DNA entry gate serves for regulated loading of the eukaryotic replicative helicase MCM2–7 onto DNA. *Genes Dev.* **2014**, *28*, 1653–1666. [CrossRef] [PubMed]

61. Boos, D.; Frigola, J.; Diffley, J.F. Activation of the replicative DNA helicase: Breaking up is hard to do. *Curr. Opin. Cell Biol.* **2012**, *24*, 423–430. [CrossRef] [PubMed]

62. Trakselis, M.A. Structural Mechanisms of Hexameric Helicase Loading, Assembly, and Unwinding. *F1000Res* **2016**, *5*. [CrossRef] [PubMed]

63. Fox, D.T.; Duronio, R.J. Endoreplication and polyploidy: Insights into development and disease. *Development* **2013**, *140*, 3–12. [CrossRef] [PubMed]

64. Lee, H.O.; Davidson, J.M.; Duronio, R.J. Endoreplication: Polyploidy with purpose. *Genes Dev.* **2009**, *23*, 2461–2477. [CrossRef] [PubMed]

65. Nordman, J.; Orr-Weaver, T.L. Regulation of DNA replication during development. *Development* **2012**, *139*, 455–464. [CrossRef] [PubMed]

66. Green, B.M.; Finn, K.J.; Li, J.J. Loss of DNA replication control is a potent inducer of gene amplification. *Science* **2010**, *329*, 943–946. [CrossRef] [PubMed]

67. Klotz-Noack, K.; McIntosh, D.; Schurch, N.; Pratt, N.; Blow, J.J. Re-replication induced by geminin depletion occurs from G2 and is enhanced by checkpoint activation. *J. Cell Sci.* **2012**, *125*, 2436–2445. [CrossRef] [PubMed]

68. Vaziri, C.; Saxena, S.; Jeon, Y.; Lee, C.; Murata, K.; Machida, Y.; Wagle, N.; Hwang, D.S.; Dutta, A. A p53-dependent checkpoint pathway prevents rereplication. *Mol. Cell* **2003**, *11*, 997–1008. [CrossRef]

69. Green, B.M.; Li, J.J. Loss of rereplication control in *Saccharomyces cerevisiae* results in extensive DNA damage. *Mol. Biol. Cell.* **2005**, *16*, 421–432. [CrossRef] [PubMed]

70. Melixetian, M.; Ballabeni, A.; Masiero, L.; Gasparini, P.; Zamponi, R.; Bartek, J.; Lukas, J.; Helin, K. Loss of Geminin induces rereplication in the presence of functional p53. *J. Cell Biol.* **2004**, *165*, 473–482. [CrossRef] [PubMed]

71. Davidson, I.F.; Li, A.; Blow, J.J. Deregulated replication licensing causes DNA fragmentation consistent with head-to-tail fork collision. *Mol. Cell* **2006**, *24*, 433–443. [CrossRef] [PubMed]

72. Drury, L.S.; Perkins, G.; Diffley, J.F. The cyclin-dependent kinase Cdc28p regulates distinct modes of Cdc6p proteolysis during the budding yeast cell cycle. *Curr. Biol.* **2000**, *10*, 231–240. [CrossRef]

73. Petersen, B.O.; Lukas, J.; Sorensen, C.S.; Bartek, J.; Helin, K. Phosphorylation of mammalian CDC6 by cyclin A/CDK2 regulates its subcellular localization. *EMBO J.* **1999**, *18*, 396–410. [CrossRef] [PubMed]

74. Elsasser, S.; Chi, Y.; Yang, P.; Campbell, J.L. Phosphorylation controls timing of Cdc6p destruction: A biochemical analysis. *Mol. Biol. Cell.* **1999**, *10*, 3263–3277. [CrossRef] [PubMed]

75. Nguyen, V.Q.; Co, C.; Irie, K.; Li, J.J. Clb/Cdc28 kinases promote nuclear export of the replication initiator proteins Mcm2–7. *Curr. Biol.* **2000**, *10*, 195–205. [CrossRef]

76. Makise, M.; Takehara, M.; Kuniyasu, A.; Matsui, N.; Nakayama, H.; Mizushima, T. Linkage between phosphorylation of the origin recognition complex and its ATP binding activity in *Saccharomyces cerevisiae*. *J. Biol. Chem.* **2009**, *284*, 3396–3407. [CrossRef] [PubMed]

77. Chen, S.; Bell, S.P. CDK prevents Mcm2–7 helicase loading by inhibiting Cdt1 interaction with Orc6. *Genes Dev.* **2011**, *25*, 363–372. [CrossRef] [PubMed]

78. Rowles, A.; Tada, S.; Blow, J.J. Changes in association of the Xenopus origin recognition complex with chromatin on licensing of replication origins. *J. Cell Sci.* **1999**, *112 Pt 12*, 2011–2018. [PubMed]

79. Mendez, J.; Zou-Yang, X.H.; Kim, S.Y.; Hidaka, M.; Tansey, W.P.; Stillman, B. Human origin recognition complex large subunit is degraded by ubiquitin-mediated proteolysis after initiation of DNA replication. *Mol. Cell* **2002**, *9*, 481–491. [CrossRef]

80. Li, C.J.; Vassilev, A.; DePamphilis, M.L. Role for Cdk1 (Cdc2)/cyclin A in preventing the mammalian origin recognition complex's largest subunit (Orc1) from binding to chromatin during mitosis. *Mol. Cell. Biol.* **2004**, *24*, 5875–5886. [CrossRef] [PubMed]

81. Nevis, K.R.; Cordeiro-Stone, M.; Cook, J.G. Origin licensing and p53 status regulate Cdk2 activity during G(1). *Cell Cycle* **2009**, *8*, 1952–1963. [CrossRef] [PubMed]

82. Shreeram, S.; Sparks, A.; Lane, D.P.; Blow, J.J. Cell type-specific responses of human cells to inhibition of replication licensing. *Oncogene* **2002**, *21*, 6624–6632. [CrossRef] [PubMed]

83. Machida, Y.J.; Teer, J.K.; Dutta, A. Acute reduction of an origin recognition complex (ORC) subunit in human cells reveals a requirement of ORC for Cdk2 activation. *J. Biol. Chem.* **2005**, *280*, 27624–27630. [CrossRef] [PubMed]

84. Moreno, A.; Carrington, J.T.; Albergante, L.; Al Mamun, M.; Haagensen, E.J.; Komseli, E.S.; Gorgoulis, V.G.; Newman, T.J.; Blow, J.J. Unreplicated DNA remaining from unperturbed S phases passes through mitosis for resolution in daughter cells. *Proc. Natl. Acad. Sci. USA* **2016**, *113*, E5757–E5764. [CrossRef] [PubMed]

85. Berbenetz, N.M.; Nislow, C.; Brown, G.W. Diversity of eukaryotic DNA replication origins revealed by genome-wide analysis of chromatin structure. *PLoS Genet.* **2010**, *6*, e1001092. [CrossRef] [PubMed]

86. Kuo, A.J.; Song, J.; Cheung, P.; Ishibe-Murakami, S.; Yamazoe, S.; Chen, J.K.; Patel, D.J.; Gozani, O. The BAH domain of ORC1 links H4K20me2 to DNA replication licensing and Meier-Gorlin syndrome. *Nature* **2012**, *484*, 115–119. [CrossRef] [PubMed]

87. Muller, P.; Park, S.; Shor, E.; Huebert, D.J.; Warren, C.L.; Ansari, A.Z.; Weinreich, M.; Eaton, M.L.; MacAlpine, D.M.; Fox, C.A. The conserved bromo-adjacent homology domain of yeast Orc1 functions in the selection of DNA replication origins within chromatin. *Genes Dev.* **2010**, *24*, 1418–1433. [CrossRef] [PubMed]

88. Burke, T.W.; Cook, J.G.; Asano, M.; Nevins, J.R. Replication factors MCM2 and ORC1 interact with the histone acetyltransferase HBO1. *J. Biol. Chem.* **2001**, *276*, 15397–15408. [CrossRef] [PubMed]

89. Iizuka, M.; Stillman, B. Histone acetyltransferase HBO1 interacts with the ORC1 subunit of the human initiator protein. *J. Biol. Chem.* **1999**, *274*, 23027–23034. [CrossRef] [PubMed]

90. Miotto, B.; Struhl, K. HBO1 histone acetylase is a coactivator of the replication licensing factor Cdt1. *Genes Dev.* **2008**, *22*, 2633–2638. [CrossRef] [PubMed]

91. Zou, Y.; Bi, X. Positive roles of SAS2 in DNA replication and transcriptional silencing in yeast. *Nucleic Acids Res.* **2008**, *36*, 5189–5200. [CrossRef] [PubMed]

92. Aggarwal, B.D.; Calvi, B.R. Chromatin regulates origin activity in Drosophila follicle cells. *Nature* **2004**, *430*, 372–376. [CrossRef] [PubMed]

93. Saksouk, N.; Avvakumov, N.; Champagne, K.S.; Hung, T.; Doyon, Y.; Cayrou, C.; Paquet, E.; Ullah, M.; Landry, A.J.; Cote, V.; et al. HBO1 HAT complexes target chromatin throughout gene coding regions via multiple PHD finger interactions with histone H3 tail. *Mol. Cell* **2009**, *33*, 257–265. [CrossRef] [PubMed]

94. Wong, P.G.; Glozak, M.A.; Cao, T.V.; Vaziri, C.; Seto, E.; Alexandrow, M. Chromatin unfolding by Cdt1 regulates MCM loading via opposing functions of HBO1 and HDAC11-geminin. *Cell Cycle* **2010**, *9*, 4351–4363. [CrossRef] [PubMed]

95. Sugimoto, N.; Maehara, K.; Yoshida, K.; Yasukouchi, S.; Osano, S.; Watanabe, S.; Aizawa, M.; Yugawa, T.; Kiyono, T.; Kurumizaka, H.; et al. Cdt1-binding protein GRWD1 is a novel histone-binding protein that facilitates MCM loading through its influence on chromatin architecture. *Nucleic Acids Res.* **2015**, *43*, 5898–5911. [CrossRef] [PubMed]

96. Sugimoto, N.; Yugawa, T.; Iizuka, M.; Kiyono, T.; Fujita, M. Chromatin remodeler sucrose nonfermenting 2 homolog (SNF2H) is recruited onto DNA replication origins through interaction with Cdc10 protein-dependent transcript 1 (Cdt1) and promotes pre-replication complex formation. *J. Biol. Chem.* **2011**, *286*, 39200–39210. [CrossRef] [PubMed]

97. Glozak, M.A.; Seto, E. Acetylation/deacetylation modulates the stability of DNA replication licensing factor Cdt1. *J. Biol. Chem.* **2009**, *284*, 11446–11453. [CrossRef] [PubMed]

98. Hofmann, J.F.; Beach, D. Cdt1 is an essential target of the Cdc10/Sct1 transcription factor: Requirement for DNA replication and inhibition of mitosis. *EMBO J.* **1994**, *13*, 425–434. [PubMed]

99. Wang, H.W.; Long, S.; Ciferri, C.; Westermann, S.; Drubin, D.; Barnes, G.; Nogales, E. Architecture and flexibility of the yeast Ndc80 kinetochore complex. *J. Mol. Biol.* **2008**, *383*, 894–903. [CrossRef] [PubMed]

100. Alushin, G.M.; Ramey, V.H.; Pasqualato, S.; Ball, D.A.; Grigorieff, N.; Musacchio, A.; Nogales, E. The Ndc80 kinetochore complex forms oligomeric arrays along microtubules. *Nature* **2010**, *467*, 805–810. [CrossRef] [PubMed]

101. Wan, X.; O'Quinn, R.P.; Pierce, H.L.; Joglekar, A.P.; Gall, W.E.; DeLuca, J.G.; Carroll, C.W.; Liu, S.T.; Yen, T.J.; McEwen, B.F.; et al. Protein architecture of the human kinetochore microtubule attachment site. *Cell* **2009**, *137*, 672–684. [CrossRef] [PubMed]

102. Jee, J.; Mizuno, T.; Kamada, K.; Tochio, H.; Chiba, Y.; Yanagi, K.; Yasuda, G.; Hiroaki, H.; Hanaoka, F.; Shirakawa, M. Structure and mutagenesis studies of the C-terminal region of licensing factor Cdt1 enable the identification of key residues for binding to replicative helicase Mcm proteins. *J. Biol. Chem.* **2010**, *285*, 15931–15940. [CrossRef] [PubMed]

103. Khayrutdinov, B.I.; Bae, W.J.; Yun, Y.M.; Lee, J.H.; Tsuyama, T.; Kim, J.J.; Hwang, E.; Ryu, K.S.; Cheong, H.K.; Cheong, C.; et al. Structure of the Cdt1 C-terminal domain: Conservation of the winged helix fold in replication licensing factors. *Protein Sci.* **2009**, *18*, 2252–2264. [CrossRef] [PubMed]

104. Hornbeck, P.V.; Zhang, B.; Murray, B.; Kornhauser, J.M.; Latham, V.; Skrzypek, E. PhosphoSitePlus, 2014: Mutations, PTMs and recalibrations. *Nucleic Acids Res.* **2015**, *43*, D512–D520. [CrossRef] [PubMed]

105. Lee, C.; Hong, B.; Choi, J.M.; Kim, Y.; Watanabe, S.; Ishimi, Y.; Enomoto, T.; Tada, S.; Kim, Y.; Cho, Y. Structural basis for inhibition of the replication licensing factor Cdt1 by geminin. *Nature* **2004**, *430*, 913–917. [CrossRef] [PubMed]

106. De Marco, V.; Gillespie, P.J.; Li, A.; Karantzelis, N.; Christodoulou, E.; Klompmaker, R.; van Gerwen, S.; Fish, A.; Petoukhov, M.V.; Iliou, M.S.; et al. Quaternary structure of the human Cdt1-Geminin complex regulates DNA replication licensing. *Proc. Natl. Acad. Sci. USA* **2009**, *106*, 19807–19812. [CrossRef] [PubMed]

107. Zhang, T.; Faraggi, E.; Li, Z.; Zhou, Y. Intrinsic Disorder and Semi-disorder Prediction by SPINE-D. *Methods Mol. Biol.* **2017**, *1484*, 159–174. [PubMed]

108. Peng, Z.; Kurgan, L. High-throughput prediction of RNA, DNA and protein binding regions mediated by intrinsic disorder. *Nucleic Acids Res.* **2015**, *43*, e121. [CrossRef] [PubMed]

109. Holm, L.; Rosenstrom, P. Dali server: Conservation mapping in 3D. *Nucleic Acids Res.* **2010**, *38*, W545–W549. [CrossRef] [PubMed]

110. Wei, Z.; Liu, C.; Wu, X.; Xu, N.; Zhou, B.; Liang, C.; Zhu, G. Characterization and structure determination of the Cdt1 binding domain of human minichromosome maintenance (Mcm) 6. *J. Biol. Chem.* **2010**, *285*, 12469–12473. [CrossRef] [PubMed]

111. Liu, C.; Wu, R.; Zhou, B.; Wang, J.; Wei, Z.; Tye, B.K.; Liang, C.; Zhu, G. Structural insights into the Cdt1-mediated MCM2–7 chromatin loading. *Nucleic Acids Res.* **2012**, *40*, 3208–3217. [CrossRef] [PubMed]

112. Coulombe, P.; Gregoire, D.; Tsanov, N.; Mechali, M. A spontaneous Cdt1 mutation in 129 mouse strains reveals a regulatory domain restraining replication licensing. *Nat. Commun.* **2013**, *4*, 2065. [CrossRef] [PubMed]

113. Ballabeni, A.; Melixetian, M.; Zamponi, R.; Masiero, L.; Marinoni, F.; Helin, K. Human geminin promotes pre-RC formation and DNA replication by stabilizing CDT1 in mitosis. *EMBO J.* **2004**, *23*, 3122–3132. [CrossRef] [PubMed]

114. Whitehall, S.; Stacey, P.; Dawson, K.; Jones, N. Cell cycle-regulated transcription in fission yeast: Cdc10-Res protein interactions during the cell cycle and domains required for regulated transcription. *Mol. Biol. Cell.* **1999**, *10*, 3705–3715. [CrossRef] [PubMed]

115. Duronio, R.J.; Xiong, Y. Signaling pathways that control cell proliferation. *Cold Spring Harb. Perspect. Biol.* **2013**, *5*, a008904. [CrossRef] [PubMed]

116. Cross, F.R.; Buchler, N.E.; Skotheim, J.M. Evolution of networks and sequences in eukaryotic cell cycle control. *Philos. Trans. R. Soc. Lond. B Biol. Sci.* **2011**, *366*, 3532–3544. [CrossRef] [PubMed]

117. Yoshida, K.; Inoue, I. Regulation of Geminin and Cdt1 expression by E2F transcription factors. *Oncogene* **2004**, *23*, 3802–3812. [CrossRef] [PubMed]

118. Valovka, T.; Schonfeld, M.; Raffeiner, P.; Breuker, K.; Dunzendorfer-Matt, T.; Hartl, M.; Bister, K. Transcriptional control of DNA replication licensing by Myc. *Sci. Rep.* **2013**, *3*, 3444. [CrossRef] [PubMed]

119. Zhang, R.; Wu, J.; Ferrandon, S.; Glowacki, K.J.; Houghton, J.A. Targeting GLI by GANT61 involves mechanisms dependent on inhibition of both transcription and DNA licensing. *Oncotarget* **2016**. [CrossRef] [PubMed]

120. Petroski, M.D.; Deshaies, R.J. Function and regulation of cullin-RING ubiquitin ligases. *Nat. Rev. Mol. Cell. Biol.* **2005**, *6*, 9–20. [CrossRef] [PubMed]

121. Havens, C.G.; Walter, J.C. Mechanism of CRL4(Cdt2), a PCNA-dependent E3 ubiquitin ligase. *Genes Dev.* **2011**, *25*, 1568–1582. [CrossRef] [PubMed]

122. Abbas, T.; Dutta, A. CRL4(Cdt2): Master coordinator of cell cycle progression and genome stability. *Cell Cycle* **2011**, *10*, 241–249. [CrossRef] [PubMed]

123. Li, X.; Zhao, Q.; Liao, R.; Sun, P.; Wu, X. The SCF(Skp2) ubiquitin ligase complex interacts with the human replication licensing factor Cdt1 and regulates Cdt1 degradation. *J. Biol. Chem.* **2003**, *278*, 30854–30858. [CrossRef] [PubMed]

124. Liu, E.; Li, X.; Yan, F.; Zhao, Q.; Wu, X. Cyclin-dependent kinases phosphorylate human Cdt1 and induce its degradation. *J. Biol. Chem.* **2004**, *279*, 17283–17288. [CrossRef] [PubMed]

125. Sugimoto, N.; Tatsumi, Y.; Tsurumi, T.; Matsukage, A.; Kiyono, T.; Nishitani, H.; Fujita, M. Cdt1 phosphorylation by cyclin A-dependent kinases negatively regulates its function without affecting geminin binding. *J. Biol. Chem.* **2004**, *279*, 19691–19697. [CrossRef] [PubMed]

126. Nishitani, H.; Sugimoto, N.; Roukos, V.; Nakanishi, Y.; Saijo, M.; Obuse, C.; Tsurimoto, T.; Nakayama, K.I.; Nakayama, K.; Fujita, M.; et al. Two E3 ubiquitin ligases, SCF-Skp2 and DDB1-Cul4, target human Cdt1 for proteolysis. *EMBO J.* **2006**, *25*, 1126–1136. [CrossRef] [PubMed]

127. Havens, C.G.; Walter, J.C. Docking of a specialized PIP Box onto chromatin-bound PCNA creates a degron for the ubiquitin ligase CRL4Cdt2. *Mol. Cell* **2009**, *35*, 93–104. [CrossRef] [PubMed]

128. Higa, L.A.; Mihaylov, I.S.; Banks, D.P.; Zheng, J.; Zhang, H. Radiation-mediated proteolysis of CDT1 by CUL4-ROC1 and CSN complexes constitutes a new checkpoint. *Nat. Cell Biol.* **2003**, *5*, 1008–1015. [CrossRef] [PubMed]

129. Nishitani, H.; Lygerou, Z.; Nishimoto, T. Proteolysis of DNA replication licensing factor Cdt1 in S-phase is performed independently of geminin through its N-terminal region. *J. Biol. Chem.* **2004**, *279*, 30807–30816. [CrossRef] [PubMed]

130. Rizzardi, L.F.; Coleman, K.E.; Varma, D.; Matson, J.P.; Oh, S.; Cook, J.G. CDK1-dependent inhibition of the E3 ubiquitin ligase CRL4(CDT2) ensures robust transition from S Phase to Mitosis. *J. Biol. Chem.* **2015**, *290*, 556–567. [CrossRef] [PubMed]

131. Hernandez-Perez, S.; Cabrera, E.; Amoedo, H.; Rodriguez-Acebes, S.; Koundrioukoff, S.; Debatisse, M.; Mendez, J.; Freire, R. USP37 deubiquitinates Cdt1 and contributes to regulate DNA replication. *Mol. Oncol.* **2016**, *10*, 1196–1206. [CrossRef] [PubMed]

132. Coleman, K.E.; Grant, G.D.; Haggerty, R.A.; Brantley, K.; Shibata, E.; Workman, B.D.; Dutta, A.; Varma, D.; Purvis, J.E.; Cook, J.G. Sequential replication-coupled destruction at G1/S ensures genome stability. *Genes Dev.* **2015**, *29*, 1734–1746. [CrossRef] [PubMed]

133. Whittaker, A.J.; Royzman, I.; Orr-Weaver, T.L. Drosophila double parked: A conserved, essential replication protein that colocalizes with the origin recognition complex and links DNA replication with mitosis and the down-regulation of S phase transcripts. *Genes Dev.* **2000**, *14*, 1765–1776. [PubMed]

134. Wohlschlegel, J.A.; Dwyer, B.T.; Dhar, S.K.; Cvetic, C.; Walter, J.C.; Dutta, A. Inhibition of eukaryotic DNA replication by geminin binding to Cdt1. *Science* **2000**, *290*, 2309–2312. [CrossRef] [PubMed]

135. McGarry, T.J.; Kirschner, M.W. Geminin, an inhibitor of DNA replication, is degraded during mitosis. *Cell* **1998**, *93*, 1043–1053. [CrossRef]

136. Min, M.; Lindon, C. Substrate targeting by the ubiquitin-proteasome system in mitosis. *Semin. Cell Dev. Biol.* **2012**, *23*, 482–491. [CrossRef] [PubMed]

137. McLean, J.R.; Chaix, D.; Ohi, M.D.; Gould, K.L. State of the APC/C: Organization, function, and structure. *Crit. Rev. Biochem. Mol. Biol.* **2011**, *46*, 118–136. [CrossRef] [PubMed]

138. You, Z.; Ode, K.L.; Shindo, M.; Takisawa, H.; Masai, H. Characterization of conserved arginine residues on Cdt1 that affect licensing activity and interaction with Geminin or Mcm complex. *Cell Cycle* **2016**, *15*, 1213–1226. [CrossRef] [PubMed]

139. Luo, L.; Yang, X.; Takihara, Y.; Knoetgen, H.; Kessel, M. The cell-cycle regulator geminin inhibits Hox function through direct and polycomb-mediated interactions. *Nature* **2004**, *427*, 749–753. [CrossRef] [PubMed]

140. Del Bene, F.; Tessmar-Raible, K.; Wittbrodt, J. Direct interaction of geminin and Six3 in eye development. *Nature* **2004**, *427*, 745–749. [CrossRef] [PubMed]

141. Yang, V.S.; Carter, S.A.; Hyland, S.J.; Tachibana-Konwalski, K.; Laskey, R.A.; Gonzalez, M.A. Geminin escapes degradation in G1 of mouse pluripotent cells and mediates the expression of Oct4, Sox2, and Nanog. *Curr. Biol.* **2011**, *21*, 692–699. [CrossRef] [PubMed]

142. Yanagi, K.; Mizuno, T.; You, Z.; Hanaoka, F. Mouse geminin inhibits not only Cdt1-MCM6 interactions but also a novel intrinsic Cdt1 DNA binding activity. *J. Biol. Chem.* **2002**, *277*, 40871–40880. [CrossRef] [PubMed]

143. Miotto, B.; Struhl, K. JNK1 phosphorylation of Cdt1 inhibits recruitment of HBO1 histone acetylase and blocks replication licensing in response to stress. *Mol. Cell* **2011**, *44*, 62–71. [CrossRef] [PubMed]

144. Chandrasekaran, S.; Tan, T.X.; Hall, J.R.; Cook, J.G. Stress-stimulated mitogen-activated protein kinases control the stability and activity of the Cdt1 DNA replication licensing factor. *Mol. Cell. Biol.* **2011**, *31*, 4405–4416. [CrossRef] [PubMed]

145. Takeda, D.Y.; Parvin, J.D.; Dutta, A. Degradation of Cdt1 during S phase is Skp2-independent and is required for efficient progression of mammalian cells through S phase. *J. Biol. Chem.* **2005**, *280*, 23416–23423. [CrossRef] [PubMed]

146. Thornton, T.M.; Rincon, M. Non-classical p38 map kinase functions: Cell cycle checkpoints and survival. *Int. J. Biol. Sci.* **2009**, *5*, 44–51. [CrossRef] [PubMed]

147. Tsunematsu, T.; Takihara, Y.; Ishimaru, N.; Pagano, M.; Takata, T.; Kudo, Y. Aurora-A controls pre-replicative complex assembly and DNA replication by stabilizing geminin in mitosis. *Nat. Commun.* **2013**, *4*, 1885. [CrossRef] [PubMed]

148. Hall, J.R.; Lee, H.O.; Bunker, B.D.; Dorn, E.S.; Rogers, G.C.; Duronio, R.J.; Cook, J.G. Cdt1 and Cdc6 are destabilized by rereplication-induced DNA damage. *J. Biol. Chem.* **2008**, *283*, 25356–25363. [CrossRef]

149. Fujita, M. Cdt1 revisited: Complex and tight regulation during the cell cycle and consequences of deregulation in mammalian cells. *Cell Div.* **2006**, *1*, 22. [CrossRef] [PubMed]

150. Seo, J.; Chung, Y.S.; Sharma, G.G.; Moon, E.; Burack, W.R.; Pandita, T.K.; Choi, K. Cdt1 transgenic mice develop lymphoblastic lymphoma in the absence of p53. *Oncogene* **2005**, *24*, 8176–8186. [CrossRef] [PubMed]

151. Petropoulou, C.; Kotantaki, P.; Karamitros, D.; Taraviras, S. Cdt1 and Geminin in cancer: Markers or triggers of malignant transformation? *Front. Biosci.* **2008**, *13*, 4485–4494. [CrossRef] [PubMed]

152. Galanos, P.; Vougas, K.; Walter, D.; Polyzos, A.; Maya-Mendoza, A.; Haagensen, E.J.; Kokkalis, A.; Roumelioti, F.M.; Gagos, S.; Tzetis, M.; et al. Chronic p53-independent p21 expression causes genomic instability by deregulating replication licensing. *Nat. Cell Biol.* **2016**, *18*, 777–789. [CrossRef] [PubMed]

153. Johnson, J.; Thijssen, B.; McDermott, U.; Garnett, M.; Wessels, L.F.; Bernards, R. Targeting the RB-E2F pathway in breast cancer. *Oncogene* **2016**, *35*, 4829–4835. [CrossRef] [PubMed]

154. Dyson, N.J. RB1: A prototype tumor suppressor and an enigma. *Genes Dev.* **2016**, *30*, 1492–1502. [CrossRef] [PubMed]

155. Karavias, D.; Maroulis, I.; Papadaki, H.; Gogos, C.; Kakkos, S.; Karavias, D.; Bravou, V. Overexpression of CDT1 Is a Predictor of Poor Survival in Patients with Hepatocellular Carcinoma. *J. Gastrointest. Surg.* **2016**, *20*, 568–579. [CrossRef] [PubMed]

156. Bicknell, L.S.; Bongers, E.M.; Leitch, A.; Brown, S.; Schoots, J.; Harley, M.E.; Aftimos, S.; Al-Aama, J.Y.; Bober, M.; Brown, P.A.; et al. Mutations in the pre-replication complex cause Meier-Gorlin syndrome. *Nat. Genet.* **2011**, *43*, 356–359. [CrossRef] [PubMed]

157. Bicknell, L.S.; Walker, S.; Klingseisen, A.; Stiff, T.; Leitch, A.; Kerzendorfer, C.; Martin, C.A.; Yeyati, P.; Al Sanna, N.; Bober, M.; et al. Mutations in ORC1, encoding the largest subunit of the origin recognition complex, cause microcephalic primordial dwarfism resembling Meier-Gorlin syndrome. *Nat. Genet.* **2011**, *43*, 350–355. [CrossRef] [PubMed]

158. de Munnik, S.A.; Bicknell, L.S.; Aftimos, S.; Al-Aama, J.Y.; van Bever, Y.; Bober, M.B.; Clayton-Smith, J.; Edrees, A.Y.; Feingold, M.; Fryer, A.; et al. Meier-Gorlin syndrome genotype-phenotype studies: 35 individuals with pre-replication complex gene mutations and 10 without molecular diagnosis. *Eur. J. Hum. Genet.* **2012**, *20*, 598–606. [CrossRef] [PubMed]

159. Burrage, L.C.; Charng, W.L.; Eldomery, M.K.; Willer, J.R.; Davis, E.E.; Lugtenberg, D.; Zhu, W.; Leduc, M.S.; Akdemir, Z.C.; Azamian, M.; et al. De Novo GMNN Mutations Cause Autosomal-Dominant Primordial Dwarfism Associated with Meier-Gorlin Syndrome. *Am. J. Hum. Genet.* **2015**, *97*, 904–913. [CrossRef] [PubMed]

160. Fenwick, A.L.; Kliszczak, M.; Cooper, F.; Murray, J.; Sanchez-Pulido, L.; Twigg, S.R.; Goriely, A.; McGowan, S.J.; Miller, K.A.; Taylor, I.B.; et al. Mutations in CDC45, Encoding an Essential Component of the Pre-initiation Complex, Cause Meier-Gorlin Syndrome and Craniosynostosis. *Am. J. Hum. Genet.* **2016**, *99*, 125–138. [CrossRef] [PubMed]

161. Shima, N.; Alcaraz, A.; Liachko, I.; Buske, T.R.; Andrews, C.A.; Munroe, R.J.; Hartford, S.A.; Tye, B.K.; Schimenti, J.C. A viable allele of Mcm4 causes chromosome instability and mammary adenocarcinomas in mice. *Nat. Genet.* **2007**, *39*, 93–98. [CrossRef] [PubMed]

162. Guernsey, D.L.; Matsuoka, M.; Jiang, H.; Evans, S.; Macgillivray, C.; Nightingale, M.; Perry, S.; Ferguson, M.; LeBlanc, M.; Paquette, J.; et al. Mutations in origin recognition complex gene ORC4 cause Meier-Gorlin syndrome. *Nat. Genet.* **2011**, *43*, 360–364. [CrossRef] [PubMed]

163. Prasanth, S.G.; Prasanth, K.V.; Stillman, B. Orc6 involved in DNA replication, chromosome segregation, and cytokinesis. *Science* **2002**, *297*, 1026–1031. [CrossRef] [PubMed]

164. Prasanth, S.G.; Prasanth, K.V.; Siddiqui, K.; Spector, D.L.; Stillman, B. Human Orc2 localizes to centrosomes, centromeres and heterochromatin during chromosome inheritance. *EMBO J.* **2004**, *23*, 2651–2663. [CrossRef] [PubMed]

165. Hemerly, A.S.; Prasanth, S.G.; Siddiqui, K.; Stillman, B. Orc1 controls centriole and centrosome copy number in human cells. *Science* **2009**, *323*, 789–793. [CrossRef] [PubMed]

Review

Elaborated Action of the Human Primosome

Andrey G. Baranovskiy and Tahir H. Tahirov *

Eppley Institute for Research in Cancer and Allied Diseases, Fred & Pamela Buffett Cancer Center, University of Nebraska Medical Center, Omaha, NE 68198, USA; abaranovskiy@unmc.edu
* Correspondence: ttahirov@unmc.edu; Tel.: +1-402-559-7607

Academic Editor: Eishi Noguchi
Received: 1 December 2016; Accepted: 4 February 2017; Published: 8 February 2017

Abstract: The human primosome is a 340-kilodalton complex of primase (DNA-dependent RNA polymerase) and DNA polymerase α, which initiates genome replication by synthesizing chimeric RNA-DNA primers for DNA polymerases δ and ε. Accumulated biochemical and structural data reveal the complex mechanism of concerted primer synthesis by two catalytic centers. First, primase generates an RNA primer through three steps: initiation, consisting of dinucleotide synthesis from two nucleotide triphosphates; elongation, resulting in dinucleotide extension; and termination, owing to primase inhibition by a mature 9-mer primer. Then Polα, which works equally well on DNA:RNA and DNA:DNA double helices, intramolecularly catches the template primed by a 9-mer RNA and extends the primer with dNTPs. All primosome transactions are highly coordinated by autoregulation through the alternating activation/inhibition of the catalytic centers. This coordination is mediated by the small C-terminal domain of the primase accessory subunit, which forms a tight complex with the template:primer, shuttles between the primase and DNA polymerase active sites, and determines their access to the substrate.

Keywords: DNA replication; human; primosome; primase; DNA polymerase α; protein-DNA interaction; RNA synthesis; initiation; termination; steric hindrance

1. Introduction

In all eukaryotic organisms, genome replication depends on activity of the primosome, a four-subunit complex of DNA primase and DNA polymerase α (Polα) [1]. The primosome initiates synthesis of both the leading and lagging strands by making chimeric RNA-DNA primers, which are required for the loading of replication factor C (RFC), proliferating cell nuclear antigen (PCNA), and replicative DNA polymerases δ and ε [2,3]. At each origin, the primosome is involved only once for leading strand initiation, while it starts every Okazaki fragment on the discontinuously synthesized lagging strand. Given the sizes of Okazaki fragments (165-bp) and chimeric primers (30–35 nucleotides), the primosome synthesizes up to 20% of the lagging strand and, therefore, approximately 10% of the genome [4,5]. During maturation of the Okazaki fragments, both the RNA and a significant portion of the DNA track of a chimeric primer are being deleted [6]. As a result, DNA synthesized by Polα comprises approximately 1.5% of the mature genome [7]. These regions are mutation hotspots because Polα has relatively low fidelity due to the absence of proofreading activity. Thus, despite low retention of Polα-synthesized DNA tracks in the mature genome, the primosome has a large impact on genome stability and evolution. Recently, it has been shown that the primosome is responsible for generation of RNA-DNA fragments in the cytosol and that it regulates the activation of type I interferons [8].

The primosome synthesizes chimeric primers in a highly coordinated fashion. RNA primer synthesis by primase involves three steps: initiation, elongation, and termination [9,10]. During the initiation step, primase binds the DNA template and two cognate rNTPs (one at the initiation site and the other at the elongation [catalytic] site) and catalyzes the formation of a dinucleotide [11,12].

Extension of the RNA is restricted due to the intrinsic property of primase to terminate synthesis at a strictly defined point [13]. Then Polα intramolecularly captures the mature RNA primer for subsequent extension by dNTPs [11,14,15]. Recent breakthroughs in structural studies of the human primosome [13] and its components [16–22] (Table 1) allow for accurate modeling of the primosome conformations during all stages of chimeric primer synthesis.

Table 1. List of the high-resolution structures of the human primosome and its domains.

PDB ID	Resolution, (Å)	Protein Construct	Structural Metals	Cofactors	Deposition Date	Reference
3L9Q	1.7	p58(272–464)	4Fe-4S		5 January 2010	[21]
3Q36	2.5	p58(266–457)	4Fe-4S		21 December 2010	[22]
4BPU	2.7	p49 [a]/p58(1–253)	Zn		28 May 2013	[16]
4BPW	3.0	p49 [a]/p58(1–253)	Zn	UTP, Mg	28 May 2013	[16]
4BPX	3.4	p49 [a]/p58(19–253) [b]/ p180(1445–1462)	Zn		28 May 2013	[16]
4LIK	1.7	p49(1–390) [c]	Zn		2 July 2013	[23]
4LIL	2.6	p49(1–390) [c]	Zn	UTP, Mn	2 July 2013	[23]
4MHQ	2.2	p49	Zn		30 August 2013	
4QCL	2.2	p180(335–1257) [d]	Zn	DNA:RNA, dCTP, Mg	12 May 2014	[17]
4Q5V	2.52	p180(335–1257) [d]		DNA:RNA, aphidicolin	17 April 2014	[17]
4RR2	2.65	p49/p58	Zn, 4Fe-4S		5 November 2014	[18]
4Y97	2.51	p180(1265–1444)/p70	Zn		17 February 2015	[19]
5DQO	2.3	p58(272–464) [e]	4Fe-4S		15 September 2015	
5EXR	3.6	p49/p58/p180 [d]/p70	Zn, 4Fe-4S		24 November 2015	[13]
5F0Q	2.2	p58(266–456)	4Fe-4S	DNA:RNA, Mg	28 November 2015	[13]
5F0S	3.0	p58(266–456)	4Fe-4S	DNA:RNA, Mn	28 November 2015	[13]
5IUD	3.3	p180(338–1255)		DNA:DNA	17 March 2016	[20]

[a] Mutations Lys-72-Ala and Met-73-Ala; [b] N-terminus of p58 is fused to the primase-binding peptide of p180 via a 15 amino acid linker; [c] Residues 360–379 and 409–420 are deleted; [d] Mutation Val-516-Ala; [e] Mutation Tyr-347-Phe.

2. Organization of the Human Primosome

Human Polα belongs to the B family of DNA Pols and is comprised of a 166-kDa catalytic subunit (p180) and a 66-kDa accessory subunit (p70). The catalytic domain of p180 (p180core) possesses DNA-polymerizing activity but has no proofreading exonuclease activity, in contrast to other replicative DNA Pols, δ and ε. The C-terminal domain of p180 (p180$_C$) is flexibly connected to a catalytic core by a 15-residue-long linker, and it contains two conserved zinc-binding modules, Zn1 and Zn2 (Figure 1), where each zinc is coordinated by four cysteines [19,24]. Zn2 and the helical region between the two zinc-binding modules provide the extended interaction interface (~4000 Å^2) with p70, while the short peptide (1447–1455) mediates the interaction between Polα and primase [13,19]. The N-terminus of p180 is predicted to be poorly folded and has no conserved motifs required for primosome function. The structural information for this region is limited to a small peptide in the catalytic subunit of yeast Polα (residues 140–147) that mediates interaction with the replisome [25]. The accessory B subunit (p70; also known as p68) consists of a globular N-terminal domain (NTD or p70$_N$), a catalytically dead phosphodiesterase domain (PDE), and an oligonucleotide/oligosaccharide-binding (OB) domain. The OB domain is embedded into the PDE domain, representing the common feature of B-family DNA Pols [19,26]. The globular NTD is attached to the PDE via a long flexible linker and participates in interactions with other DNA replication proteins [19,27,28].

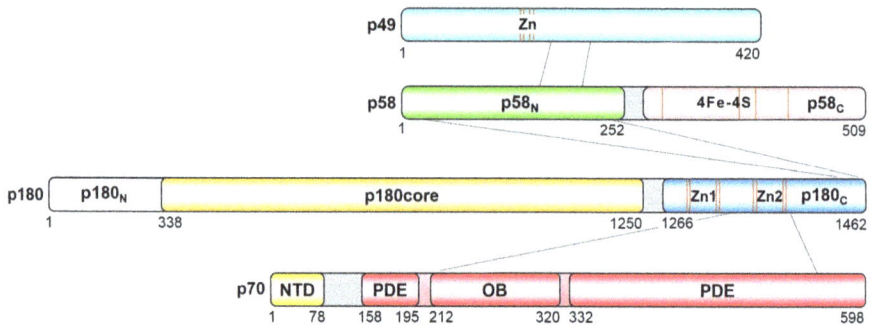

Figure 1. Schematic representation of the domain organization in the human primosome. The borders of the regions participating in intersubunit interactions are designated by dotted lines. Positions of the conserved cysteines coordinating zinc or [4Fe-4S] cluster are indicated by orange lines. The linkers responsible for flexible connections between domains are colored gray.

Human primase consists of a 50-kda catalytic subunit (p49; also known as p48, PRIM1, Pri1, and PriS) and a 59-kDa regulatory subunit (p58; also known as PRIM2, Pri2, and PriL) (Figure 1). Eukaryotic and archaeal primases have a similar structural organization, which indicates a common evolutionary ancestor [29]. In contrast to prokaryotic primases, the zinc-binding motif of eukaryotic/archaeal primases is integrated into the "prim" fold of the catalytic subunit and probably plays only a structural role [16,18,30–32]. p58 has two distinct domains: the N-terminal domain (p58$_N$) with a mixed α/β-fold and the all-helical C-terminal domain (p58$_C$), connected by an 18-residue linker (253–270) [18]. Similar to yeast primase [33], four cysteines of p58$_C$ coordinate an iron-sulfur cluster ([4Fe-4S]) which is buried inside of the domain and is important for p58$_C$ folding [21,22,34,35].

There was one report claiming that all four *Saccharomyces cerevisiae* B-family DNA polymerases coordinate the [4Fe-4S] cluster at the second cysteine-rich module (referred to here as Zn2) of the C-terminal domain of the catalytic subunits (CTD, analog of p180$_C$) [36]. However, the provided experimental evidence was uncertain for Polα. For example, Polα CTD purified under anaerobic conditions contained only 0.1 mol non-heme iron and acid-labile sulfide per mol CTD, while CTDs of other B-family DNA polymerases (δ, ε, and ζ) contained 2.0 to 2.6 Fe and S per monomer. Coordination of the [4Fe-4S] cluster by Polα CTD has not been confirmed in subsequent studies where high-purity stoichiometric Polα complexes have been obtained [37,38]. Structural studies of yeast and human Polα do not support the presence of an iron-sulfur cluster in Polα CTD; only two zinc ions coordinated by Zn1 and Zn2 modules were seen [13,19,39]. Zn2 is important for interaction between Polα subunits and snugly fits the docking site on the OB domain. Coordination of an [4Fe-4S] cluster by the Zn2 module would certainly change its shape and disrupt the interaction between the catalytic and B subunits. It was also shown that partially purified Polε CTD contained significant levels of iron, whereas its complex with the B subunit was iron-free [37]. These data support the idea that the CTDs of Polα and Polε with an inadvertently misincorporated iron-sulfur cluster cannot form stable complexes with the corresponding B subunits. It is worth noting, that placing an affinity tag on the B subunit is crucial for obtaining stoichiometric complexes of B-family DNA polymerases, because it prevents the contamination of preparations with a free catalytic subunit.

Substantial conformational changes in the primosome are essential for seamlessly carrying out the entire cycle of RNA-DNA primer synthesis. The primosome has three functional centers: the RNA- and DNA-polymerizing centers, located on p49 and p180core, respectively [11,40,41], and regulatory p58$_C$, which is responsible for template:primer binding and translocation from primase to Polα [42,43]. The structure of the human primosome reveals an elongated platform p49-p58$_N$-p180$_C$-p70 (Figure 2) that can hold p180core and p58$_C$ either stationary, by docking in inactive form, or flexibly, by linkers during various stages of primer synthesis [13]. Interestingly, the points of the linker's attachment to the platform are fairly close despite their origination from different subunits. The platform itself has limited flexibility because $p58_N$ subdomains were shown to oscillate by several degrees relative each other [13,16,18]. $p58_N$ could be considered as a core of the platform; its smaller subdomain interacts with p49, while the larger, α-helical subdomain interacts with $p180_C$ and is connected by the linker to $p58_C$ (Figure 2). Such organization of the primosome provides significant freedom for the functional centers in their movement relative to each other.

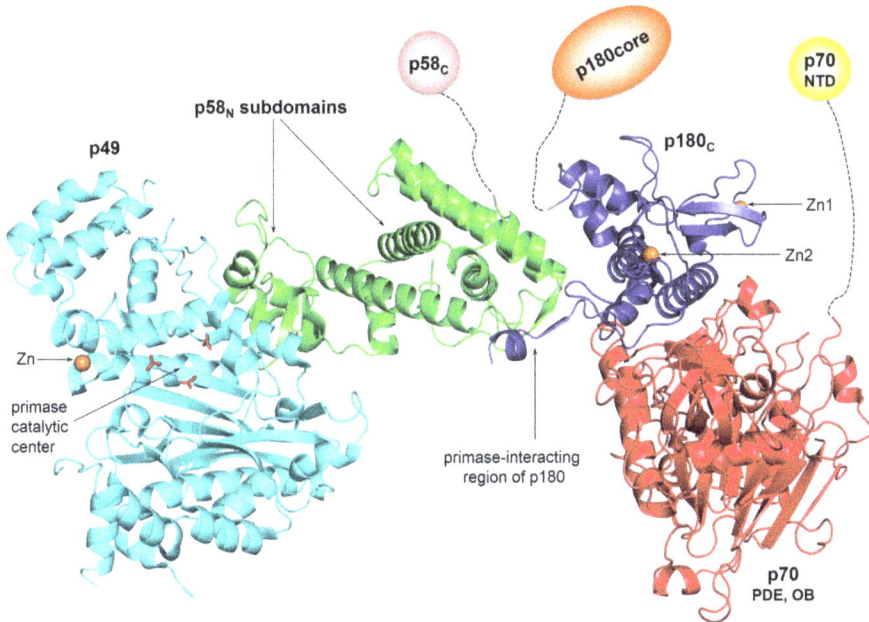

Figure 2. The platform of the human primosome. Coordinates of the human primosome (PDB ID 5EXR) were used to represent the platform structure. The color scheme for domains is the same as in Figure 1. The positions of $p58_C$ and p180core, as well as the linkers connecting them to the platform, vary depending on the primer synthesis step. For space-saving purposes, $p58_C$, p180core, and p70-NTD are shown at reduced scale relative to the platform. All figures were prepared using the PyMOL Molecular Graphics System (version 1.8, Schrödinger, LLC).

3. Interaction of Human Primase with a Template:Primer

Recent biochemical and structural studies finally unveiled the mechanism of human primase interaction with a DNA template and an RNA primer, where $p58_C$ firmly holds the DNA:RNA duplex while p49 catalyzes the attachment of rNTPs to the 3′-end of the primer [13,43]. $p58_C$ specifically recognizes the junction at the 5′-end of the RNA primer, which contains the 5′-triphosphate group (Figure 3A). The β- and γ-phosphates of the triphosphate moiety make six hydrogen bonds with $p58_C$, explaining the critical role of these phosphates in primase activity and their affinity for the DNA:RNA substrate [43–45]. Moreover, recognition of the 5′-triphosphate prevents $p58_C$ rotation around the duplex, thereby strictly determining the position and orientation of $p58_C$ relative to the platform and p180core during all primosome transactions. Coordination of a divalent metal probably stabilizes the conformation of the triphosphate group and its complex with $p58_C$. Arg-306 interacts with both the β- and γ-phosphates and is critical for primase activity, especially during dinucleotide synthesis [16,42]. There are no other contacts between $p58_C$ and the RNA primer except for stacking between His-303 and the base of the 5′-GTP (Figure 3B).

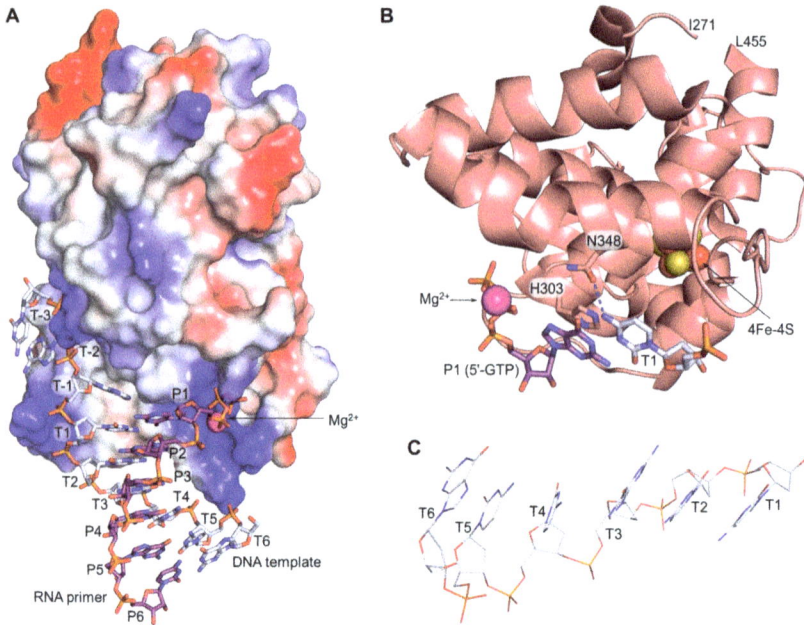

Figure 3. Interaction of p58$_C$ with a DNA template primed by RNA. (**A**) p58$_C$ specifically recognizes the DNA:RNA junction at the primer 5′-end containing the triphosphate. The p58$_C$ surface is represented by the vacuum electrostatic potential at 20% transparency; (**B**) mechanism of p58$_C$ specificity to a purine at the initiation site. The hydrogen bond is depicted by dashed blue line; (**C**) DNA template bends between T3 and T4. All parts of the figure were drawn using the coordinates of the p58$_C$/DNA:RNA complex (PDB ID 5F0Q).

The structure of p58$_C$/DNA:RNA revealed the location and organization of the initiation site with bound initiating GTP which forms the 5′-end of the nascent dinucleotide [13]. The critical role of p58$_C$ in binding the initiating nucleotide explains why p49 is able to extend RNA fragments but cannot initiate synthesis from two rNTPs [11]. The relatively weak coordination of the initiating rNTP by only six hydrogen bonds explains the low affinity of this site (K_m(ATP) = 3 mM), which is 11-fold lower compared to the elongation site [11]. Human primase has no obvious sequence specificity except the well-known preference of the initiation site for GTP/ATP [9,46], which is probably due to the cumulative effect of two factors. First, His-303 demonstrates good stacking with the initiating purine, while its ring would only partially overlap with a pyrimidine base (Figure 2B). Secondly, Asn-348 can use its carbonyl or amino group to form a hydrogen bond with N4 or O4 of the templating cytidine or thymine, respectively.

p58$_C$ forms 13 hydrogen bonds with the template, the majority of which are located near the junction. The presence of 19 hydrogen bonds between p58$_C$ and DNA:RNA results in a stable complex with a K_d of 32.7 nM [43]. For comparison, the catalytic core domains of human Polε and Polα bind the template:primer with 2.4-fold and 10-fold lower affinity, respectively [20,47]. The intact human primase and p58$_C$ have similar affinities for DNA:RNA, supporting the idea that p58$_C$ is a major DNA-binding domain in the primosome [43]. The primer 5′-triphosphate and the template 3′-overhang exhibit a synergistic effect on duplex binding by primase and its RNA-polymerizing activity [43]. The dependence of p58$_C$ affinity on the stability of the DNA:RNA duplex explains the abortive character of RNA synthesis at the beginning of the elongation stage and in the case of AT-rich templates [12,18]. The structure of p58$_C$/DNA:RNA complex explains why the His-401-Arg mutation in yeast primase leads to lethality [34]. The bulky side chain of the arginine in place of His-351 (corresponds to His-401 in yeast primase) disrupts the interaction with DNA:RNA because of steric hindrance with the template and/or the DNA-interacting loop containing residues 355–366. p58$_C$ affects the template conformation in the DNA:RNA duplex: it maintains the B-DNA conformation of the template deoxyriboses that are in contact with p58$_C$ (T1–T3), while three nucleotides at the 5′-end (T4–T5) are in the A-DNA conformation (Figure 3C).

4. Mechanisms of RNA Synthesis Initiation, Elongation, and Termination

The structure of p58$_C$/DNA:RNA (PDB ID 5F0Q) together with the structures of p49–p58 (PDB ID 4RR2) [18] and p49–p58(1–253)/UTP (PDB ID 4BPW) [16] allows for obtaining accurate models of primase during all steps of RNA synthesis [13]. Structure-based modeling by superimposition of the second nucleotide of the primer from the p58$_C$/DNA:RNA complex with UTP bound at the elongation site of p49 reveals the compact initiation complex (Figure 4) with good shape complementarity and eight potential hydrogen bonds between p49 and p58$_C$ [13]. This organization of the initiation complex where the active site is shared by p49 and p58$_C$ results in cooperative binding of four template nucleotides and initiating rNTP (Figure 5). The active site is able to accommodate only three template nucleotides which are placed between Tyr-54 of p49 at the 5′-end and Met-307 of p58$_C$ at the 3′-end. p49 can make only six hydrogen bonds with a template because of its shallow DNA-binding interface (Figure 5). The active site elements accommodated by two flanking β-sheets of p49 are adopted for the common mechanism of nucleic acids synthesis through the coordination of two divalent metals [48].

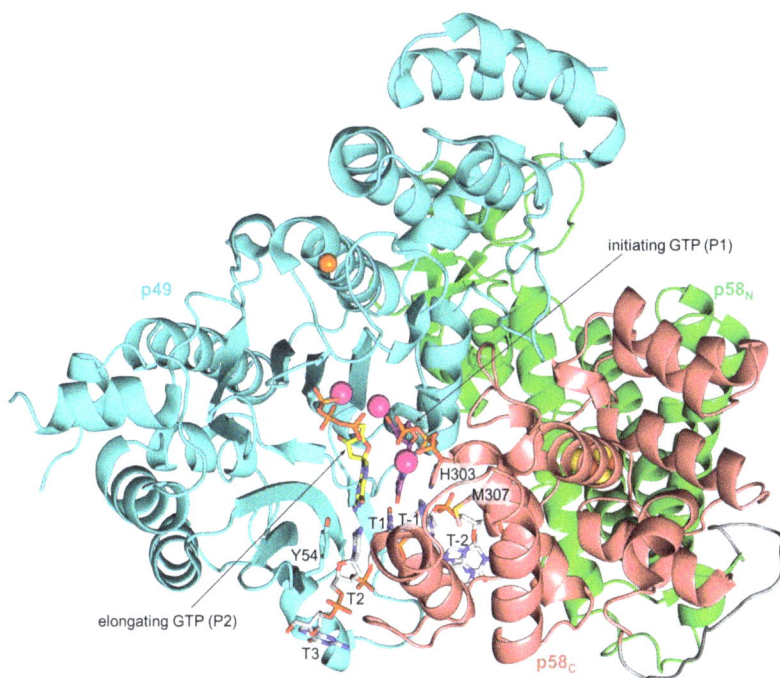

Figure 4. The model of human primase in the initiation complex with a DNA template and two GTP molecules. The linker between p58$_N$ and p58$_C$ colored gray is shown for reference purposes only. The carbons of the DNA template, initiating GTP, and elongating GTP are colored gray, purple, and yellow, respectively. The atoms of zinc, magnesium, iron, and sulfur are represented as spheres and colored orange, magenta, red, and yellow, respectively.

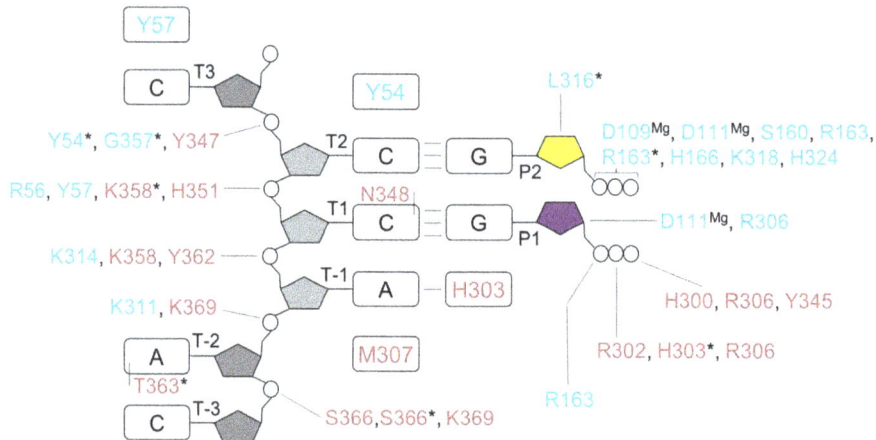

Figure 5. Interaction of human primase with a DNA template and rNTPs during RNA synthesis initiation. The color scheme is the same as in Figure 4. The residues of p49 interacting with the DNA template and the initiating GTP are identified from the model of the initiation complex. The asterisk indicates that a main-chain atom of the amino acid forms a hydrogen bond with a nucleotide. Amino acids participating in stacking interactions with nucleotides are shown in rectangular boxes. Interactions of aspartates 109 and 111 with both rNTPs are mediated by the Mg^{2+} ions.

The model of the initiation complex revealed that p49 participates in pre-catalytic positioning of the initiating GTP by making three hydrogen bonds: Arg-163 with the α-phosphate, Asp-306 with the O2′ of a ribose, and the bond between Asp-111-coordinated Mg^{2+} and the O3′ of a ribose (Figure 5). During the elongation stage of RNA synthesis, the initiation site disintegrates due to the growing distance between its structural elements provided by both subunits: $p58_C$ continues holding the 5′-end of the primer, while p49 is establishing the above-described three hydrogen bonds with the growing 3′-end, because during primer extension the 3′-terminal nucleotide occupies the same space on p49 as the initiating rNTP. The interaction between the O2′ of the initiating GTP and Asp-306 of p49 explains the strict preference for ribonucleotides at the initiation step [46]. Consistently, the primase is also sensitive to the presence of the O2′ at the primer 3′-terminus during its extension [38,49]. Replacement of Asp-306 by Ala severely affects primase activity, but to a lesser extent compared to alanine substitutions of Asp-109 or Asp-111 which coordinate the catalytic Mg^{2+} ions [41]. In contrast, the elongation site demonstrates low selectivity for rNTPs [38,49], compensated for by a 10- to 130-fold higher cellular concentration of rNTPs versus dNTPs [50]. Therefore, the probability of dNTP insertion, which works as a chain terminator for primase, is a rare event in vivo. Selectivity of the initiation site to ribose, mediated by the hydrogen bond between Asp-306 and the O2′, is probably due to the requirement for accurate positioning of the O3′, which is deprotonated by Mg^{2+} for the nucleophilic attack on the α-phosphate of the incoming NTP. Moreover, such selectivity potentially prevents the primase from extending DNA tracks made by Polα or other DNA Pols. It is quite possible that primase binds all three substrates before formation of the initiation complex, which works as a locking mechanism and fixes the substrates in catalytically proficient position.

Modeling [13,18] and mutational [16] studies indicate that p49 employs the same amino acids for interactions with the DNA template during the initiation and elongation steps of primer synthesis. The weak interaction between p49 and the template-primer [43] suggests the mechanism of primase translocation along the template: p49 dissociates from DNA:RNA, held by $p58_C$, after each round of nucleotide incorporation and quickly rebinds it by placing the 3′-terminal nucleotide of the primer at the binding site for the initiating nucleotide or, more exactly, to its section located on the catalytic subunit. In accordance with biochemical data [43], the model of the elongation complex (Figure 6) revealed a lack of interaction between human primase and the emerging RNA strand, except for the same contacts as found in the initiation complex (Figure 5). The open architecture of the primase/DNA:RNA complex, where contacts with both the minor and major grooves are absent, explains the ability of DNA primases to extend mispaired primer termini and perform translesion synthesis [51,52].

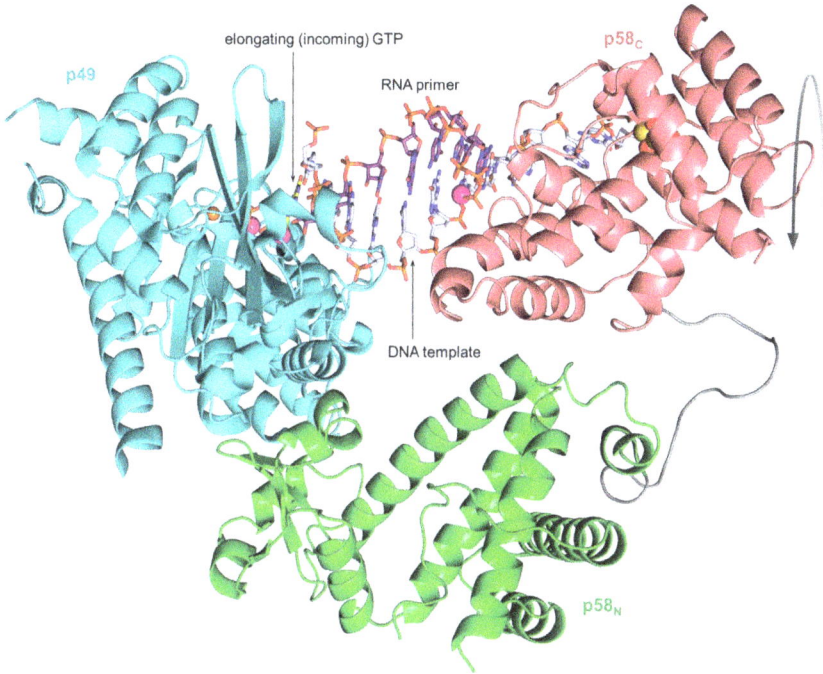

Figure 6. The model of human primase in elongation complex with a DNA template, primed by 7-mer RNA, and an incoming GTP. The curved arrow shows the direction of $p58_C$ rotation relative to $p49$-$p58_N$ during primer extension. The atoms of zinc, magnesium, iron, and sulfur are represented as spheres and colored orange, magenta, red, and yellow, respectively.

Due to the tight association with the template:primer junction [43], $p58_C$ must move away from $p49$ during primer extension, by following the helical path of the growing DNA:RNA duplex [13]. Probably, such spiral movement of $p58_C$ defines the mechanism of the primase counting phenomenon, which results in primer synthesis termination [12,18,43,53]. The model of the elongation complex, where primase is ready to generate an 8-mer primer, demonstrates that $p58_C$ is in proximity to the helical subdomain of $p58_N$ (Figure 6). Extension of the 8-mer primer would be complicated because of the emerging steric hindrance between the two p58 domains, which compromises the pre-catalytic alignment of the O3′ of a primer and the α-phosphate of an incoming NTP. The plasticity of $p58_N$ allows primase to overcome steric hindrance during synthesis of the 9-mer primer but not during the following extension step [13]. Due to this plasticity, the intra-subunit steric hindrance works as a molecular brake to stop primase, which results in an RNA primer with a well-defined length optimal for utilization by Polα. The linker between $p58_N$ and $p58_C$ is not important for RNA synthesis termination because its shortening did not reduce the size of RNA products [18]. In contrast, primase pausing is dependent on the strength of the $p58_C$/DNA:RNA complex; that is why its disturbance by changes in $p58_C$ sequence [42] and the template:primer structure [43] attenuates the counting effect.

Salt, the type of divalent metal, and the metal's concentration affect the distribution of RNA synthesis products [43]. Moreover, the *de novo* assay masks the effect of synthesis termination on templates, forming stable duplexes with 9-mer RNA primers, due to a 6000-fold lower primase affinity for single-stranded DNA versus a primed one [18,43]. On the other hand, 9-bp AT-rich DNA-RNA duplexes are not stable at common reaction conditions (30–35 °C), which significantly reduces the probability of RNA synthesis restart. Modeling of elongation complexes with 9 to 11-mer primers indicates that the steric hindrance is predominant only upon synthesis of 10- and 11-mer RNA [13]. If Polα is absent in the reaction, primase occasionally bypasses this barrier, using DNA:RNA substrates dissociated from p58$_C$, which results in the accumulation of longer products upon extended incubation.

5. Mechanism of RNA Primer Transfer to Polα and Its Extension with dNTPs

According to biochemical data, upon completion of RNA primer synthesis p58$_C$ continues to hold the template-primer until Polα captures it [11,12]. Recent structural data support this observation by showing that the predominant length of RNA primers is nine nucleotides and the optimal substrate for Polα is a 9-bp DNA:RNA duplex [17,43]. These data indicate that p58$_C$ and p180core will form a switch complex before Polα starts an extension of the RNA primer with dNTPs. The model of this complex revealed the concurrent binding of a 9-bp DNA:RNA duplex and shape complementarity between both subunits (Figure 7). According to this model, p58$_C$ will not allow Polα to extend shorter duplexes because the 3′-end of the primer does not reach the active site. Finally, biochemical experiments confirmed the idea that Polα in the primosome extends only the mature 9-mer RNA primers [13].

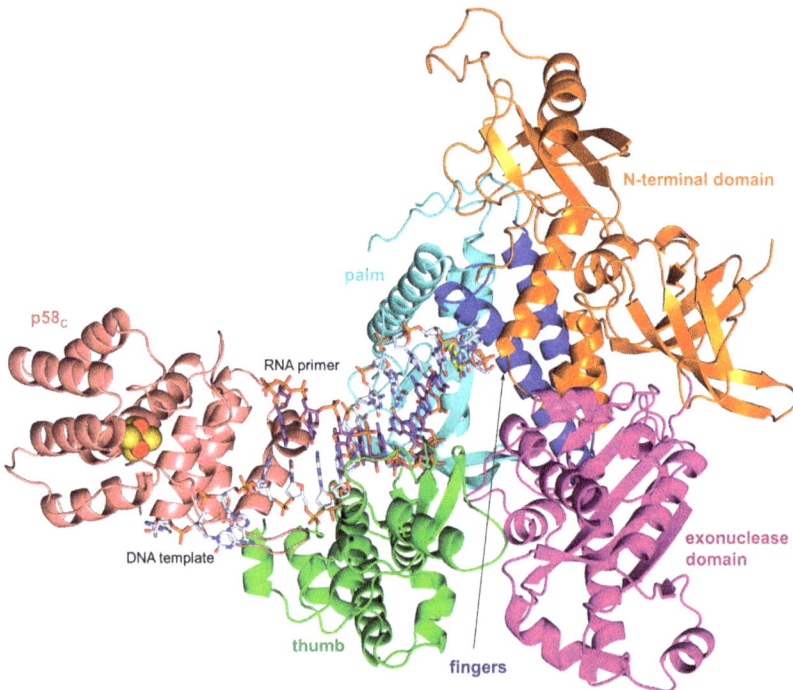

Figure 7. The model of the switch complex containing p180core, p58$_C$, a DNA template primed by a 9-mer RNA, and incoming dCTP. p180core subdomains are shown in different colors. The carbons of the DNA template, RNA primer, and incoming dCTP are colored gray, purple, and yellow, respectively. This model was made using the coordinates of the p180core/DNA:RNA/dCTP complex (PDB ID 4QCL) and p58$_C$/DNA:RNA complex (PDB ID 5F0Q).

Similar to other B-family DNA polymerases and their prototypes from viruses, bacteriophages and bacteria, p180core has a "right-hand" fold: an active site formed by a "palm" holding the catalytic residues and making a set of interactions with three base pairs of the DNA double helix at the 3'-end of a primer, a "thumb" that secures the polymerase grip onto the template-primer helix, and "fingers" providing the induced-fit closure of the active site after binding of the cognate dNTP (Figure 7). Polα cannot correct its own mistakes during DNA copying because of evolutionary substitution of the catalytic amino acid residues in the exonuclease active site [54].

Polα possesses an interesting feature of binding and extending DNA:RNA and DNA:DNA duplexes with similar efficiency [20,38,55]. Structural data for p180core in ternary complex with DNA:RNA/dCTP and in binary complex with DNA:DNA indicate that Polα binds the DNA and hybrid duplexes in a similar way [17,20]. There are no significant conformational changes in p180core to accommodate different duplexes; instead, Polα imposes the A-DNA conformation on the DNA primer [20] and bends the RNA primer [17,56] to keep the same contacts with the sugar-phosphate backbone. It is probable that the requirement for similar binding of both types of duplexes explains a smaller footprint of Polα on the template:primer and a less extensive network of contacts, which results in a low affinity with a K_d of ~320 nM for the RNA:DNA helix [20]. Its relatively weak interaction with the template:primer explains the high sensitivity of Polα to unconventional DNA structures, which is manifested by DNA synthesis abrogation on the certain templates [38,56,57]. It is likely that the limited Polα processivity on poly-dT templates is due to DNA bending and/or the triplex formation between the DNA:DNA duplex and the template's 5'-tail [57,58], rather than to the intrinsic ability of Polα to count the amount of incorporated dNMPs [56]. Moreover, no Polα pausing was observed on DNA templates of random sequence [38,55].

6. Polα Inhibition by Aphidicolin

Aphidicolin, an antimitotic metabolite of the mold *Cephalosporium aphidicola*, is a potent inhibitor of DNA replication in a variety of organisms [59,60]. It specifically inhibits B-family DNA polymerases, with Polα being the most sensitive to it [61]. Aphidicolin demonstrated potent growth-inhibitory and cytotoxic activities against human tumor cell lines cultured in vitro, but the absence of structural information hampered the improvement of its inhibitory properties [62–64]. The structure of p180core in ternary complex with a DNA:RNA duplex and aphidicolin revealed the mechanism of Polα inhibition and provided the structural rationale for design of a new generation of drugs with superior solubility, stability, and inhibitory activity [17]. Aphidicolin binds Polα at the active site by occupying the hydrophobic pocket for a nascent base pair (Figure 8). The interaction between aphidicolin and Polα is mediated by an extensive pattern of hydrophobic contacts as well as by the hydrogen bonds between two oxygens and the main-chain nitrogens. Accommodation of the bulky "potato" shape of the inhibitor results in the fingers opening and *syn* conformation of the templating guanine due to the base rotation by 118° around the N-glycosidic bond. The preference of aphidicolin for purine at this position is due to stabilization of the *syn* conformation of a purine mediated by stacking with a side chain of Arg-784, by the hydrogen bond between N7 and Oγ of Ser-955, and by several van der Waals interactions. In contrast to the imidazole ring of a purine base, the larger pyrimidine ring would hardly fit the pocket formed mainly by a second α-helix of the fingers domain.

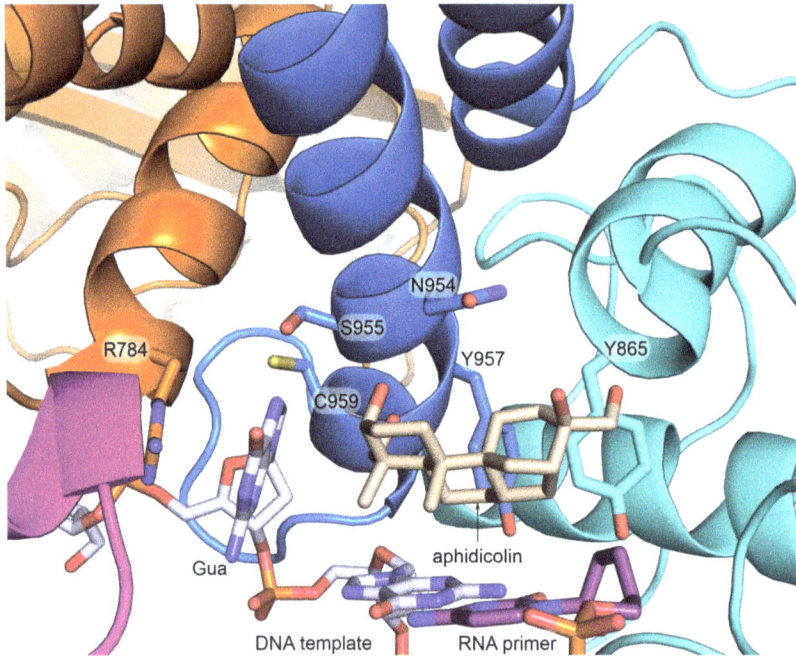

Figure 8. Close-up view of the Polα active site with bound aphidicolin and the DNA:RNA duplex. The color scheme for p180core subdomains is same as in Figure 7. The carbons of aphidicolin are colored wheat. Side chains of the key residues, participating in hydrophobic interactions with aphidicolin and in stabilization of the *syn* conformation of the templating guanine, are shown as sticks. RNA primer contains a dideoxy-cytidine at the 3′-end. This figure was drawn using the coordinates of the p180core/DNA:RNA/aphidicolin complex (PDB ID 4Q5V).

7. Mechanism of Concerted RNA-DNA Primer Synthesis by the Human Primosome

The accumulated structural data allow for visualization of all key steps of the chimeric primer synthesis (Figure 9 and movie provided in [13]). The structure of the primosome in apo-form revealed the autoinhibited state of Polα due to p180core docking on the platform where the Zn2 module of p180$_C$ and the OB domain of p70 are wedged into the template:primer-binding cleft of Polα [13]. During the initiation of RNA synthesis, p58$_C$ binds the template and initiating rNTP and moves toward the active site of p49 residing on the platform. In the presence of the cognate, elongating rNTP at the catalytic site, the initiation complex is stabilized and proceeds toward the dinucleotide formation. While p58$_C$ is important for primosome loading on early replication origins [65], it has low affinity for single-stranded DNA [43]. Presumably, other replication factors, like RPA, facilitate p58$_C$ loading on the template [21]. During the RNA elongation step, p58$_C$ moves toward p180core and pushes it to dissociate from the platform, resulting in Polα activation. The following primer extension results in a clash between p58$_C$ and the platform that is responsible for RNA synthesis termination. At this step the interaction of p49 with a 9-bp DNA:RNA held by p58$_C$ is compromised, leading to flotation of p58$_C$/DNA:RNA and its capture by p180core floating nearby that results in the template-primer loading to the Polα active site. p58$_C$ and p180core have an additional level of freedom relative to each other because they are independently connected with a platform by long linkers. According to modeling studies, these linkers allow Polα to generate a DNA track up to 20 nucleotides long, with p58$_C$ holding the 5′-end of the primer. The weak grip of Polα on the DNA double helix could facilitate its displacement from the template:primer by RFC/PCNA or Polε.

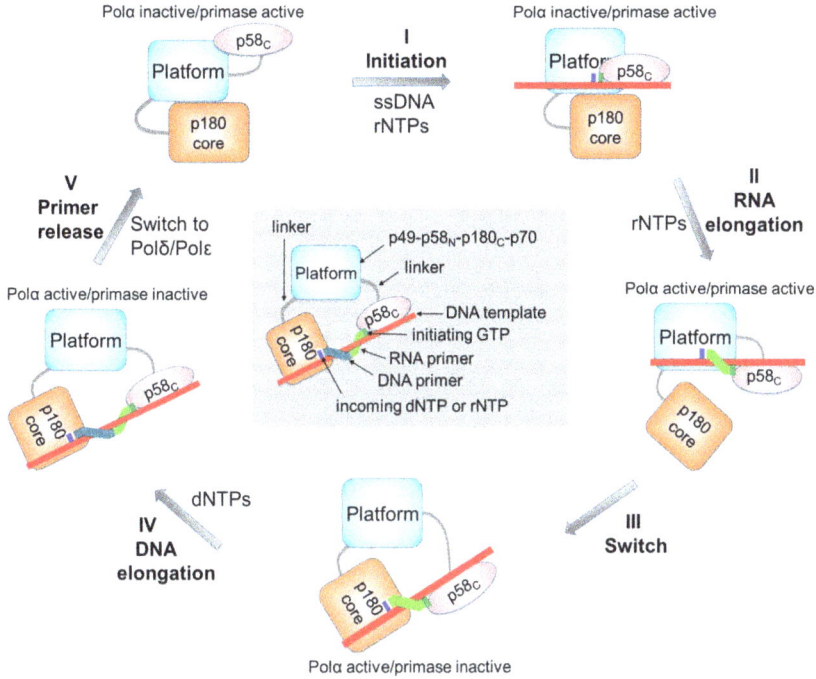

Figure 9. Schematic representation of conformational changes in the primosome during chimeric primer synthesis. At the first step (steps are labeled by roman numerals), $p58_C$ moves toward p49 to initiate RNA synthesis. During the second step, $p58_C$ moves toward p180core and pushes it to dissociate from the platform. Additionally, when RNA primer length is nine nucleotides, $p58_C$ makes a steric hindrance with the platform, which prevents primer extension by p49. At the third step, $p58_C$ rotates and loads the template:primer to the Polα active site. At the fourth step, Polα extends the RNA primer with dNTPs. At the fifth step, the primosome is replaced by Polε or Polδ.

Structural and biochemical data indicate that $p58_C$ is a central mediator of all primosome transactions [13,42,43]. $p58_C$ shuttles between the RNA- and DNA-polymerizing centers in the primosome, playing the role of the universal template:primer loader and regulator of primase and Polα. The linker between $p58_N$ and $p58_C$ allows $p58_C$ to form the initiation complex with p49 during dinucleotide synthesis, to move away together with the 5′-end of the primer during its extension, and, finally, to intramolecularly transfer and load the template primed by a 9-mer RNA to the Polα active site. To perform these multiple duties, the $p58_C$ shape conforms to several topological requirements: it is complementary to p49 during initiation and to p180core during the switch, and clashes with $p58_N$ during RNA synthesis termination.

8. Concluding Remarks

The eukaryotic primosome was discovered more than 30 years ago [46,66–68] but its intricate mechanism of RNA-DNA primer synthesis has become clear only recently, owing to thrilling progress in structural studies. Comprehensive understanding of all primosome transactions, including initiation, elongation, and accurate termination of RNA synthesis, as well as primer transfer from primase to Polα, requires the crystal structures of the primosome in complex with a variety of substrates. Crystallization of these complexes is extremely challenging due to the size of the primosome and its significant flexibility. Fortunately, several key structures allowed for obtaining plausible three-dimensional models for all steps of chimeric primer synthesis. These structures include the human primosome in apo-form [13], the ternary complex p180core/DNA:RNA/dCTP [17], the binary complex p58$_C$/DNA:RNA [13], complexes of p49–p58(19–253) or p49(1–390) with UTP [16,23], and full-length primase in apo-form [18]. Precise regulation of the concerted action of the two catalytic centers in the primosome is mainly based on the shape complementarity or the steric hindrance between its three components: a platform and two mobile domains, p58$_C$ and p180core [13].

Further studies are required to understand the mechanism of primosome integration into the replisome and its regulation by other replicative factors. Studies in yeast have shown that trimeric Ctf4 links the N-terminal domain of the Polα catalytic subunit to the GINS complex, which is a part of the CMG helicase also containing Cdc45 and Mcm2–7 [25,69]. The helical N-terminal domain of p70 connected with the primosome by an 80-residue-long linker is a potential candidate for interaction with the replisome or regulatory proteins. It interacts with the hexameric helicase of SV40 large T antigen and activates the viral replisome [27,28]. Moreover, the N-terminal domain of the B subunit of Polε has a similar structure and interacts with the replisome [70,71]. A recent model of the replisome organization in *Saccharomyces cerevisiae* obtained from electron microscopy studies indicates that Polα is located behind the helicase, in proximity to both unwound parental strands [72]. High-resolution structural data are needed to build accurate replisome models (human-system models are more desirable) showing the primosome orientation and conformation during priming of the leading and lagging strands.

Acknowledgments: We thank Kelly Jordan and Youri Pavlov for critical reading and editing of this manuscript. This work was supported by the National Institute of General Medical Sciences (NIGMS) grant GM101167 to Tahir H. Tahirov.

Conflicts of Interest: The authors declare no conflicts of interest.

References

1. Muzi-Falconi, M.; Giannattasio, M.; Foiani, M.; Plevani, P. The DNA polymerase α-primase complex: Multiple functions and interactions. *Sci. World J.* **2003**, *3*, 21–33. [CrossRef] [PubMed]
2. Pavlov, Y.I.; Shcherbakova, P.V. DNA polymerases at the eukaryotic fork-20 years later. *Mutat. Res.* **2010**, *685*, 45–53. [CrossRef] [PubMed]
3. Lujan, S.A.; Williams, J.S.; Kunkel, T.A. DNA polymerases divide the labor of genome replication. *Trends Cell Biol.* **2016**, *26*, 640–654. [CrossRef] [PubMed]
4. Smith, D.J.; Whitehouse, I. Intrinsic coupling of lagging-strand synthesis to chromatin assembly. *Nature* **2012**, *483*, 434–438. [CrossRef] [PubMed]
5. Pellegrini, L. The pol α-primase complex. *Subcell. Biochem.* **2012**, *62*, 157–169. [PubMed]
6. MacNeill, S. Composition and dynamics of the eukaryotic replisome: A brief overview. *Subcell. Biochem.* **2012**, *62*, 1–17. [PubMed]
7. Reijns, M.A.; Kemp, H.; Ding, J.; de Proce, S.M.; Jackson, A.P.; Taylor, M.S. Lagging-strand replication shapes the mutational landscape of the genome. *Nature* **2015**, *518*, 502–506. [CrossRef] [PubMed]
8. Starokadomskyy, P.; Gemelli, T.; Rios, J.J.; Xing, C.; Wang, R.C.; Li, H.; Pokatayev, V.; Dozmorov, I.; Khan, S.; Miyata, N.; et al. DNA polymerase-α regulates the activation of type I interferons through cytosolic rna:DNA synthesis. *Nat. Immunol.* **2016**, *17*, 495–504. [CrossRef] [PubMed]
9. Kuchta, R.D.; Stengel, G. Mechanism and evolution of DNA primases. *Biochim. Biophys. Acta* **2010**, *1804*, 1180–1189. [CrossRef] [PubMed]
10. Frick, D.N.; Richardson, C.C. DNA primases. *Annu. Rev. Biochem.* **2001**, *70*, 39–80. [CrossRef]
11. Copeland, W.C.; Wang, T.S. Enzymatic characterization of the individual mammalian primase subunits reveals a biphasic mechanism for initiation of DNA replication. *J. Biol. Chem.* **1993**, *268*, 26179–26189. [PubMed]
12. Sheaff, R.J.; Kuchta, R.D. Mechanism of calf thymus DNA primase: Slow initiation, rapid polymerization, and intelligent termination. *Biochemistry* **1993**, *32*, 3027–3037. [CrossRef] [PubMed]

13. Baranovskiy, A.G.; Babayeva, N.D.; Zhang, Y.; Gu, J.; Suwa, Y.; Pavlov, Y.I.; Tahirov, T.H. Mechanism of concerted RNA-DNA primer synthesis by the human primosome. *J. Biol. Chem.* **2016**, *291*, 10006–10020. [CrossRef] [PubMed]

14. Sheaff, R.J.; Kuchta, R.D.; Ilsley, D. Calf thymus DNA polymerase α-primase: "Communication" and primer-template movement between the two active sites. *Biochemistry* **1994**, *33*, 2247–2254. [CrossRef] [PubMed]

15. Kuchta, R.D.; Reid, B.; Chang, L.M. DNA primase. Processivity and the primase to polymerase α activity switch. *J. Biol. Chem.* **1990**, *265*, 16158–16165. [PubMed]

16. Kilkenny, M.L.; Longo, M.A.; Perera, R.L.; Pellegrini, L. Structures of human primase reveal design of nucleotide elongation site and mode of pol α tethering. *Proc. Natl. Acad. Sci. USA* **2013**, *110*, 15961–15966. [CrossRef] [PubMed]

17. Baranovskiy, A.G.; Babayeva, N.D.; Suwa, Y.; Gu, J.; Pavlov, Y.I.; Tahirov, T.H. Structural basis for inhibition of DNA replication by aphidicolin. *Nucleic Acids Res.* **2014**, *42*, 14013–14021. [CrossRef] [PubMed]

18. Baranovskiy, A.G.; Zhang, Y.; Suwa, Y.; Babayeva, N.D.; Gu, J.; Pavlov, Y.I.; Tahirov, T.H. Crystal structure of the human primase. *J. Biol. Chem.* **2015**, *290*, 5635–5646. [CrossRef] [PubMed]

19. Suwa, Y.; Gu, J.; Baranovskiy, A.G.; Babayeva, N.D.; Pavlov, Y.I.; Tahirov, T.H. Crystal structure of the human Pol α B subunit in complex with the C-terminal domain of the catalytic subunit. *J. Biol. Chem.* **2015**, *290*, 14328–14337. [CrossRef] [PubMed]

20. Coloma, J.; Johnson, R.E.; Prakash, L.; Prakash, S.; Aggarwal, A.K. Human DNA polymerase α in binary complex with a DNA:DNA template-primer. *Sci. Rep.* **2016**, *6*, 1–10. [CrossRef] [PubMed]

21. Vaithiyalingam, S.; Warren, E.M.; Eichman, B.F.; Chazin, W.J. Insights into eukaryotic DNA priming from the structure and functional interactions of the 4Fe-4S cluster domain of human DNA primase. *Proc. Natl. Acad. Sci. USA* **2010**, *107*, 13684–13689. [CrossRef] [PubMed]

22. Agarkar, V.B.; Babayeva, N.D.; Pavlov, Y.I.; Tahirov, T.H. Crystal structure of the C-terminal domain of human DNA primase large subunit: Implications for the mechanism of the primase-polymerase α switch. *Cell Cycle* **2011**, *10*, 926–931. [CrossRef] [PubMed]

23. Vaithiyalingam, S.; Arnett, D.R.; Aggarwal, A.; Eichman, B.F.; Fanning, E.; Chazin, W.J. Insights into eukaryotic primer synthesis from structures of the p48 subunit of human DNA primase. *J. Mol. Biol.* **2014**, *426*, 558–569. [CrossRef] [PubMed]

24. Mizuno, T.; Yamagishi, K.; Miyazawa, H.; Hanaoka, F. Molecular architecture of the mouse DNA polymerase α-primase complex. *Mol. Cell. Biol.* **1999**, *19*, 7886–7896. [CrossRef] [PubMed]

25. Simon, A.C.; Zhou, J.C.; Perera, R.L.; van Deursen, F.; Evrin, C.; Ivanova, M.E.; Kilkenny, M.L.; Renault, L.; Kjaer, S.; Matak-Vinkovic, D.; et al. A Ctf4 trimer couples the cmg helicase to DNA polymerase α in the eukaryotic replisome. *Nature* **2014**, *510*, 293–297. [CrossRef] [PubMed]

26. Baranovskiy, A.G.; Babayeva, N.D.; Liston, V.G.; Rogozin, I.B.; Koonin, E.V.; Pavlov, Y.I.; Vassylyev, D.G.; Tahirov, T.H. X-ray structure of the complex of regulatory subunits of human DNA polymerase delta. *Cell Cycle* **2008**, *7*, 3026–3036. [CrossRef] [PubMed]

27. Zhou, B.; Arnett, D.R.; Yu, X.; Brewster, A.; Sowd, G.A.; Xie, C.L.; Vila, S.; Gai, D.; Fanning, E.; Chen, X.S. Structural basis for the interaction of a hexameric replicative helicase with the regulatory subunit of human DNA polymerase α-primase. *J. Biol. Chem.* **2012**, *287*, 26854–26866. [CrossRef] [PubMed]

28. Huang, H.; Zhao, K.; Arnett, D.R.; Fanning, E. A specific docking site for DNA polymerase {α}-primase on the SV40 helicase is required for viral primosome activity, but helicase activity is dispensable. *J. Biol. Chem.* **2010**, *285*, 33475–33484. [CrossRef] [PubMed]

29. Leipe, D.D.; Aravind, L.; Koonin, E.V. Did DNA replication evolve twice independently? *Nucleic Acids Res.* **1999**, *27*, 3389–3401. [CrossRef] [PubMed]

30. Ito, N.; Nureki, O.; Shirouzu, M.; Yokoyama, S.; Hanaoka, F. Crystal structure of the *Pyrococcus horikoshii* DNA primase-UTP complex: Implications for the mechanism of primer synthesis. *Genes Cells* **2003**, *8*, 913–923. [CrossRef] [PubMed]

31. Lao-Sirieix, S.H.; Nookala, R.K.; Roversi, P.; Bell, S.D.; Pellegrini, L. Structure of the heterodimeric core primase. *Nat. Struct. Mol. Biol.* **2005**, *12*, 1137–1144. [CrossRef] [PubMed]

32. Augustin, M.A.; Huber, R.; Kaiser, J.T. Crystal structure of a DNA-dependent RNA polymerase (DNA primase). *Nat. Struct. Biol.* **2001**, *8*, 57–61. [CrossRef] [PubMed]

33. Klinge, S.; Hirst, J.; Maman, J.D.; Krude, T.; Pellegrini, L. An iron-sulfur domain of the eukaryotic primase is essential for RNA primer synthesis. *Nat. Struct. Mol. Biol.* **2007**, *14*, 875–877. [CrossRef] [PubMed]

34. Francesconi, S.; Longhese, M.P.; Piseri, A.; Santocanale, C.; Lucchini, G.; Plevani, P. Mutations in conserved yeast DNA primase domains impair DNA replication in vivo. *Proc. Natl. Acad. Sci. USA* **1991**, *88*, 3877–3881. [CrossRef] [PubMed]

35. Weiner, B.E.; Huang, H.; Dattilo, B.M.; Nilges, M.J.; Fanning, E.; Chazin, W.J. An iron-sulfur cluster in the C-terminal domain of the p58 subunit of human DNA primase. *J. Biol. Chem.* **2007**, *282*, 33444–33451. [CrossRef] [PubMed]

36. Netz, D.J.; Stith, C.M.; Stumpfig, M.; Kopf, G.; Vogel, D.; Genau, H.M.; Stodola, J.L.; Lill, R.; Burgers, P.M.; Pierik, A.J. Eukaryotic DNA polymerases require an iron-sulfur cluster for the formation of active complexes. *Nat. Chem. Biol.* **2011**, *8*, 125–132. [CrossRef] [PubMed]

37. Baranovskiy, A.G.; Lada, A.G.; Siebler, H.M.; Zhang, Y.; Pavlov, Y.I.; Tahirov, T.H. DNA polymerase delta and zeta switch by sharing accessory subunits of DNA polymerase delta. *J. Biol. Chem.* **2012**, *287*, 17281–17287. [CrossRef] [PubMed]

38. Zhang, Y.; Baranovskiy, A.G.; Tahirov, T.H.; Pavlov, Y.I. The C-terminal domain of the DNA polymerase catalytic subunit regulates the primase and polymerase activities of the human DNA polymerase α-primase complex. *J. Biol. Chem.* **2014**, *289*, 22021–22034. [CrossRef]

39. Klinge, S.; Nunez-Ramirez, R.; Llorca, O.; Pellegrini, L. 3D architecture of DNA pol α reveals the functional core of multi-subunit replicative polymerases. *EMBO J.* **2009**, *28*, 1978–1987. [CrossRef] [PubMed]

40. Copeland, W.C.; Wang, T.S. Mutational analysis of the human DNA polymerase α. The most conserved region in α-like DNA polymerases is involved in metal-specific catalysis. *J. Biol. Chem.* **1993**, *268*, 11028–11040. [PubMed]

41. Copeland, W.C.; Tan, X. Active site mapping of the catalytic mouse primase subunit by alanine scanning mutagenesis. *J. Biol. Chem.* **1995**, *270*, 3905–3913. [PubMed]

42. Zerbe, L.K.; Kuchta, R.D. The p58 subunit of human DNA primase is important for primer initiation, elongation, and counting. *Biochemistry* **2002**, *41*, 4891–4900. [CrossRef] [PubMed]

43. Baranovskiy, A.G.; Zhang, Y.; Suwa, Y.; Gu, J.; Babayeva, N.D.; Pavlov, Y.I.; Tahirov, T.H. Insight into the human DNA primase interaction with template-primer. *J. Biol. Chem.* **2016**, *291*, 4793–4802. [CrossRef] [PubMed]

44. Podust, V.N.; Vladimirova, O.V.; Manakova, E.N.; Lavrik, O.I. Eukaryotic DNA primase appears to act as oligomer in DNA-polymerase-α–primase complex. *Eur. J. Biochem.* **1992**, *206*, 7–13. [CrossRef] [PubMed]

45. Kirk, B.W.; Kuchta, R.D. Arg304 of human DNA primase is a key contributor to catalysis and NTP binding: Primase and the family X polymerases share significant sequence homology. *Biochemistry* **1999**, *38*, 7727–7736. [CrossRef] [PubMed]

46. Grosse, F.; Krauss, G. The primase activity of DNA polymerase α from calf thymus. *J. Biol. Chem.* **1985**, *260*, 1881–1888. [PubMed]

47. Zahurancik, W.J.; Klein, S.J.; Suo, Z. Kinetic mechanism of DNA polymerization catalyzed by human DNA polymerase epsilon. *Biochemistry* **2013**, *52*, 7041–7049. [CrossRef] [PubMed]

48. Yang, W.; Lee, J.Y.; Nowotny, M. Making and breaking nucleic acids: Two-Mg^{2+}-ion catalysis and substrate specificity. *Mol. Cell* **2006**, *22*, 5–13. [CrossRef] [PubMed]

49. Kuchta, R.D.; Ilsley, D.; Kravig, K.D.; Schubert, S.; Harris, B. Inhibition of DNA primase and polymerase α by arabinofuranosylnucleoside triphosphates and related compounds. *Biochemistry* **1992**, *31*, 4720–4728. [CrossRef] [PubMed]

50. Traut, T.W. Physiological concentrations of purines and pyrimidines. *Mol. Cell. Biochem.* **1994**, *140*, 1–22. [CrossRef] [PubMed]

51. Urban, M.; Joubert, N.; Purse, B.W.; Hocek, M.; Kuchta, R.D. Mechanisms by which human DNA primase chooses to polymerize a nucleoside triphosphate. *Biochemistry* **2010**, *49*, 727–735. [CrossRef] [PubMed]

52. Jozwiakowski, S.K.; Borazjani Gholami, F.; Doherty, A.J. Archaeal replicative primases can perform translesion DNA synthesis. *Proc. Natl. Acad. Sci. USA* **2015**, *112*, E633–E638. [CrossRef] [PubMed]

53. Ogawa, T.; Okazaki, T. Discontinuous DNA replication. *Ann. Rev. Biochem.* **1980**, *19*, 421–457. [CrossRef] [PubMed]

54. Pavlov, Y.I.; Shcherbakova, P.V.; Rogozin, I.B. Roles of DNA polymerases in replication, repair, and recombination in eukaryotes. *Int. Rev. Cytol.* **2006**, *255*, 41–132.

55. Thompson, H.C.; Sheaff, R.J.; Kuchta, R.D. Interactions of calf thymus DNA polymerase α with primer/templates. *Nucleic Acids Res.* **1995**, *23*, 4109–4115. [CrossRef] [PubMed]

56. Perera, R.L.; Torella, R.; Klinge, S.; Kilkenny, M.L.; Maman, J.D.; Pellegrini, L. Mechanism for priming DNA synthesis by yeast DNA polymerase α. *Elife* **2013**, *2*, e00482. [CrossRef] [PubMed]

57. Zhang, Y.; Baranovskiy, A.G.; Tahirov, E.T.; Tahirov, T.H.; Pavlov, Y.I. Divalent ions attenuate DNA synthesis by human DNA polymerase α by changing the structure of the template/primer or by perturbing the polymerase reaction. *DNA Repair* **2016**, *43*, 24–33. [CrossRef] [PubMed]

58. Nadeau, J.G.; Crothers, D.M. Structural basis for DNA bending. *Proc. Natl. Acad. Sci. USA* **1989**, *86*, 2622–2626. [CrossRef] [PubMed]

59. Ikegami, S.; Taguchi, T.; Ohashi, M.; Oguro, M.; Nagano, H.; Mano, Y. Aphidicolin prevents mitotic cell division by interfering with the activity of DNA polymerase-α. *Nature* **1978**, *275*, 458–460. [CrossRef] [PubMed]

60. Huberman, J.A. New views of the biochemistry of eucaryotic DNA replication revealed by aphidicolin, an unusual inhibitor of DNA polymerase α. *Cell* **1981**, *23*, 647–648. [CrossRef]

61. Sheaff, R.; Ilsley, D.; Kuchta, R. Mechanism of DNA polymerase α inhibition by aphidicolin. *Biochemistry* **1991**, *30*, 8590–8597. [CrossRef] [PubMed]

62. Prasad, G.; Edelson, R.A.; Gorycki, P.D.; Macdonald, T.L. Structure-activity relationships for the inhibition of DNA polymerase α by aphidicolin derivatives. *Nucleic Acids Res.* **1989**, *17*, 6339–6348. [CrossRef] [PubMed]

63. Pedrali-Noy, G.; Belvedere, M.; Crepaldi, T.; Focher, F.; Spadari, S. Inhibition of DNA replication and growth of several human and murine neoplastic cells by aphidicolin without detectable effect upon synthesis of immunoglobulins and HLA antigens. *Cancer Res.* **1982**, *42*, 3810–3813. [PubMed]

64. Cinatl, J., Jr.; Cinatl, J.; Kotchetkov, R.; Driever, P.H.; Bertels, S.; Siems, K.; Jas, G.; Bindseil, K.; Rabenau, H.F.; Pouckova, P.; et al. Aphidicolin glycinate inhibits human neuroblastoma cell growth in vivo. *Oncol. Rep.* **1999**, *6*, 563–568. [CrossRef] [PubMed]

65. Liu, L.; Huang, M. Essential role of the iron-sulfur cluster binding domain of the primase regulatory subunit Pri2 in DNA replication initiation. *Protein Cell* **2015**, *6*, 194–210. [CrossRef] [PubMed]

66. Conaway, R.C.; Lehman, I.R. A DNA primase activity associated with DNA polymerase α from *Drosophila melanogaster* embryos. *Proc. Natl. Acad. Sci. USA* **1982**, *79*, 2523–2527. [CrossRef] [PubMed]

67. Hubscher, U. The mammalian primase is part of a high molecular weight DNA polymerase α polypeptide. *EMBO J.* **1983**, *2*, 133–136. [PubMed]

68. Plevani, P.; Badaracco, G.; Augl, C.; Chang, L.M. DNA polymerase I and DNA primase complex in yeast. *J. Biol. Chem.* **1984**, *259*, 7532–7539.

69. Villa, F.; Simon, A.C.; Ortiz Bazan, M.A.; Kilkenny, M.L.; Wirthensohn, D.; Wightman, M.; Matak-Vinkovic, D.; Pellegrini, L.; Labib, K. Ctf4 is a hub in the eukaryotic replisome that links multiple CIP-Box proteins to the CMG helicase. *Mol. Cell* **2016**, *63*, 385–396. [CrossRef] [PubMed]

70. Sengupta, S.; van Deursen, F.; de Piccoli, G.; Labib, K. Dpb2 integrates the leading-strand DNA polymerase into the eukaryotic replisome. *Curr. Biol.* **2013**, *23*, 543–552. [CrossRef] [PubMed]

71. Langston, L.D.; Zhang, D.; Yurieva, O.; Georgescu, R.E.; Finkelstein, J.; Yao, N.Y.; Indiani, C.; O'Donnell, M.E. Cmg helicase and DNA polymerase epsilon form a functional 15-subunit holoenzyme for eukaryotic leading-strand DNA replication. *Proc. Natl. Acad. Sci. USA* **2014**, *111*, 15390–15395. [CrossRef] [PubMed]

72. Sun, J.; Shi, Y.; Georgescu, R.E.; Yuan, Z.; Chait, B.T.; Li, H.; O'Donnell, M.E. The architecture of a eukaryotic replisome. *Nat. Struct. Mol. Biol.* **2015**, *22*, 976–982. [CrossRef] [PubMed]

genes

MDPI

Review

Mechanisms Governing DDK Regulation of the Initiation of DNA Replication

Larasati and Bernard P. Duncker *

Department of Biology, University of Waterloo, 200 University Avenue West, Waterloo, ON N2L3G1, Canada; llarasat@uwaterloo.ca
* Correspondence: bduncker@uwaterloo.ca; Tel.: +1-519-888-4567 (ext. 33957)

Academic Editor: Eishi Noguchi
Received: 1 November 2016; Accepted: 16 December 2016; Published: 22 December 2016

Abstract: The budding yeast Dbf4-dependent kinase (DDK) complex—comprised of cell division cycle (Cdc7) kinase and its regulatory subunit dumbbell former 4 (Dbf4)—is required to trigger the initiation of DNA replication through the phosphorylation of multiple minichromosome maintenance complex subunits 2-7 (Mcm2-7). DDK is also a target of the radiation sensitive 53 (Rad53) checkpoint kinase in response to replication stress. Numerous investigations have determined mechanistic details, including the regions of Mcm2, Mcm4, and Mcm6 phosphorylated by DDK, and a number of DDK docking sites. Similarly, the way in which the Rad53 forkhead-associated 1 (FHA1) domain binds to DDK—involving both canonical and non-canonical interactions—has been elucidated. Recent work has revealed mutual promotion of DDK and synthetic lethal with *dpb11-1* 3 (Sld3) roles. While DDK phosphorylation of Mcm2-7 subunits facilitates their interaction with Sld3 at origins, Sld3 in turn stimulates DDK phosphorylation of Mcm2. Details of a mutually antagonistic relationship between DDK and Rap1-interacting factor 1 (Rif1) have also recently come to light. While Rif1 is able to reverse DDK-mediated Mcm2-7 complex phosphorylation by targeting the protein phosphatase glycogen 7 (Glc7) to origins, there is evidence to suggest that DDK can counteract this activity by binding to and phosphorylating Rif1.

Keywords: DNA replication; DDK; Dbf4; Cdc7; MCM; Rad53; cell cycle checkpoint; Rif1; Sld3

1. Introduction

In an unperturbed cell cycle, budding yeast Dbf4-dependent kinase (DDK) complex triggers the initiation of DNA replication mainly through the phosphorylation of minichromosome maintenance complex subunits 2-7 (Mcm2-7) (reviewed in [1]). When DNA damage or dNTP depletion results in checkpoint activation, the normal role of DDK is opposed by radiation sensitive 53 (Rad53) kinase, which phosphorylates DDK, leading to its dissociation from chromatin [2–6]. Recently, a much better understanding of the way in which DDK associates with both Mcm2-7 and Rad53 (structurally and functionally) has been gained. This review will focus on genetic and molecular studies that have identified and characterized the subunits of Mcm2-7 which mediate the binding of DDK, and those that are the critical targets of DDK phosphorylation. Similarly, crucial mechanistic details of both canonical and non-canonical ways in which the Rad53 forkhead-associated 1 domain (FHA1) interacts with DDK have been determined. Recently, roles for additional protein factors in regulating DDK stimulation have also been uncovered. These include synthetic lethal with *dpb11-1* 3 (Sld3), which both stimulates DDK phosphorylation of Mcm2 and binds to DDK-phosphorylated Mcm4 and Mcm6; and Rap1-interacting factor 1 (Rif1), which counteracts DDK activity by recruiting the protein phosphatase glycogen 7 (Glc7) to dephosphorylate Mcm4. Finally, evidence supporting a role for DDK in coordinating the initiation of DNA replication with sister chromatid cohesion will be discussed.

2. Insights into DDK Interactions with Mcm2-7

One of the essential players in the initiation of eukaryotic DNA replication is the DDK complex, comprised of the serine-threonine kinase cell division cycle 7 (Cdc7), and its regulatory subunit, dumbbell former 4 (Dbf4). In the budding yeast *Saccharomyces cerevisiae*, each protein is encoded by a single gene, and deletion of either in a haploid strain is lethal [7]. Recently developed in vitro systems which recapitulate the molecular events culminating in origin firing have further demonstrated that the inclusion of DDK is absolutely required for the initiation of DNA replication [8–11]. The crucial function of DDK is to phosphorylate the Mcm2-7 ring, part of the larger CMG (Cdc45-Mcm2-7-go-ichi-ni-san (GINS)) replicative DNA helicase complex formed at origins of DNA replication (reviewed in [1]). The onset of these events is triggered by a rise in Dbf4 levels at the end of G1-phase, which fall after mitosis as Dbf4 is degraded in an anaphase promoting complex (APC)-dependent manner [12–15]. The high levels of active DDK at the end of G1-phase are also important for overcoming Rif1-Glc7 activity (discussed below) [16,17]. In recent years, a much higher-resolution understanding of these mechanisms has been obtained (summarized in Figure 1).

It has been known for some time that DDK is essential for DNA replication in vivo, likely due to its phosphorylation of Mcm2-7 inducing a conformational change, thereby favoring interaction with other firing factors. A P83L mutation in Mcm5 encoded by the *mcm5-bob1* allele can bypass DDK's requirement for viability, presumably mimicking a conformational change that facilitates DNA replication [18,19]. Similarly, some initial insight as to which residues of the Mcm2-7 subunits are the critical DDK targets was provided through a report that pointed to the N-terminal serine/threonine-rich domain (NSD) of Mcm4 as being a target of DDK as well as being required for cell growth and S-phase progression [20]. To test the hypothesis that the NSD is inhibitory to the activation of origins, an allele of *MCM4* lacking the NSD was transformed into temperature-sensitive *cdc7-4* and *dbf4-1* budding yeast strains and—reminiscent of *mcm5-bob1*—found to complement the growth defects at non-permissive temperatures [21]. Further examination of the NSD revealed that it could be functionally divided into overlapping proximal (amino acids 74–174) and distal (amino acids 2–145) regions. The proximal region inhibits origin activation, as demonstrated by a comparison of wild-type *MCM4* and *mcm4Δ74-174* strains. When both were exposed to the ribonucleotide reductase inhibitor hydroxyurea (HU, which synchronizes cells in early S-phase by provoking a checkpoint response), the *MCM4* cells only allowed origins that are normally active in early S-phase to fire, whereas with the *mcm4Δ74-174* strain, both early- and late-firing origins were activated. In contrast, the distal region was found to restrict the rate of replication fork progression [22,23].

Mcm2 has also been identified as an important DDK target, and is phosphorylated at serines 164 and 170 [24–26]. Plasmid-based expression of an allele where sequences encoding the two serines were changed to specify alanines, *mcm2-2A*, acted in a dominant negative fashion in an *MCM2* wild-type strain, resulting in severe growth defects. When the same *mcm2-2A* mutant was expressed at wild-type levels from a plasmid in a temperature-activated degron (td)-tagged *mcm2* strain at 37 °C (a temperature at which the td-tagged Mcm2 is degraded), again severe growth defects were observed, and fluorescence-activated cell sorting (FACS) analysis revealed impaired S-phase progression. Interestingly, in both cases, the defects could be partially suppressed by the *mcm5-bob1* mutation [26]. Mcm2 and Mcm5 lie adjacent to one another in the Mcm2-7 ring, and disruption of the interaction between the two of them leads to an opening, which allows for loading onto double-stranded DNA [27–29]. Insight as to the possible biological role of Mcm2 modification by DDK was provided by the observation that DDK-phosphorylated Mcm2 dissociates from Mcm5 and triggers opening of the Mcm2-7 ring [26] to allow extrusion of single-stranded DNA generated from origin melting. Electron microscopy analysis of *Drosophila melanogaster* Mcm2-7 suggests that the Mcm2-Mcm5 gap is later sealed through the interaction of the Mcm2-7 ring with Cdc45 and GINS [30]. As is the case for Mcm2 and Mcm4, Mcm6 has an unstructured N-terminal domain including several DDK target sites [29], and is phosphorylated by this kinase complex in vitro [31,32]. Recently, both DDK-phosphorylated Mcm4 and Mcm6 were shown to bind Sld3, which in turn recruits Cdc45 to origins (discussed below).

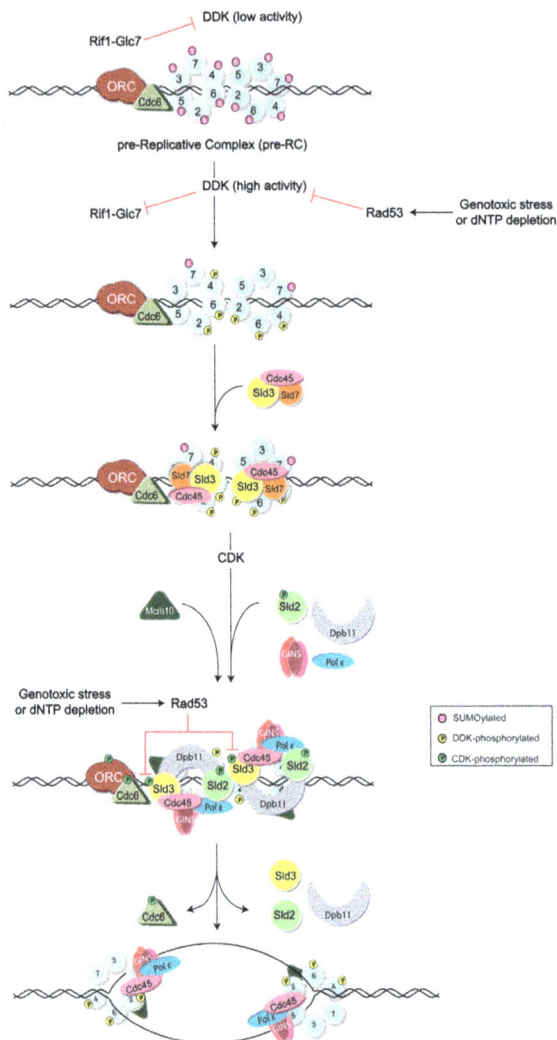

Figure 1. Model of DNA replication initiation. DNA replication is initiated by the assembly of the pre-Replicative Complex (pre-RC) at G1 phase, which is then followed by a series of phosphorylation events carried out by Dbf4-dependent kinase (DDK) and cyclin-dependent kinase (CDK) to generate the active form of the CMG (Cdc45-Mcm2-7-go-ichi-ni-san(GINS)) helicase. Normally, DDK activity is low until the end of G1 phase, as Dbf4—the regulatory subunit of DDK—is degraded in an anaphase promoting complex (APC)-dependent manner [12–15]. However, some Dbf4 that has escaped this process can provide residual DDK activity, contributing to potential premature Mcm2-7 complex phosphorylation. To avoid this, Rif1 recruits the protein phosphatase Glc7 to dephosphorylate the DDK targets. High activity of DDK in late G1 phase is proposed as a mechanism to inhibit Rif1-Glc7 activity [16,17]. DDK activity is also inhibited by the S-phase checkpoint kinase, Rad53, during exposure to genotoxic agents or dNTP depletion. Rad53 binds to and phosphorylates Dbf4 to remove DDK from chromatin and prevent subsequent origin firing [2–6]. Rad53 also phosphorylates an essential target of CDK, Sld3, to ensure the inhibition of DNA replication during replication stress [3,4], ORC: Origin Recognition Complex.

In addition to characterizing the regions of MCM subunits that are phosphorylated, insight has been gained regarding the way in which the DDK complex is targeted to Mcm2-7. Sequential analysis of each MCM subunit's ability to bind the DDK components through both two-hybrid assays and co-immunoprecipitation analysis revealed that Dbf4 and Cdc7 bind to Mcm2 and Mcm4, respectively [33], and DDK docking regions have been uncovered in these two MCM subunits [20,24,33]. In the case of Mcm4, a region comprising amino acids 175–333 was found to mediate binding by DDK [20], while two different regions on Mcm2 are required, including amino acids 2–63 [33] and 204–278 [24]. Interestingly, while structural studies have shown that Mcm2 and Mcm4 are not in close proximity in a single Mcm2-7 hexamer, the situation is different with the double hexameric form known to be loaded onto origins of DNA replication, where these subunits lie adjacent to one another, forming a bipartite DDK binding site, consistent with the finding that the double hexamer is a preferred DDK substrate over the single hexamer [34,35]. Previous work has revealed that DDK interacts with Mcm2 through the conserved Dbf4 N- and C-motifs [36,37], however little is known about the Cdc7 region that interacts with Mcm4.

While Mcm10 does not share sequence homology with Mcm2-7 [38] and is not included in the Mcm2-7 ring, it is nevertheless indispensable for DNA replication [11,38]. A recent study showed that both DDK subunits associate with Mcm10 in vitro, with Dbf4 binding more strongly than Cdc7 [39], which is consistent with an earlier finding that Cdc23 (homolog of Mcm10 in fission yeast *Schizosaccharomyces pombe*) binds to Dfp1 (homolog of Dbf4 in fission yeast) [40]. Mcm10 also interacts with Mcm2-7 [38,41–44], and the strength of this interaction is increased in the presence of DDK and cyclin-dependent kinase (CDK) [45], which may facilitate double hexamer separation [46]. Moreover, Mcm10 increases DDK phosphorylation of Mcm2 [39,40] and the Mcm2-7 complex as a whole [40] in vitro.

3. Regulation of DDK Activity by Rad53

The ability of DDK to phosphorylate MCM subunits can be impeded by the checkpoint kinase Rad53, which is known to bind Dbf4 primarily through its FHA1 domain [47]. Under conditions where DNA is damaged or cellular dNTP pools are depleted, Dbf4 is a target of Rad53, which results in removal of the DDK complex from chromatin [2], thereby inhibiting further origin firing [3–6]. Furthermore, in vitro phosphorylation of the DDK complex by Rad53 has been found to inhibit the phosphorylation of Mcm2 by DDK [48]. Numerous Rad53 phosphorylation sites have been identified in Dbf4, and mutation of four of these to alanines in a strain for which Rad53 phosphorylation sites in Sld3 were similarly mutated resulted in late origin firing, despite exposure to HU [3]. More recently, characterization of the Dbf4 region required for binding Rad53 revealed that a stretch including amino acid residues 105–221 is both necessary and sufficient for the interaction of Dbf4 and Rad53. A crystal structure was subsequently obtained, confirming a BRCA1 C terminus (BRCT)fold, but with an additional N-terminal alpha-helix required for FHA1 binding, and was therefore designated the H-BRCT domain [49]. As FHA domains are known to bind phosphothreonine-containing motifs, each H-BRCT threonine was systematically mutated, but none of these changes resulted in an abrogation of the interaction with FHA1. Subsequently, a combination of bioinformatics, nuclear magnetic resonance (NMR) spectroscopy, and two-hybrid analysis uncovered a non-canonical lateral surface patch on Rad53 FHA1 that binds to Dbf4 H-BRCT, distinct from its phosphothreonine epitope-binding domain [50]. Importantly, the Rad53 FHA1 domain is able to simultaneously engage Dbf4 H-BRCT and a Cdc7 phosphoepitope known to be recognized by Rad53 [50,51], suggesting a bipartite mode of interaction with the DDK complex. Indeed, this has now been confirmed through the elucidation of the crystal structure of Rad53 FHA1 simultaneously bound to Dbf4 and the phosphorylated Cdc7 peptide [52]. A requirement for FHA1 interaction with both DDK subunits may serve to simultaneously ensure that this only occurs during a checkpoint response (canonical phosphothreonine interaction with Cdc7), and exhibits substrate specificity (non-canonical interaction with Dbf4).

4. Mutual Promotion of Sld3 and DDK Activities

Sld3 is a key factor in the initiation of DNA replication and represents an essential target of CDK at this point in the cell cycle [53–55]. Sld3 associates with early-firing origins in G1 phase and late-firing origins in late S-phase [56], consistent with it being one of the limiting factors that differentiate early and late origins [55,57]. It binds to both Mcm2-7 and Cdc45, thus serving to recruit the latter to origins [56,58]. In recruiting Cdc45, Sld3 forms a complex with Sld7 [55], which acts to reduce Sld3's affinity to Cdc45 [59], likely helping Sld3 to dissociate from the origin while Cdc45 remains and eventually moves with the replication fork as a part of the CMG helicase. GINS may also help to displace Sld3 from origins, as they compete with each other for Mcm2-7 binding [58]. Like Dbf4, Sld3 is targeted by Rad53 phosphorylation as a mechanism to inhibit origin firing in response to DNA damage [3,4].

For some time, it has been known that Sld3's association with origins of DNA replication is DDK-dependent [55,60], but the molecular mechanisms involved have been uncovered more recently. Sld3 binds to Mcm2-7 [58], which facilitates its recruitment to origins. An in vitro replication system comprised of origin DNA attached to magnetic beads supplemented with purified budding yeast replication proteins was used to show that Sld3 binds loaded Mcm2-7 in a manner dependent upon DDK [61]. Further analysis revealed that Sld3 amino acids 510–545 mediate this interaction [61]. Interestingly, this region includes many of the sites that Rad53 phosphorylates to inhibit origin firing [3], and preincubation of Sld3 with Rad53 prevented it from binding MCM in the presence of DDK [61], in much the same way as Rad53 phosphorylation prevents Sld3 from interacting with scaffold protein Dpb11 (see Figure 1) [3]. This same system was further used to examine MCM subunit binding, and revealed that Sld3 specifically interacts with DDK-phosphorylated Mcm4 and Mcm6 [61]. To test whether the binding of Sld3 to Mcm4 and Mcm6 represents the essential function of DDK, Mcm4 and Mcm6 phosphomimic mutants were generated for which N-terminal DDK-targeted serine and threonine residues were substituted with negatively charged aspartate residues (Mcm4-25D, Mcm6-25D), and were able to support roughly 60% the wild-type level of DNA replication in the absence of DDK. Two recent studies have also reported DDK-dependent interactions between Sld3 and Mcm2 through pull-down, co-immunoprecipitation, and two hybrid assays [10,62]. Intriguingly, the crystal structure of Sld3 uncovered two conserved basic patches close to one another with the potential of mediating interactions with phosphorylated Mcm2-7 subunits [63]. One of these (amino acids 301–330) has been found to act as a Cdc45-binding interface [63], while mutation of the second patch for the *sld3-4E* mutant (K188E, R192E, K404E, K405E) resulted in disrupted interactions with Mcm2 and Mcm6, but not Cdc45 [62]. Importantly, a similar mutation of this region (K181E, R186E, R192E, K404E, K405E) also maintained an interaction with Cdc45, but displayed a severe growth inhibition phenotype, and this was mirrored by a failure of *sld3-4E* to support growth in place of the wild-type *SLD3* allele [62,63]. As with Mcm4 and Mcm6, the N-terminus of Mcm2 has proven to be crucial for Sld3 binding, as Mcm2 amino acids 1–390 are sufficient for this interaction, but amino acids 1–299 are not [62]. Confirmation of the physiological importance of Sld3 interactions with the Mcm2 and Mcm6 N-termini, was obtained through in vivo complementation assays, in which deletion mutants with disrupted Sld3 binding for Mcm2 (Δ300–390) or Mcm6 (Δ1–122) failed to support growth or S-phase progression in the absence of wild-type Mcm2 or Mcm6 expression, respectively [62]. Furthermore, quantitative PCR analysis of chromatin immunoprecipitation samples (ChIP-qPCR) revealed that the *mcm6Δ1-122* mutant is deficient in recruiting both Sld3 and the single-stranded DNA binding protein replication protein A (RPA) to early-firing origin ARS607, consistent with a defect in replication initiation [62].

Interestingly, there is some evidence to suggest that—in addition to DDK facilitating the association of Sld3 with origins of DNA replication—Sld3 in turn may aid DDK in carrying out one of its roles. As mentioned above, DDK phosphorylates Mcm2 at serines 164 and 170 [24–26]. In vitro, the addition of either full-length Sld3 or its C-terminus alone was able to substantially enhance DDK phosphorylation of Mcm2 [64]. Further evidence for the importance of this stimulatory role was obtained by generating a *SLD3* mutant, *sld3-m16* (Sld3-S556A, H557A, S558A, T559A), defective in aiding DDK with Mcm2 phosphorylation, but competent with respect to other functions, including binding to Dpb11, Mcm2-7, Cdc45, and T-rich single-stranded origin DNA. Reminiscent of the *mcm2-2A* mutant, expression of *sld3-m16* resulted in a dominant-negative growth defect phenotype, and a decrease in association between Mcm2-7 and GINS was observed, pointing to a defect in CMG helicase assembly [64]. This mutually stimulatory relationship between DDK and Sld3 activities likely represents an important positive feedback loop that helps push origins past the threshold of CMG formation required for origin firing.

5. Opposing Activities of SUMOylation, Rif1, and DDK

Rif1 was initially identified as a regulator of telomeric length [65], but has more recently been implicated in the regulation of DNA replication in budding yeast, fission yeast, and mammalian cells [66–70]. More specifically, several lines of evidence point towards an important role for Rif1 in opposing the MCM phosphorylation activity of DDK. For example, temperature-sensitive *cdc7-1* cells can typically be synchronized at the G1-S boundary at 37 °C, yet failed to arrest at this temperature in a *cdc7-1 rif1Δ* strain [16]. Furthermore, deletion of *RIF1* was found to increase the proportion of hyperphosphorylated Mcm4 in budding yeast whole cell extracts, as judged by slower mobility in sodium dodecyl sulfate polyacrylamide gel electrophoresis (SDS-PAGE) immunoblots [16,17,71]. To promote dephosphorylation, Rif1 possesses two conserved motifs for the docking of Glc7—the sole budding yeast protein phosphatase 1 [16,17,71]. Mutation of the Rif1 Glc7 docking domains was able to suppress *cdc7-4* and *dbf4-1* growth defects, consistent with it normally reversing the MCM phosphorylation carried out by DDK [17]. Further evidence for such a role was provided by the observation that a Rif1 Glc7 docking domain mutant resulted in increased Mcm4 phosphorylation, which could be reduced or prevented altogether in a *cdc7-1* background at permissive and non-permissive temperatures, respectively [71]. The idea of Rif1 targeting protein phosphatases to origin-bound MCM complexes is further supported by ChIP analysis carried out in both *S. cerevisiae* and the fission yeast *S. pombe*, which showed a reduction of Glc7 and *S. pombe* protein phosphatase 1 Sds21 and Dis2 at late-firing origins in the absence of Rif1 or with mutation of its protein phosphatase 1-binding motifs [71].

Intriguingly, a hint of another mechanistic layer in the opposing actions of Rif1 and Dbf4 has been provided by the key finding that Dbf4 can itself bind to Rif1 through the latter's C terminus (amino acids 1790–1916) [16,17]. It is tempting to speculate that Rif1 may thus directly counteract DDK activity, however, the ability of DDK to phosphorylate Mcm4 in vitro was not inhibited by the addition of the purified Rif1 C terminus [16]. The inverse may also be true—namely, that DDK binding and potential phosphorylation of Rif1 hinders the latter's ability to target Glc7 to origins. Indeed, putative conserved DDK and CDK phosphorylation sites are found adjacent to the protein phosphatase 1 docking domains in both *S. cerevisiae* and *S. pombe* Rif1. A *S. cerevisiae* Rif1 mutant for which nine of these serines were changed to alanine enhanced the temperature-sensitivity of *cdc7-1*, while changing them to aspartic acid as a phosphomimic had the opposite effect, reminiscent of what was observed when the docking sites themselves were mutated, and equivalent results were observed with similar *S. pombe* mutants [16,71].

Bringing things full circle, one further role of Rif1 is to potentially counteract DDK phosphorylation of Sld3. Although it has been clearly established that Sld3 is a crucial target for CDK phosphorylation, a significant phos-tag gel mobility shift has been observed for Sld3 in G1 phase *rif1Δ* cells, consistent with phosphorylation, and this shift is prevented in *cdc7-4* cells at non-permissive (37 °C) temperature [17].

Interestingly, a recent report has uncovered a potential additional mechanism for DDK-mediated promotion of MCM phosphorylation [72]. SUMOylation of chromatin-bound Mcm2-6 subunits was detected, peaking in G1 phase after MCM loading, declining during S-phase, then rising again in M phase. Mcm7 showed a slightly different pattern, with SUMOylation persisting through most of S-phase, before declining at the end of S-phase. SUMOylation of Mcm6 was shown to increase its interaction with Glc7 [72], promoting the dephosphorylation of Mcm2-7 [16,17,71]. When DDK was inactivated, Mcm2-6 SUMOylation was no longer lost as cells transitioned from G1 to S-phase, suggesting that DDK mediates this process. As SUMOylated forms of Mcm4 did not appear to be phosphorylated by DDK, the authors speculated that DDK might instead act on deSUMOylation enzymes, although this remains to be investigated [72].

6. Targeting of DDK to Early Replicating Centromeric Origins of DNA Replication

Initiation events at budding yeast origins of DNA replication are temporally regulated, with individual origins characteristically firing in early, mid, or late S-phase [73]. DDK activity is limiting for DNA replication, as Dbf4 is present at low abundance and is required throughout S-phase to promote new initiation events [57]. DDK is therefore one of the determinants of which origins fire first in S-phase; however, only recently have some of the underlying mechanistic details been uncovered. Among the earliest origins to fire in S-phase are those associated with the 16 centromeric regions of *S. cerevisiae* chromosomes [73]. Similar findings have been obtained with other yeast species [74–76], *Trypanosoma brucei* [77], and *D. melanogaster* [78], suggesting that this is a conserved aspect of eukaryotic cell cycles. Interestingly, live cell imaging in *S. cerevisiae* has revealed that both Dbf4 and Cdc7 accumulate near spindle pole bodies and kinetochores in late M and early G1 phase [79]. The centromeric localization of Dbf4 was confirmed by ChIP-qPCR for cells arrested in G1 phase, but was strongly impaired when the genes encoding either chromosome transmission fidelity 19 (Ctf19) or chromomsome loss 4 (Chl4) (both kinetochore constituents) were deleted. This effect was specific for Dbf4 association at centromeres, as Dbf4 association with early-firing origins *ARS606* and *ARS607* was not altered in *ctf19Δ* or *chl4Δ* cells. The discovery that Dbf4 Myc-tagged at its C terminus is impaired for association with centromeres, but not with replication origins allowed researchers to determine that abrogation of DDK targeting results in a specific reduction in Sld3-Sld7 origin association and a delay in replication timing at centromeric regions [79]. Importantly, the recruitment of DDK to kinetochores also appears to promote sister chromatid cohesion by targeting the sister chromatid cohesion protein 2 (Scc2)-Scc4 cohesin loader to centromeres in G1 phase, which has also been observed in *Xenopus laevis* [80]. Thus, DDK likely plays a central role in coordinating S-phase onset with sister chromatid cohesion. Recently, Dbf4 localization at centromeres has also been observed in human cells, and DDK was implicated in regulating the recruitment of topoisomerase 2-alpha (TOP2A), which is required for chromosome condensation and sister chromatid separation [81]. Although the timing of Dbf4 centromere association was not coincident with the onset of DNA replication, this study involved the overexpression of tagged Dbf4. Thus, it would be interesting to observe if a stronger correlation is found with normal levels of Dbf4 expression.

7. Conclusions and Perspectives

To initiate DNA replication, DDK binds to and phosphorylates its essential target—the Mcm2-7 ring. This phosphorylation leads to gate opening between Mcm2 and Mcm5, allowing extrusion of single stranded DNA generated by origin melting. DDK also facilitates the association of one of the essential firing factors, Sld3, with origin DNA. A key feature of this DDK-dependent recruitment is that Sld3 interacts with DDK targets Mcm2, 4, and 6. Sld3 in turn targets Cdc45 to origins, thereby facilitating the formation of the CMG replicative helicase. As many of these mechanistic details have been determined through the use of in vitro systems, the additional construction of mutant strains will be required to confirm that they hold true in vivo. The recruitment of DDK to yeast centromeric sequences in G1 phase promotes early S-phase replication of these regions, and likely ensures proper coordination with sister chromatid cohesion through Scc2-Scc4 targeting to the same loci. Similar findings in other eukaryotes merit further investigation to establish the degree of mechanistic conservation.

Negative regulation of DNA replication by opposing DDK activity can occur via two distinct mechanisms. The checkpoint kinase Rad53 impedes DDK activity during S-phase replication stress. How Rad53 binds DDK to facilitate its phosphorylation has been characterized, exposing two FHA domain-mediated binding modes, one canonical and the other non-canonical. Rif1 and Mcm2-7 SUMOylation can each counteract DDK activity to prevent precocious DNA replication initiation in G1 phase by targeting Glc7 to dephosphorylate MCM subunits, yet exactly how Rif1 is itself recruited to origins of DNA replication, and the precise mechanism of Mcm2-7 SUMOylation, are open questions that remain to be investigated.

Acknowledgments: B.P.D. is funded by Natural Sciences and Engineering Research Council of Canada grant RGPIN 238392. L. is funded by a Lembaga Pengelola Dana Pendidikan studentship from the Indonesian Government.

Conflicts of Interest: The authors declare no conflict of interest.

References

1. Labib, K. How Do Cdc7 and Cyclin-Dependent Kinases Trigger the Initiation of Chromosome Replication in Eukaryotic Cells? *Genes Dev.* **2010**, *24*, 1208–1219. [CrossRef] [PubMed]
2. Pasero, P.; Duncker, B.P.; Schwob, E.; Gasser, S.M. A Role for the Cdc7 Kinase Regulatory Subunit Dbf4p in the Formation of Initiation-Competent Origins of Replication. *Genes Dev.* **1999**, *13*, 2159–2176.
3. Zegerman, P.; Diffley, J.F.X. Checkpoint-Dependent Inhibition of DNA Replication Initiation by Sld3 and Dbf4 Phosphorylation. *Nature* **2010**, *467*, 474–478. [CrossRef] [PubMed]
4. Lopez-Mosqueda, J.; Maas, N.L.; Jonsson, Z.O.; DeFazio-Eli, L.G.; Wohlschlegel, J.; Toczyski, D.P. Damage-Induced Phosphorylation of Sld3 Is Important to Block Late Origin Firing. *Nature* **2010**, *467*, 479–483. [CrossRef] [PubMed]
5. Duch, A.; Palou, G.; Jonsson, Z.O.; Palou, R.; Calvo, E.; Wohlschlegel, J.; Quintana, D.G. A Dbf4 Mutant Contributes to Bypassing the Rad53-Mediated Block of Origins of Replication in Response to Genotoxic Stress. *J. Biol. Chem.* **2011**, *286*, 2486–2491. [CrossRef] [PubMed]
6. Chen, Y.-C.; Kenworthy, J.; Gabrielse, C.; Hänni, C.; Zegerman, P.; Weinreich, M. DNA Replication Checkpoint Signaling Depends on a Rad53–Dbf4 N-Terminal Interaction in *Saccharomyces cerevisiae*. *Genetics* **2013**, *194*, 389–401. [CrossRef] [PubMed]
7. Kitada, K.; Johnston, L.H.; Sugino, T.; Sugino, A. Temperature-Sensitive cdc7 Mutations of *Saccharomyces cerevisiae* Are Suppressed by the DBF4 Gene, Which Is Required for the G1/S Cell Cycle Transition. *Genetics* **1992**, *131*, 21–29. [PubMed]
8. Heller, R.C.; Kang, S.; Lam, W.M.; Chen, S.; Chan, C.S.; Bell, S.P. Eukaryotic Origin-Dependent DNA Replication In Vitro Reveals Sequential Action of DDK and S-CDK Kinases. *Cell* **2011**, *146*, 80–91.
9. On, K.F.; Beuron, F.; Frith, D.; Snijders, A.P.; Morris, E.P.; Diffley, J.F.X. Prereplicative Complexes Assembled In Vitro Support Origin-Dependent and Independent DNA Replication. *EMBO J.* **2014**, *33*, 605–620.
10. Herrera, M.C.; Tognetti, S.; Riera, A.; Zech, J.; Clarke, P.; Fernández-Cid, A.; Speck, C. A Reconstituted System Reveals How Activating and Inhibitory Interactions Control DDK Dependent Assembly of the Eukaryotic Replicative Helicase. *Nucleic Acids Res.* **2015**, *43*, 10238–10250. [CrossRef] [PubMed]
11. Yeeles, J.T.P.; Deegan, T.D.; Janska, A.; Early, A.; Diffley, J.F.X. Regulated Eukaryotic DNA Replication Origin Firing with Purified Proteins. *Nature* **2015**, *519*, 431–435. [CrossRef] [PubMed]
12. Cheng, L.; Collyer, T.; Hardy, C.F.J. Cell Cycle Regulation of DNA Replication Initiator Factor Dbf4p. *Mol. Cell. Biol.* **1999**, *19*, 4270–4278. [CrossRef] [PubMed]
13. Oshiro, G.; Owens, J.C.; Shellman, Y.; Sclafani, R.A.; Li, J.J. Cell Cycle Control of Cdc7p Kinase Activity through Regulation of Dbf4p Stability. *Mol. Cell. Biol.* **1999**, *19*, 4888–4896. [CrossRef] [PubMed]
14. Weinreich, M.; Stillman, B. Cdc7p-Dbf4p Kinase Binds to Chromatin during S Phase and Is Regulated by Both the APC and the RAD53 Checkpoint Pathway. *EMBO J.* **1999**, *18*, 5334–5346. [CrossRef] [PubMed]
15. Ferreira, M.F.; Santocanale, C.; Drury, L.S.; Diffley, J.F. Dbf4p, an Essential S Phase-Promoting Factor, Is Targeted for Degradation by the Anaphase-Promoting Complex. *Mol. Cell. Biol.* **2000**, *20*, 242–248.
16. Hiraga, S.; Alvino, G.M.; Chang, F.; Lian, H.; Sridhar, A.; Kubota, T.; Brewer, B.J.; Weinreich, M.; Raghuraman, M.K.; Donaldson, A.D. Rif1 Controls DNA Replication by Directing Protein Phosphatase 1 to Reverse Cdc7-Mediated Phosphorylation of the MCM Complex. *Genes Dev.* **2014**, *28*, 372–383.

17. Mattarocci, S.; Shyian, M.; Lemmens, L.; Damay, P.; Altintas, D.M.; Shi, T.; Bartholomew, C.R.; Thomä, N.H.; Hardy, C.F.J.; Shore, D. Rif1 Controls DNA Replication Timing in Yeast through the PP1 Phosphatase Glc7. *Cell Rep.* **2014**, *7*, 62–69. [CrossRef] [PubMed]

18. Hardy, C.F.; Dryga, O.; Seematter, S.; Pahl, P.M.; Sclafani, R.A. *mcm5/cdc46-bob1* Bypasses the Requirement for the S Phase Activator Cdc7p. *Proc. Natl. Acad. Sci. USA* **1997**, *94*, 3151–3155. [CrossRef] [PubMed]

19. Hoang, M.L.; Leon, R.P.; Pessoa-Brandao, L.; Hunt, S.; Raghuraman, M.K.; Fangman, W.L.; Brewer, B.J.; Sclafani, R.A. Structural Changes in Mcm5 Protein Bypass Cdc7-Dbf4 Function and Reduce Replication Origin Efficiency in *Saccharomyces cerevisiae*. *Mol. Cell. Biol.* **2007**, *27*, 7594–7602. [CrossRef] [PubMed]

20. Sheu, Y.-J.; Stillman, B. Cdc7-Dbf4 Phosphorylates MCM Proteins via a Docking Site-Mediated Mechanism to Promote S Phase Progression. *Mol. Cell* **2006**, *24*, 101–113. [CrossRef] [PubMed]

21. Sheu, Y.-J.; Stillman, B. The Dbf4-Cdc7 Kinase Promotes S Phase by Alleviating an Inhibitory Activity in Mcm4. *Nature* **2010**, *463*, 113–117. [CrossRef] [PubMed]

22. Sheu, Y.-J.; Kinney, J.B.; Lengronne, A.; Pasero, P.; Stillman, B. Domain within the Helicase Subunit Mcm4 Integrates Multiple Kinase Signals to Control DNA Replication Initiation and Fork Progression. *Proc. Natl. Acad. Sci. USA* **2014**, *111*, E1899–E1908. [CrossRef] [PubMed]

23. Sheu, Y.-J.; Kinney, J.B.; Stillman, B. Concerted Activities of Mcm4, Sld3, and Dbf4 in Control of Origin Activation and DNA Replication Fork Progression. *Genome Res.* **2016**, *26*, 315–330. [CrossRef] [PubMed]

24. Bruck, I.; Kaplan, D. Dbf4-Cdc7 Phosphorylation of Mcm2 Is Required for Cell Growth. *J. Biol. Chem.* **2009**, *284*, 28823–28831. [CrossRef] [PubMed]

25. Stead, B.E.; Brandl, C.J.; Davey, M.J. Phosphorylation of Mcm2 Modulates Mcm2-7 Activity and Affects the Cell's Response to DNA Damage. *Nucleic Acids Res.* **2011**, *39*, 6998–7008. [CrossRef] [PubMed]

26. Bruck, I.; Kaplan, D.L. The Dbf4-Cdc7 Kinase Promotes Mcm2-7 Ring Opening to Allow for Single-Stranded DNA Extrusion and Helicase Assembly. *J. Biol. Chem.* **2015**, *290*, 1210–1221. [CrossRef] [PubMed]

27. Bochman, M.L.; Schwacha, A. The *Saccharomyces cerevisiae* Mcm6/2 and Mcm5/3 ATPase Active Sites Contribute to the Function of the Putative Mcm2-7 "Gate". *Nucleic Acids Res.* **2010**, *38*, 6078–6088.

28. Samel, S.A.; Fernández-Cid, A.; Sun, J.; Riera, A.; Tognetti, S.; Herrera, M.C.; Li, H.; Speck, C. A Unique DNA Entry Gate Serves for Regulated Loading of the Eukaryotic Replicative Helicase MCM2-7 onto DNA. *Genes Dev.* **2014**, *28*, 1653–1666. [CrossRef] [PubMed]

29. Li, N.; Zhai, Y.; Zhang, Y.; Li, W.; Yang, M.; Lei, J.; Tye, B.-K.; Gao, N. Structure of the Eukaryotic MCM Complex at 3.8 Å. *Nature* **2015**, *524*, 186–191. [CrossRef] [PubMed]

30. Costa, A.; Ilves, I.; Tamberg, N.; Petojevic, T.; Nogales, E.; Botchan, M.R.; Berger, J.M. The Structural Basis for MCM2-7 Helicase Activation by GINS and Cdc45. *Nat. Struct. Mol. Biol.* **2011**, *18*, 471–477.

31. Lei, M.; Kawasaki, Y.; Young, M.R.; Kihara, M.; Sugino, A.; Tye, B.K. Mcm2 Is a Target of Regulation by Cdc7-Dbf4 during the Initiation of DNA Synthesis. *Genes Dev.* **1997**, *11*, 3365–3374. [CrossRef] [PubMed]

32. Randell, J.C.W.; Fan, A.; Chan, C.; Francis, L.I.; Heller, R.C.; Galani, K.; Bell, S.P. Mec1 Is One of Multiple Kinases That Prime the Mcm2-7 Helicase for Phosphorylation by Cdc7. *Mol. Cell* **2010**, *40*, 353–363.

33. Ramer, M.D.; Suman, E.S.; Richter, H.; Stanger, K.; Spranger, M.; Bieberstein, N.; Duncker, B.P. Dbf4 and Cdc7 Proteins Promote DNA Replication through Interactions with Distinct Mcm2-7 Protein Subunits. *J. Biol. Chem.* **2013**, *288*, 14926–14935. [CrossRef] [PubMed]

34. Francis, L.I.; Randell, J.C.W.; Takara, T.J.; Uchima, L.; Bell, S.P. Incorporation into the Prereplicative Complex Activates the Mcm2-7 Helicase for Cdc7-Dbf4 Phosphorylation. *Genes Dev.* **2009**, *23*, 643–654.

35. Sun, J.; Fernandez-Cid, A.; Riera, A.; Tognetti, S.; Yuan, Z.; Stillman, B.; Speck, C.; Li, H. Structural and Mechanistic Insights into Mcm2-7 Double-Hexamer Assembly and Function. *Genes Dev.* **2014**, *28*, 2291–2303.

36. Varrin, A.E.; Prasad, A.A.; Scholz, R.-P.; Ramer, M.D.; Duncker, B.P. A Mutation in Dbf4 Motif M Impairs Interactions with DNA Replication Factors and Confers Increased Resistance to Genotoxic Agents. *Mol. Cell. Biol.* **2005**, *25*, 7494–7504. [CrossRef] [PubMed]

37. Jones, D.R.; Prasad, A.A.; Chan, P.K.; Duncker, B.P. The Dbf4 Motif C Zinc Finger Promotes DNA Replication and Mediates Resistance to Genotoxic Stress. *Cell Cycle* **2010**, *9*, 2018–2026. [CrossRef] [PubMed]

38. Merchant, A.M.; Kawasaki, Y.; Chen, Y.; Lei, M.; Tye, B.K. A Lesion in the DNA Replication Initiation Factor Mcm10 Induces Pausing of Elongation Forks through Chromosomal Replication Origins in *Saccharomyces cerevisiae*. *Mol. Cell. Biol.* **1997**, *17*, 3261–3271. [CrossRef] [PubMed]

39. Perez-Arnaiz, P.; Bruck, I.; Kaplan, D.L. Mcm10 Coordinates the Timely Assembly and Activation of the Replication Fork Helicase. *Nucleic Acids Res.* **2016**, *44*, 315–329. [CrossRef] [PubMed]

40. Lee, J.-K.; Seo, Y.-S.; Hurwitz, J. The Cdc23 (Mcm10) Protein Is Required for the Phosphorylation of Minichromosome Maintenance Complex by the Dfp1-Hsk1 Kinase. *Proc. Natl. Acad. Sci. USA* **2003**, *100*, 2334–2339. [CrossRef] [PubMed]

41. Homesley, L.; Lei, M.; Kawasaki, Y.; Sawyer, S.; Christensen, T.; Tye, B.K. Mcm10 and the MCM2-7 Complex Interact to Initiate DNA Synthesis and to Release Replication Factors from Origins. *Genes Dev.* **2000**, *14*, 913–926. [PubMed]

42. Izumi, M.; Yanagi, K.; Mizuno, T.; Yokoi, M.; Kawasaki, Y.; Moon, K.-Y.; Hurwitz, J.; Yatagai, F.; Hanaoka, F. The Human Homolog of *Saccharomyces cerevisiae* Mcm10 Interacts with Replication Factors and Dissociates from Nuclease-Resistant Nuclear Structures in G2 Phase. *Nucleic Acids Res.* **2000**, *28*, 4769–4777.

43. Liachko, I.; Tye, B.K. Mcm10 Mediates the Interaction between DNA Replication and Silencing Machineries. *Genetics* **2009**, *181*, 379–391. [CrossRef] [PubMed]

44. Van Deursen, F.; Sengupta, S.; De Piccoli, G.; Sanchez-Diaz, A.; Labib, K. Mcm10 Associates with the Loaded DNA Helicase at Replication Origins and Defines a Novel Step in Its Activation. *EMBO J.* **2012**, *31*, 2195–2206.

45. Douglas, M.E.; Diffley, J.F.X. Recruitment of Mcm10 to Sites of Replication Initiation Requires Direct Binding to the Minichromosome Maintenance (MCM) Complex. *J. Biol. Chem.* **2016**, *291*, 5879–5888.

46. Quan, Y.; Xia, Y.; Liu, L.; Cui, J.; Li, Z.; Cao, Q.; Chen, X.S.; Campbell, J.L.; Lou, H. Cell-Cycle-Regulated Interaction between Mcm10 and Double Hexameric Mcm2-7 Is Required for Helicase Splitting and Activation during S Phase. *Cell Rep.* **2015**, *13*, 2576–2586. [CrossRef] [PubMed]

47. Duncker, B.P.; Shimada, K.; Tsai-Pflugfelder, M.; Pasero, P.; Gasser, S.M. An N-Terminal Domain of Dbf4p Mediates Interaction with Both Origin Recognition Complex (ORC) and Rad53p and Can Deregulate Late Origin Firing. *Proc. Natl. Acad. Sci. USA* **2002**, *99*, 16087–16092. [CrossRef] [PubMed]

48. Kihara, M.; Nakai, W.; Asano, S.; Suzuki, A.; Kitada, K.; Kawasaki, Y.; Johnston, L.H.; Sugino, A. Characterization of the Yeast Cdc7p/Dbf4p Complex Purified from Insect Cells. Its Protein Kinase Activity Is Regulated by Rad53p. *J. Biol. Chem.* **2000**, *275*, 35051–35062. [CrossRef] [PubMed]

49. Matthews, L.A.; Jones, D.R.; Prasad, A.A.; Duncker, B.P.; Guarné, A. *Saccharomyces cerevisiae* Dbf4 Has Unique Fold Necessary for Interaction with Rad53 Kinase. *J. Biol. Chem.* **2012**, *287*, 2378–2387. [CrossRef] [PubMed]

50. Matthews, L.A.; Selvaratnam, R.; Jones, D.R.; Akimoto, M.; McConkey, B.J.; Melacini, G.; Duncker, B.P.; Guarné, A. A Novel Non-Canonical Forkhead-Associated (FHA) Domain-Binding Interface Mediates the Interaction between Rad53 and Dbf4 Proteins. *J. Biol. Chem.* **2014**, *289*, 2589–2599. [CrossRef] [PubMed]

51. Aucher, W.; Becker, E.; Ma, E.; Miron, S.; Martel, A.; Ochsenbein, F.; Marsolier-Kergoat, M.-C.; Guerois, R. A Strategy for Interaction Site Prediction between Phospho-Binding Modules and Their Partners Identified from Proteomic Data. *Mol. Cell. Proteom.* **2010**, *9*, 2745–2759. [CrossRef] [PubMed]

52. Almawi, A.W.; Matthews, L.A.; Larasati; Myrox, P.; Boulton, S.; Lai, C.; Moraes, T.; Melacini, G.; Ghirlando, R.; Duncker, B.P.; et al. "AND" Logic Gates at Work: Crystal Structure of Rad53 Bound to Dbf4 and Cdc7. *Sci. Rep.* **2016**, *6*, 34237. [CrossRef] [PubMed]

53. Tanaka, S.; Umemori, T.; Hirai, K.; Muramatsu, S.; Kamimura, Y.; Araki, H. CDK-Dependent Phosphorylation of Sld2 and Sld3 Initiates DNA Replication in Budding Yeast. *Nature* **2007**, *445*, 328–332. [CrossRef] [PubMed]

54. Zegerman, P.; Diffley, J.F.X. Phosphorylation of Sld2 and Sld3 by Cyclin-Dependent Kinases Promotes DNA Replication in Budding Yeast. *Nature* **2007**, *445*, 281–285. [CrossRef] [PubMed]

55. Tanaka, S.; Nakato, R.; Katou, Y.; Shirahige, K.; Araki, H. Origin Association of Sld3, Sld7, and Cdc45 Proteins Is a Key Step for Determination of Origin-Firing Timing. *Curr. Biol.* **2011**, *21*, 2055–2063.

56. Kamimura, Y.; Tak, Y.-S.; Sugino, A.; Araki, H. Sld3, Which Interacts with Cdc45 (Sld4), Functions for Chromosomal DNA Replication in *Saccharomyces cerevisiae*. *EMBO J.* **2001**, *20*, 2097–2107.

57. Mantiero, D.; Mackenzie, A.; Donaldson, A.; Zegerman, P. Limiting Replication Initiation Factors Execute the Temporal Programme of Origin Firing in Budding Yeast. *EMBO J.* **2011**, *30*, 4805–4814. [CrossRef] [PubMed]

58. Bruck, I.; Kaplan, D.L. GINS and Sld3 Compete with One Another for Mcm2-7 and Cdc45 Binding. *J. Biol. Chem.* **2011**, *286*, 14157–14167. [CrossRef] [PubMed]

59. Tanaka, T.; Umemori, T.; Endo, S.; Muramatsu, S.; Kanemaki, M.; Kamimura, Y.; Obuse, C.; Araki, H. Sld7, an Sld3-associated Protein Required for Efficient Chromosomal DNA Replication in Budding Yeast. *EMBO J.* **2011**, *30*, 2019–2030. [CrossRef] [PubMed]

60. Yabuuchi, H.; Yamada, Y.; Uchida, T.; Sunathvanichkul, T.; Nakagawa, T.; Masukata, H. Ordered Assembly of Sld3, GINS and Cdc45 Is Distinctly Regulated by DDK and CDK for Activation of Replication Origins. *EMBO J.* **2006**, *25*, 4663–4674. [CrossRef] [PubMed]

61. Deegan, T.D.; Yeeles, J.T.; Diffley, J.F. Phosphopeptide Binding by Sld3 Links Dbf4-Dependent Kinase to MCM Replicative Helicase Activation. *EMBO J.* **2016**, *35*, 961–973. [CrossRef] [PubMed]

62. Fang, D.; Cao, Q.; Lou, H. Sld3-MCM Interaction Facilitated by Dbf4-Dependent Kinase Defines an Essential Step in Eukaryotic DNA Replication Initiation. *Front. Microbiol.* **2016**, *7*, 885. [CrossRef] [PubMed]

63. Itou, H.; Muramatsu, S.; Shirakihara, Y.; Araki, H. Crystal Structure of the Homology Domain of the Eukaryotic DNA Replication Proteins Sld3/Treslin. *Structure* **2014**, *22*, 1341–1347. [CrossRef] [PubMed]

64. Bruck, I.; Kaplan, D.L. Conserved Mechanism for Coordinating Replication Fork Helicase Assembly with Phosphorylation of the Helicase. *Proc. Natl. Acad. Sci. USA* **2015**, *112*, 11223–11228. [CrossRef] [PubMed]

65. Hardy, C.F.; Sussel, L.; Shore, D. A RAP1-Interacting Protein Involved in Transcriptional Silencing and Telomere Length Regulation. *Genes Dev.* **1992**, *6*, 801–814. [CrossRef] [PubMed]

66. Lian, H.-Y.; Robertson, E.D.; Hiraga, S.; Alvino, G.M.; Collingwood, D.; McCune, H.J.; Sridhar, A.; Brewer, B.J.; Raghuraman, M.K.; Donaldson, A.D. The Effect of Ku on Telomere Replication Time Is Mediated by Telomere Length but Is Independent of Histone Tail Acetylation. *Mol. Biol. Cell* **2011**, *22*, 1753–1765.

67. Cornacchia, D.; Dileep, V.; Quivy, J.-P.; Foti, R.; Tili, F.; Santarella-Mellwig, R.; Antony, C.; Almouzni, G.; Gilbert, D.M.; Buonomo, S.B.C. Mouse Rif1 Is a Key Regulator of the Replication-Timing Programme in Mammalian Cells. *EMBO J.* **2012**, *31*, 3678–3690. [CrossRef] [PubMed]

68. Hayano, M.; Kanoh, Y.; Matsumoto, S.; Renard-Guillet, C.; Shirahige, K.; Masai, H. Rif1 Is a Global Regulator of Timing of Replication Origin Firing in Fission Yeast. *Genes Dev.* **2012**, *26*, 137–150. [CrossRef] [PubMed]

69. Yamazaki, S.; Ishii, A.; Kanoh, Y.; Oda, M.; Nishito, Y.; Masai, H. Rif1 Regulates the Replication Timing Domains on the Human Genome. *EMBO J.* **2012**, *31*, 3667–3677. [CrossRef] [PubMed]

70. Foti, R.; Gnan, S.; Cornacchia, D.; Dileep, V.; Bulut-Karslioglu, A.; Diehl, S.; Buness, A.; Klein, F.A.; Huber, W.; Johnstone, E.; et al. Nuclear Architecture Organized by Rif1 Underpins the Replication-Timing Program. *Mol. Cell* **2016**, *61*, 260–273. [CrossRef] [PubMed]

71. Davé, A.; Cooley, C.; Garg, M.; Bianchi, A. Protein Phosphatase 1 Recruitment by Rif1 Regulates DNA Replication Origin Firing by Counteracting DDK Activity. *Cell Rep.* **2014**, *7*, 53–61. [CrossRef] [PubMed]

72. Wei, L.; Zhao, X. A New MCM Modification Cycle Regulates DNA Replication Initiation. *Nat. Struct. Mol. Biol.* **2016**, *23*, 209–216. [CrossRef] [PubMed]

73. Raghuraman, M.K.; Winzeler, E.A.; Collingwood, D.; Hunt, S.; Wodicka, L.; Conway, A.; Lockhart, D.J.; Davis, R.W.; Brewer, B.J.; Fangman, W.L. Replication Dynamics of the Yeast Genome. *Science* **2001**, *294*, 115–121. [CrossRef] [PubMed]

74. Kim, S.-M.; Dubey, D.D.; Huberman, J.A. Early-Replicating Heterochromatin. *Genes Dev.* **2003**, *17*, 330–335.

75. Koren, A.; Tsai, H.-J.; Tirosh, I.; Burrack, L.S.; Barkai, N.; Berman, J. Epigenetically-Inherited Centromere and Neocentromere DNA Replicates Earliest in S-Phase. *PLoS Genet.* **2010**, *6*, e1001068. [CrossRef] [PubMed]

76. Müller, C.A.; Nieduszynski, C.A. Conservation of Replication Timing Reveals Global and Local Regulation of Replication Origin Activity. *Genome Res.* **2012**, *22*, 1953–1962. [CrossRef] [PubMed]

77. Tiengwe, C.; Marcello, L.; Farr, H.; Dickens, N.; Kelly, S.; Swiderski, M.; Vaughan, D.; Gull, K.; Barry, J.D.; Bell, S.D.; et al. Genome-Wide Analysis Reveals Extensive Functional Interaction between DNA Replication Initiation and Transcription in the Genome of *Trypanosoma brucei*. *Cell Rep.* **2012**, *2*, 185–197.

78. Ahmad, K.; Henikoff, S. Centromeres Are Specialized Replication Domains in Heterochromatin. *J. Cell Biol.* **2001**, *153*, 101–110. [CrossRef] [PubMed]

79. Natsume, T.; Müller, C.A.; Katou, Y.; Retkute, R.; Gierliński, M.; Araki, H.; Blow, J.J.; Shirahige, K.; Nieduszynski, C.A.; Tanaka, T.U. Kinetochores Coordinate Pericentromeric Cohesion and Early DNA Replication by Cdc7-Dbf4 Kinase Recruitment. *Mol. Cell* **2013**, *50*, 661–674. [CrossRef] [PubMed]

80. Takahashi, T.S.; Basu, A.; Bermudez, V.; Hurwitz, J.; Walter, J.C. Cdc7-Drf1 Kinase Links Chromosome Cohesion to the Initiation of DNA Replication in *Xenopus* Egg Extracts. *Genes Dev.* **2008**, *22*, 1894–1905.

81. Wu, K.Z.L.; Wang, G.-N.; Fitzgerald, J.; Quachthithu, H.; Rainey, M.D.; Cattaneo, A.; Bachi, A.; Santocanale, C. DDK Dependent Regulation of TOP2A at Centromeres Revealed by a Chemical Genetics Approach. *Nucleic Acids Res.* **2016**, *44*, 8786–8798. [CrossRef] [PubMed]

![genes logo](GCAT TACG GCAT *genes*)

MDPI

Review

Origin DNA Melting—An Essential Process with Divergent Mechanisms

Matthew P. Martinez †, John M. Jones †, Irina Bruck and Daniel L. Kaplan *

Department of Biomedical Sciences, Florida State University College of Medicine, 1115 W. Call St., Tallahassee, FL 32306, USA; mm14d@my.fsu.edu (M.P.M.); jmj13b@my.fsu.edu (J.M.J.); irina.bruck@med.fsu.edu (I.B.)
* Correspondence: Daniel.Kaplan@med.fsu.edu; Tel.: +1-850-645-0237
† These two authors contributed equally to this work.

Academic Editor: Eishi Noguchi
Received: 21 November 2016; Accepted: 3 January 2017; Published: 11 January 2017

Abstract: Origin DNA melting is an essential process in the various domains of life. The replication fork helicase unwinds DNA ahead of the replication fork, providing single-stranded DNA templates for the replicative polymerases. The replication fork helicase is a ring shaped-assembly that unwinds DNA by a steric exclusion mechanism in most DNA replication systems. While one strand of DNA passes through the central channel of the helicase ring, the second DNA strand is excluded from the central channel. Thus, the origin, or initiation site for DNA replication, must melt during the initiation of DNA replication to allow for the helicase to surround a single-DNA strand. While this process is largely understood for bacteria and eukaryotic viruses, less is known about how origin DNA is melted at eukaryotic cellular origins. This review describes the current state of knowledge of how genomic DNA is melted at a replication origin in bacteria and eukaryotes. We propose that although the process of origin melting is essential for the various domains of life, the mechanism for origin melting may be quite different among the different DNA replication initiation systems.

Keywords: DNA helicase; DNA replication; initiation; protein-DNA interaction; DnaA; Large T antigen; E1 helicase; Mcm2–7; melting

---

1. Review of Bacterial Replication Initiator DnaA

Like every other organism, bacteria must replicate their DNA in order to produce viable offspring. However, bacteria cannot infinitely replicate, meaning there must be a tight regulation of this process. The fact that replication does not just start and pause indicates that there is a lot of regulation on the initiation of chromosome replication. DnaA, the key initiator protein among almost all bacteria, is a highly conserved protein and is the driver of the system in which DNA replication initiation is regulated. This protein has been studied extensively and understood through the *Escherichia coli* model.

2. DnaA-Orisome Structure

DnaA is a key protein in the initiation of bacterial replication (Figure 1). Bound to high- and low-affinity sites at the initiation sequence, *oriC*, DnaA is a highly conserved protein among all bacteria that comprises the DNA-protein complex termed the orisome, which triggers the initiation of chromosome replication. OriC DNA is not bare throughout the cell cycle, but instead has bound DnaA to three high-affinity sites (left to right: R1, R2, R4). These three DnaA sites, along with oriC bending protein Fis, set a nucleosome-like conformation in the origin that has been suggested to prevent replication initiation (Figure 2) [1]. Fis is not necessary for viability, however, the lack of Fis binding results in asynchronous replication in rapidly growing cells. This is due to the binding of DnaA to low affinity sites at a lower concentration than what is normally required, since there is no Fis protein to inhibit DnaA binding [2]. Additionally, this conformation keeps the DNA double-stranded until the appropriate replication-promoting proteins bind and separate the two strands. The review by Leonard and Grimwade [1] discusses that these replication-promoting proteins include additional DnaA and another DNA bending protein, Integration Host Factor (IHF). Upon accumulation of a sufficient level of DnaA-ATP, the active form of DnaA, Fis will be displaced and IHF will bind, along with DnaA, to low affinity sites between R1 and R2, and R2 and R4 [3]. IHF has been shown to be nonessential for the assembly of a functional orisome, however, this loss of IHF results in perturbed replication initiation [1]. The viability of cells lacking IHF binding is most likely due to the flexibility of the DNA between R1 and R5M.

Although the exact mechanism of the displacement of this initiator inhibition is unclear, a recent study has shown that ATP-bound DnaA, as opposed to ADP-bound DnaA, experiences a conformational change within domains I–III that enhances its ability to bind to low affinity sites within oriC as well as cooperatively bind to already bound DnaA molecules [4]. Once a threshold concentration of DnaA-ATP is achieved in the cell, Fis can successfully be displaced and the inhibitory complex can progress to an active one. DNase I footprinting studies have suggested that DNA wraps around the DnaA oligomer once bound [5]. As illustrated in Figure 1, the function of each domain has been determined via reverse genetics: DnaA recruitment (I), DNA binding (IV), oligomerization (I, III), ATP binding (III), and helicase loading (I, III) [1]. Between domains III and IV is an amphipathic region that is involved in binding to the inner membrane of the cell [6]. Additionally, domain II serves as a flexible linker, aligning domain I with domains III + IV [4].

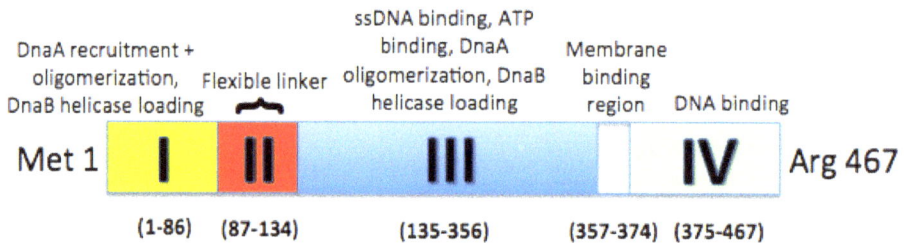

Figure 1. A schematic map of the four domains of DnaA. ssDNA: single-stranded DNA.

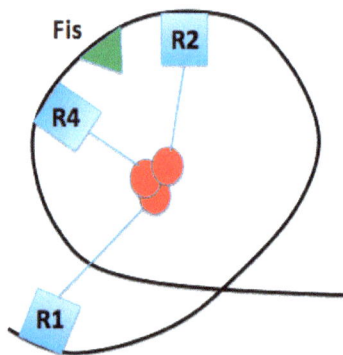

Figure 2. Proposed loop conformation of inactive oriC, constrained by DnaA bound to high-affinity sites R1, R2, and R4 via domain I N-terminus interactions. This conformation is facilitated by Fis.

3. DNA Conformation

The oriC DNA contains multiple sites of DnaA binding in which specific binding is required for duplex unwinding (Figure 3). Between the three high affinity sites mentioned in the above paragraph are low affinity DnaA sites (R5M, τ2, I1, and I2, respectively, between R1 and R2; C3, C2, I3, and C1, respectively, between R2 and R4) [3], which become DnaA bound just before origin melting. The left half of oriC (R1–I2) and right half of oriC (R2–R4) have opposite orientations, with both oriented inward of oriC (towards each other) [3]. Kaur et al. demonstrated that the loss of any two high affinity sites resulted in the loss of oriC function, while the loss of any single high affinity site resulted in a functional oriC with perturbed initiation timing, with an R4 mutation being the most significant [3]. The loss of R2 showed the least significant impact, implying that R2 may be a redundant site or may stabilize the oligomers from R1 and R4. The loss of any single high affinity site rendered the cell dependent on both Fis and IHF binding for a functional oriC. When either R1 or R4 was deleted, R2 was shown to be capable of nucleating a DnaA oligomer, although a higher concentration of DnaA was required. Less DnaA was detected in the right half of oriC in R4 mutants, supporting the importance of R4 [2]. Additionally, it was shown that *E. coli* mutants with a deletion in the entire right half or oriC (R2–R4) are still viable under slow growth conditions. However, with sensitivity to rich media and other rapid growth conditions, it is possible that the right half of oriC has evolved to support multi-forked replication [7]. With these data on alternative methods for pre-RC formation, the minimum requirements for origin melting can be further investigated and understood with greater complexity.

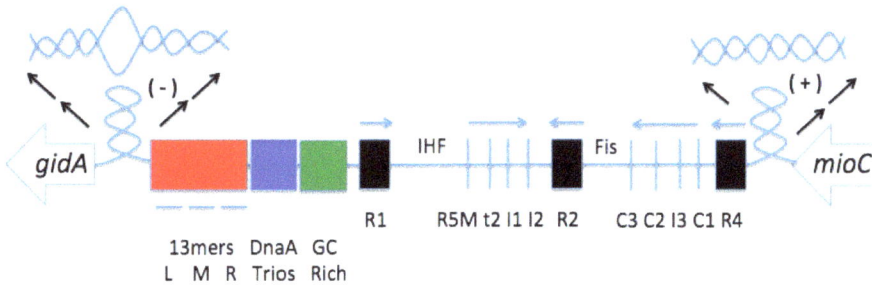

Figure 3. The origin of replication in *Escherichia coli*, oriC. This 245-bp sequence consists of the 13-mer DNA unwinding element (red), DnaA-trio motifs (blue), and binding sites for DnaA, Integration Host Factor (IHF), and Fis. Additionally, flanking genes *gidA* and *mioC* are shown. The arrows represent the transcription direction of the flanking genes (large, hollow arrows) and directionality of DnaA filament formation (small arrows above DnaA boxes). The black arrows help visualize each type of supercoiling, shown above the oriC.

The DNA unwinding element is an AT-rich region towards the left of oriC that has less helical stability than the rest of oriC DNA. DNA unwinding element (DUE) consists of three regions (L, M, R) 13-mer repeats [1]. The DUE is the first piece of DNA to unwind in replication initiation [8], with evidence supporting initiation of melting beginning with the L-region [9]. Kowalski and Eddy have demonstrated that by deleting the l-13mer and replacing it with a dissimilar sequence, its helical instability, rather than its specific sequence, is essential for origin function and duplex unwinding. Meanwhile, the sequence of the r-13mer is the most evolutionarily conserved of the three segments, suggesting a role for the r-13mer in specific protein recognition [10].

An increase in net negative supercoiling (a general undertwist in the DNA has been shown in more efficient *E. coli* initiation, indicating that this chromosomal topology is preferred for replication initiation [11]. The flanking gene *gidA* introduces negative supercoiling to the left of the DUE, which helps further destabilize the already less thermodynamically stable AT-rich DUE [11]. Supporting this, maximal *gidA* transcription occurs before initiation. Additionally, Magnan and Bates discuss in their review the importance of positive supercoiling in regulating oriC transcription [11]. The positive supercoiling to the right of DUE is regulated by the flanking gene *mioC*, with maximal transcription immediately after initiation. While *gidA* and *mioC* are both dispensable, it is possible that they help drive initiation under suboptimal conditions [11]. Kaur et al. tested for the conformation of oriC pre-melting, and developed a model in which oriC forms a constrained loop by interactions of the N-termini of high affinity-bound DnaA, and this loop and repression of active low affinity sites is assisted by Fis binding (Figure 2) [3]. It is possible that this pre-initiation complex causes a reduction of negative supercoiling adjacent to the DUE, and further research is needed to support this.

Recent research has found a DnaA-trio, which consists of a repeating trinucleotide motif, beginning with 3′-GAT-5′, which lies between the AT-rich DUE and the GC rich region (which is adjacent to the DnaA boxes) (Figure 3) [12]. These newer findings will be discussed in greater detail later on.

4. Initiator Mechanism

DnaA contains various AAA+ (ATPases Associated with various cellular Activities) motif sequences which provide a range of functions, including DnaA-DnaA binding [13] and DnaA-ssDNA (single-stranded DNA) binding [14]. DnaA bound to oriC high affinity DNA boxes, via its domain IV helix-turn-helix motif [4], nucleates by binding ATP-DnaA at adjacent low affinity sites. Interestingly, one method of regulation of this step is through a chromosome-membrane protein tether. Bound to an array of operator sequences on the chromosome up to 1 Mb away from oriC, this tether is proposed to inhibit DnaA binding to DNA by reducing the net negative supercoiling [15], although this mechanism is not quite yet understood. DnaA-ATP is required for effective binding to low affinity sites and DnaA oligomer formation [4], yet DnaA cannot always be bound to ATP. Examining the crystal structure of DnaA bound to ssDNA revealed four DnaA protomers per oligomer, forming a right hand spiral around a single strand of the duplex DNA [14].

The DnaA oligomer formation from R1 and R4 inwards towards R2 [3] is mediated by the Arg285 residue within domain III, which is oriented inward towards R2 for both the right and left half of oriC [13]. These Arg285 fingers stimulate subcomplex formation by binding the ATP nucleotide of the next DnaA monomer, eventually forming a DnaA oligomer. This study also found that the Arg285 finger of R1-box-bound DnaA is crucial for DUE unwinding and single-stranded DNA unwinding element (ssDUE) binding, where the same residue of R4-box-bound DnaA plays a necessary role in DnaB helicase loading.

The interaction between DnaA monomers facilitates a conformational change in the bound strand of DNA, stretching the contacted strand and disrupting the base pairs of the thermodynamically unstable DUE [14]. Once this region of oriC unwinds, origin melting is enhanced by binding of the DnaA box-bound DnaA filaments to the partially melted region of oriC DNA. DnaA forms a helical filament around the ssDNA, where each protomer binds three nucleotides via two pairs of helices, $\alpha 3/\alpha 4$ and $\alpha 5/\alpha 6$, which line the inner channel of this protein assembly [14]. Additionally, this conformation prevents reannealing of the two strands of DNA. Once this conformation is set, as visualized in Figure 4, DnaA stabilizes the partially melted origin by nucleating from the already bound DnaA, forming dynamic filaments on the ssDNA monomer by monomer in a 3′-5′ directionality [16].

While the details of this mechanism have been widely unknown, a recent study has identified a DnaA-trio motif within the ssDNA, which is recognized by the DnaA box-bound DnaA [12], facilitating filament formation on the ssDNA via the domain III Initiator Specific Motif (ISM) Initiator Specific Motif [14]. According to the study conducted by Richardson et al., the box-bound DnaA recognizes a 3′-GAT-5′ sequence, with some variability between the first and third nucleotide, but a highly conserved second adenine nucleotide [12]. At this point, DnaA will nucleate across the next few DnaA-trios and into the DUE. Upon filament formation and further duplex melting, DnaA will load DnaB via domain I and domain III interactions, initiating the formation of the prepriming complex (Figure 4, [17]).

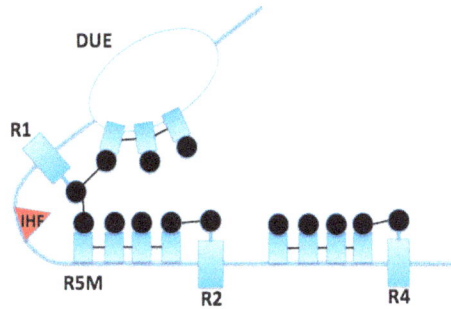

Figure 4. Active oriC conformation, showing interactions between R1 and R5M facilitated by IHF, and interactions between R1 DnaA and DNA unwinding element (DUE)-bound DnaA, which facilitates filament formation on ssDNA. The thicker blue line represents double-stranded oriC DNA, and the thin lines of the "bubble" represent the single-stranded DNA of the melted DUE.

5. Large T-Antigen and E1 Helicases

Mechanisms of origin melting can be derived from the structural analysis of the DNA tumor virus Simian virus 40 (SV40) and papillomavirus. Specifically, SV40 utilizes its Large T antigen (LTag) to initially separate and continually unwind double-stranded DNA (dsDNA) in host cells, and the papillomavirus enlists E1 to do the same. Due to eukaryotic similarities, such as homohexameric domains and beta hairpin loops, results derived from these models may be applicable to the understanding of eukaryotic melting processes. Unlike replication in eukaryotes, melting with initiators SV40 and E1 is performed through cooperation of only a handful of protein domains compared to the variety of protein complexes often necessary to facilitate eukaryotic DNA replication. This lack of complexity yet abundance of shared homology has allowed recent studying of SV40 and E1 to elucidate potential mechanisms for eukaryotic origin melting.

6. Structure of LTag and the Core Ori

Melting of eukaryotic DNA is thought to require a variety of protein factors which work together to manipulate dsDNA, ultimately separating the two strands via mechanical force. Due to the complexity of the eukaryotic cellular machinery, researchers have turned to more simplistic models of initiation, such as the Large T antigen. LTag is a double hexameric protein complex produced by the SV40 virus which is solely responsible for melting of SV40 viral DNA origins, as well as helicase activity once replication forks have been established. Three distinguishable domains compartmentalize these actions, the first of which is known as the origin binding domain (OBD). The OBD of LTag has been shown to bind both dsDNA and ssDNA [18] much like the DNA binding domains of DnaA [12]. Many similarly structured DNA binding domains (DBDs) of eukaryotic and prokaryotic replication machinery bind ssDNA specifically, such as eukaryotic replication protein A (RPA), and bacterial *E. coli* single-stranded DNA-binding proteins (EcoSSB) [19–21]. The second and third domains are the Zn domains, and AAA+ domains, respectively [22]. The three domains can be found in Figure 5A. To initiate replication, these domains seek out designated binding sites on viral DNA along a segment known as the core origin of DNA.

The SV40 core origin for DNA replication (core ori) is composed of four pentanucleotide GAGGC sequences, an AT-rich region (AT), and an early palindromic sequence (EP). From 5′-3′ the ori is composed of the EP, the four pentanucleotides, and the AT region (Figure 5B). Due to the double hexameric nature of LTag, and the asymmetry of the core ori, each hexamer is bound to two GAGGC sequences and either an EP or AT. Once each hexamer is bound, the double hexamer is complete, and completion of the double hexamer is associated with ori melting [23]. The GAGGC sequences themselves are recognized by the OBDs of LTag at major grooves [24,25], while AAA+ regions were found to utilize histidine residues at the tips of beta hairpin loops to interact with ori DNA electrostatically at minor grooves (Figure 6B) [22]. Because proteins often use arginine residues to orient themselves into narrow minor grooves of DNA [26], histidine's role in the AAA+ domains of LTag was originally thought to be the same as that of arginine elsewhere (i.e., as a DNA recognition element) [22]. However, research into the role of these histidines and their respective beta hairpins has suggested unique models for melting discussed below.

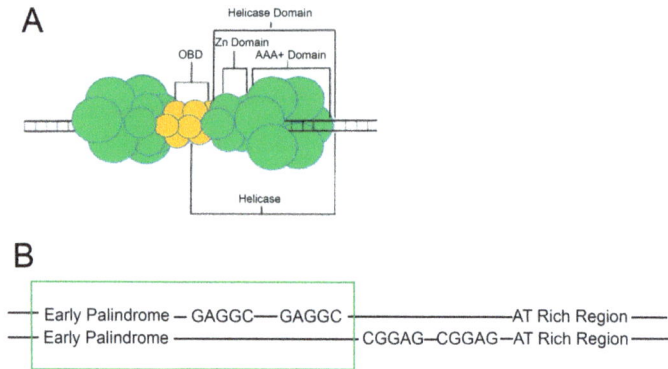

Figure 5. Structure of the Large T Antigen (LTag) and the Simian Virus 40 (SV40) Core Ori. (**A**) A cartoon model illustrating the double hexameric LTag complex and its relevant subdivisions. A single hexamer is noted to contain a portion of the origin binding domain (OBD) and a helicase domain, which itself includes a Zn and AAA+ domain; (**B**) Depiction of the Core Ori of SV40 viral double-stranded DNA (dsDNA) including the four GAGGC pentamers and the flanking AT-rich (AT) and early palindromic sequence (EP) regions. The box around two pentamers and the EP region indicates what portion of the core ori a single hexamer of LTag would occupy.

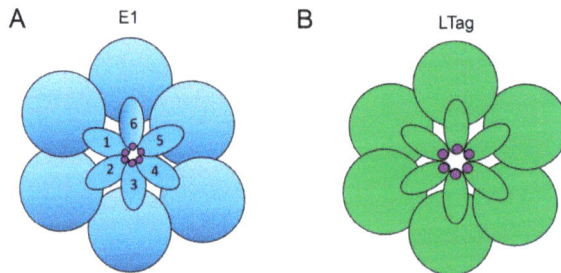

Figure 6. E1 vs. LTag Beta Hairpin Structure. (**A**) A cartoon model depicting the central channel of an E1 helicase domain from a down-the-barrel point of view. The outer circles represent helicase subunits while the structures numbered 1–6 designate the beta hairpin loops. These loops overlap to create a "staircase" pattern. The foot of the beta hairpin staircase is numbered 1. The increasing numbers correspond to higher steps in the staircase. Hairpin loop 2 sits higher than hairpin 1, while 3 overlaps 2, 4 overlaps 3, and so on in an ascending pattern characteristic of E1 hairpins. The histidine residues employed in the untwisting mechanism of melting are denoted in purple at the tip of each hairpin; (**B**) A cartoon model depicting the central channel of an LTag hexameric complex. The six circular domains signify the six helicase subunits while the six oval structures represent beta hairpin loops. The hairpin loops are organized into a planar arrangement characteristic of LTag helicase domains, a distinct organizational method not found in E1 that may contribute to unique melting mechanisms. Histidine residues at the tip of each hairpin are marked in purple [27].

7. Mechanisms of Melting with LTag

The identified histidine residue is a component of beta hairpin loops which the AAA+ domain utilizes to interact with DNA (Figure 6B). Because LTag is a double hexamer, the dodecahedric complex contains two AAA+ domains with a total of twelve beta hairpin loops, and therefore twelve interactive histidines [28]. Only a single pair of histidines, one imidazole ring from each AAA+ domain, were found to lie in the same minor groove, as well as in the same plane, and within 2.7 angstroms of each other, suggesting the presence of hydrogen bonds to provide enhanced stabilization of the LTag dimer [22]. Portions of the core ori, at which these histidine anchors were found, have been confirmed to be melted after double hexamer assembly [29]. Further mutagenesis of these beta hairpin structures has confirmed their necessity during melting of regions flanking the central pentameric sequences of the core ori [30]. Each set of six beta hairpins are arranged in a planar pattern ultimately creating a ring with a central, positively charged channel (Figure 6B) [31]. This channel is between 7–15 angstroms in diameter [32], making it incredibly unlikely for dsDNA to be thread through, but highly likely for ssDNA [30,33]. For comparison, a hexameric helicase that has been shown to envelop dsDNA, known as RuvB, has a central channel diameter of 30 angstroms [34]. The SV40 distant homolog, E1, utilizes helicase domains determined to envelop solely ssDNA (Figure 7D), and it contains a central channel 17 angstroms in diameter as a result [35]. It is therefore likely that after initial melting, the ssDNA will become engulfed in the central channel as the helicase domains translocate down the DNA, separating the double helix via steric exclusion principles. The steric exclusion model of strand separation occurs when one ssDNA strand, from the duplex that was melted, is enclosed by a hexameric helicase channel so that when the other strand remains outside of the channel, the duplex may be pried apart further by helicase progression down the ssDNA [36].

Crystal structures of LTag-DNA complexes have elucidated that each of the OBDs of the double hexamer are oriented 180 degrees to each other when bound to DNA, potentially as a result of a twisting motion which could have generated mechanical force to melt the ori DNA [22]. Since hairpin histidines act as the anchor for LTag's AAA+ domains, and the minor grooves in which they anchor were subsequently melted, it is feasible that this twisting motion would provide enough force to disrupt hydrogen bonds between base pairs of ori nucleotides, similar to the "untwisting" mechanism utilized by the LTag homolog, E1 (Figure 7). However, LTag-ori-DNA crystal structures showed no significant deformations of DNA [22]. Because of proposals of E1 utilizing trimers in the "untwisting" mechanism before construction of the E1 double hexamers homologous to LTag (Figure 7) [37], it has been proposed that an intermediate LTag structure is formed as well, which melts the ori before the final LTag double hexamer is assembled for translocation [22].

8. Structure of the E1 Double Hexamer and Double Trimer

Much like SV40's LTag, papillomavirus's E1 is a homohexameric protein complex responsible for both the initiation of melting and the successive unwinding of DNA. E1 recognizes its unique origin of replication (ori) through DBDs which work to recognize four E1 binding sites, in a nature homologous to SV40's use of OBDs to bind four GAGGC sequences. From left to right, the E1 protein complex consists of an N-terminal domain, a DBD, an oligomerization domain, a helicase domain, and an acidic C-terminal tail [37]. The DBD is oriented between the two helicase domains which are arranged facing each other. The DBD binds to the E1 binding sites at the center of the ori, while the neighboring helicase domains bind to flanking regions of DNA. The helicase domains of the double hexamer (DH) arrange their beta hairpins in a staircase manner as opposed to the planar formation characteristic of LTag helicase domains (Figure 6).

Unique to studies of E1, formation of an E1 double trimer (DT) has been identified before formation of a double hexamer. Although the exact structure of the DT has not been identified, it is accepted that the DBD of the trimer is oriented between helicase domains, and that the DBD binds the center of the origin while the helicase domains remain bound to flanking regions of DNA. The DT arises when E1 interacts with a dsDNA ori probe in the presence of nucleotides, while the DH subsequently forms in the presence of ATP [37]. The DT has been shown to recognize the origin of replication, and ultimately convert into a double hexamer on ssDNA derived from a melted origin [37].

9. Mechanisms of Melting with E1

Although a single E1 trimer does not maintain helicase abilities [38], the DT has conclusively demonstrated an ability to melt dsDNA into ssDNA so that the resulting ssDNA may be used as a template for DH assembly [37]. The identity of the melting complex as DT and not DH, or an intermediate between the two, was concluded through time-course experimentation [39]. The determination of the DT as the melting machinery of E1 has led to recent extensive kinetic and biochemical analyses with the goal of identifying the DT melting mechanism. Plasmid untwisting assays have supported the hypothesis that initial melting is performed via an untwisting mechanism of ori DNA by the DT (Figure 7) [39]. It is proposed that by hydrolyzing ATP, the DT manages to utilize Histidine residues (H507) in the beta hairpins of the helicase domains to initiate melting (Figure 6A). Because the helicase domains themselves remain on the flanks of the E1 binding sites, the histidine interactions with DNA are thought to melt the central portion indirectly through structural deformations of the flanks which propagate through the center binding sites via an untwisting mechanism [39]. If this mechanism were to occur, then mechanical force must be transmitted from the flanks of the ori through the central binding sites. Therefore, nicks in the DNA should inhibit ori melting as a result of interrupting the path of force transference. This is precisely what Shuck and Stenlund found during nicking experiments of ori DNA [39].

The untwisting mechanism has become a widely accepted proposal. However, a "squeeze-to-open" model has been suggested, in which dsDNA is enveloped by DH and ultimately compressed in the central channel of the helicase domains until base pairs are separated [40]. The "squeeze-to-open" model is supported by evidence of melting occurring simultaneously as LTag assembly occurs [29]. Since E1 has only demonstrated central channels capable of enveloping ssDNA, a model involving a larger central channel proves more promising for LTag structures, because they have been shown to undergo conformational changes promoting slight dilations of their central channels [32].

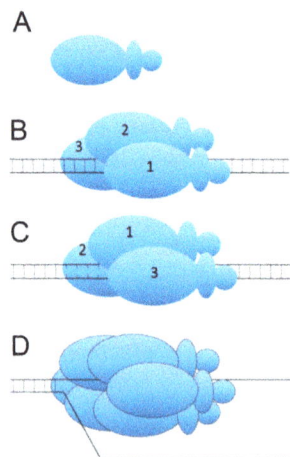

Figure 7. The E1 "Unwinding" Mechanism of Origin Melting. (**A**) An illustration of an E1 monomer. Twelve of these constitute an E1 double hexameric complex shown to unwind DNA after initial melting; (**B**) Pre-twist: Assembly of a single trimer of E1 monomers around dsDNA, and the insertion of histidine residues into the dsDNA; The numbers 1, 2, and 3 demark the three subunits of the trimer. (**C**) Post-twist: The slight rotation, or "twist", has resulted in a melted origin and reorientation of the three subunits as a result; (**D**) Assembly of a single hexamer of E1 onto ssDNA post melting.

10. MCM2-7 Helicase

In eukaryotic cells, it has not yet been determined what melts replication origin DNA. The MCM2–7 helicase [41] and the origin recognition complex (ORC) [42] assemblies are the most likely candidates, since these complexes hydrolyze ATP, and energy is required for origin melting. The MCM2–7 helicase is related to Large T and E1 helicase proteins, suggesting conservation of mechanism [43]. However, MCM2–7 lacks much of the machinery present in the viral counterparts, suggesting that the mechanism for origin melting is different for MCM2–7 compared to Large T and E1 [43]. Furthermore, the MCM2–7 helicase is very weak on its own [44], and MCM2–7 requires Cdc45 and GINS attachment for full helicase activity [45]. The CMG (CDC45-MCM2–7-GINS) helicase is conserved in archaea as well [46].

The MCM2–7 has an N-terminal domain, required for double hexamer attachment, and a C-terminal AAA+ domain, required for ATPase activity [41,47]. The double-hexamer interface is active during late M and G_1 phase, when the MCM2–7 is loaded as a double hexamer [47,48]. However, during S phase, when the replication fork helicase is activated, the MCM2–7 double hexamers dissociate, and the resulting CMG helicases unwind bidirectionally from the origin [49,50]. The MCM2–7 helicase also has DNA binding regions within the N-terminal and AAA+ domains [51,52]. It is generally agreed that the CDC45-MCM2–7-GINS assembly, the fully-active helicase, unwinds DNA by a steric exclusion mechanism [49,53–55]. In this model, the leading strand passes through the central channel of CMG [49,53–55]. The excluded lagging strand may pass through a side channel of the CMG, or alternatively the lagging strand may pass completely outside the CMG [49,53–55]. In either event, the double-stranded origin DNA must be melted to activate CMG unwinding.

What is the mechanism for replication fork unwinding by the CMG? According to the rotary model, the ssDNA lying inside the central channel of CMG is passed from one AAA+ domain to another in a sequential manner [35]. This model is derived mainly from homology to the Large T and E1 viral helicase systems, for which a rotary model is proposed [35]. A second model, based upon recent electron microscopy structures, proposes that the ssDNA binding regions of the AAA+ domain hands-off the ssDNA to the ssDNA binding region within the N-terminal region [49,53–55]. Future studies may reveal which one of these two models reflects the CMG mechanism for unwinding DNA in vivo.

The origin dsDNA encircled by MCM2–7 must be converted from dsDNA to ssDNA during replication initiation. In budding yeast, the origins are AT-rich, similar to the origins of bacteria and eukaryotic viral origins, suggesting that this may be conserved to promote initial melting of the origin, since AT-rich regions are inherently prone to melting. The MCM2–7 may open to promote exclusion of the lagging strand during the replication initiation. However, the mechanism for MCM2–7 ring opening is currently not known, but it may occur at the MCM2–MCM5 interface because this interaction surface is inherently weak [44,56,57]. Future studies may reveal how the MCM2–7 ring opens during S phase to allow for origin melting, and future studies may also reveal whether ring opening occurs before or after MCM2–7 double hexamer dissociation.

Additional ssDNA binding proteins may participate in the origin melting process. Proteins that bind origin ssDNA in budding yeast include MCM10 [58,59], SLD3 [60], SLD2 [61], DPB11 [62], and RPA [63], the eukaryotic single-stranded binding protein. These proteins do not hydrolyze ATP, and therefore their contribution to origin melting lies in their ability to bind ssDNA and stabilize the melted state. Interestingly, mutating the ssDNA binding residues of MCM10, SLD2, SLD3, and DPB11 results in decreased replication initiation and diminished recruitment of RPA to replication origins [62,64–66]. These data suggest that one or more of these initiation factors may be required to stabilize melted origin ssDNA, and perhaps even hand off melted origin DNA to RPA. However, little is known regarding the mechanism for how the initiation factors melt origin DNA, and little is known how the initiation factors hand off ssDNA to RPA. The human homologs of MCM10 (human MCM10) [67,68], SLD3 (Treslin) [66], and SLD2, RECQL4 [69], have also been shown to bind ssDNA, suggesting that the function may be conserved from budding yeast to human.

A replication initiation assay has recently been reconstituted for budding yeast using only purified proteins [70]. Furthermore, methods exist in budding yeast for the induced-degradation of essential genes, with phenotypic scoring of the mutant phenotype [71,72]. In addition, the ssDNA binding residues of the initiation factors have now been identified for budding yeast [62,64–66]. Thus, through a combination of in vitro reconstitution assays and in vivo experiments, a mechanistic understanding of how origin DNA is melted, stabilized, and transferred to RPA will soon be revealed for this model eukaryotic organism.

11. Concluding Remarks

A key step in replication initiation in all organisms may be the melting of origin DNA, since replication fork helicases in all systems seem to unwind DNA by a steric exclusion mechanism. In bacteria, the DnaA protein may be responsible for melting origin DNA, and also for loading the helicase onto the melted ssDNA. For eukaryotic viruses, the Large T and E1 helicases are competent to melt the origin DNA and subsequently unwind the DNA by steric exclusion. For the cellular eukaryotic replication initiation machinery, it appears that essential initiation factors, including MCM10, SLD3, SLD2, and DPB11, may be responsible for stabilizing the melted origin DNA, and these proteins may also participate in the hand-off of melted origin ssDNA to RPA. Thus, while origin melting is common for all domains of life, the mechanism for origin melting may be quite different for each DNA replication initiation system.

Acknowledgments: This work has been supported by the National Institute of General Medical Sciences of the National Institutes of Health under Award Number R15GM113167.

Author Contributions: M.P.M., J.M.J., I.B., and D.L.K. wrote the manuscript.

Conflicts of Interest: The authors declare no conflict of interest.

References

1. Leonard, A.; Grimwade, J. The orisome: Structure and function. *Front. Microbiol.* **2015**, *6*, 1–13. [CrossRef] [PubMed]
2. Ryan, V.T.; Grimwade, J.; Camara, J.E.; Crooke, E.; Leonard, A. Escherichia coli prereplication complex assembly is regulated by dynamic interplay among Fis, IHF and DnaA. *Mol. Microbiol.* **2004**, *51*, 1347–1359. [CrossRef] [PubMed]
3. Kaur, G.; Vora, M.; Czerwonka, C.; Rozgaja, T.; Grimwade, J.; Leonard, A. Building the bacterial orisome: High affinity DnaA recognition plays a role in setting the conformation of oriC DNA. *Mol. Microbiol.* **2014**, *91*, 1148–1163. [CrossRef] [PubMed]
4. Saxena, R.; Vasudevan, S.; Patil, D.; Ashoura, N.; Grimwade, J.; Crooke, E. Nucleotide-induced conformational changes in *Escherichia coli* DnaA protein are required for bacterial ORC to pre-RC conversion at the chromosomal origin. *Int. J. Mol. Sci.* **2015**, *16*, 27897–27911. [CrossRef] [PubMed]
5. Fuller, R.S.; Funnell, B.E.; Kornberg, A. The DnaA protein complex with the *E. coli* chromosomal replication origin (oriC) and other DNA sites. *Cell* **1984**, *38*, 889–900. [CrossRef]
6. Periasamy, V. Co-Ordination of Replication Initiation with Transcriptional Regulation in *Escherichia coli*. Ph.D. Thesis, University of Buffalo, Buffalo, NY, USA, 2015.
7. Stepankiw, N.; Kaidow, A.; Boye, E.; Bates, D. The right half of the *Escherichia coli* replication origin is not essential for viability, but facilitates multi-forked replication. *Mol. Microbiol.* **2009**, *74*, 467–479. [CrossRef] [PubMed]
8. Gille, H.; Messer, W. Localized DNA melting and structural perturbations in the origin of replication, oriC, of *Escherichia coli* in vitro and in vivo. *EMBO J.* **1991**, *10*, 1579–1584. [PubMed]
9. González-Soltero, R.; Botello, E.; Jiménez-Sánchez, A. Initiation of heat-induced replication requires DnaA and the L-13-mer of oriC. *J. Bacteriol.* **2006**, *188*, 8294–8298. [CrossRef] [PubMed]
10. Kowalski, D.; Eddy, M. The DNA unwinding element: A novel, *cis*-acting component that facilitates opening of the *Escherichia coli* replication origin. *EMBO J.* **1989**, *8*, 4335–4344. [PubMed]

11. Magnan, D.; Bates, D. Regulation of DNA replication initiation by chromosome structure. *J. Bacteriol.* **2015**, *197*, 3370–3377. [CrossRef] [PubMed]

12. Richardson, T.; Harran, O.; Murray, H. The bacterial DnaA-trio replication origin element specifies single-stranded DNA initiator binding. *Nature* **2016**, *534*, 412–416. [CrossRef] [PubMed]

13. Noguchi, Y.; Sakiyama, Y.; Kawakami, H.; Katayama, T. The Arg fingers of key DnaA promoters are oriented inward of the replication origin oriC and stimulate DnaA subcomplexes in the initiation complex. *J. Biol. Chem.* **2015**, *290*, 20295–20312. [CrossRef] [PubMed]

14. Duderstadt, K.; Chuang, K.; Berger, J. DNA stretching by bacterial initiators promotes replication origin melting. *Nature* **2012**, *478*, 209–213. [CrossRef] [PubMed]

15. Magnan, D.; Joshi, M.; Barker, A.; Visser, B.; Bates, D. DNA replication initiation is blocked by a distant chromosome-membrane attachment. *Curr. Biol.* **2015**, *25*, 2143–2149. [CrossRef] [PubMed]

16. Cheng, H.; Gröger, P.; Hartann, A.; Schlierf, M. Bacterial initiators form dynamic filaments on single-stranded DNA monomer by monomer. *Nucleic Acids Res.* **2015**, *43*, 396–405. [CrossRef] [PubMed]

17. Chodavarapu, S.; Felczak, M.; Yaniv, J.; Kaguni, J. *Escherichia coli* DnaA interacts with HU in initiation at the *E. coli* replication origin. *Mol. Microbiol.* **2008**, *67*, 781–792. [CrossRef] [PubMed]

18. Titolo, S.; Welchner, E.; White, P.W.; Archambault, J. Characterization of the DNA-binding properties of the origin-binding domain of simian virus 40 large T antigen by fluorescence anisotropy. *J. Virol.* **2003**, *77*, 5512–5518. [CrossRef] [PubMed]

19. Bochkarev, A.; Pfuetzner, R.A.; Edwards, A.M.; Frappier, L. Single Stranded Dna-Binding Domain of Human Replication Protein A Bound to Single Stranded DNA, RpA70 Subunit, Residues 183–420. *Nature* **1997**, *385*, 176–181. [CrossRef] [PubMed]

20. Bhat, K.P.; Betous, R.; Cortez, D. High-affinity DNA-binding Domains of Replication Protein A (RPA) Direct SMARCAL1-dependent Replication Fork Remodeling. *J. Biol. Chem.* **2014**, *290*, 4110–4117. [CrossRef] [PubMed]

21. Sigal, N.; Delius, H.; Kornberg, T.; Gefter, M.L.; Alberts, B. A DNA unwinding protein isolated from *Escherichia coli*: Its interaction with DNA and DNA polymerase. *Proc. Natl. Acad. Sci. USA* **1972**, *69*, 3537–3541. [CrossRef] [PubMed]

22. Chang, Y.; Xu, M.; Machado, A.; Yu, X.; Rohs, R.; Chen, X. Mechanism of origin DNA recognition and assembly of an initiator-helicase complex by SV40 large tumor antigen. *Cell Rep.* **2013**, *3*, 1117–1127. [CrossRef] [PubMed]

23. Valle, M.; Chen, X.; Donate, L.; Fanning, E.; Carazo, J. Structural basis for the cooperative assembly of large T antigen on the origin of replication. *J. Mol. Biol.* **2006**, *357*, 1295–1305. [CrossRef] [PubMed]

24. Bochkareva, E.; Martynowski, D.; Seitova, A.; Bochkarev, A. Structure of the origin-binding domain of simian virus 40 large T antigen bound to DNA. *EMBO J.* **2006**, *25*, 5961–5969. [CrossRef] [PubMed]

25. Luo, X.; Sanford, D.; Bullock, P.; Bachovchin, W. Solution structure of the origin DNA-binding domain of SV40 T-antigen. *Nat. Struct Biol.* **1996**, *3*, 1034–1039. [CrossRef] [PubMed]

26. Rohs, R.; West, S.; Sosinsky, A.; Liu, P.; Mann, R.; Honig, B. The role of DNA shape in protein-DNA recognition. *Nature* **2009**, *461*, 1248–1253. [CrossRef] [PubMed]

27. Erzberger, J.; Mott, M.; Berger, J. Structural basis for ATP-dependent DnaA assembly and replication-origin remodeling. *Nat. Struct. Mol. Biol.* **2006**, *13*, 676–683. [CrossRef] [PubMed]

28. Shen, J.; Gai, D.; Patrick, A.; Greenleaf, W.; Chen, X. The roles of the residues on the channel beta-hairpin and loop structures of simian virus 40 hexameric helicase. *Proc. Natl. Acad. Sci. USA* **2005**, *102*, 11248–11253. [CrossRef] [PubMed]

29. Borowiec, J.A.; Hurwitz, J. Localized melting and structural changes in the SV40 origin of replication induced by T-antigen. *EMBO J.* **1988**, *7*, 3149–3158. [PubMed]

30. Kumar, A.; Meinke, G.; Reese, D.; Moine, S.; Phelan, P.; Fradet-Turcotte, A.; Archambault, J.; Bohm, A.; Bullock, P. Model for T-antigen-dependent melting of the simian virus 40 core origin based on studies of the interaction of the beta-hairpin with DNA. *J. Virol.* **2007**, *81*, 4808–4818. [CrossRef] [PubMed]

31. Lilyestrom, W.; Klein, M.; Zhang, R.; Joachimiak, A.; Chen, X. Crystal structure of SV40 large T-antigen bound to p53: Interplay between a viral oncoprotein and a cellular tumor suppressor. *Genes Dev.* **2006**, *20*, 2373–2382. [CrossRef] [PubMed]

32. Gai, D.; Zhao, R.; Li, D.; Finkielstein, C.; Chen, X. Mechanisms of conformational change for a replicative hexameric helicase of SV40 large tumor antigen. *Cell* **2004**, *119*, 47–60. [CrossRef] [PubMed]

33. Meinke, G.; Bullock, P.; Bohm, A. Crystal structure of the simian virus 40 large T-antigen origin-binding domain. *J. Virol.* **2006**, *80*, 4304–4312. [CrossRef] [PubMed]

34. Miyata, T.; Yamada, K.; Iwasaki, H.; Shinagawa, H.; Morikawa, K.; Mayanagi, K. Two different oligomeric states of the RuvB branch migration motor protein as revealed by electron microscopy. *J. Struct. Biol.* **2000**, *131*, 83–89. [CrossRef] [PubMed]

35. Enemark, E.; Joshua-Tor, L. Mechanism of DNA translocation in a replicative hexameric helicase. *Nature* **2006**, *442*, 270–275. [CrossRef] [PubMed]

36. Hacker, K.; Kenneth, J. A Hexameric Helicase Encircles One DNA Strand and Excludes the Other during DNA Unwinding. *Am. Chem. Soc.* **1997**, *46*, 14080–14087. [CrossRef] [PubMed]

37. Schuck, S.; Stenlund, A. Assembly of a double hexameric helicase. *Mol. Cell* **2005**, *20*, 377–389. [CrossRef] [PubMed]

38. Lee, S.; Syed, S.; Enemark, E.; Schuck, S.; Stenlund, A.; Ha, T.; Joshua-Tor, L. Dynamic look at DNA unwinding by a replicative helicase. *Proc. Natl. Acad. Sci. USA* **2014**, *111*, E827–E835. [CrossRef] [PubMed]

39. Schuck, S.; Stenlund, A. Mechanistic analysis of local ori melting and helicase assembly by the papillomavirus E1 protein. *Mol. Cell* **2011**, *43*, 776–787. [CrossRef] [PubMed]

40. Gai, D.; Chang, Y.; Chen, X. Origin DNA melting and unwinding in DNA replication. *Curr. Opin. Struct. Biol.* **2010**, *20*, 756–762. [CrossRef] [PubMed]

41. Schwacha, A.; Bell, S.P. Interactions between two catalytically distinct MCM subgroups are essential for coordinated ATP hydrolysis and DNA replication. *Mol. Cell* **2001**, *8*, 1093–1104. [CrossRef]

42. Randell, J.; Bowers, J.; Rodriguez, H.; Bell, S. Sequential ATP Hydrolysis by Cdc6 and ORC Directs Loading of the MCM2–7 Helicase. *Mol. Cell* **2006**, *21*, 29–39. [CrossRef] [PubMed]

43. Yao, N.; O'Donnell, M. Evolution of replication machines. *Crit. Rev. Biochem. Mol. Biol.* **2016**, *51*, 135–149. [CrossRef] [PubMed]

44. Bochman, M.; Schwacha, A. The MCM2–7 complex has in vitro helicase activity. *Mol. Cell* **2008**, *31*, 287–293. [CrossRef] [PubMed]

45. Ilves, I.; Petojevic, T.; Pesavento, J.; Botchan, M. Activation of the MCM2–7 Helicase by Association with Cdc45 and GINS Proteins. *Mol. Cell* **2010**, *37*, 247–258. [CrossRef] [PubMed]

46. Xu, Y.; Gristwood, T.; Hodgson, B.; Trinidad, J.; Albers, S.; Bell, S. Archaeal orthologs of CDC45 and GINS form a stable complex that stimulates the helicase activity of MCM. *Proc. Natl. Acad. Sci. USA* **2016**, *113*, 13390–13395. [CrossRef] [PubMed]

47. Remus, D.; Beuron, F.; Tolun, G.; Griffith, J.; Morris, E.; Diffley, J. Concerted loading of MCM2–7 double hexamers around DNA during DNA replication origin licensing. *Cell* **2009**, *139*, 719–730. [CrossRef] [PubMed]

48. Evrin, C.; Clarke, P.; Zech, J.; Lurz, R.; Sun, J.; Uhle, S.; Li, H.; Stillman, B.; Speck, C. A double-hexameric MCM2–7 complex is loaded onto origin DNA during licensing of eukaryotic DNA replication. *Proc. Natl. Acad. Sci. USA* **2009**, *106*, 20240–20245. [CrossRef] [PubMed]

49. Fu, Y.; Yardimci, H.; Long, D.; Ho, T.; Guainazzi, A.; Bermudez, V.; Hurwitz, J.; van Oijen, A.; Schärer, O.; Walter, J. Selective bypass of a lagging strand roadblock by the eukaryotic replicative DNA helicase. *Cell* **2011**, *146*, 931–941. [CrossRef] [PubMed]

50. Yardimci, H.; Loveland, A.; Habuchi, S.; van Oijen, A.; Walter, J. Uncoupling of Sister Replisomes during Eukaryotic DNA Replication. *Mol. Cell* **2010**, *40*, 834–840. [CrossRef] [PubMed]

51. Froelich, C.; Kang, S.; Epling, L.; Bell, S.; Enemark, E. A conserved MCM single-stranded DNA binding element is essential for replication initiation. *eLife* **2014**, *3*, e019993. [CrossRef] [PubMed]

52. McGeoch, A.; Trakselis, M.; Laskey, R.; Bell, S.D. Organization of the archaeal MCM complex on DNA and implications for the helicase mechanism. *Nat. Struct. Mol. Biol.* **2005**, *12*, 756–762. [CrossRef] [PubMed]

53. Sun, J.; Shi, Y.; Georgescu, R.; Yuan, Z.; Chait, B.; Li, H.; O'Donnell, M. The architecture of a eukaryotic replisome. *Nat. Struct. Mol. Biol.* **2015**, *22*, 976–982. [CrossRef] [PubMed]

54. Yuan, Z.; Bai, L.; Sun, J.; Georgescu, R.; Liu, J.; O'Donnell, M.; Li, H. Structure of the eukaryotic replicative CMG helicase suggests a pumpjack motion for translocation. *Nat. Struct. Mol. Biol.* **2016**, *23*, 217–224. [CrossRef] [PubMed]

55. Pellegrini, L.; Costa, A. New insights inot the mechanism of DNA duplicaiton by the eukaryotic replisome. *Trends Biochem. Sci.* **2016**, *41*, 859–871. [CrossRef] [PubMed]

56. Bochman, M.; Schwacha, A. The Saccharomyces cerevisiae MCM6/2 and MCM5/3 ATPase active sites contribute to the function of the putative MCM2–7 'gate'. *Nucleic Acids Res.* **2010**, *38*, 6078–6088. [CrossRef] [PubMed]

57. Bruck, I.; Kaplan, D.L. The DBF4-CDC7 kinase promotes MCM2–7 ring opening to allow for single-stranded DNA extrusion and helicase assembly. *J. Biol. Chem.* **2015**, *290*, 1210–1221. [CrossRef] [PubMed]

58. Warren, E.; Vaithiyalingam, S.; Haworth, J.; Greer, B.; Bielinsky, A.; Chazin, W.; Eichman, B. Structural basis for DNA binding by replication initiator MCM10. *Structure* **2008**, *16*, 1892–1901. [CrossRef] [PubMed]

59. Eisenberg, S.; Korza, G.; Carson, J.; Liachko, I.; Tye, B. Novel DNA binding properties of the Mcm10 protein from Saccharomyces cerevisiae. *J. Biol. Chem.* **2009**, *284*, 25412–25420. [CrossRef] [PubMed]

60. Bruck, I.; Kaplan, D. Origin Single-stranded DNA Releases SLD3 Protein from the MCM2–7 Complex, Allowing the GINS Tetramer to Bind the MCM2–7 Complex. *J. Biol. Chem.* **2011**, *286*, 18602–18613. [CrossRef] [PubMed]

61. Kanter, D.; Kaplan, D. SLD2 binds to origin single-stranded DNA and stimulates DNA annealing. *Nucleic Acids Res.* **2011**, *39*, 2580–2592. [CrossRef] [PubMed]

62. Dhingra, N.; Bruck, I.; Smith, S.; Ning, B.; Kaplan, D. DPB11 helps control assembly of the CDC45-MCM2–7-GINS replication fork helicase. *J. Biol. Chem.* **2015**, *290*, 7586–7601. [CrossRef] [PubMed]

63. Chen, R.; Wold, M. Replication protein A: Single-stranded DNA's first responder: Dynamic DNA-interactions allow replication protein A to direct single-strand DNA intermediates into different pathways for synthesis or repair. *Bioessays* **2014**, *36*, 1156–1161. [CrossRef] [PubMed]

64. Perez-Arnaiz, P.; Kaplan, D. An Mcm10 mutant defective in ssDNA binding shows defects in DNA replication initiation. *J. Mol. Biol.* **2016**, *428*, 4608–4625. [CrossRef] [PubMed]

65. Bruck, I.; Kaplan, D. The replication initiation protein sld2 regulates helicase assembly. *J. Biol. Chem.* **2014**, *289*, 1948–1959. [CrossRef] [PubMed]

66. Bruck, I.; Kaplan, D. The replication initiation protein SLD3/Treslin orchestrates the assembly of the replication fork helicase during S phase. *J. Biol. Chem.* **2015**, *290*, 27414–27424. [CrossRef] [PubMed]

67. Tye, B.K. MCM proteins in DNA replication. *Annu. Rev. Biochem.* **1999**, *68*, 649–686. [CrossRef] [PubMed]

68. Thu, Y.; Bielinsky, A. Enigmatic rles of Mcm10 in DNA replication. *Trends Biochem. Sci.* **2013**, *38*, 184–194. [CrossRef] [PubMed]

69. Ohlenschläger, O.; Kuhnert, A.; Schneider, A.; Haumann, S.; Bellstedt, P.; Keller, H.; Saluz, H.-P.; Hortschansky, P.; Hänel, F.; Grosse, F.; et al. The N-terminus of the human RECQL4 helicase is a homeodomain-like DNA interaction motif. *Nucleic Acids Res.* **2012**, *40*, 8309–8324. [CrossRef] [PubMed]

70. Yeeles, J.; Deegan, T.; Janska, A.; Early, A.; Diffley, J. Regulated eukaryotic DNA replication origin firing with purified proteins. *Nature* **2015**, *519*, 431–435. [CrossRef] [PubMed]

71. Tanaka, S.; Miyazawa-Onami, M.; Iida, T.; Araki, H. iAID: An improved auxin-inducible degron system for the construction of a 'tight' conditional mutant in the budding yeast *Saccharomyces cerevisiae*. *Yeast* **2015**, *32*, 567–581. [CrossRef] [PubMed]

72. Labib, K.; Tercero, J.A.; Diffley, J.F.X. Uninterrupted MCM2–7 function required for DNA replication fork progression. *Science* **2000**, *288*, 1643–1647. [CrossRef] [PubMed]

genes

MDPI

Review

The Role of the N-Terminal Domains of Bacterial Initiator DnaA in the Assembly and Regulation of the Bacterial Replication Initiation Complex

Anna Zawilak-Pawlik [1,*], Małgorzata Nowaczyk [1] and Jolanta Zakrzewska-Czerwińska [1,2]

1   Hirszfeld Institute of Immunology and Experimental Therapy, Polish Academy of Sciences, Weigla 12,
    Wroclaw 53-114, Poland; malgorzata.nowaczyk@iitd.pan.wroc.pl (M.N.);
    jolanta.zakrzewska@uni.wroc.pl (J.Z.-C.)
2   Department of Molecular Microbiology, Faculty of Biotechnology, University of Wrocław,
    ul. Joliot-Curie 14A, Wrocław 50-383, Poland
*   Correspondence: zawilak@iitd.pan.wroc.pl; Tel.: +48-71-370-9949

Academic Editor: Eishi Noguchi
Received: 23 March 2017; Accepted: 4 May 2017; Published: 10 May 2017

Abstract: The primary role of the bacterial protein DnaA is to initiate chromosomal replication. The DnaA protein binds to DNA at the origin of chromosomal replication (*oriC*) and assembles into a filament that unwinds double-stranded DNA. Through interaction with various other proteins, DnaA also controls the frequency and/or timing of chromosomal replication at the initiation step. *Escherichia coli* DnaA also recruits DnaB helicase, which is present in unwound single-stranded DNA and in turn recruits other protein machinery for replication. Additionally, DnaA regulates the expression of certain genes in *E. coli* and a few other species. Acting as a multifunctional factor, DnaA is composed of four domains that have distinct, mutually dependent roles. For example, C-terminal domain IV interacts with double-stranded DnaA boxes. Domain III drives ATP-dependent oligomerization, allowing the protein to form a filament that unwinds DNA and subsequently binds to and stabilizes single-stranded DNA in the initial replication bubble; this domain also interacts with multiple proteins that control oligomerization. Domain II constitutes a flexible linker between C-terminal domains III–IV and N-terminal domain I, which mediates intermolecular interactions between DnaA and binds to other proteins that affect DnaA activity and/or formation of the initiation complex. Of these four domains, the role of the N-terminus (domains I–II) in the assembly of the initiation complex is the least understood and appears to be the most species-dependent region of the protein. Thus, in this review, we focus on the function of the N-terminus of DnaA in orisome formation and the regulation of its activity in the initiation complex in different bacteria.

Keywords: DnaA; N-terminus of DnaA; *oriC*; chromosomal replication; orisome; HobA; DiaA; SirA; Hda; Dps; DnaB

---

1. Introduction

Chromosomal replication is a key step in cell cycle progression in all organisms of the three domains of life: Bacteria, Archaea, and Eukaryota. This process begins by the assembly of a multiprotein complex at a predefined locus (multiple loci in Archaea and Eukaryota) on a chromosome, which is called the origin(s) of chromosomal replication (*ori*, in bacteria called *oriC*) [1,2]. The main roles of these nucleoprotein initiation complexes are to recognize the *ori* site, to distort the double helix, and to provide a platform for the assembly of the multiprotein replication machinery, termed the replisome, that will synthesize the nascent chromosome [3,4]. Chromosomal replication is highly regulated, mainly at the first step (initiation), to ensure that DNA replication does not begin under conditions that prevent the cell from completing the process, thus preventing the cell from dividing and producing a viable offspring cell [5,6].

The general mechanism of replication initiation is similar in all organisms. However, the number of initiation complexes per chromosome, initiation complex composition, protein-protein and protein-DNA interactions between initiation complex components, and check-point steps vary among organisms, with greater differences occurring among more unrelated taxonomic groups [3,4]. It is assumed that the molecular mechanism of replication initiation and its control are simplest in bacteria and most complex in Eukaryota. Indeed, the composition of the initiation complex in bacteria is less intricate than in organisms from the other two domains of life [1]. Nonetheless, the bacterial initiator protein DnaA is highly specialized, such that it can perform the functions of distinct subunits of Archaeal and Eukaryotic initiation complexes. For example, all initiators, including bacterial DnaA, Archaeal Orc1/Cdc6, or Eukaryotic Orc1-Orc6 origin recognition complex (ORC), recognize *ori* sites. However, in contrast to the last two, which are unable to melt DNA, only DnaA unwinds DNA and recruits other replisome proteins, especially the replicative helicase DnaB, to the newly formed single-stranded replication eye [7,8]. The DnaA protein and *oriC* are also the main factors controlling the assembly of the initiation complex or are subjected to control mechanisms that restrict the number of replications to one per cell cycle [6,9,10]. It is noteworthy that in some species, e.g., *Escherichia coli* or *Bacillus subtilis*, DnaA also serves as a transcription factor [11,12]. Thus, DnaA is a multifunctional protein, which is reflected by its complex structure and structure-function related activities.

2. Bacterial DnaA—General Overview of the Structure and Function

To form a bacterial initiation complex, often called an orisome, DnaA binds to DNA at *oriC* and employs protein-protein interactions between protomers to assemble into a helical filament that is capable of opening double-stranded DNA (dsDNA) at the DNA unwinding element (DUE) [13]. DnaA is encoded by the *dnaA* gene, which is found in nearly all bacterial species. Exceptions include a few endosymbiotic bacteria, such as *Azolla filiculoides*, *Blochmannia floridanus*, and *Wigglesworthia glossinidia*, which lack a functional *dnaA* gene. In these bacteria, the initiator protein and mechanisms of initiation of chromosomal replication remain unidentified [14–16]. The DnaA proteins in bacteria characterized thus far vary in molecular weight between 47 kDa and 73 kDa (399-amino acid *Aquifex aeolicus* DnaA and 656-amino acid *Streptomyces coelicolor* DnaA, respectively). DnaA is composed of four structural and functional domains (Figure 1). The C-terminal domain IV encompasses approx. 120 amino acids (~13 kDa) and, together with domain III (approx. 230 amino acids, ~25 kDa), constitutes the most conserved part of DnaA with regard to structure and function. Domain II, which links domain III and domain I, is the most diverse domain between species with respect to sequence and length, varying between approx. 20 amino acids (~2 kDa) in *Helicobacter pylori* and approx. 250 amino acids (~28 kDa) in *S. coelicolor*. However, it should be noted that some DnaA proteins, such as the *A. aeolicus* initiator protein, appear to lack domain II (Figure 2) [17]. N-terminal domain I is composed of approx. 75–110 amino acids (~8–12 kD) (74 amino acids in *A. aeolicus* DnaA, 90 amino acids in *E. coli* DnaA, 108 amino acids in *Mycobacterium tuberculosis* DnaA), and in contrast to a well-conserved secondary structure, its sequence is poorly conserved among unrelated bacterial species.

Domain IV is responsible for DNA binding via a helix-turn-helix motif (Figure 1). The domain recognizes 9-mer, non-palindromic DNA sequences called DnaA boxes that are clustered at *oriC* (*E. coli* consensus sequence: 5′-TTATNCACA-3′). Domain III belongs to the ATPases Associated with diverse cellular Activities (AAA+) class of proteins; upon interaction with adenosine triphosphate (ATP), but not adenosine diphosphate (ADP), domain III changes conformation to enable the protein to properly oligomerize into a filament. The structure of such a filament bound to dsDNA and the means by which DnaA melts *oriC* is not fully understood. Nonetheless, the interaction between DnaA monomers within the filament introduce a conformational change in the bound DNA to melt its double-stranded structure at the DUE [18–20]. Subsequently, multiple domain III's of the filament bind to and stabilize single-stranded DNA (ssDNA) via initiator-specific motifs (ISMs) [18,21–24]. *E. coli* DnaA domain III, together with domain I, recruits DnaB helicase to an open complex and helps position the helicase onto the ssDNA [25,26]; however, DnaA interactions with DnaB helicase and helicase loaders vary among species [27–30]. It should be noted that filamentation is mediated by domain III and controlled by other proteins that interact directly with this domain, such as a complex of the beta subunit of the DNA polymerase III (β-clamp) and the protein homologous to DnaA (Hda) (β-clamp-Hda- complex) in *E. coli* and possibly in *Caulobacter crescentus* or the sporulation initiation inhibitor protein Soj and the initiation-control protein YabA in *B. subtilis* [31–35]. Interestingly, as shown for *E. coli* DnaA, domains III and IV are sufficient in vitro for opening the *oriC* region; i.e., proteins that lack domains I and II unwind *oriC* in vitro in a manner similar to that of the full-length protein [36]. However, N-terminally truncated DnaA does not support DNA replication in vitro and is not viable in vivo, which indicates that the N-terminal part of *E. coli* DnaA is required to maintain its function in bacterial cells. Indeed, it has been shown that DnaA domain I, similar to domain III, mediates interactions between DnaA monomers and interacts with other proteins, including the helicase DnaB (see below).

Although the N-terminal domain is crucial for DnaA activity in vivo, its role in orisome formation is the least understood of the four domains. The reason for that is, in part, related to the lack of structure of full-length DnaA. The structure of the N-terminal portion of DnaA [37,38], which consists of a largely unstructured domain II and independently solved structures of domains III–IV [13,17,22], does not allow us to predict how the N-terminal domain is positioned within the orisome and how domain I is oriented with regard to the C-terminal domains III and IV. Due to the flexible domain II, DnaA domain I appears to be structurally detached from domains III–IV; however, it does affect DnaA activity in the orisome. Moreover, domain I is sensitive to regulation by cellular proteins (Figure 1B) that appear to coordinate DnaA activity with the bacterial growth phase or cell cycle, stress, or unknown stimuli. Domain I possibly controls the transition from the initiation phase to the elongation phase in *E. coli* through mutually exclusive interactions with regulatory proteins and DnaB. Altogether, the findings indicate that domain I is important for the activity of DnaA at the orisome.

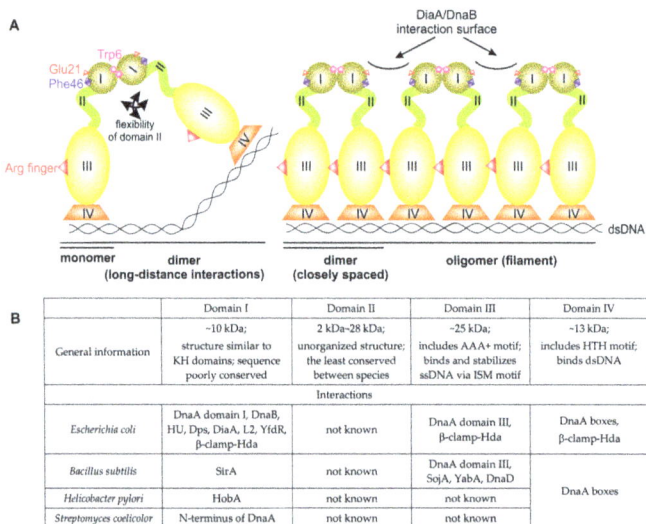

	Domain I	Domain II	Domain III	Domain IV
General information	~10 kDa; structure similar to KH domains; sequence poorly conserved	2 kDa~28 kDa; unorganized structure; the least conserved between species	~25 kDa; includes AAA+ motif; binds and stabilizes ssDNA via ISM motif	~13 kDa; includes HTH motif; binds dsDNA
			Interactions	
Escherichia coli	DnaA domain I, DnaB, HU, Dps, DiaA, L2, YfdR, β-clamp-Hda	not known	DnaA domain III, β-clamp-Hda	DnaA boxes, β-clamp-Hda
Bacillus subtilis	SirA	not known	DnaA domain III, SojA, YabA, DnaD	DnaA boxes
Helicobacter pylori	HobA	not known	not known	
Streptomyces coelicolor	N-terminus of DnaA	not known	not known	

Figure 1. Domain structure of bacterial initiator protein DnaA. (**A**) A schematic overview of DnaA domains and their activities in orisome formation. Crucial residues involved in domain I dimerization (*E. coli* Trp6) and DnaB binding (*E. coli* Glu21 and Phe46) are marked. An arginine finger (*E. coli* Arg285), an ATPases Associated with diverse cellular Activities (AAA+) family-specific motif that recognizes ATP bound to an adjacent subunit in a multimeric complex, is also depicted. (**B**) General information about motifs, activities, and interacting partners of DnaA domains.

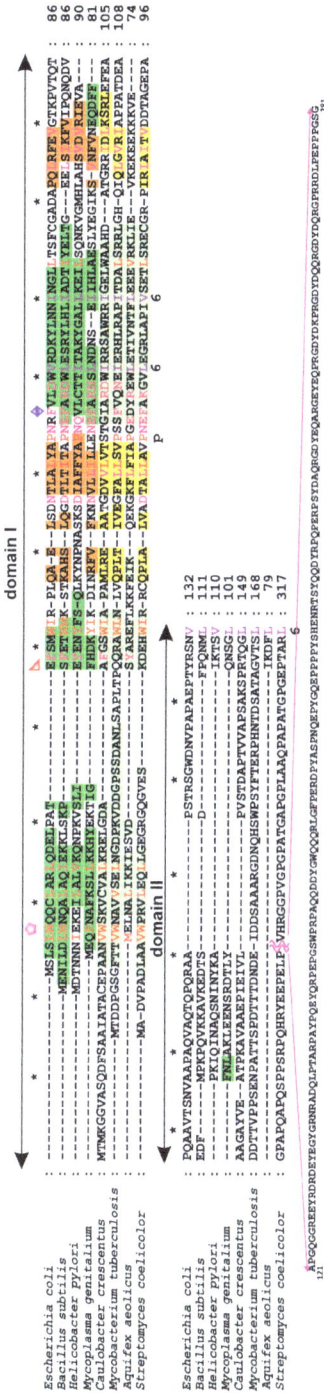

Figure 2. Sequence alignment of DnaA domains I–II from selected bacterial genera. The sequences were aligned using Profile ALIgNmEnt (PRALINE) [39]. Secondary elements of domains I–II are marked in green (α-helices) and brown (β-strands). Dark green and dark brown correspond to experimentally resolved structures of *H. pylori* (pdb 2WP0), *E. coli* (pdb 2E0G), *B. subtilis* (pdb 4TPS), and *M. genitalium* (pdb 2JMP) DnaAs; light green and light brown correspond to predicted secondary structures of *C. crescentus*, *M. tuberculosis*, and *A. aeolicus* DnaAs. The coloured fonts indicate conserved residues in domain I (violet, red, and pink from highest to lowest conservation, respectively); non-conserved residues are shown in black. Conserved residues involved in domain I dimerization (*E. coli* W6 (Trp6)). DnaB binding (*E. coli* E21 (Glu21), and F46 (Phe46)) are marked by a pink pentagon, red triangle, and violet peen, respectively; these symbols correspond to Figure 1.

3. N-Terminus of Bacterial DnaA

3.1. Structures of Bacterial DnaA Domains I and II

The structures of *E. coli*, *B. subtilis*, *H. pylori*, and *Mycoplasma genitalium* DnaA domain I have been solved; for the last, however, no functional analyses have been performed to date. Despite high sequence diversity (Figure 2), domain I is structurally conserved and consists of α-helices and β-strands (Figure 3). *E. coli* domain I is composed of 3 α-helices and 3 β-strands in the order of α1-α2-β1-β2-α3-β3 [37,38]. *H. pylori* DnaA is missing one β-strand between α1 and α2 [40], and *B. subtilis* DnaA contains an extra α4 helix between α3 and β3 [41]; *M. genitalium* contains two additional α-helices in the order of α1-α2-β1-β2-α3-α-β3-α [38] (Figure 3). Structurally, the α-helices and β-strands form distinct surfaces; an exception is for *M. genitalium*, in which the β-strands are packed between helices α1-α2 and α3-α4 at one site and α5 at the other. The β-strands comprise a β-sheet; however, the functional roles of the individual β-strands and entire β-sheet in domain I are unknown. The α helices are involved in different protein-protein interactions, and α1 of *E. coli* DnaA, together with a loop between β1-β2, forms a hydrophobic patch that engages in intermolecular interactions between the N-termini of DnaA monomers [37,42,43]. Nonetheless, this hydrophobic patch is not conserved among all DnaAs; for example, it is not present in *H. pylori* DnaA, and the N-terminus of this DnaA does not dimerize [40]. The α2 and α3 helices of *E. coli*, *H. pylori*, and *B. subtilis* DnaAs interact with other proteins (the DnaA initiator-associating factor DiaA and DnaB [37,44], the *Helicobacter* orisome binding protein A (HobA) [40], and the sporulation inhibitor of replication SirA [41], respectively), and despite a lack of sequence conservation, they are proposed to form structurally conserved protein-protein interaction surfaces utilized by regulatory proteins to control DnaA activity (see below) [41,44].

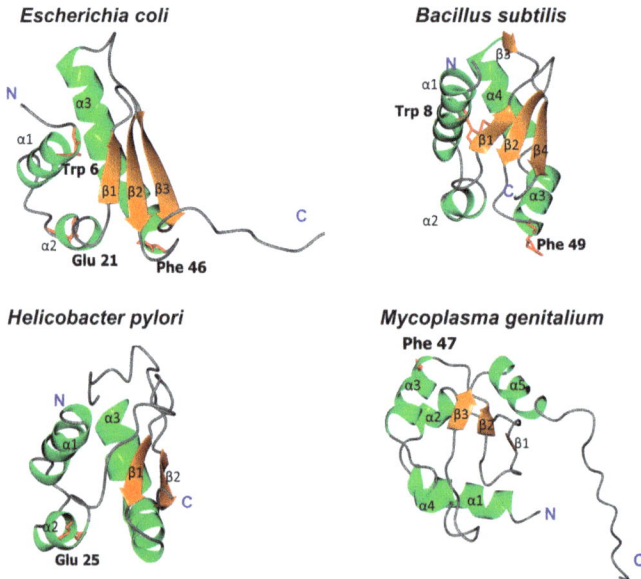

Figure 3. Ribbon diagrams of DnaA domain I in *E. coli* (pdb 2E0G), *B. subtilis* (pdb 4TPS), *H. pylori* (pdb 2WP0), and *M. genitalium* (pdb 2JMP). Residues involved in *E. coli* domain I dimerization (Trp) and DnaB binding (Glu, Phe) are marked (if conserved).

It has been reported that the structure of DnaA domain I is similar to the K homology domain (KH domain) [37,40]. KH domains interact with RNA and ssDNA nucleic acids, and affinity toward ssDNA or RNA is increased by the presence of multiple KH domains [45]. Additionally, the N-terminus of *E. coli* DnaA weakly interacts with ssDNA [37], though DnaA lacking domain I is able to unwind DNA and stabilize ssDNA via the ISM motif located in domain III [21,36,46]. Therefore, it remains unknown whether the KH motif plays any role in ssDNA binding upon unwinding of DNA by DnaA.

Domain II is unstructured and the most variable in sequence (Figure 2). Accordingly, there is little information about the possible motifs in regions that function in overall DnaA structure or function, especially within the context of mutual interdependence between domain I and domains III–IV.

3.2. Escherichia Coli DnaA Domain I

E. coli is a gram-negative, non-sporulating, facultatively anaerobic bacterium. Although *E. coli* constitutes a natural microflora in the lower intestine of warm-blooded organisms, including humans, some strains are pathogenic. This bacterium can survive and multiply outside of its host despite a decline in growth over time. The genomes of natural isolates of *E. coli* range from 4.5 to 6.0 Mb and encode approx. 4200–6500 genes. The bacterium has been used as a model organism for studying bacterial processes, including chromosomal replication and the cell cycle. Therefore, *E. coli* DnaA is one of the best characterized initiator proteins, especially within the context of structure-function relationships. In fact, studies on *E. coli* DnaA pioneered work on other initiators, including those in Archaea and Eukaryota. The resolved structure of *E. coli* DnaA domain I (1–86 aa) complements comprehensive biochemical data collected to date. It has been shown that domain I is engaged in numerous protein-protein interactions that include other DnaA monomers, as well as proteins that regulate DnaA activity at the orisome (DiaA, the histone-like protein HU, the ribosomal protein L2, the DNA-binding proteins from starved cells Dps, cryptic prophage protein YfdR, the β-clamp-Hda complex). Domain I of *E. coli* DnaA also participates in recruiting the replisome protein DnaB helicase; thus, it is important for the transition between the initiation and DNA synthesis (elongation) phases of replication.

The amino acids important for domain I head-to-head dimerization have been mapped to a patch formed by helix α1 and the loop between β1 and β2 (Figures 2 and 3; amino acids leucine 5 (Leu5), tryptophan 6 (Trp6), glutamine 8 (Gln8), cysteine 9 (Cys9), Leu10, and Leu33) [37,42,43,47,48]. Regardless, how these interactions impact the structure and function of the entire DnaA protein, especially within the context of the assembled orisome, is still not fully understood. It has been suggested that N-terminal domains of *E. coli* DnaA, possibly due to dimerization of domain I, mediate long-distance interactions between DnaA monomers (Figure 1), similar to *S. coelicolor* (see below), and that this interaction facilitates or stabilizes DnaA binding to distantly located DnaA binding sites [49,50]. Dimerization might also be important to facilitate cooperativity of DnaA binding to closely spaced DnaA boxes, particularly for those with low affinity [49,51,52]. Indeed, domain I promotes DnaA oligomerization at *oriC*, possibly by bringing DnaA monomers into a closer contact so they can make a filament via domain III (Figure 1) [42,43]. The N-terminal domain is also required for DnaB loading [43]; DnaA defective in dimerisation via domain I (e.g., DnaA lacking the N-terminal domain or DnaA mutated at the amino acid Trp6, which is critical for domain I dimerization), is not able to load DnaB onto an open complex despite the fact that it can unwind DNA and bind to DnaB via a second interaction surface located at domain III [36,43,53]. It was suggested that dimerized domain I of DnaA oligomers at *oriC* provides an array of sites that, together with domain III, stably bind to DnaB and help load helicase onto ssDNA (Figure 1) [23,37]. Indeed, DnaB interacts with DnaA domain I via the amino acids glutamic acid 21 (Glu21) and phenylalanine 46 (Phe46), which are located on helix α2 and α3, respectively, i.e., at the region opposite from the α1 dimerization surface (Figures 3 and 4) [36,37,53]. Such localization of surface interaction allows domain I to simultaneously dimerize and interact with DnaB.

As they are also engaged in interactions with DiaA and Hda regulatory proteins, DnaA helices α2 and α3 exposed to protein surfaces appear to be a hot spot for protein-protein interactions. DiaA is found in many bacterial species [54,55]. Although *E. coli* DiaA is not essential in vivo, it stimulates chromosomal replication, controls synchrony of initiation events, and ensures that the process is coordinated with the cell cycle [56]. Upon orisome formation, the DiaA tetramer simultaneously binds to multiple DnaA molecules and stimulates the assembly of DnaA onto *oriC*, which in turn facilitates the unwinding of the *oriC* duplex DNA [55]. In particular, amino acids Glu21 and Trp25 on α2 and asparagine 44 (Asn44), Phe46, and Trp50 on α3 are important for DiaA binding (Figure 4) [31,44,55]. Moreover, it has been shown that DiaA and DnaB compete for binding to DnaA and that DiaA bound to DnaA inhibits the DnaA-DnaB interaction and DnaB loading onto DnaA multimers at *oriC* [44]. These results demonstrate that DiaA controls DnaB loading [44,57]. The possible mechanism that regulates DiaA binding to DnaA is not known; however, it has been suggested that unknown cellular factors control DnaA-DiaA interactions [44].

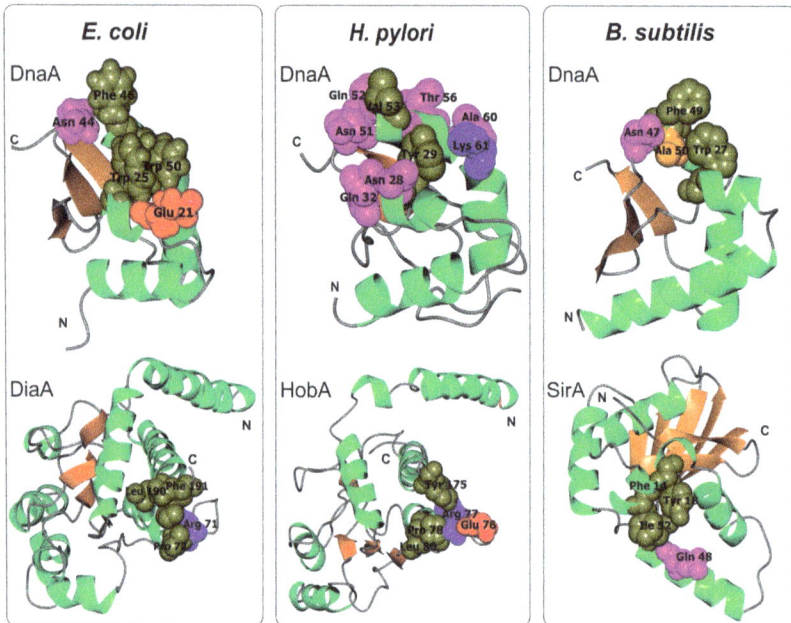

Figure 4. Ribbon diagrams of *E. coli*, *H. pylori*, and *B. subtilis* DnaA domain I and cognate interacting partners: DiaA (pdb 4U6N), HobA (pdb 2WP0), SirA (pdb 4PTS), respectively. Residues most important for complex formation are indicated by color-coded spheres (magenta—polar, orange—small non-polar, olive green—hydrophobic, red—negative charged, blue—positive charged).

Hda plays a pivotal role in regulating DnaA activity via a mechanism called RIDA (regulatory inactivation of DnaA). Hda consists of an N-terminal β-clamp-binding consensus sequence and the AAA+ domain, which shares homology with DnaA domain III. Hda-ADP in a complex with a β-clamp of DNA polymerase III interacts with DnaA domains I, III, and IV shortly after initiation [31,58], and inter-AAA+ interactions between domain III of *E. coli* DnaA and Hda stimulate the hydrolysis of ATP bound to DnaA [31,59]. DnaA-ADP is not able to properly oligomerize and unwind DNA; thus, it is inactive for initiation until it becomes reactivated into DnaA-ATP, which occurs either by DnaA de novo synthesis or by the interaction of DnaA-ADP with DnaA-reactivating sequences (DARS) or phospholipids (see below) [6,60,61]. Interactions between domains I and IV with Hda likely stabilize the complex and promote interactions between the AAA+ domains. In particular, DnaA mutated at Asn44 or lysine 54 (Lys54) located on helix α3 is insensitive to RIDA in vitro and in vivo [31]. Interestingly, *E. coli* domain I has also been proposed to participate in the transition of DnaA-ADP into DnaA-ATP, which is able to initiate replication [62]. Such an exchange of nucleotides, called rejuvenation, is promoted by the interaction between DnaA domain III and acidic phospholipids in the cell membrane [61]. However, it has recently been demonstrated that this process strongly depends on DnaA protein membrane occupancy, which affects the functional state of DnaA [62,63]. It was proposed that domain I is particularly important for rejuvenation associated with DnaA density-driven, cooperative oligomerization [62].

The molecular mechanisms of DnaA domain I interactions with HU, Dps, L2, and YfdR, and their roles in the initiation of chromosomal replication are much less understood than those described above. The HU protein is a DNA-binding protein that functions in compaction of the bacterial chromosome (by inducing DNA bends) and regulates DNA-related processes, including replication and transcription [64]. HU is composed of two subunits, α and β, that can form homo- and heterodimers. HU is known to stimulate in vitro DNA unwinding by DnaA, though the mechanism remains obscure [7,65]. Recently, it was shown that HU directly interacts with DnaA and that this interaction stabilizes DnaA oligomers assembled at *oriC* [66]. In particular, DnaA domain I preferentially binds to the α subunit of HU, either as an $\alpha2$ or $\alpha\beta$ dimer. In vitro, the $\alpha2$ homodimer stimulates DNA replication more efficiently than $\alpha\beta$ or $\beta2$. In vivo, the composition of the subunits in a dimer changes with the growth phase: the $\alpha2$ dimer predominates during early log-phase growth but decreases to only approx. 5% of HU in the stationary phase [67]. Moreover, inactivation of the α but not the β subunit perturbs coordination between the initiation of DNA replication and the cell cycle. These findings suggest that HU facilitates initiation of chromosomal replication in *E. coli* during logarithmic growth.

In contrast to HU, proteins Dps, L2, and YfdR inhibit initiation [68–70]. Dps is synthesized upon exposure to environmental stress (e.g., oxidation, starvation) and protects DNA from oxidative stress via three intrinsic activities: DNA binding, iron sequestration, and ferroxidase enzymatic activity [71]. In vitro, Dps weakly inhibits DnaA-dependent replication of plasmids; however, the protein significantly (but not completely) inhibits chromosomal replication in vivo [68]. Interestingly, Dps synthesis is especially induced in oxygen-stressed cells during the logarithmic phase of growth. Under these conditions, Dps might be especially important for protecting replicating DNA and for inhibiting new rounds of DNA synthesis. However, it has been suggested that incomplete inhibition of replication initiation might allow for the synthesis of nascent DNA with mutations and, as a consequence, an increase in genetic variation within a population in response to oxidative stress [68].

L2 is a ribosomal protein that has recently been shown to interact with the N-terminus of DnaA [70]. In vitro, L2 and its truncated form, which lacks 59 N-terminal amino acids, destabilizes DnaA oligomers at *oriC* and thus inhibits DnaA-dependent DUE unwinding. Thus, L2 interferes with prepriming complex formation because it precludes DnaB loading, which is required for further replisome assembly. It has been suggested that L2 coordinates replication with transcription under specific, yet unknown, conditions.

YfdR, a protein encoded by a set of genes of the cryptic phage CPS-53, binds to domain I of *E. coli* DnaA in a Phe46-dependent manner [69]. Consistently, YfdR inhibits the binding of other Phe46-dependent proteins, DiaA and DnaB, to DnaA. YfdR also reduces the initiation of plasmid replication in vitro. Although the exact role of the YfdR protein is still not clarified, it has been suggested that the protein may regulate replication under specific stress conditions because the cryptic phage CPS-53 is involved in response to oxidative and acid stresses.

3.3. Bacillus Subtilis DnaA Domain I

B. subtilis is a gram-positive soil bacterium that sporulates under suboptimal growth conditions [72,73]. The genomes of natural isolates of *B. subtilis* range from 4.0 to 4.3 Mb and encode approx. 4000–4500 genes. Many *B. subtilis* cellular processes, including chromosomal replication, adjust to environmental conditions to promote vegetative growth, sporulation, or spore germination. Accordingly, a master Spo0A regulator, which is responsible for entry into sporulation, directly controls the activity of *oriC* [74,75] and indirectly regulates DnaA (see below). *B. subtilis oriC* is bipartite, i.e., it contains two clusters of DnaA boxes separated by a *dnaA* gene; both clusters are required for the initiation of chromosomal replication in vivo [76,77]. In vitro, DnaA binds to both sub-regions, acting as a bridge and looping out the *dnaA* gene [78]. *B. subtilis* DnaA-ATP has been shown to interact with *oriC* in a manner characteristic of AAA+ proteins; upon orisome assembly, DnaA-ATP forms a helix-like structure that unwinds DNA and binds to ssDNA [33,46]. Domain III of *B. subtilis* DnaA has a predominant role in DnaA filament assembly and is thus a target for binding numerous regulatory proteins, such as Soj, YabA, and the primosomal protein DnaD, none of which is found in *E. coli* [33,34,79]. In fact, *B. subtilis* DnaA domain III is the best characterized domain of the entire DnaA protein, whereas the roles of the other domains in the formation and activity of the initiation complex are much less understood. Knowledge of the role of the *B. subtilis* N-terminal domains (1–86 aa domain I, 87–111 aa domain II) in orisome assembly is particularly scarce. It is known that the N-terminal domains are not required for filament formation and ssDNA binding by *B. subtilis* DnaA in vitro [46], though it remains unclear whether *B. subtilis* DnaA domain I dimerizes. Most residues involved in the dimerization of *E. coli* DnaA domain I are conserved in *B. subtilis* DnaA (Figures 2 and 3), and 22 amino acids of the N-terminus of the latter can functionally replace the 20 N-terminal residues of the former (i.e., helix α1) [48]. Such a hybrid protein complements the temperature-sensitive (Ts) growth phenotype of the dnaA46 mutant strain WM2063, though *E. coli* DnaA lacking 23 N-terminal amino acids is unable to complement this Ts strain. This suggests that the interaction between molecules of *B. subtilis* DnaA via domain I may occur and play a role in formation of the DnaA-*oriC* complex. This hypothesis is supported by the fact that SirA, which interacts with domain I of *B. subtilis* DnaA, displaces the initiator protein from *oriC* when incubated with the DnaA-*oriC* complex [80]. In vivo, SirA is produced under Spo0A~P regulation and inhibits new rounds of replication prior to sporulation [80,81]. SirA forms a heterodimer with domain I of DnaA via interaction with initiator protein α2 and α3 helices. In addition, certain amino acids in domain I (Trp27, Asn47, Phe49, and alanine 50 (Ala50)) were shown to be especially important for interaction with SirA [41,82] (Figure 4). It is noteworthy that SirA also interacts with domain III [83] and, together with domain III-binding Soj and *oriC*-interacting Spo0A, controls *B. subtilis* chromosomal replication and coordinates replication during the transition from a vegetative to dormant state [74,83,84].

Unlike in *E. coli*, *B. subtilis* DnaA domain I appears to play no role in helicase recruitment into an open complex. Thus far, no interactions between *B. subtilis* DnaA domain I and helicase DnaC or helicase loading proteins (a loader—DnaI, a co-loader—DnaB, and an assisting protein—DnaD; please note the differences in helicase-related nomenclature; DnaD interacts with domain III of DnaA) have been reported [29,85]. Moreover, *B. subtilis* helicase is loaded onto ssDNA via a "ring-making" mechanism, which is different from the "ring-breaking" mechanism in *E. coli* [86,87]. Thus, distinct protein-protein interactions might be involved in helicase assembly into an open complex.

3.4. Helicobacter Pylori DnaA Domain I

H. pylori is a gram-negative pathogenic bacterium that resides in the human stomach, a relatively stable, albeit hostile, ecological niche [88,89]. The genomes of natural isolates of *H. pylori* range from 1.5 to 1.7 Mb and encode approx. 1400–1800 genes, with only a few regulatory proteins controlling cellular processes [90,91]. *H. pylori oriC* resembles *B. subtilis oriC*, i.e., it is bipartite and consists of two clusters of DnaA boxes, *oriC*1 and *oriC*2, separated by a *dnaA* gene [92]. The structure of *H. pylori oriC* and DnaA-DNA interactions have recently been well characterized [92–95], but there are limited biochemical data for *H. pylori* DnaA, particularly concerning domain III. For instance, it is not known whether *H. pylori* is regulated by ATP binding and hydrolysis, and no protein homologous to Hda has been found in *H. pylori*. Moreover, no proteins interacting with domain III of *H. pylori* DnaA have been identified thus far. As domain III is highly homologous among species, it likely forms a filament that is typical of DnaA. The N-terminus of *H. pylori* DnaA has been relatively well characterized. It comprises 110 amino acids (1–90 amino acids domain I, 91–110 amino acids domain II) and does not self-associate [40], possibly due to structural obstacles that may preclude dimerization. These obstacles include a shorter helix α1, a lack of conserved Trp6, and a positively charged (non-hydrophobic) area of interaction. *H. pylori* DnaA domain I interacts with HobA, a protein essential for *H. pylori* survival. To date, HobA is the only known protein that interacts with DnaA, and it influences DnaA assembly at *oriC* [96,97]. Indeed, HobA binding to DnaA stimulates DnaA oligomerization at *oriC*1 [54]. Despite low sequence homology, HobA is a structural and functional homologue of *E. coli* DiaA [54,98]. Similar to DiaA and SirA, HobA interacts with DnaA helices α2 and α3 [40], and residues tyrosine 29 (Tyr29), Asn28, and Gln32 on α2, and Lys61, valine 53 (Val53), Gln52, Asn51, Thr56, and Ala60 on α3 have been shown to be involved in interactions with HobA (Figure 4). However, DiaA and HobA cannot substitute for each other in vitro or in vivo because DiaA–*E. coli* DnaA and HobA–*H. pylori* DnaA interaction surfaces co-evolved [54]. Despite the high functional homology between DiaA and HobA, the dynamics of HobA/DiaA-stimulated oligomerization differ. HobA enhances and accelerates *H. pylori* DnaA binding to *oriC*, whereas DiaA increases but decelerates *E. coli* DnaA binding to *oriC*. Interestingly, the kinetics of responses involving domains III–IV do not depend on the stimulating protein (DiaA or HobA). In a hybrid system in which *E. coli* domain I was fused to domains II–IV of *H. pylori* DnaA (Ec^{I}Hp^{II-IV}DnaA), DiaA stimulated Ec^{I}Hp^{II-IV}DnaA in a manner similar to that of HobA stimulation of *H. pylori* DnaA, though with a sensitivity characteristic of DiaA [54]. This suggests that HobA or DiaA binding to cognate DnaA stimulates subsequent interaction, possibly between domain III, and that an induced response depends on domain III, the activity of which apparently differs slightly between these species.

It is not known whether the N-terminus of *H. pylori* DnaA or any domain of the DnaA protein participates in helicase loading onto an open complex because no DnaA-DnaB interactions, either between isolated proteins or within an orisome, have been shown thus far. Glu21, which is important for interactions of *E. coli* DnaA with *E. coli* DnaB, is present in *H. pylori* (Glu 25), but Phe46 is missing. It should be noted that *H. pylori* DnaB helicase is atypical, and unlike bacterial hexameric helicases, it forms a dodecamer that dissociates into hexamers upon interaction with DnaG primase [99,100]. Regardless, the mechanism for DnaB loading onto an open complex is still unknown.

3.5. Streptomyces Coelicolor DnaA Domain I

S. coelicolor is a gram-positive soil bacterium. It possesses a large, 9 Mb chromosome encoding approx. 8300 genes, which is almost twice as large as the *E. coli* or *B. subtilis* chromosome. *S. coelicolor* grows as substrate mycelia, which differentiate into an aerial mycelium and spores upon nutrient depletion. The key elements of the initiation of *S. coelicolor* chromosomal replication, DnaA and *oriC*, have been identified, and their interactions have been characterized [101–106]. *S. coelicolor* *oriC* contains two clusters of DnaA boxes separated by a short spacer DNA [103]; in total, there are 19 DnaA boxes spread over nearly 1000 bp. The DnaA-DNA complexes formed on both sides of the DNA spacer interact with each other to form a hairpin-like structure [106]. Although this resembles DnaA binding to bipartite origins in *B. subtilis* and *H. pylori*, the number of distinct nucleoprotein complexes is higher in *S. coelicolor* (up to 4 complexes per hairpin) than in the other two bacteria (1 complex per loop), as visualized by electron microscopy [78,92,106]. *S. coelicolor* DnaA is one of the largest known DnaA proteins (656 amino acids) due to the presence of a long domain II, which comprises an additional stretch (approx. 150 amino acids) of predominantly acidic amino acids. Such an exceptionally large domain II should enable DnaA dimers or oligomers to interact with distantly located DnaA boxes to establish a functional nucleoprotein complex. Domain I of the *S. coelicolor* DnaA protein dimerises [106], and together with domain III it participates in DnaA oligomerization [105,106]. It is possible that domain I mediates interactions between DnaA bound to distal DnaA boxes, whereas domain III mediates interactions between closely spaced boxes [106]. In addition, DnaA lacking domain I aggregates strongly upon DNA binding; thus, domain I should support the correct DnaA structure upon orisome formation [106]. Nonetheless, there is no detailed information concerning possible interaction surfaces or amino acids that participate in domain I intermolecular interactions, and there are no known proteins that interact with *S. coelicolor* DnaA. Thus, further studies are required to gain insight into protein-protein interactions that lead to assembly or regulation of a functional *S. coelicolor* orisome.

3.6. DnaA Domain II

Domain II was initially regarded as only a flexible linker that joins domain I with domains III–IV. However, it has been suggested that "nonessential" regions of domain II may be transiently involved in DnaB recruitment, and this domain, similar to DiaA, is presumably required to promote optimal helicase loading [107]. Moreover, domain II can be extended, and it tolerates the insertion of structured fragments. This was shown in *E. coli*, whereby green fluorescent protein (GFP) of 238 amino acids was inserted into domain II or into the C-terminal region of domain I (right after β3), without the loss of DnaA functionality in vivo [108,109]. In fact, it was the only location of GFP in DnaA that was tolerated by the *E. coli* protein. In addition, comprehensive deletion analysis within domain II of *E. coli* DnaA showed that at least 21–27 residues are required to sustain the correct conformation of the entire protein, possibly because they properly align domain I with domains III–IV [110]. Furthermore, deletions shortening *E. coli* domain II resulted in an under-initiation phenotype [107,111], which raises the question of how domain I and domains III–IV are aligned in proteins that have almost no existing domain II. Because domain I plays an important role in the cooperative binding of DnaA molecules at *oriC*, it is tempting to speculate that the length of domain II is adjusted according to the spacing between DnaA boxes. Regarding this hypothesis, the *S. coelicolor* DnaA protein can bind to widely spaced DnaA boxes due to the presence of a long domain II, whereas the *H. pylori* DnaA protein, with a relatively short domain II, binds to closely spaced *H. pylori* DnaA boxes [3]. It should reminded here, that the N-terminal domain I of *H. pylori* DnaA does not dimerise (Section 3.4, see also below), however, the direct interactions between the N-terminal domains of DnaA might be substituted by not-direct, HobA mediated, tetramerisation of DnaA [40,112].

4. Conclusions and Perspectives

The N-terminal domains of bacterial DnaAs are essential for full protein activity upon initiation of chromosomal replication, ensuring cooperativity of the protein in DNA binding and correct spatial assembly at *oriC*. This, in turn, is required for proper control of orisome activity with respect to further replisome assembly (e.g., DnaB loading) and the transition from the initiation to the DNA synthesis step. The N-terminal domains are also engaged in coordinating chromosomal replication with the cell cycle (e.g., sporulation) and other cellular processes (e.g., transcription) or environmental conditions (e.g., oxidative stress).

It should be noted that the N-terminal domains exhibit the least conserved sequence (Figure 2), and accordingly, it has been shown that the N-termini of DnaA from various species have different activities or interactions (Figure 1). The N-terminal domains likely evolved to meet the requirements of species that reflect differences in the structures of *oriCs*, the mechanisms of replisome assembly and the strategies of regulating DnaA activity. However, there are relatively few experimental data that assert the general features of the N-terminal domains with respect to the structure-function relationship of orisomes in different species. Nonetheless, dimerization and interaction with other proteins are the most conservative features of domain I. Domain II serves as a linker that coordinates the function of largely independent domains I, III, and IV.

It was experimentally shown that domain I in *E. coli* and *S. coelicolor* DnaAs dimerize. Helix α1 is crucial for dimerization in *E. coli*, but amino acids and interaction surfaces involved in *S. coelicolor* DnaA dimerization are unknown. In contrast, *H. pylori* DnaA domain I was shown not to interact, and there are no data regarding the dimerization of *B. subtilis* DnaA domain I. It was proposed that domain I dimerization and a sufficiently long, flexible domain II help to establish long-distance interactions. Thus, it was suggested that for some orisomes, domain I dimerization is not important when DnaA boxes are closely spaced at *oriC*, such as for *H. pylori oriC* [93,95,113]. However, *H. pylori* DnaA participates in long-distance interactions between DnaA-*oriC*1 and DnaA-*oriC*2 subcomplexes [92], raising the question of which domain (or domains) mediates the interactions between subcomplexes in *H. pylori*, *B. subtilis*, and other bipartite orisomes (e.g., mollicutes or Epsilonproteobacteria) [9,85,114].

Interaction of DnaA domain I with other proteins (*E. coli* DiaA, *H. pylori* HobA, and *B. subtilis* SirA) is mediated by helices α2 and α3, which likely comprise a common interface for protein-protein interactions (Figure 4). Interactions with DiaA and HobA are species specific, i.e., one protein cannot be substituted with another for interaction with DnaA in other species. Although it is not known whether SirA-DnaA interaction is also species specific, the amino acid sequence within the *B. subtilis* DnaA α2-α3 interface is quite different from that of *E. coli* and *H. pylori* DnaAs (Figure 4). In the structure-function relationship, it appears that proteins that bind multiple DnaA molecules, such as DiaA or HobA, stimulate DnaA oligomerization, whereas proteins that bind only a single DnaA protomer, such as SirA, destabilize DnaA oligomers. Multimerization of domain I might be important for cooperative binding of DnaA with DnaA boxes or for assembly of the multi-protomer interface for protein-protein interactions. When this interaction interface is released by DiaA/HobA, it can be further utilized by other proteins, such as when it is used by *E. coli* DnaB. However, proteins such as SirA might destabilize dimerization or the multi-protomer interface and thus preclude cooperative DNA binding or inhibit the loading of other proteins. It would be interesting to analyse how SirA affects oligomerization of hybrid DnaAs (*E. coli* (Bs^{I}Ec^{II-IV}DnaA) or *H. pylori* (Bs^{I}Hp^{II-IV}DnaA)), in which domain I is swapped for *B. subtilis* domain I. Such proteins should be able to interact with SirA, and this interaction could possibly destabilize orisomes formed by chimeric DnaAs.

Interaction between DnaA domain I and the helicase has only been demonstrated for *E. coli*. However, the interaction between DnaA domain III and helicase loader/loader assisting proteins appears to be more common in bacteria (DnaC binds to *A. aeolicus* DnaA [27], and DnaD interacts with *B. subtilis* DnaA [34,79]). It is reasonable to assume that by participating in helicase loading and activation, DnaA might be a key factor controlling the transition from initiation to elongation. More studies are required to reveal whether the binding between helicase and domain I of DnaA depends on the helicase loading mechanism (ring-making in *E. coli* vs. ring-breaking in *B. subtilis*), the loading proteins (*E. coli* DnaC, *B. subtilis* DnaI, or recently discovered DciA [30]), the *oriC* structure (*E. coli* mono- vs. *B. subtilis* bipartite), or other species-specific factors.

As mentioned above, domain I has various activities and has a different number and variety of interacting partners. The fact that there is a large discrepancy between the known activities exhibited by *E. coli* DnaA and initiators from other species is especially puzzling. Within this context, the N-terminus of *E. coli* DnaA appears to be an omnipotent domain. However, within the context of environmental challenges, physiology, and genetics, *E. coli* is not that different from other species, particularly *B. subtilis* or *S. coelicolor*. This makes it difficult to justify such an increase or decrease in the properties or interaction partners (seven, one, and zero DnaA interacting partners have been discovered thus far in *E. coli*, *B. subtilis*, and *S. coelicolor*, respectively—Figure 1). Nonetheless, these species have different life cycles. Thus, for example, because *E. coli* is unable to sporulate, it may require additional or different regulatory proteins to control chromosomal replication, whereas *B. subtilis* and *S. coelicolor* enter a dormant state under similar unfavourable conditions. Indeed, the initiation of *B. subtilis* chromosomal replication is controlled by Spo0A, SojA, and SirA, which are proteins associated with sporulation cycle control. Nonetheless, information is likely missing for many proteins that can interact with the N-terminal domain of DnaAs from other species, which, in turn, may regulate the initiation of chromosomal replication. For example, no interacting partners are known for *C. crescentus*, *S. coelicolor*, and *M. tuberculosis* DnaAs. It should be noted that in some bacteria, the number of proteins that regulate replication might be very low. For example, in *H. pylori*, a bacterium known for an overall limited number of regulatory proteins (compare approx. 30 proteins involved in signal transduction in *H. pylori* with approx. 300 and 1000 proteins in *E. coli*/*B. subtilis* and *S. coelicolor*, respectively [115]), the number of DnaA-interacting proteins might not be much higher than has been identified thus far. However, it is also possible that alternative pathways have been developed to control DnaA activity in *B. subtilis*, *S coelicolor*, *H. pylori*, and other bacteria. For example, it appears that *B. subtilis* DnaA is controlled primarily at domain III, whereas *C. crescentus* DnaA is primarily controlled at the levels of expression and proteolysis [116].

Functional and structural studies on *E. coli* DnaA-DiaA and *H. pylori* DnaA-HobA heterocomplexes have revealed relatively high specificity of interactions between initiation proteins [54]. This finding opens new possibilities for selective pathogen eradication by targeting essential protein-protein interactions involved in the initiation of chromosomal replication. Indeed, replication proteins are increasingly being considered as drug targets [117,118], among which species-specific domain I interactions appear promising. Thus, further studies will be important to increase our knowledge about the role of the N-terminus in controlling the initiation of bacterial chromosomal replication.

Acknowledgments: This research was supported by research grant SONATA BIS3 from the National Science Centre, Poland, (DEC-2013/10/E/NZ1/00718). The cost of publication was supported by the Wroclaw Centre of Biotechnology under the Leading National Research Centre (KNOW) program for years 2014-2018.

Conflicts of Interest: The authors declare no conflict of interest. The funding sponsors had no role in the writing of the manuscript.

References

1. Leonard, A.C.; Méchali, M. DNA replication origins. *Cold Spring Harb. Perspect. Biol.* **2013**, *5*, a010116. [CrossRef] [PubMed]
2. Kawakami, H.; Katayama, T. DnaA, ORC, and Cdc6: Similarity beyond the domains of life and diversity. *Biochem. Cell Biol. Biochim. Biol. Cell.* **2010**, *88*, 49–62. [CrossRef] [PubMed]
3. Costa, A.; Hood, I.V.; Berger, J.M. Mechanisms for initiating cellular DNA replication. *Annu. Rev. Biochem.* **2013**, *82*, 25–54. [CrossRef] [PubMed]
4. O'Donnell, M.; Langston, L.; Stillman, B. Principles and concepts of DNA replication in bacteria, archaea, and eukarya. *Cold Spring Harb. Perspect. Biol.* **2013**, *5*. [CrossRef] [PubMed]
5. Deegan, T.D.; Diffley, J.F. MCM: One ring to rule them all. *Curr. Opin. Struct. Biol.* **2016**, *37*, 145–151.
6. Katayama, T.; Ozaki, S.; Keyamura, K.; Fujimitsu, K. Regulation of the replication cycle: Conserved and diverse regulatory systems for DnaA and *oriC*. *Nat. Rev. Microbiol.* **2010**, *8*, 163–170. [CrossRef] [PubMed]
7. Hwang, D.S.; Kornberg, A. Opening of the replication origin of Escherichia coli by DnaA protein with protein HU or IHF. *J. Biol. Chem.* **1992**, *267*, 23083–23086. [PubMed]
8. Mott, M.L.; Berger, J.M. DNA replication initiation: Mechanisms and regulation in bacteria. *Nat. Rev. Microbiol.* **2007**, *5*, 343–354. [CrossRef] [PubMed]
9. Wolański, M.; Donczew, R.; Zawilak-Pawlik, A.; Zakrzewska-Czerwińska, J. *oriC*-encoded instructions for the initiation of bacterial chromosome replication. *Front. Microbiol.* **2014**, *5*, 735. [PubMed]
10. Chodavarapu, S.; Kaguni, J.M. Replication Initiation in Bacteria. *The Enzymes* **2016**, *39*, 1–30. [PubMed]
11. Smith, J.L.; Grossman, A.D. In Vitro Whole Genome DNA Binding Analysis of the Bacterial Replication Initiator and Transcription Factor DnaA. *PLoS Genet.* **2015**, *11*, e1005258. [CrossRef] [PubMed]
12. Messer, W.; Weigel, C. DnaA as a transcription regulator. *Methods Enzymol.* **2003**, *370*, 338–349. [PubMed]
13. Erzberger, J.P.; Mott, M.L.; Berger, J.M. Structural basis for ATP-dependent DnaA assembly and replication-origin remodeling. *Nat. Struct. Mol. Biol.* **2006**, *13*, 676–683. [CrossRef] [PubMed]
14. Ran, L.; Larsson, J.; Vigil-Stenman, T.; Nylander, J.A.A.; Ininbergs, K.; Zheng, W.-W.; Lapidus, A.; Lowry, S.; Haselkorn, R.; Bergman, B. Genome Erosion in a Nitrogen-Fixing Vertically Transmitted Endosymbiotic Multicellular Cyanobacterium. *PLoS ONE* **2010**, *5*, e11486. [CrossRef]
15. Akman, L.; Yamashita, A.; Watanabe, H.; Oshima, K.; Shiba, T.; Hattori, M.; Aksoy, S. Genome sequence of the endocellular obligate symbiont of tsetse flies, *Wigglesworthia glossinidia*. *Nat. Genet.* **2002**, *32*, 402–407.
16. Gil, R.; Silva, F.J.; Zientz, E.; Delmotte, F.; González-Candelas, F.; Latorre, A.; Rausell, C.; Kamerbeek, J.; Gadau, J.; Hölldobler, B.; et al. The genome sequence of *Blochmannia floridanus*: comparative analysis of reduced genomes. *Proc. Natl. Acad. Sci. USA* **2003**, *100*, 9388–9393. [CrossRef] [PubMed]
17. Erzberger, J.P.; Pirruccello, M.M.; Berger, J.M. The structure of bacterial DnaA: implications for general mechanisms underlying DNA replication initiation. *EMBO J.* **2002**, *21*, 4763–4773. [CrossRef] [PubMed]

18. Duderstadt, K.E.; Berger, J.M. A structural framework for replication origin opening by AAA+ initiation factors. *Curr. Opin. Struct. Biol.* **2013**, *23*, 144–153. [CrossRef] [PubMed]

19. Martinez, M.P.; Jones, J.M.; Bruck, I.; Kaplan, D.L. Origin DNA Melting-An Essential Process with Divergent Mechanisms. *Genes* **2017**, *8*, 26. [CrossRef] [PubMed]

20. Shimizu, M.; Noguchi, Y.; Sakiyama, Y.; Kawakami, H.; Katayama, T.; Takada, S. Near-atomic structural model for bacterial DNA replication initiation complex and its functional insights. *Proc. Natl. Acad. Sci. USA* **2016**, *113*, E8021–E8030. [CrossRef] [PubMed]

21. Duderstadt, K.E.; Chuang, K.; Berger, J.M. DNA stretching by bacterial initiators promotes replication origin opening. *Nature* **2011**, *478*, 209–213. [CrossRef] [PubMed]

22. Ozaki, S.; Kawakami, H.; Nakamura, K.; Fujikawa, N.; Kagawa, W.; Park, S.-Y.; Yokoyama, S.; Kurumizaka, H.; Katayama, T. A common mechanism for the ATP-DnaA-dependent formation of open complexes at the replication origin. *J. Biol. Chem.* **2008**, *283*, 8351–8362. [CrossRef] [PubMed]

23. Ozaki, S.; Katayama, T. Highly organized DnaA-*oriC* complexes recruit the single-stranded DNA for replication initiation. *Nucleic Acids Res.* **2012**, *40*, 1648–1665. [CrossRef] [PubMed]

24. Ozaki, S.; Noguchi, Y.; Hayashi, Y.; Miyazaki, E.; Katayama, T. Differentiation of the DnaA-*oriC* Subcomplex for DNA Unwinding in a Replication Initiation Complex. *J. Biol. Chem.* **2012**, *287*, 37458–37471. [CrossRef] [PubMed]

25. Carr, K.M.; Kaguni, J.M. Stoichiometry of DnaA and DnaB protein in initiation at the *Escherichia coli* chromosomal origin. *J. Biol. Chem.* **2001**, *276*, 44919–44925. [CrossRef] [PubMed]

26. Carr, K.M.; Kaguni, J.M. *Escherichia coli* DnaA protein loads a single DnaB helicase at a DnaA box hairpin. *J. Biol. Chem.* **2002**, *277*, 39815–39822. [CrossRef] [PubMed]

27. Mott, M.L.; Erzberger, J.P.; Coons, M.M.; Berger, J.M. Structural synergy and molecular crosstalk between bacterial helicase loaders and replication initiators. *Cell* **2008**, *135*, 623–634. [CrossRef] [PubMed]

28. Smits, W.K.; Goranov, A.I.; Grossman, A.D. Ordered association of helicase loader proteins with the *Bacillus subtilis* origin of replication in vivo. *Mol. Microbiol.* **2010**, *75*, 452–461. [CrossRef] [PubMed]

29. Soultanas, P. Loading mechanisms of ring helicases at replication origins: Helicase loading. *Mol. Microbiol.* **2012**, *84*, 6–16. [CrossRef] [PubMed]

30. Brézellec, P.; Vallet-Gely, I.; Possoz, C.; Quevillon-Cheruel, S.; Ferat, J.-L. DciA is an ancestral replicative helicase operator essential for bacterial replication initiation. *Nat. Commun.* **2016**, *7*, 13271.

31. Su'etsugu, M.; Harada, Y.; Keyamura, K.; Matsunaga, C.; Kasho, K.; Abe, Y.; Ueda, T.; Katayama, T. The DnaA N-terminal domain interacts with Hda to facilitate replicase clamp-mediated inactivation of DnaA. *Environ. Microbiol.* **2013**, *15*, 3183–3195. [CrossRef] [PubMed]

32. Wargachuk, R.; Marczynski, G.T. The Caulobacter crescentus Homolog of DnaA (HdaA) Also Regulates the Proteolysis of the Replication Initiator Protein DnaA. *J. Bacteriol.* **2015**, *197*, 3521–3532. [CrossRef] [PubMed]

33. Scholefield, G.; Errington, J.; Murray, H. Soj/ParA stalls DNA replication by inhibiting helix formation of the initiator protein DnaA. *EMBO J.* **2012**, *31*, 1542–1555. [CrossRef] [PubMed]

34. Scholefield, G.; Murray, H. YabA and DnaD inhibit helix assembly of the DNA replication initiation protein DnaA. *Mol. Microbiol.* **2013**, *90*, 147–159. [CrossRef] [PubMed]

35. Cho, E.; Ogasawara, N.; Ishikawa, S. The functional analysis of YabA, which interacts with DnaA and regulates initiation of chromosome replication in *Bacillus subtils*. *Genes Genet. Syst.* **2008**, *83*, 111–125.

36. Sutton, M.D.; Carr, K.M.; Vicente, M.; Kaguni, J.M. Escherichia coli DnaA protein. The N-terminal domain and loading of DnaB helicase at the *E. coli* chromosomal origin. *J. Biol. Chem.* **1998**, *273*, 34255–34262.

37. Abe, Y.; Jo, T.; Matsuda, Y.; Matsunaga, C.; Katayama, T.; Ueda, T. Structure and function of DnaA N-terminal domains: Specific sites and mechanisms in inter-DnaA interaction and in DnaB helicase loading on *oriC*. *J. Biol. Chem.* **2007**, *282*, 17816–17827. [CrossRef] [PubMed]

38. Lowery, T.J.; Pelton, J.G.; Chandonia, J.-M.; Kim, R.; Yokota, H.; Wemmer, D.E. NMR structure of the N-terminal domain of the replication initiator protein DnaA. *J. Struct. Funct. Genomics* **2007**, *8*, 11–17.

39. Simossis, V.A.; Heringa, J. PRALINE: A multiple sequence alignment toolbox that integrates homology-extended and secondary structure information. *Nucleic Acids Res.* **2005**, *33*, W289–W294.

40. Natrajan, G.; Noirot-Gros, M.F.; Zawilak-Pawlik, A.; Kapp, U.; Terradot, L. The structure of a DnaA/HobA complex from *Helicobacter pylori* provides insight into regulation of DNA replication in bacteria. *Proc. Natl. Acad. Sci. USA* **2009**, *106*, 21115–21120. [CrossRef] [PubMed]

41. Jameson, K.H.; Rostami, N.; Fogg, M.J.; Turkenburg, J.P.; Grahl, A.; Murray, H.; Wilkinson, A.J. Structure and interactions of the *Bacillus subtilis* sporulation inhibitor of DNA replication, SirA, with domain I of DnaA. *Mol. Microbiol.* **2014**, *93*, 975–991. [CrossRef] [PubMed]

42. Simmons, L.A.; Felczak, M.; Kaguni, J.M. DnaA Protein of Escherichia coli: oligomerization at the *E. coli* chromosomal origin is required for initiation and involves specific N-terminal amino acids. *Mol. Microbiol.* **2003**, *49*, 849–858. [CrossRef] [PubMed]

43. Felczak, M.M.; Simmons, L.A.; Kaguni, J.M. An essential tryptophan of *Escherichia coli* DnaA protein functions in oligomerization at the *E. coli* replication origin. *J. Biol. Chem.* **2005**, *280*, 24627–24633. [CrossRef] [PubMed]

44. Keyamura, K.; Abe, Y.; Higashi, M.; Ueda, T.; Katayama, T. DiaA dynamics are coupled with changes in initial origin complexes leading to helicase loading. *J. Biol. Chem.* **2009**, *284*, 25038–25050. [CrossRef] [PubMed]

45. Valverde, R.; Edwards, L.; Regan, L. Structure and function of KH domains. *FEBS J.* **2008**, *275*, 2712–2726. [CrossRef] [PubMed]

46. Richardson, T.T.; Harran, O.; Murray, H. The bacterial DnaA-trio replication origin element specifies single-stranded DNA initiator binding. *Nature* **2016**, *534*, 412–416. [CrossRef] [PubMed]

47. Sutton, M.D.; Kaguni, J.M. Novel alleles of the Escherichia coli dnaA gene. *J. Mol. Biol.* **1997**, *271*, 693–703. [CrossRef] [PubMed]

48. Weigel, C.; Schmidt, A.; Seitz, H.; Tüngler, D.; Welzeck, M.; Messer, W. The N-terminus promotes oligomerization of the *Escherichia coli* initiator protein DnaA. *Mol. Microbiol.* **1999**, *34*, 53–66. [CrossRef] [PubMed]

49. Messer, W.; Blaesing, F.; Majka, J.; Nardmann, J.; Schaper, S.; Schmidt, A.; Seitz, H.; Speck, C.; Tüngler, D.; Wegrzyn, G.; Weigel, C.; Welzeck, M.; Zakrzewska-Czerwinska, J. Functional domains of DnaA proteins. *Biochimie* **1999**, *81*, 819–825. [CrossRef]

50. Messer, W. The bacterial replication initiator DnaA. DnaA and *oriC*, the bacterial mode to initiate DNA replication. *FEMS Microbiol. Rev.* **2002**, *26*, 355–374. [PubMed]

51. Messer, W.; Blaesing, F.; Jakimowicz, D.; Krause, M.; Majka, J.; Nardmann, J.; Schaper, S.; Seitz, H.; Speck, C.; Weigel, C.; et al. Bacterial replication initiator DnaA. Rules for DnaA binding and roles of DnaA in origin unwinding and helicase loading. *Biochimie* **2001**, *83*, 5–12. [CrossRef]

52. Miller, D.T.; Grimwade, J.E.; Betteridge, T.; Rozgaja, T.; Torgue, J.J.-C.; Leonard, A.C. Bacterial origin recognition complexes direct assembly of higher-order DnaA oligomeric structures. *Proc. Natl. Acad. Sci. USA* **2009**, *106*, 18479–18484. [CrossRef] [PubMed]

53. Seitz, H.; Weigel, C.; Messer, W. The interaction domains of the DnaA and DnaB replication proteins of *Escherichia coli*. *Mol. Microbiol.* **2000**, *37*, 1270–1279. [CrossRef] [PubMed]

54. Zawilak-Pawlik, A.; Donczew, R.; Szafrański, S.; Mackiewicz, P.; Terradot, L.; Zakrzewska-Czerwińska, J. DiaA/HobA and DnaA: A pair of proteins co-evolved to cooperate during bacterial orisome assembly. *J. Mol. Biol.* **2011**, *408*, 238–251. [CrossRef] [PubMed]

55. Keyamura, K.; Fujikawa, N.; Ishida, T.; Ozaki, S.; Su'etsugu, M.; Fujimitsu, K.; Kagawa, W.; Yokoyama, S.; Kurumizaka, H.; Katayama, T. The interaction of DiaA and DnaA regulates the replication cycle in *E. coli* by directly promoting ATP DnaA-specific initiation complexes. *Genes Dev.* **2007**, *21*, 2083–2099. [CrossRef] [PubMed]

56. Ishida, T.; Akimitsu, N.; Kashioka, T.; Hatano, M.; Kubota, T.; Ogata, Y.; Sekimizu, K.; Katayama, T. DiaA, a novel DnaA-binding protein, ensures the timely initiation of *Escherichia coli* chromosome replication. *J. Biol. Chem.* **2004**, *279*, 45546–45555. [CrossRef] [PubMed]

57. Kaguni, J.M. Replication initiation at the *Escherichia coli* chromosomal origin. *Curr. Opin. Chem. Biol.* **2011**, *15*, 606–613. [CrossRef] [PubMed]

58. Su'etsugu, M.; Shimuta, T.-R.; Ishida, T.; Kawakami, H.; Katayama, T. Protein associations in DnaA-ATP hydrolysis mediated by the Hda-replicase clamp complex. *J. Biol. Chem.* **2005**, *280*, 6528–6536. [CrossRef] [PubMed]

59. Keyamura, K.; Katayama, T. DnaA protein DNA-binding domain binds to Hda protein to promote inter-AAA+ domain interaction involved in regulatory inactivation of DnaA. *J. Biol. Chem.* **2011**, *286*, 29336–29346. [CrossRef] [PubMed]

60. Fujimitsu, K.; Senriuchi, T.; Katayama, T. Specific genomic sequences of *E. coli* promote replicational initiation by directly reactivating ADP-DnaA. *Genes Dev.* **2009**, *23*, 1221–1233. [CrossRef] [PubMed]

61. Saxena, R.; Fingland, N.; Patil, D.; Sharma, A.K.; Crooke, E. Crosstalk between DnaA Protein, the Initiator of *Escherichia coli* Chromosomal Replication, and Acidic Phospholipids Present in Bacterial Membranes. *Int. J. Mol. Sci.* **2013**, *14*, 8517–8537. [CrossRef] [PubMed]

62. Aranovich, A.; Braier-Marcovitz, S.; Ansbacher, E.; Granek, R.; Parola, A.H.; Fishov, I. N-terminal-mediated oligomerization of DnaA drives the occupancy-dependent rejuvenation of the protein on the membrane. *Biosci. Rep.* **2015**, *35*. [CrossRef] [PubMed]

63. Aranovich, A.; Gdalevsky, G.Y.; Cohen-Luria, R.; Fishov, I.; Parola, A.H. Membrane-catalyzed nucleotide exchange on DnaA. Effect of surface molecular crowding. *J. Biol. Chem.* **2006**, *281*, 12526–12534. [CrossRef] [PubMed]

64. Macvanin, M.; Adhya, S. Architectural organization in *E. coli* nucleoid. *Biochim. Biophys. Acta* **2012**, *1819*, 830–835. [CrossRef] [PubMed]

65. Ryan, V.T.; Grimwade, J.E.; Nievera, C.J.; Leonard, A.C. IHF and HU stimulate assembly of pre-replication complexes at *Escherichia coli oriC* by two different mechanisms. *Mol. Microbiol.* **2002**, *46*, 113–124. [CrossRef] [PubMed]

66. Chodavarapu, S.; Felczak, M.M.; Yaniv, J.R.; Kaguni, J.M. *Escherichia coli* DnaA interacts with HU in initiation at the *E. coli* replication origin. *Mol. Microbiol.* **2008**, *67*, 781–792. [CrossRef] [PubMed]

67. Claret, L.; Rouviere-Yaniv, J. Variation in HU composition during growth of *Escherichia coli*: The heterodimer is required for long term survival. *J. Mol. Biol.* **1997**, *273*, 93–104. [CrossRef] [PubMed]

68. Chodavarapu, S.; Gomez, R.; Vicente, M.; Kaguni, J.M. *Escherichia coli* Dps interacts with DnaA protein to impede initiation: A model of adaptive mutation. *Mol. Microbiol.* **2008**, *67*, 1331–1346. [CrossRef] [PubMed]

69. Noguchi, Y.; Katayama, T. The *Escherichia coli* Cryptic Prophage Protein YfdR Binds to DnaA and Initiation of Chromosomal Replication Is Inhibited by Overexpression of the Gene Cluster yfdQ-yfdR-yfdS-yfdT. *Front. Microbiol.* **2016**, *7*, 239. [CrossRef] [PubMed]

70. Chodavarapu, S.; Felczak, M.M.; Kaguni, J.M. Two forms of ribosomal protein L2 of *Escherichia coli* that inhibit DnaA in DNA replication. *Nucleic Acids Res.* **2011**, *39*, 4180–4191. [CrossRef] [PubMed]

71. Calhoun, L.N.; Kwon, Y.M. Structure, function and regulation of the DNA-binding protein Dps and its role in acid and oxidative stress resistance in *Escherichia coli*: A review. *J. Appl. Microbiol.* **2011**, *110*, 375–386. [CrossRef] [PubMed]

72. Higgins, D.; Dworkin, J. Recent progress in *Bacillus subtilis* sporulation. *FEMS Microbiol. Rev.* **2012**, *36*, 131–148. [CrossRef] [PubMed]

73. Tan, I.S.; Ramamurthi, K.S. Spore formation in *Bacillus subtilis*. *Environ. Microbiol. Rep.* **2014**, *6*, 212–225. [CrossRef] [PubMed]

74. Boonstra, M.; de Jong, I.G.; Scholefield, G.; Murray, H.; Kuipers, O.P.; Veening, J.-W. Spo0A regulates chromosome copy number during sporulation by directly binding to the origin of replication in *Bacillus subtilis*. *Mol. Microbiol.* **2013**, *87*, 925–938. [CrossRef] [PubMed]

75. Castilla-Llorente, V.; Muñoz-Espín, D.; Villar, L.; Salas, M.; Meijer, W.J.J. Spo0A, the key transcriptional regulator for entrance into sporulation, is an inhibitor of DNA replication. *EMBO J.* **2006**, *25*, 3890–3899. [CrossRef] [PubMed]

76. Moriya, S.; Atlung, T.; Hansen, F.G.; Yoshikawa, H.; Ogasawara, N. Cloning of an autonomously replicating sequence (ars) from the *Bacillus subtilis* chromosome. *Mol. Microbiol.* **1992**, *6*, 309–315. [CrossRef] [PubMed]

77. Moriya, S.; Imai, Y.; Hassan, A.K.; Ogasawara, N. Regulation of initiation of *Bacillus subtilis* chromosome replication. *Plasmid* **1999**, *41*, 17–29. [CrossRef] [PubMed]

78. Krause, M.; Rückert, B.; Lurz, R.; Messer, W. Complexes at the replication origin of *Bacillus subtilis* with homologous and heterologous DnaA protein. *J. Mol. Biol.* **1997**, *274*, 365–380. [CrossRef] [PubMed]

79. shigo-Oka, D.; Ogasawara, N.; Moriya, S. DnaD protein of *Bacillus subtilis* interacts with DnaA, the initiator protein of replication. *J. Bacteriol.* **2001**, *183*, 2148–2150. [CrossRef] [PubMed]

80. Wagner, J.K.; Marquis, K.A.; Rudner, D.Z. SirA enforces diploidy by inhibiting the replication initiator DnaA during spore formation in *Bacillus subtilis*. *Mol. Microbiol.* **2009**, *73*, 963–974. [CrossRef] [PubMed]

81. Rahn-Lee, L.; Gorbatyuk, B.; Skovgaard, O.; Losick, R. The conserved sporulation protein YneE inhibits DNA replication in *Bacillus subtilis*. *J. Bacteriol.* **2009**, *191*, 3736–3739. [CrossRef] [PubMed]

82. Rahn-Lee, L.; Merrikh, H.; Grossman, A.D.; Losick, R. The sporulation protein SirA inhibits the binding of DnaA to the origin of replication by contacting a patch of clustered amino acids. *J. Bacteriol.* **2011**, *193*, 1302–1307. [CrossRef] [PubMed]

83. Duan, Y.; Huey, J.D.; Herman, J.K. The DnaA inhibitor SirA acts in the same pathway as Soj (ParA) to facilitate *oriC* segregation during *Bacillus subtilis* sporulation. *Mol. Microbiol.* **2016**, *102*, 530–544. [CrossRef] [PubMed]

84. Xenopoulos, P.; Piggot, P.J. Regulation of growth of the mother cell and chromosome replication during sporulation of *Bacillus subtilis*. *J. Bacteriol.* **2011**, *193*, 3117–3126. [CrossRef] [PubMed]

85. Briggs, G.S.; Smits, W.K.; Soultanas, P. Chromosomal replication initiation machinery of low-g+c-content firmicutes. *J. Bacteriol.* **2012**, *194*, 5162–5170. [CrossRef] [PubMed]

86. Velten, M.; McGovern, S.; Marsin, S.; Ehrlich, S.D.; Noirot, P.; Polard, P. A two-protein strategy for the functional loading of a cellular replicative DNA helicase. *Mol. Cell* **2003**, *11*, 1009–1020. [CrossRef]

87. Davey, M.J.; O'Donnell, M. Replicative helicase loaders: Ring breakers and ring makers. *Curr. Biol.* **2003**, *13*, R594–R596. [CrossRef]

88. Atherton, J.C.; Blaser, M.J. Coadaptation of *Helicobacter pylori* and humans: Ancient history, modern implications. *J. Clin. Investig.* **2009**, *119*, 2475–2487. [CrossRef] [PubMed]

89. Tan, S.; Tompkins, L.S.; Amieva, M.R. *Helicobacter pylori* usurps cell polarity to turn the cell surface into a replicative niche. *PLoS Pathog.* **2009**, *5*, e1000407. [CrossRef] [PubMed]

90. Tomb, J.F.; White, O.; Kerlavage, A.R.; Clayton, R.A.; Sutton, G.G.; Fleischmann, R.D.; Ketchum, K.A.; Klenk, H.P.; Gill, S.; Dougherty, B.A.; et al. The complete genome sequence of the gastric pathogen *Helicobacter pylori*. *Nature* **1997**, *388*, 539–547. [CrossRef] [PubMed]

91. Danielli, A.; Scarlato, V. Regulatory circuits in *Helicobacter pylori*: Network motifs and regulators involved in metal-dependent responses. *FEMS Microbiol. Rev.* **2010**, *34*, 738–752. [CrossRef] [PubMed]

92. Donczew, R.; Weigel, C.; Lurz, R.; Zakrzewska-Czerwinska, J.; Zawilak-Pawlik, A. *Helicobacter pylori* oriC—The first bipartite origin of chromosome replication in Gram-negative bacteria. *Nucleic Acids Res.* **2012**, *40*, 9647–9660. [CrossRef] [PubMed]

93. Donczew, R.; Mielke, T.; Jaworski, P.; Zakrzewska-Czerwińska, J.; Zawilak-Pawlik, A. Assembly of *Helicobacter pylori* initiation complex is determined by sequence-specific and topology-sensitive DnaA-oriC interactions. *J. Mol. Biol.* **2014**, *426*, 2769–2782. [CrossRef] [PubMed]

94. Zawilak, A.; Durrant, M.C.; Jakimowicz, P.; Backert, S.; Zakrzewska-Czerwińska, J. DNA binding specificity of the replication initiator protein, DnaA from *Helicobacter pylori*. *J. Mol. Biol.* **2003**, *334*, 933–947. [CrossRef] [PubMed]

95. Zawilak-Pawlik, A.; Kois, A.; Majka, J.; Jakimowicz, D.; Smulczyk-Krawczyszyn, A.; Messer, W.; Zakrzewska-Czerwińska, J. Architecture of bacterial replication initiation complexes: orisomes from four unrelated bacteria. *Biochem. J.* **2005**, *389*, 471–481. [CrossRef] [PubMed]

96. Rain, J.C.; Selig, L.; De Reuse, H.; Battaglia, V.; Reverdy, C.; Simon, S.; Lenzen, G.; Petel, F.; Wojcik, J.; Schächter, V.; Chemama, Y.; Labigne, A.; Legrain, P. The protein-protein interaction map of *Helicobacter pylori*. *Nature* **2001**, *409*, 211–215. [CrossRef] [PubMed]

97. Zawilak-Pawlik, A.; Kois, A.; Stingl, K.; Boneca, I.G.; Skrobuk, P.; Piotr, J.; Lurz, R.; Zakrzewska-Czerwińska, J.; Labigne, A. HobA—A novel protein involved in initiation of chromosomal replication in *Helicobacter pylori*. *Mol. Microbiol.* **2007**, *65*, 979–994. [CrossRef] [PubMed]

98. Natrajan, G.; Hall, D.R.; Thompson, A.C.; Gutsche, I.; Terradot, L. Structural similarity between the DnaA-binding proteins HobA (HP1230) from *Helicobacter pylori* and DiaA from *Escherichia coli*. *Mol. Microbiol.* **2007**, *65*, 995–1005. [CrossRef] [PubMed]

99. Bazin, A.; Cherrier, M.V.; Gutsche, I.; Timmins, J.; Terradot, L. Structure and primase-mediated activation of a bacterial dodecameric replicative helicase. *Nucleic Acids Res.* **2015**, *43*, 8564–8576. [CrossRef] [PubMed]

100. Stelter, M.; Gutsche, I.; Kapp, U.; Bazin, A.; Bajic, G.; Goret, G.; Jamin, M.; Timmins, J.; Terradot, L. Architecture of a dodecameric bacterial replicative helicase. *Struct. Lond. Engl. 1993* **2012**, *20*, 554–564.

101. Zakrzewska-Czerwińska, J.; Schrempf, H. Characterization of an autonomously replicating region from the *Streptomyces lividans* chromosome. *J. Bacteriol.* **1992**, *174*, 2688–2693. [CrossRef] [PubMed]

102. Majka, J.; Messer, W.; Schrempf, H.; Zakrzewska-Czerwińska, J. Purification and characterization of the *Streptomyces lividans* initiator protein DnaA. *J. Bacteriol.* **1997**, *179*, 2426–2432. [CrossRef] [PubMed]

103. Jakimowicz, D.; Majka, J.; Messer, W.; Speck, C.; Fernandez, M.; Martin, M.C.; Sanchez, J.; Schauwecker, F.; Keller, U.; Schrempf, H.; Zakrzewska-Czerwińska, J. Structural elements of the *Streptomyces oriC* region and their interactions with the DnaA protein. *Microbiol. Read. Engl.* **1998**, *144* (Pt 5), 1281–1290. [CrossRef] [PubMed]

104. Majka, J.; Jakimowicz, D.; Messer, W.; Schrempf, H.; Lisowski, M.; Zakrzewska-Czerwińska, J. Interactions of the *Streptomyces lividans* initiator protein DnaA with its target. *Eur. J. Biochem. FEBS* **1999**, *260*, 325–335. [CrossRef]

105. Majka, J.; Zakrzewska-Czerwiñska, J.; Messer, W. Sequence recognition, cooperative interaction, and dimerization of the initiator protein DnaA of *Streptomyces*. *J. Biol. Chem.* **2001**, *276*, 6243–6252. [CrossRef] [PubMed]

106. Jakimowicz, D.; Majkadagger, J.; Konopa, G.; Wegrzyn, G.; Messer, W.; Schrempf, H.; Zakrzewska-Czerwińska, J. Architecture of the *Streptomyces lividans* DnaA protein-replication origin complexes. *J. Mol. Biol.* **2000**, *298*, 351–364. [CrossRef] [PubMed]

107. Molt, K.L.; Sutera, V.A.; Moore, K.K.; Lovett, S.T. A role for nonessential domain II of initiator protein, DnaA, in replication control. *Genetics* **2009**, *183*, 39–49. [CrossRef] [PubMed]

108. Boeneman, K.; Fossum, S.; Yang, Y.; Fingland, N.; Skarstad, K.; Crooke, E. *Escherichia coli* DnaA forms helical structures along the longitudinal cell axis distinct from MreB filaments. *Mol. Microbiol.* **2009**, *72*, 645–657. [CrossRef] [PubMed]

109. Nozaki, S.; Niki, H.; Ogawa, T. Replication initiator DnaA of *Escherichia coli* changes its assembly form on the replication origin during the cell cycle. *J. Bacteriol.* **2009**, *191*, 4807–4814. [CrossRef] [PubMed]

110. Nozaki, S.; Ogawa, T. Determination of the minimum domain II size of *Escherichia coli* DnaA protein essential for cell viability. *Microbiol. Read. Engl.* **2008**, *154*, 3379–3384. [CrossRef] [PubMed]

111. Leonard, A.C.; Grimwade, J.E. The orisome: Structure and function. *Front. Microbiol.* **2015**, *6*, 545. [CrossRef] [PubMed]

112. Terradot, L.; Zawilak-Pawlik, A. Structural insight into *Helicobacter pylori* DNA replication initiation. *Gut Microbes* **2010**, *1*, 330–334. [CrossRef] [PubMed]

113. Zawilak, A.; Cebrat, S.; Mackiewicz, P.; Król-Hulewicz, A.; Jakimowicz, D.; Messer, W.; Gosciniak, G.; Zakrzewska-Czerwinska, J. Identification of a putative chromosomal replication origin from *Helicobacter pylori* and its interaction with the initiator protein DnaA. *Nucleic Acids Res.* **2001**, *29*, 2251–2259. [CrossRef] [PubMed]

114. Jaworski, P.; Donczew, R.; Mielke, T.; Thiel, M.; Oldziej, S.; Weigel, C.; Pawlik, A.M. Unique and universal features of Epsilonproteobacterial origins of chromosome replication and DnaA-DnaA box interactions. *Evol. Genomic Microbiol.* **2016**, *7*, 1555. [CrossRef] [PubMed]

115. Ulrich, L.E.; Zhulin, I.B. The MiST2 database: A comprehensive genomics resource on microbial signal transduction. *Nucleic Acids Res.* **2010**, *38*, D401–D407. [CrossRef] [PubMed]

116. Collier, J. Regulation of chromosomal replication in *Caulobacter crescentus*. *Plasmid* **2012**, *67*, 76–87. [CrossRef] [PubMed]

117. Robinson, A.; Causer, R.J.; Dixon, N.E. Architecture and conservation of the bacterial DNA replication machinery, an underexploited drug target. *Curr. Drug Targets* **2012**, *13*, 352–372. [CrossRef] [PubMed]

118. van Eijk, E.; Wittekoek, B.; Kuijper, E.J.; Smits, W.K. DNA replication proteins as potential targets for antimicrobials in drug-resistant bacterial pathogens. *J. Antimicrob. Chemother.* **2017**, *72*, 1275–1284. [CrossRef] [PubMed]

Review

Control of Initiation of DNA Replication in *Bacillus subtilis* and *Escherichia coli*

Katie H. Jameson[1] and Anthony J. Wilkinson[2,*]

[1] Institute of Integrative Biology, University of Liverpool, Crown Street, Liverpool L69 7ZB, UK; katie.jameson@liverpool.ac.uk

[2] Structural Biology Laboratory, Department of Chemistry, University of York, York YO10 5DD, UK

* Correspondence: tony.wilkinson@york.ac.uk; Tel.: +44-1904-328-261

Academic Editor: Eishi Noguchi

Received: 12 November 2016; Accepted: 20 December 2016; Published: 10 January 2017

Abstract: Initiation of DNA Replication is tightly regulated in all cells since imbalances in chromosomal copy number are deleterious and often lethal. In bacteria such as *Bacillus subtilis* and *Escherichia coli*, at the point of cytokinesis, there must be two complete copies of the chromosome to partition into the daughter cells following division at mid-cell during vegetative growth. Under conditions of rapid growth, when the time taken to replicate the chromosome exceeds the doubling time of the cells, there will be multiple initiations per cell cycle and daughter cells will inherit chromosomes that are already undergoing replication. In contrast, cells entering the sporulation pathway in *B. subtilis* can do so only during a short interval in the cell cycle when there are two, and only two, chromosomes per cell, one destined for the spore and one for the mother cell. Here, we briefly describe the overall process of DNA replication in bacteria before reviewing initiation of DNA replication in detail. The review covers DnaA-directed assembly of the replisome at *oriC* and the multitude of mechanisms of regulation of initiation, with a focus on the similarities and differences between *E. coli* and *B. subtilis*.

Keywords: initiation of DNA replication; DnaA; *oriC*; regulation of DNA replication; *Bacillus subtilis*; sporulation

1. Introduction

The initiation of DNA replication is highly regulated and tightly coupled to the progression of the cell cycle to ensure that the frequency of initiation appropriately matches that of cell division. In this way, cells maintain correct chromosome copy number and ensure success in reproduction [1–3]. Under-replication leads to cells likely to be missing essential genetic information, whilst over-replication is highly disruptive of genetic regulatory processes and is frequently associated with disease and cell death.

Regulation of DNA replication is exerted primarily at the initiation step when an initiator protein binds to the origin of replication and promotes the assembly of a nucleoprotein complex from which replication forks diverge [4]. Much of our current understanding of DNA replication and its regulatory control in bacteria is derived from studies of the Gram-negative organism *Escherichia coli*, in which the initiator protein is DnaA and the origin is *oriC*. It is now clear that while the principles underlying the regulation of DNA replication initiation in *E. coli* apply to many other bacteria, the regulatory components are somewhat restricted in their distribution [2,5]. Thus the Gram-positive organism *Bacillus subtilis* has no known DNA replication regulators in common with *E. coli*, moreover, its bipartite origin of replication is strikingly different in arrangement to the continuous origin of *E. coli* [6]. Furthermore, when starved of nutrients, additional layers of DNA replication control are exerted in *B. subtilis* as it enters into the pathway of sporulation which is characterized by asymmetric cell division, and compartment-specific gene expression.

This review describes our current understanding of DNA replication initiation and its regulation in *B. subtilis*. As bacterial DNA replication is best understood in *E. coli*, we provide an overview of the replication phases of initiation, elongation and termination in this organism before highlighting differences that are known in *Bacillus*. This is followed by an in-depth coverage of initiation of DNA replication including the initiation machinery and the mechanisms of DnaA assembly at the origin, with particular emphasis on the roles of the *Bacillus*-specific components, DnaB and DnaD, in replisome assembly. Next, we discuss the activities of the regulators, YabA and Soj/Spo0J, during growth and Spo0A/Sda and SirA during sporulation. Finally their mechanisms of action are compared with those of the *E. coli* regulatory components. This review is concerned with the regulatory mechanisms of DNA replication initiation in *B. subtilis* and *E. coli*—it is not intended as a comprehensive review of the DNA replication mechanisms of all bacterial species.

2. DNA Replication

The process of DNA replication can be separated into three distinct phases: initiation, elongation and termination. During the initiation phase, a nucleoprotein complex assembles at the origin of replication. This induces localized DNA unwinding leading to helicase loading and recruitment of a full complement of replisome machinery. In the elongation phase, this replication machinery carries out template-directed DNA synthesis. This is continuous and processive on the leading strand, but discontinuous on the lagging strand where a more complex cycle of primer synthesis, strand elongation and fragment ligation takes place. Finally, during termination, DNA polymerization is halted at a specific termination site. Regulation of DNA replication occurs principally at the initiation stage, during or prior to the recruitment of the replication machinery.

2.1. Initiation of DNA Replication

In bacteria, DNA replication is initiated by the binding of a protein initiator, DnaA, to the origin of replication, *oriC* (Figure 1). DnaA is understood to form a right-handed helical oligomer on the DNA [7,8] directed by its binding to a series of recognition sites within the origin termed DnaA-boxes [9]. The formation of this oligomer induces a localized unwinding of the DNA duplex within the origin at an AT-rich region termed the DUE (*DNA Unwinding Element*) [10,11]. DnaA then plays a role in recruiting the processive DNA helicase, named DnaB in *E. coli* or DnaC in *B. subtilis* [12], which is loaded onto the unwound single-stranded DNA (ssDNA) by a helicase loader, named DnaC and DnaI respectively [13] (Table 1).

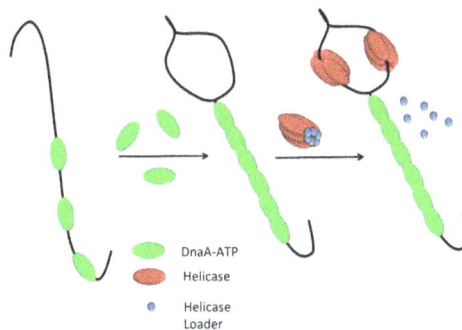

Figure 1. DNA replication initiation at *oriC*: DnaA (**green**) recognizes binding sites on *oriC*, forming a nucleoprotein complex which induces unwinding at the DNA unwinding element (DUE). The helicase loader then facilitates binding of the DNA helicase (**red**) as a prelude to recruitment and assembly of other components of the replication machinery. Figure inspired by [4].

Table 1. The essential DNA replication initiation machinery of *Bacillus subtilis* and *Escherichia. coli*.

Role in DNA Replication Initiation	B. subtilis	E. coli
Initiator	DnaA	DnaA
Helicase	DnaC	DnaB
DNA Remodelling	DnaB, DnaD	–
Helicase Loader	DnaI	DnaC
Primase	DnaG	DnaG

The helicase subsequently recruits the primase, DnaG, and the polymerase β-clamp, DnaN, which in turn recruits other components of the replication machinery in readiness for de novo DNA strand synthesis [14]. In *B. subtilis*, initiation requires two additional essential proteins, DnaD and DnaB [15], both of which possess DNA remodelling activities [16] and bind to the origin prior to helicase loading [17]. DnaD is thought to play a role in double-stranded DNA (dsDNA) melting, while DnaB appears to have a role in helicase loading. The essential components of the *B. subtilis* DNA replication machinery and their *E. coli* equivalents are listed in Table 1.

2.2. The Elongation Phase

During the elongation phase of DNA replication, DNA is synthesized processively by the action of a large multi-subunit complex known as the replisome (Figure 2A). Based on single molecule biophysics studies in *E. coli*, the replisome consists of three DNA polymerase complexes, a hexameric DNA helicase, DNA primase (assumed from structural studies to comprise three subunits [18]), three processivity clamps, DnaN (two of which are associated with the core replisome), and a pentameric clamp loader complex [19].

The helicase forms a homohexameric ring that is understood to sit at the head of the replication fork on the lagging strand of the template DNA. The helicase mechanically separates dsDNA by translocating along the lagging template strand in a process driven by ATP-hydrolysis. Separated DNA strands are coated in single-stranded DNA binding protein, SSB, which prevents the strands from re-annealing and offers protection to the ssDNA from nucleases [20–22].

The primase, DnaG, contains three functional domains; an N-terminal zinc-binding domain (ZBD), a central RNA polymerase domain (RPD) and a C-terminal helicase binding domain. Three DnaG molecules associate with the N-terminal domains of the helicase, positioned such that the primase captures the ssDNA which has been newly unwound by the helicase, ready for primer synthesis [23,24] (Figure 2B). The primase contains a groove that is thought to interact non-specifically with ssDNA, allowing the primase to track along the ssDNA and orientate it correctly for entry into the active site in the RPD, where primers are synthesized from available ribonucleoside tri-phosphate (rNTPs). The newly synthesized primer is extruded on the outside of the DnaB-DnaG complex, ready for handoff to SSB and DNA polymerase [23]. Whilst the RPD contains the catalytic site for RNA primer synthesis, the ZBD is responsible for modulating the activity of the RPD. Interestingly, the ZBD of DnaG regulates the RPD of an adjacent subunit in *trans* [25]. The RPD and ZBD from separate chains recognize the ssDNA template and initiate primer synthesis at specific trinucleotide recognition sites; with the ZBD increasing the catalytic activity of the *trans* RPD, as well as restricting processivity and primer length [25].

Strand extension in *E. coli* (Figure 2A) is carried out by DNA polymerase III (Pol III), which has an αεθ structure, where α is the catalytic subunit, ε is responsible for proofreading and θ is a non-essential subunit thought to stimulate the activity of ε. Pol III extends the primer with the assistance of the processivity clamp, DnaN (also known as the β-clamp). DnaN sits directly behind Pol III, as a closed ring on the DNA formed from two C-shaped subunits. DnaN binds across, rather than within, the major and minor grooves of duplex DNA, allowing the protein to slide along the DNA. In this way, the β-clamp enables the polymerase to synthesize up to 1000 bases a second [20–22]. The synthesis of each lagging strand Okazaki fragment requires the loading of a new β-clamp; thus the clamp loader complex forms part of the replisome machinery. The clamp loader is a pentameric complex with a subunit structure $\tau_3\delta\delta'$. The τ subunit, the product of the gene *dnaX*, interacts with both the DNA helicase and Pol III—it is thought to play an architectural role at the replisome and couple DNA unwinding and DNA extension [20–22].

Elongation in *B. subtilis* occurs by a similar mechanism; however it uses two different, but related, replicative DNA polymerases, PolC and DnaE (Figure 2A). DnaE is more closely related to *E. coli* Pol III than PolC [26]. Both polymerases have been shown to be essential for lagging strand synthesis, whilst PolC is required for leading strand synthesis [27]. Each can extend DNA primers, but DnaE alone is able to extend the RNA primers produced by DNA primase. It is thus thought that DnaE extends the RNA primers with DNA before handing over to PolC for further strand synthesis [27]. This is analogous to systems in eukaryotes where DNA polymerase α extends RNA primers with DNA, before handing over to the lagging strand polymerase δ [20].

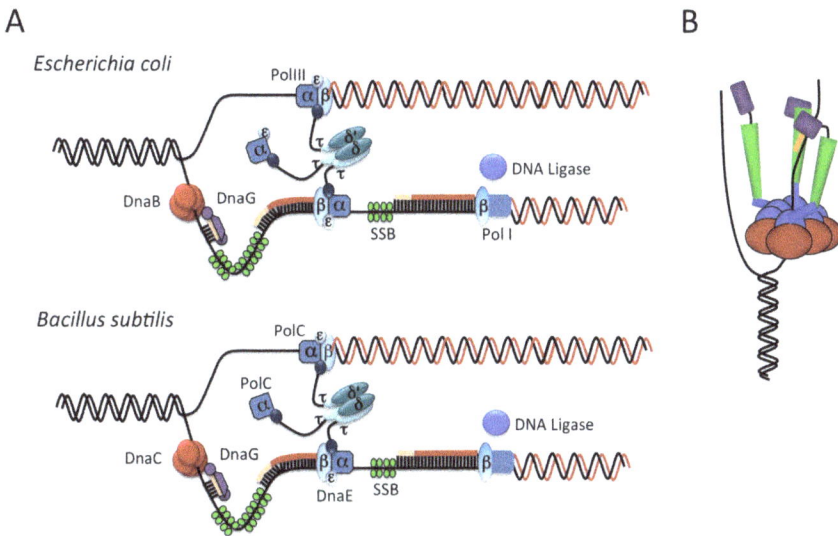

Figure 2. (**A**) Schematic representation of the *E. coli* and *B. subtilis* replisomes showing locations of the helicase, primase, DNA polymerase, the β-clamp and the clamp loader ($\tau_3\delta\delta'$) at the replication fork. Figure adapted from [20] (**B**) Schematic of primase function. The helicase (**red**) unwinds the parental DNA, positioning a single strand ready for primer formation. The RNA polymerase binding domain (**green**) of one primase molecule forms a complex with the Zn binding domain (**purple**) of another primase molecule and single-stranded DNA (ssDNA) in order to synthesize the primer (**orange**). The C-terminal helicase-binding domain of the primase is shown in blue. Adapted from [24]

2.3. Termination of DNA Replication

The termination of DNA replication occurs at a termination locus positioned directly opposite *oriC*. In both *B. subtilis* and *E. coli*, replication termination is controlled by a polar mechanism in which the *Ter* site can be approached from a 'permissive' or 'non-permissive' direction. However, different mechanisms have evolved in each species.

In *E. coli*, the locus directly opposite *oriC* is flanked on either side by five non-palindromic 23-bp sites, *Ter*A-J (Figure 3), which bind the monomeric protein Tus (terminator utilisation substance) [28–30]. The orientation of these *Ter* sites dictates whether or not a travelling replication fork is able to pass the site or is halted in DNA replication [28,30]. Thus, a replication fork can bypass a *Ter* site unimpeded when travelling in the permissive direction, but is blocked when travelling in the non-permissive direction. For example, in Figure 3A, a replication fork travelling clockwise would bypass *TerH*, *TerI*, *TerE*, *TerD* and *TerA*, but would be halted at *TerC* (or failing that, at *TerB*, *TerF*, *TerG* or *TerJ*). Tus is a 36 kDa protein which specifically binds *Ter* sites in an asymmetric manner [31] (Figure 3B). Collision with the DNA helicase DnaB approaching from the permissive direction, causes Tus to rapidly dissociate. In contrast, when the approach is from the non-permissive direction, Tus-*Ter* forms a roadblock which prevents the translocation of DnaB and the associated replication fork [32]. Tus functions like a 'molecular mousetrap' at *Ter*. The trap is set by asymmetric binding of Tus to dsDNA in the non-permissive orientation, such that strand unwinding by the oncoming replication machinery 'triggers' the trap causing a specific cytosine base at position 6 of the *Ter* site to flip into a binding site on Tus. This gives rise to a 'locked' Tus-*Ter* complex (Figure 3B) which presents a roadblock to the progression of the replication fork [32,33].

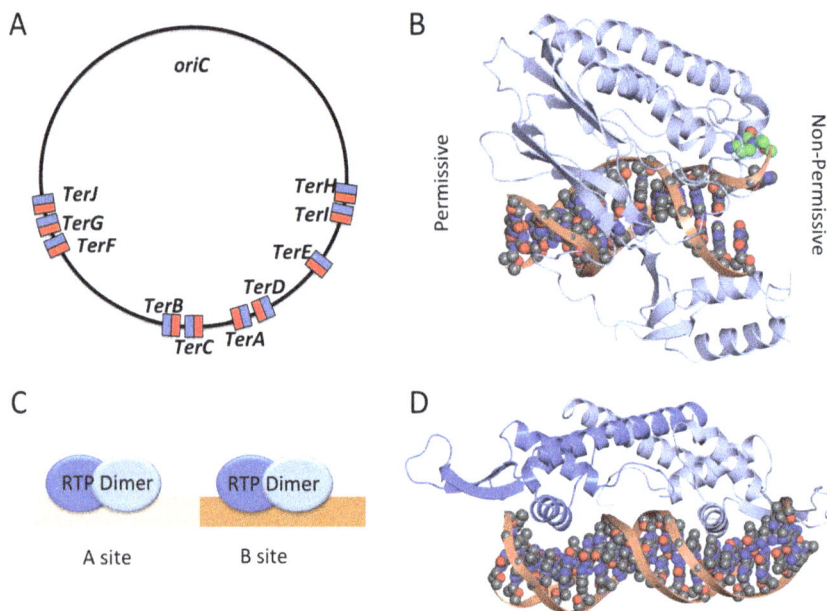

Figure 3. (**A**) Location and orientation of *Ter* sites in *E. coli*: permissive face shown in blue, non-permissive face shown in red; (**B**) Structure of the Tus-*Ter* complex (PDB code: 2EWJ) showing the permissive face (**left**) and the non-permissive face (**right**). On the non-permissive face a specific cytosine base (**green**) flips into Tus when double-stranded DNA (dsDNA) is unwound by the oncoming replication fork, creating a 'locked' complex; (**C**) Schematic image of two RTP dimers binding at the A and B sites of the *Bacillus* terminus region; (**D**) Structure of an RTP dimer bound to dsDNA (PDB code: 2EFW) with the sequence of a B-site region; one molecule displays a 'wing up' conformation (adjacent to the A site) and the other a 'wing down' conformation. (**B**), (**D**) and subsequent structural figures were rendered in CCP4MG [34].

In *B. subtilis*, the binding of two homodimers of the replication termination protein (RTP) at 'A' and 'B' sites within the *Ter* region is required to arrest replication (Figure 3C) [35,36]. The approach of the replication machinery from the 'B' site results in termination of replication (non-permissive direction) whilst approach from the 'A' site allows replication to continue (permissive direction). The crystal structure of a single RTP dimer bound to the native 'B' site has been shown to display asymmetry in the 'wing' region of the winged-helix domain [37] (Figure 3D). The protomer that lies proximal to the A-site shows a 'wing-up' conformation, while the other protomer displays a 'wing down' conformation, each making different contacts with the dsDNA. It is possible that this asymmetry gives rise to the 'permissive' and 'non-permissive' directions. However, A-site binding is also required to block replication fork progression, and A-site binding by RTP is co-operative following B-site binding [38]. The structural consequences of A-site binding are unknown and therefore the molecular basis of RTP action in replication termination remains unknown. Although the details of the *E. coli* and *B. subtilis* replication termination mechanisms vary, they appear to have evolved conceptually similar mechanisms for terminating replication in a direction specific manner.

3. Initiation of DNA Replication

3.1. Replication Origins

Replication origins have formed the topic of comprehensive recent reviews [39,40]. Knowledge of replication origins and how they encode DnaA-origin binding is key to the understanding of initiation mechanisms and how they are regulated. All origins harbor sequences that direct the formation of replication complexes, DNA unwinding, and species-specific regulatory activities. Conserved features of all bacterial replication origins include DnaA-box clusters and an AT-rich DUE. However across species, origins vary significantly in organization and length, including the number and spacing of DnaA-boxes and DnaA-box location with respect to the DNA unwinding elements. Of particular relevance to this discussion are two key differences between the origins of *B. subtilis* and *E. coli*: the genomic context of the origin and the number of intergenic regions that constitute *oriC*.

3.1.1. Genetic Context of Replication Origins

The location of the replication origin and its gene context are well conserved across bacterial species, with most flanked by, or containing, the *dnaA* gene [6,41]. The genes surrounding *oriC* and *dnaA* are also well conserved, consisting of the gene cluster *rnpA-rpmH-dnaA-dnaN-recF-gyrB-gyrA* with *oriC* residing in one or two intergenic regions adjacent to *dnaA* [6]. Unusually, the *E. coli* origin has undergone a major rearrangement resulting in a translocation of the origin 44 kb away from the *dnaA* gene and the *rnpA-rpmH-dnaA-dnaN-recF-gyrB-gyrA* cluster [41] so that it is instead flanked by the genes *gidA* and *mioC* [39]. Thus the origin of replication in *B. subtilis* may be more primitive than that of *E. coli*. Moreover, *B. subtilis* may provide a better model for bacterial replication origins in general [6].

3.1.2. Continuous and Bipartite Origins

Origins are described as either continuous or bipartite according to whether all of the functional elements are contained in one or two intergenic regions respectively. For example, the origin of DNA replication in *B. subtilis* (Figure 4A) is bipartite, containing two DnaA-box clusters, separated by the *dnaA* gene [42,43]. In *E. coli*, the origin of replication is a continuous ≈250 bp element. (Figure 3B). The bipartite origin in *B. subtilis* has been shown to be important for proper replication initiation [36], although it is not clear how this difference in origin structure affects the assembly and architecture of the initiation machinery at the origin. During replication initiation, *B. subtilis oriC* forms looped structures which are thought to be a consequence of the bipartite nature of its origin [44]. These looped structures can also form using *E. coli* DnaA but *E. coli* DnaA is unable to unwind the *B. subtilis* origin. This supports the idea that a mechanism of DnaA binding at the origin leading to DnaA oligomerisation is applicable across bacterial species, as might be expected given the high conservation of DnaA. However, specific assembly and regulation of initiation encoded by each origin is likely to be species-specific.

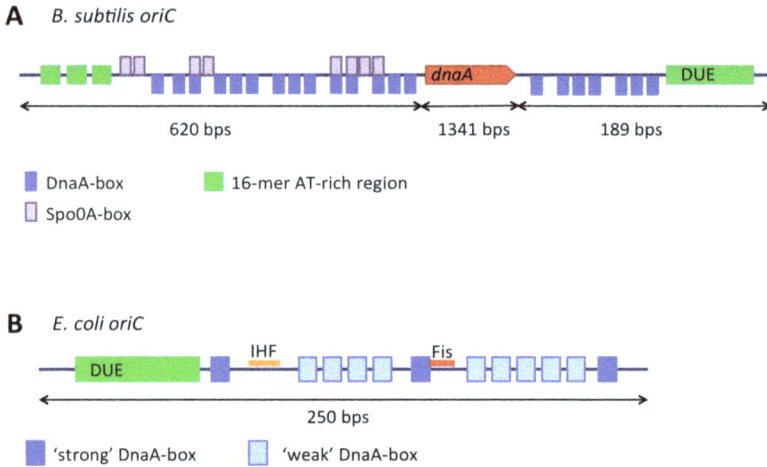

Figure 4. (**A**) *B. subtilis* origin of replication: DnaA-boxes are shown in blue, the *dnaA* gene in red, DNA unwinding element in green and Spo0A-boxes in purple; (**B**) The *E. coli* origin of replication: strong DnaA-boxes are shown in dark blue, weak DnaA-boxes in light blue, the DNA unwinding element in green and binding sites for accessory proteins integration host factor (IHF) and Fis in orange and red, respectively.

3.2. The DNA Replication Initiator, DnaA

The initator DnaA is a member of the AAA+ ATPase family (<u>A</u>TPases <u>a</u>ssociated with diverse cellular <u>a</u>ctivities) and contains four distinct domains [45,46] (Figure 5A). In the cell, DnaA exists in both ATP- and ADP-bound forms [47]. DnaA–ATP is considered to be the 'active' form of the protein as this is required for oligomerisation at the origin [48,49], an event which triggers DNA unwinding and ultimately, assembly of the replisome. The C-terminal domain IV of DnaA is a dsDNA binding domain which is responsible for DnaA-box recognition [50,51]. The adjacent domain III contains Walker A and B motifs that are involved in ATP binding and hydrolysis. This domain plays a role in self-interaction/oligomer formation and in ssDNA binding [52]. Domain II of DnaA is poorly conserved and of variable length and considered to form a flexible linker which may play a role in controlling replication efficiency [53]. Finally, the N-terminal domain I is an 'interaction domain' which has been shown to interact with various protein regulators of DnaA across different organisms [54–56]. In *E. coli*, it also interacts with the helicase, DnaB [57], and has been suggested to play a role in the self-assembly of DnaA at the origin.

3.2.1. DnaA-Box Recognition by DnaA

The DnaA-boxes within the origin of replication vary in their affinity for DnaA, according to their similarity to a consensus binding sequence, and on the adenosine nucleotide bound state of DnaA [58,59]. In *E. coli* and *B. subtilis*, the consensus DNA-box is the nine-base-pair sequence, 5'-TTATNCACA-3' [60].

An X-ray structure of DnaA domain IV bound to a consensus DnaA-box sequence revealed that DNA binding is mediated by a helix-turn-helix which interacts primarily with the major groove of the dsDNA, with additional contacts made in the adjacent minor groove [51] (Figure 5A). Base-specific interactions were observed at 8 of the 9 base pairs in the DnaA-box; the exception being the base pair at position 5, where there is no sequence preference [51]. Mutations at residues involved in base-specific interactions result in loss of DnaA-box binding specificity, or loss of DNA-binding altogether [61].

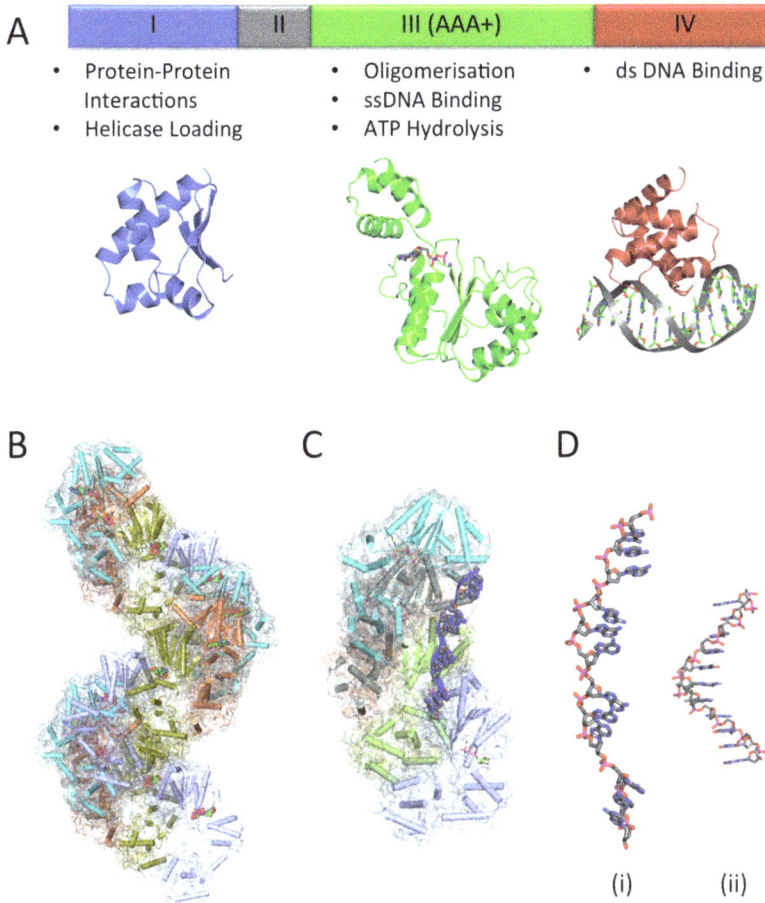

Figure 5. (**A**) Schematic of DnaA showing domain architecture and structures. Domain I (PDB code: 4TPS) is shown in blue, domain II in gray, domain III (PDB code: 1L8Q) in green and domain IV (PDB code: 1JLV) in red. Figure adapted from [2]. (**B**) DnaA Domains III–IV bound to a non-hydrolysable ATP analogue (PDB code: 2HCB) form a spiral structure that is thought to mimic DnaA oligomerisation at the origin. A repeating pattern of DnaA protomers is shown in light blue, gold, coral and cyan; (**C**) ssDNA-binding mode of DnaA domains III–IV (PDB code: 3R8F). Separate DnaA protomers are shown in light blue, green, gray and cyan; (**D**) ssDNA binding by DnaA stretches the strand into an extended form (i) compared to B-form DNA (ii).

3.2.2. Variable Affinity of DnaA-Boxes

The DnaA-boxes at the origin can be either 'strong' or 'weak'; where strong boxes bind both DnaA–ATP and DnaA–ADP with equal affinity and 'weak' boxes have a much greater relative affinity for DnaA–ATP [62]. In order for the helical DnaA oligomer to form at the origin and induce DNA unwinding, both strong and weak DnaA-boxes need to bind DnaA [63,64]. In the *E. coli* origin (Figure 4B), DnaA-boxes are distributed such that three strong boxes lie at either end of the origin and at its centre. As DnaA–ATP recruitment to the origin has been shown to be co-operative, these strong boxes are thought to form anchoring points from which the DnaA oligomer can grow [63,65]. In this model, DnaA–ATP is recruited to weak binding sites via co-operative interactions with DnaA–ATP molecules already bound to neighbouring sites [65].

3.2.3. DnaA Oligomerisation

Domains III–IV of *Aquifex aeolicus* DnaA have been shown to adopt an open spiral conformation [8] which likely mimics the right-handed helical oligomers ATP-bound DnaA forms at *oriC* [7] (Figure 5B). Adjacent protomers interact with one another via two clusters of conserved residues located on either side of the nucleotide binding pocket [8]. Significantly, DnaA–ADP cannot form this right-handed oligomer [7]; instead it appears to be monomeric [66]. The binding of ATP induces a small conformational change in the ATPase domain which allows an adjacent DnaA protomer to interact with the ATP via a conserved arginine residue known as an 'arginine finger'. This interaction is significant in stabilising the DnaA helical filament [8] and similar 'arginine finger' interactions are frequently observed in other AAA+ ATPases [67]. Significantly, these observations provide a molecular explanation for why DnaA–ATP is the 'active' form of the initiatior.

In order to reconcile the Domain IV-DnaA-box binding mode with the DnaA helical oligomer formed by DnaA domains III–IV on dsDNA, a conformational change in the linker helix between domains III and IV has been invoked [51,68]. A significant kink in the linker helix is observed in the ATP-bound structure compared to the ADP-bound form (where the helix is straight) suggesting that the two domains are conformationally uncoupled and would be able to rotate with respect to one another to allow filament formation at the origin [8].

3.2.4. DNA Unwinding and ssDNA Binding

After the DnaA oligomer has formed at the origin, localized strand unwinding occurs at the DUE [69] (Figure 1). Based on structural work carried out with *Aquifex aeolicus* DnaA, unwinding is mediated by the DnaA-oligomer, which introduces positive writhe in the bound DNA [7]. Compensatory negative writhe at the DUE would facilitate DNA unwinding [4,8]. This unwound DNA is then stabilized by binding to the ssDNA binding site of DnaA located in the ATPase domain [69,70]. ATP-bound DnaA binds ssDNA in the same open spiral conformation displayed by DnaA domains III-IV [69]. In complexes of DnaA with single-stranded poly-(dA) DNA, each DnaA protomer binds three nucleotides, making multiple interactions with the DNA phosphodiester backbone. Each nucleotide triplet displays a normal B-form DNA conformation, but the triplets are separated by gaps of approximately 10 Å creating an overall extended form of DNA [69] (Figure 5D). This strand extension has been shown to be ATP-dependent in solution and is highly reminiscent of ssDNA binding displayed by the homologous recombination protein, RecA. The third base of each triplet is rotated however, making bases in the DnaA-bound strand discontiguous; this presumably prevents re-annealing of the strand at the origin [69] (Figure 5D).

Recently identified trinucleotide sequences within bacterial origins termed 'DnaA-trios' appear to be responsible for providing specificity of binding of DnaA to ssDNA, facilitating DNA-unwinding at the origin [71]. The trinucleotide motifs have the consensus sequence 3'-G/AAT-5' and are separated from a proximal DnaA-box, or pair of boxes, by a GC-rich region. A DnaA molecule bound to the proximal DnaA-box via domain IV appears to be able to bind to the first of these DnaA-trio motifs via its AAA+ motif in domain III. Additional DnaA molecules interact with further DnaA-trio motifs, forming an oligomer on the ssDNA and facilitating DNA-unwinding [71].

3.2.5. *Bacillus* DnaA

B. subtilis DnaA has also been shown to form helical oligomers on both double and single stranded DNA [72], moreover, the DnaA–ATP form is required for co-operative binding to the origin [73]. *Bacillus anthracis* DnaA displays an ATP-dependent variable affinity for DnaA-box sequences [74]. Together these findings imply *Bacillus* DnaA functions at *oriC* in a similar manner to *E. coli* DnaA.3.2.6. The Role of DnaA Domains I–II in Initiation

DnaA domains I–II are not necessary for DnaA oligomerisation, or DnaA loading onto ssDNA [71]. Nevertheless DnaA domains I–II are required for initiation of replication [75]. DnaA domain I is known to interact with several regulators of DNA replication initiation; these include *E. coli* DiaA [76] and

H. pylori HobA [55]—structural homologues and promoters of initiation in their respective organisms—and *E. coli* Hda [77] and *B. subtilis* SirA—two negative regulators of initiation [54]. In *E. coli*, domain I also interacts with the helicase DnaB [57,78,79] where it is thought to help correctly orientate the loading of DnaB at the origin. Domain I of DnaA has also been suggested to play a role in the self-assembly of DnaA at the origin [80,81]. It has a K homology (KH)-domain fold typically found in ssDNA binding proteins [82–84] (Figure 5A). In vitro DnaA domain I binds to single-stranded *oriC* DNA, albeit weakly, suggesting a potential role in binding ssDNA at the origin [84]. However, no ssDNA binding role has yet been demonstrated for DnaA domain I in vivo.

Domain II has been shown to be unstructured, consistent with a role as a flexible tether between domains I and III. It is not completely dispensable for DnaA function, but it is poorly conserved and varies significantly in length between organisms [46]. Two studies in *E. coli* have indicated that domain II contributes to the efficiency of initiation of replication. In one study, a spontaneous deletion in domain II allowed suppression of an over-initiation phenotype, suggesting that the deletion had reduced the efficiency of DNA replication initiation [85]. In another study, when deletions longer than 17–19 residues were made from domain II, the doubling time of cells harbouring this mutation was increased compared to wild type cells, suggesting the length of domain II contributed to the efficiency of DNA replication [53]. The same study defined the minimum length of domain II in *E. coli* to be 21–27 residues [53].

3.3. Helicase Loading

Following the unwinding of the DUE, a homohexameric DNA helicase is loaded onto single stranded DNA at the replication origin by the action of a helicase loader protein. In *E. coli*, the helicase, DnaB, is loaded onto the ssDNA by the helicase loader DnaC. This occurs via a 'ring-breaking' mechanism whereby DnaC forms a spiral oligomer which remodels the hexameric DnaB ring, producing a break in the ring large enough to allow loading onto ssDNA [86]. The recruitment of the DnaB–DnaC complex to the origin occurs by an interaction between the N-terminal domain of DnaA and the helicase, DnaB [12,84,87]. This interaction is thought to orient DnaB for loading onto the bottom strand of the DNA, while an interaction between the AAA+ domains of DnaA and DnaC is thought to recruit the complex in the right orientation for DnaB loading on the upper strand (Figure 6) [86,88].

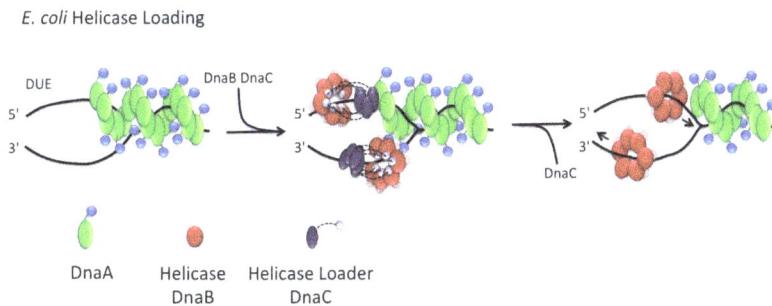

Figure 6. In *E. coli*, the initiator DnaA forms a helical oligomer during initiation which associates with the upper strand of the ssDNA. Following unwinding of the DUE, interactions between DnaA and DnaB or DnaC in the DnaC–DnaB complex are thought to correctly orientate DnaB for loading onto the bottom and top strands of DNA, respectively. Figure adapted from [88].

The primase, DnaG, is next recruited via an interaction with the N-terminal domain of DnaB. Subsequently, active primer formation appears to induce the dissociation of DnaC, in a step which is necessary for DnaB to begin to function as an active helicase. Release of DnaC appears to be dependent on the ATPase activity of DnaC which is thought to be induced by a conformational change in DnaB during primer formation [89]. DnaG interacts with the N-terminal domain of DnaB, while DnaC interacts with its C-terminal domain [90]. The loading of the helicase is important for the recruitment of the DNA polymerase clamp, DnaN. The clamp, in turn, recruits the DNA polymerase, in readiness for primer elongation [90,91].

Helicase loading in *B. subtilis* is thought to occur via a different mechanism known as 'ring assembly' [91]. In this model, the helicase loader, DnaI, facilitates the assembly of the helicase DnaC onto ssDNA [92,93]. In the presence of DnaI, pre-formed DnaC hexamers exhibit no helicase or translocase activity in contrast to monomeric DnaC which displays both helicase and translocase activities [92]. The helicase loader DnaI, like *E. coli*'s loader protein, contains an N-terminal helicase interaction domain and a C-terminal AAA+ domain [94]. The ATPase activity of the C-terminal domain of DnaI is stimulated in the presence of ssDNA, but only once inhibition by the N-terminal domain is overcome; binding of the N-terminal domain of DnaI to the helicase DnaC reveals a cryptic ssDNA binding site on the C-terminal domain [93]. It is thought that this then facilitates helicase loading onto ssDNA. Finally, the ATPase activity of the C-terminal domain may stimulate the release of DnaI once loading has occurred [93].

3.4. Bacillus Initiation Proteins DnaD and DnaB

Besides DnaA, DnaC (equivalent to *E. coli* DnaB), DnaI and DnaG, DNA replication initiation in *B. subtilis* requires the presence of two additional essential proteins, DnaD and DnaB [95,96]. A summary of their structure and function forms part of the discussion in an excellent recent review [6]. Both DnaD and DnaB are components of the replication initiation machinery at *oriC* [15] as well as components of the replication restart machinery which is DnaA-independent [97]. Both proteins exhibit DNA remodelling activities [16] and share structural similarity [96]. The *B. subtilis* initiation machinery assembles in a hierarchical manner, and DnaD and DnaB recruitment occurs between DnaA binding at *oriC* and the loading of the helicase, DnaC [17]. On binding to *oriC*, DnaD forms direct interactions with DnaA [98]. DnaD is required for the recruitment of DnaB and this, in turn, is then required for recruitment of DnaC–DnaI [17]. Together DnaB and DnaI are thought to function as a helicase loader [92].

The exact roles of DnaD and DnaB in replication initiation remain unclear. DnaD is able to untwist supercoiled DNA into an open looped form [99]. It forms tetramers which can assemble into large protein scaffolds that appear to mediate DNA loop formation and enhance melting of dsDNA [100]. The N-terminal domain of DnaD (DDBH1) is implicated in tetramer formation [101,102] with the C-terminal domain (DDBH2) involved in both double- and single-stranded DNA binding [100,102]. The full-length protein is required for DnaD to exhibit DNA looping and melting activities [100,102]. It is estimated that there are 3000–5000 DnaD molecules [103] in the cell and this relative abundance has led to the suggestion that DnaD plays a global role in DNA remodeling, beyond that required for DNA replication initiation [16]. In support of this idea, a study has shown that DNA remodeling by DnaD stimulates DNA repair by Nth endonucleases in response to DNA damage following treatment with H_2O_2 [104].

It is generally thought that DnaB acts together with DnaI to enable the loading of DnaC onto forked DNA [92]. However, studies [93,105] suggest that DnaI alone is sufficient to load the helicase onto DNA and that DnaB is required to recruit DnaC–DnaI to the origin [17] and that it acts to stimulate the helicase and translocase activities of DnaC in the presence of DnaI [92]. DnaB has also been implicated in the association of the DNA replication machinery with the cell membrane [95,106]. It has also been shown to laterally compact DNA—although it is not known how this contributes to its function [16].

Although DnaD and DnaB show little sequence similarity, a Hidden Markov Model analysis identified two shared domains known as DDBH1 and DDBH2 (DDBH2 belongs to the PFAM domain: *DnaB_2*). DnaD has a DDBH1–DDBH2 architecture, whilst DnaB has a DDBH1–DDBH2–DDBH2 organization [96] (Figure 7A). The structure of the DDBH1 domain of DnaD revealed a winged helix domain with two additional structural elements: an N-terminal helix–strand–helix and a C-terminal helix [101] (Figure 7C). The β-strand of the helix–strand–helix was found to mediate interactions between DnaD molecules in both dimer and tetramer formation (Figure 7C). The C-terminal helix has been shown to be important in higher-order oligomerisation of these tetramers [101]. These structural elements appear to be present in DnaB DDBH1 [96] which has also been shown to form tetramers mediated by its N-terminus [16], suggesting that DnaB and DnaD share similar oligomerisation properties.

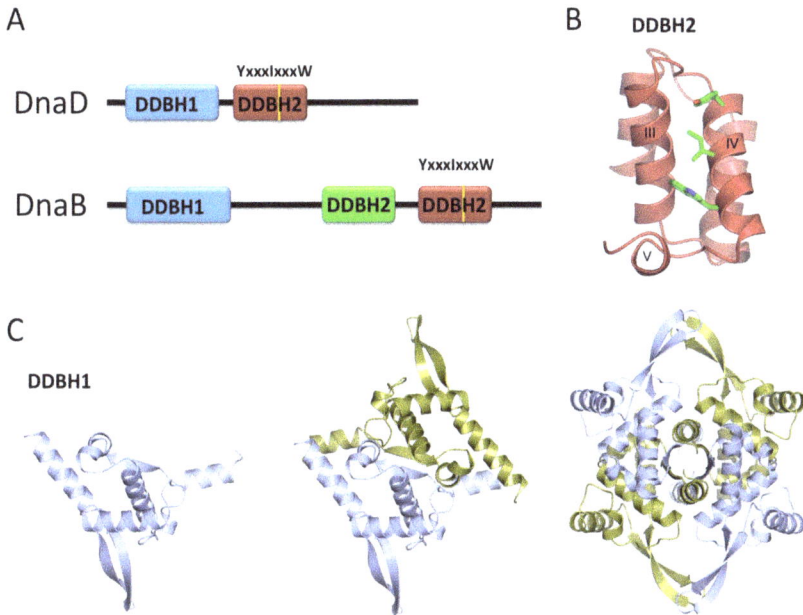

Figure 7. (**A**) Diagram showing the architecture of DnaD and DnaB. Conserved DNA binding motif YxxxIxxxW is marked on the relevant DDBH2 domain; (**B**) Ribbon diagram of the DnaD DDBH2 domain from *Streptococcus mutans* (PDB code: 2ZC2). Tyrosine, Isoleucine and Tryptophan residues of the YxxxIxxxW motif are coloured by atom (carbon in green, nitrogen in blue and oxygen in red); (**C**) Ribbon diagram of DnaD DDBH1 domain from *Bacillus subtilis* (PDB code: 2V79) showing a winged helix with additional structural elements. Monomer, dimer and tetramer architectures are shown. Dimer and tetramer interactions are mediated by the β-strand of the additional helix–strand–helix. Figure inspired by [6].

DnaD's DDBH2 domain has been shown to be involved in DNA-binding and in DNA-dependent higher-order oligomerization [102]. Two structures of the DDBH2 domain of DnaD homologues from *Streptococcus mutans* (PDB code: 2ZC2) and *Enterococcus faecalis* (PDB code: 2I5U) show a compact helical structure with four longer helices I–IV and a shorter fifth helix (V) of only 4 residues (Figure 7B). Although residues following helix V are poorly conserved across DnaD homologs, secondary structure prediction and analysis of the *B. subtilis* DnaD DDBH2 domain by NMR suggests that helix V is extended by a further seven residues [96]. Helix V is followed by a region at the C-terminus that is predicted to be disordered. A YxxxIxxxW motif residing in helix IV, the poorly conserved helix V and the C-terminal unstructured region [96] have been shown to be important for ssDNA binding. These structural elements appear to be conserved in the second of the DDBH2 domains of DnaB [96]. This domain has been implicated in dsDNA and ssDNA binding as well as in higher-order oligomerization [107]. Again, this suggests that the domains play similar roles in the respective proteins.

A DnaB (1–300) fragment encompassing DDBH1–DDBH2 (missing the C-terminal DDBH2 domain) forms tetramers and binds ssDNA [96,107]. Interestingly, C-terminally truncated cytosolic forms of DnaB have been observed during the mid–late growth phase. Full length DnaB alone is observed at *oriC*, thus proteolysis may be regulating DnaB function [107]. It is unclear whether the truncated version of the protein has a discrete function [107], however the different DNA binding capabilities of the DDBH2 domains of DnaB may be important in differentiating the functions of the full-length and truncated versions of DnaB.

3.5. Regulation of DNA Replication

3.5.1. During Vegetative Growth in *B. subtilis*

YabA

YabA is a negative regulator of DNA replication in *Bacillus subtilis*, affecting both the timing and synchrony of DNA replication in vegetatively growing cells [108]. Deletion of *yabA* causes an increased frequency of initiation events and asynchronous DNA replication [108] as well as a growth phenotype associated with increased initiation events [109,110]. YabA interacts with both the replication initiator, DnaA, and the DNA polymerase clamp, DnaN [109,110]. Mutations of YabA affecting the interaction with either DnaA or DnaN have been shown to exhibit an over-initiation phenotype similar to that in Δ*yabA* cells. This suggests that both interactions are important for replication regulation [110].

Expression of *yabA* genes encoding DnaA-loss-of-interaction or DnaN-loss-of-interaction mutations disrupts the formation of YabA foci at mid-cell, where it is assumed that YabA is co-localized with the replisome. Significantly, however, co-expression of DnaA-loss-of-interaction and DnaN-loss-of-interaction YabA-mutants restores YabA foci, presumably through a hetero-oligomer produced by the two mutants. This implies that both interactions are simultaneously required for YabA localization at the replisome [110,111].

YabA forms tetramers through interactions of N-terminal coiled-coil domains to form an intermolecular 4-helix bundle. This provides a structural scaffold from which four C-terminal Zn-binding domains project. These are connected to the N-terminal domain by a flexible linker and they appear to be independent domains [111] (Figure 8A). The determinants on YabA for DnaA and DnaN interactions lie within these C-terminal domains. Significantly, yeast three-hybrid experiments show that full-length YabA is able to interact simultaneously with DnaA and DnaN [110], whereas the C-terminal domain alone cannot [111]. Thus, the YabA tetramer organization facilitates simultaneous interactions with DnaA and DnaN.

Despite much study, the mechanism of YabA action remains elusive. YabA is not able to promote DnaA–ATP hydrolysis in vitro [112], however, it has been shown to affect the co-operative binding of DnaA to *oriC* [73], and it is capable of disrupting DnaA oligomerisation in vitro [112]. It is not clear, however, if this is its main mode of action in vivo. Two alternative models have been proposed. In the first, YabA tethers DnaA to DnaN at the replisome for most of the cell cycle (Figure 8B) [113], sequestering DnaA from the origin during ongoing rounds of replication. This model is consistent with the alternate localisations of DnaA in wild type and Δ*yabA* cells. In wild type cells, DnaA localizes at the origin in small cells (which are at early points in their cell cycle) and at mid-cell, co-incident with DnaX and therefore the replisome, in larger cells (in later stages of the cell cycle) [113]. In Δ*yabA* cells, by contrast, DnaA is localized with the origin throughout the cell cycle [113].

The alternative model proposes that YabA binds to DnaA at *oriC* so as to inhibit its cooperative binding to further DnaA molecules throughout the cell cycle up to the point where DNA replication is completed and the replisome disassembles. At this point free DnaN competitively titrates YabA away from its complex with DnaA, allowing the latter to bind cooperatively at the origin (Figure 8C) [73]. This model is consistent with evidence that the cellular level of DnaN correlates with the frequency of replication initiation, with increased DnaN levels increasing replication initiation frequency, and decreased levels, decreasing initiation frequency [73]. Additionally, in a strain replicating from a

DnaA-independent origin, *oriN*, YabA was shown to affect the cooperativity of DnaA binding at *oriC*, and increased levels of DnaN removed YabA from *oriC*, suggesting that DnaN could be controlling the binding of YabA at the origin.

Further studies to establish the dynamics and stoichiometry of the interactions between YabA, DnaA and DnaN are required to further refine and reconcile these models: bearing in mind that they are unlikely to be mutually exclusive.

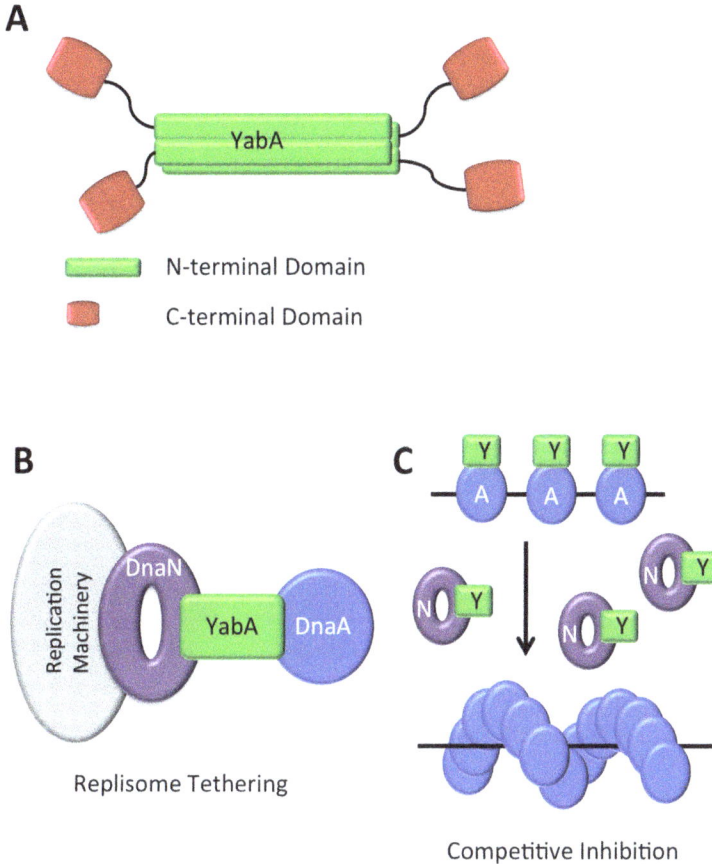

Figure 8. (**A**) Schematic of YabA tetramer structure: YabA N-terminal domains (**green**) form a 4-stranded coiled coil structure. Pseudo-monomeric C-terminal Zn-binding domains (**red**) are attached by flexible linkers; (**B**) Replisome tethering model. YabA tethers DnaA to the replisome via an interaction with both DnaA and DnaN, titrating DnaA away from the replication origin; (**C**) Co-operative inhibition model. YabA (Y) inhibits the cooperative binding of DnaA (A) at the origin during replication. When DnaN (N) is released after replication, YabA binds DnaN, releasing DnaA.

Soj/Spo0J

Soj is an ATPase which negatively and positively regulates DNA replication in *B. subtilis* [114], according to its oligomeric state [115], which is controlled by nucleotide binding. ATP-bound Soj forms dimers which co-operatively interact with DNA in a sequence unspecific manner, whilst ADP-bound Soj is monomeric [116]. Dimeric ATP-bound Soj appears to stimulate initiation of replication, whilst monomeric Soj inhibits replication [72,115]. Spo0J regulates Soj activity by stimulating its ATPase activity, thus converting the dimer back to the monomeric form [115].

Soj appears to interact with the ATPase domain (III) of DnaA, although it does not affect ATP binding or hydrolysis by DnaA [72]. Instead, it acts by inhibiting DnaA oligomerisation at *oriC*. A Soj mutant trapped in the monomeric state has been shown to inhibit DnaA oligomer formation both in vitro and in vivo [72]. Curiously, Soj trapped in the dimeric state is also able to interact with DnaA on a similar surface, without inhibiting DnaA oligomerization. Thus, it has been suggested that monomeric Soj inhibits conformational changes in DnaA that are needed to form an active initiation complex [72], whilst dimeric Soj may stabilize DnaA in this oligomerization-competent conformation [72].

Soj and Spo0J are orthologues of ParA and ParB, respectively. ParA, and ParB, along with a *cis*-acting DNA sequence parS, are components of a plasmid partitioning system found in many prokaryotic species. These systems ensure partitioning of low copy number plasmids into daughter cells. ParB binds to parS sequences on the plasmid, while ParA forms filaments on chromosomal DNA. An interaction between ParA and ParB simulates the ATPase activity of the former, which is thought to cause dissociation of the terminal ParA molecule from the filament; the plasmid can then either dissociate or translocate along the chromosomal DNA by binding to the next ParA molecule. Continuous cycles of ParA assembly and disassembly lead to equidistribution of the plasmids within the cell [117], ensuring partitioning on either side of the division plane [118].

Chromosomal orthologues of ParA and ParB and parS sites are found in some bacterial species and it is attractive to assume that they perform a role in chromosomal segregation similar to that of the plasmid partitioning proteins. In *B. subtilis*, although Spo0J-parS contributes to accurate chromosome segregation, it is not essential for this function [119]. Instead it plays a role in the recruitment of the SMC complex to the origin, and it is the SMC proteins that are responsible for proper segregation and condensation of the chromosome [120,121]. Regardless, Spo0J provides a mechanism through which *B. subtilis* may be able to co-ordinate DNA replication and chromosome segregation [121].

DnaD

DnaD has also been reported to play a role in the regulation of DNA replication initiation in *B. subtilis*. Like YabA, DnaD has been shown to inhibit the ATP-dependent cooperative binding of DnaA to *oriC* DNA [122] and to affect the formation of helical DnaA filaments in vitro [112]. It remains unclear however, how these activities can be reconciled with the role of DnaD in vivo, where it is essential for DNA replication initiation.

DnaA-Box Clusters

A *B. subtilis* deletion strain, in which six DnaA-box clusters (DBCs) found outside of the replication origin were removed, displayed an early initiation of DNA replication phenotype. This phenotype was strong only when all six clusters were removed and could be partially relieved by the re-introduction of a single DBC at various locations [123]. Nevertheless, these data suggest that *B. subtilis* DNA replication is sensitive to the amount of free DnaA in the cell, which might otherwise be bound at these sites.

3.5.2. During Sporulation in *Bacillus subtilis*

A characteristic of *B. subtilis* is its ability to differentiate under nutrient limiting conditions to form a dormant endospore. The spore is metabolically inactive and resistant to harsh conditions such as high temperatures, desiccation and ionizing radiation. When nutrients become available again, the spore can germinate, returning the cell to vegetative growth, even after thousands of years [124,125]. Unlike vegetative growth which is characterized by division at mid-cell, during sporulation the cell divides asymmetrically forming a larger mother cell compartment and smaller forespore compartment (Figure 9A). These two daughter cells each contain an identical copy of the genome, however, differential pathways of gene expression lead to dramatically different cell fates. The forespore is engulfed by the mother cell, and in the cytoplasm of the latter it matures into a resistant spore. In the final stages, the mother cell lyses to release the fully formed spore [126] (Figure 9A). Entry into the sporulation pathway is under the control of a complex signaling pathway, at the heart of which is an expanded two-component system termed the phosphorelay, which culminates in the phosphorylation of the response regulator Spo0A, the master control element of sporulation [127] (Figure 9B). Spo0A~P acts as a transcriptional regulator, controlling directly or indirectly the expression of over 500 genes [128].

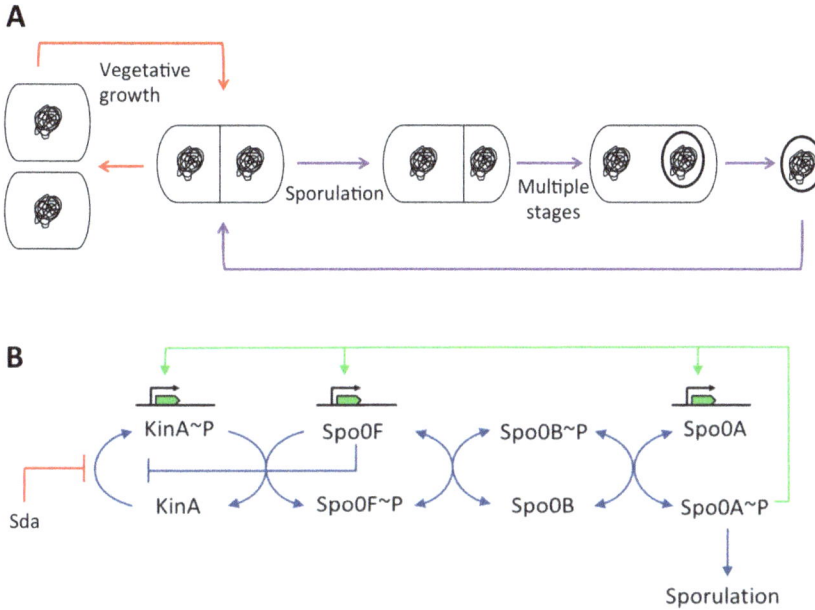

Figure 9. (**A**) Vegetative growth and sporulation in *B. subtilis*. In normal vegetative growth (**red** arrows) cells divide symmetrically, producing identical daughter cells. During sporulation (**purple** arrows) cells divide asymmetrically forming a mother cell and forespore; each receives an identical copy of the genome, and through differential gene regulation they experience different fates. The mother cell engulfs the forespore, nurturing it as it matures. In the final stages, the mother cell lyses releasing the dormant spore; (**B**) Phosphorelay leading to the induction of sporulation. A series of phosphoryl transfer reactions lead to the accumulation of threshold levels of Spo0A~P needed for entry into the sporulation pathway.

At the point of entry into sporulation, DNA replication and asymmetric cell division must be coordinated to ensure that the cell contains two, and only two, copies of the chromosome—one destined for the mother cell and the other for the forespore. Trapping of more than a single chromosome in the forespore compartment can reduce the viability of the spore and its capacity to germinate [129].

Spo0A~P Pulsing

It has long been recognized that there is a 'sensitive period' in the cell cycle when the cell can enter into the sporulation pathway. If the cell progresses beyond this point, it is committed to a new round of vegetative division [130,131]. The critical determinant is the concentration of Spo0A~P which fluctuates over the course of the cell cycle and is at its highest immediately after DNA replication is completed [129,132]. A threshold level of Spo0A~P must be reached for sporulation to be triggered. As Spo0A~P levels increase, low-threshold target genes are turned on, however, a higher threshold Spo0A~P concentration must be achieved in order to trigger the sporulation process [133,134].

Spo0A~P pulsing is linked to DNA replication, but until recently it was not known how. The cellular Spo0A~P concentration is controlled by the sporulation inhibitor protein, Sda [129]. Sda inhibits the sporulation sensor kinases KinA and KinB, which feed phosphate into the phosphorelay leading ultimately to the phosphorylation of Spo0A [135–137] (Figure 9B). Sda production is controlled at the transcriptional level by DnaA [129,135] such that *sda* expression requires the presence of replication active DnaA. Thus, Sda levels spike at the same time as, or just after, the replisome forms [129]. Sda is subsequently rapidly proteolysed [138]. This provides a feedback mechanism whereby Sda blocks phosphorylation of Spo0A and entry into sporulation, during ongoing rounds of DNA replication [129]. However, factors other than Sda influence Spo0A~P pulsing, as deletion of Sda does not prevent the pulsing of Spo0F levels (*spo0F* is a 'low-threshold' gene under the control of Spo0A~P) [139] suggesting that Spo0A~P pulsing still occurs.

The chromosomal arrangement of the phosphorelay genes *spo0F* and *kinA* is important for Spo0A~P pulsing [129]. *spo0F* is located close to the replication origin, in contrast to *kinA* which is located near the replication terminus. As a result, two copies of *spo0F* will be present in the cell during most of the period of DNA replication, alongside a single copy of *kinA* [132]. Alterations in the chromosomal positioning of *spo0F*, or induction of Spo0F from an inducible promoter, have been shown to affect Spo0A~P pulsing [132], with high Spo0F:KinA ratios inhibiting KinA phosphorylation and preventing sporulation [140,141]. As rapidly growing cells undertake multiple rounds of DNA replication simultaneously, the Spo0F:KinA ratio also provides a mechanism for inhibiting sporulation under nutrient-rich conditions. Collectively, the chromosomal arrangement of the phosphorelay genes in *B. subtilis*, together with direct inhibition of KinA activity by Sda, serve to coordinate the entry into sporulation with DNA replication.

SirA

Spo0A~P pulsing provides a mechanism for preventing replicating cells from entering into sporulation. Interestingly, cells which are artificially induced to sporulate under conditions of rapid growth are able to maintain correct chromosome copy number [142]. This is attributable to the activity of SirA, an inhibitor of DNA replication, produced under Spo0A~P control. Deletion of *sirA* results in loss of chromosome number control upon induction of sporulation during rapid growth [142]. Meanwhile, cells overproducing SirA do not form colonies on plates and in liquid culture many of these cells are elongated and anucleate, with some containing nucleoids which have been severed by division septa—a phenotype reminiscent of DnaA depletion [142,143]. SirA inhibits DNA replication through a direct interaction with DnaA [143]. A genetic screen indicated that the determinants of SirA binding reside in domain I of DnaA [54] and a later structure of a complex of SirA with DnaA domain I fully delineated this binding surface [144]. Cells harbouring alleles with *sirA* point mutations mapping to the DnaA domain I binding surface of SirA, exhibit a similar phenotype to Δ*sirA* cells. Moreover, these mutations disrupted SirA foci normally observed in sporulating cells [144]. This suggests that SirA localizes to the replisome via an interaction with DnaA during sporulation.

Intriguingly, SirA binds to a surface on DnaA domain I structurally equivalent to that used by the positive regulators of DNA replication initiation, DiaA and HobA, structural homologues found in *E. coli* and *Helicobacter pylori*, respectively (Figure 10). This raises an intriguing question about the role of DnaA domain I in replication initiation in the respective organisms. How is the same topological site used to positively and negatively regulate replication initiation?

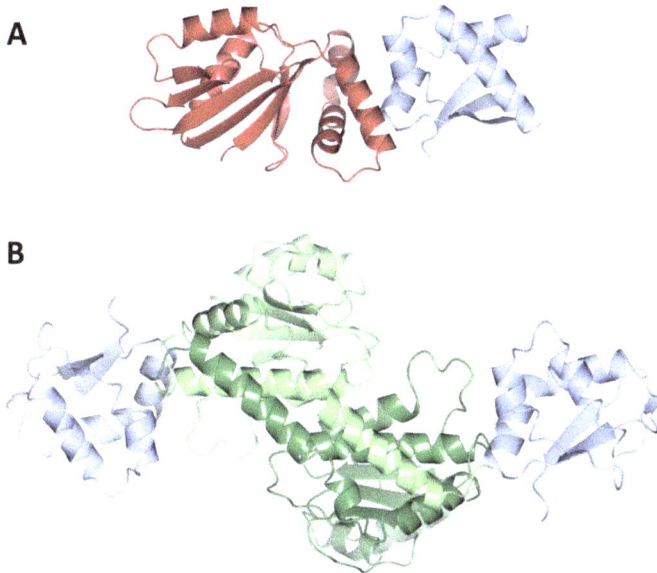

Figure 10. (**A**) SirA–DnaA-DomainI structure (PDB code: 4TPS). SirA is shown in crimson, domain I of DnaA in blue; (**B**) HobA–DnaA-DomainI structure (PDB code: 2WP0). HobA is shown in green, domain I of DnaA in blue. HobA and SirA interact on equivalent surfaces of DnaA, despite exerting different regulatory effects.

Recently, SirA was shown to facilitate chromosome segregation during sporulation, independent of its role in DNA replication regulation [145]. Newly synthesized bacterial origins are localized to a cell pole (or future pole) with high fidelity. In sporulating cells, *oriC* must be segregated or 'captured' at the respective poles in the future forespore and mother cell compartments following the onset of DNA replication. In a Δ*sirA* mutant strain, 10% of sporulating cells fail to capture *oriC*. This activity is distinct from the role of SirA in DNA replication regulation, as *sirA* mutants deficient in DNA replication inhibition were able to facilitate normal *oriC* capture. Soj has also been implicated in *oriC* capture during sporulation, with 20% of cells failing to capture *oriC* in a Δ*soj* mutant strain [120,145]. There is no further increase in the failure of cells to capture *oriC* in a Δ*soj*Δ*sirA* double mutant, implying the two proteins are acting in the same pathway. Using a gain of interaction bacterial-two-hybrid screen, a potential interaction site between the C-terminus of SirA and domain III of DnaA was identified. This site overlaps with residues previously identified in the Soj-interaction site on DnaA, suggesting this interaction may facilitate *oriC* capture [145].

A Direct Role for Spo0A~P

The *B. subtilis* replication origin contains a number of Spo0A-boxes which partially overlap with DnaA-boxes [146] (Figure 4A). Indeed, the consensus Spo0A-box sequence 5′-TGTCGAA-3′ is similar to the DnaA-box consensus sequence 5′-TGTGNATAA-3′ [147]. Spo0A~P has been shown to bind these Spo0A-boxes in vitro [147], and sequence changes that alter the resemblance to the Spo0A-box consensus, without affecting that to the DnaA-box consensus, affect Spo0A~P, but not DnaA binding to the origin [146]. The binding of Spo0A~P to *oriC* appears to play a role in chromosome copy number control in a Δ*sda*/Δ*sirA* mutant strain when sporulation is induced by starvation, or in cells induced to sporulate during rapid growth [146]. Δ*sda* and Δ*sirA* strains each show a more significant loss of copy number control than upon mutation of the Spo0A boxes, with the phenotype being more profound for Δ*sda* than Δ*sirA* [146].

4. In E. coli

4.1. Regulatory Inactivation of DnaA (RIDA)

In *E. coli*, the concentration of available 'initiation-active' DnaA–ATP is considered to be a limiting factor in the initiation of DNA replication from *oriC* [148]. Thus regulation of the availability of DnaA–ATP serves to control initiation events. The 'regulatory inactivation of DnaA' is a term given to the process of converting 'active' DnaA–ATP into 'inactive' DnaA–ADP by ATP hydrolysis. Hydrolysis is promoted by the protein Hda, and requires a complex between DnaA, Hda and the DNA-bound polymerase β-clamp, DnaN [149–151]. In this way, the regulation of initiation is coupled to elongation in DNA replication, as ATP hydrolysis becomes activated following the start of DNA synthesis [150].

Hda is homologous to DnaA domain III, with a 48% sequence similarity between their AAA+ ATPase domains [150]. Hda binds DnaN via an N-terminal clamp binding motif of sequence QL[SP]LPL [152], whilst Hda:DnaA interactions are mediated via their respective ATPase domains [153]. Strains carrying *hda* deletions, and inactivated Hda mutants or DnaA mutants unable to hydrolyse ATP, exhibit overinitiation of DNA replication and growth inhibition [48,148,149,154]. Hda-mediated hydrolysis of DnaA–ATP to DnaA–ADP requires ADP-bound Hda [155] which is monomeric. In contrast, apo-Hda appears to form homodimers and larger multimers [155] implying that Hda's oligomerisation state plays a role in its ability to promote DnaA–ATP hydrolysis [155]. A crystal structure of *Shewanella amazonensis* Hda bound to the nucleotide CDP also revealed a dimer, however because the DnaN binding motif was buried, it was assumed to represent an inactive conformation of the protein [156].

Mutations in both Hda and DnaA suggest that the proteins interact via their respective AAA+ domains, with the arginine finger residue of Hda playing an important role in DnaA–ATP hydrolysis [153,157]. Models of the DnaA–Hda interaction suggest that it may be similar to that formed between molecules of DnaA [156,157]. Hda's interaction with the β-clamp is important for Hda–DnaA binding, suggesting that the β-clamp alters the conformation of the Hda–DnaA interaction to promote ATP hydrolysis [157]. Recently interactions of Hda with DnaA domains I and IV have also been shown to be important for RIDA as mutations at specific DnaA domain I and IV residues lead to higher cellular concentrations of DnaA–ATP than seen in wildtype cells [77,158]. The Hda–DnaA model suggests domain IV makes contacts with Hda at a nucleotide interaction surface towards the C-terminus of Hda [158]. It has therefore been proposed that DnaA domain I interacts with the N-terminal portion of Hda's ATPase domain, to stabilize the DnaA–Hda interaction from both sides [77] (Figure 11). Further studies of the interactions between Hda, DnaA and the β-clamp are required to fully elucidate molecular mechanism of Hda action.

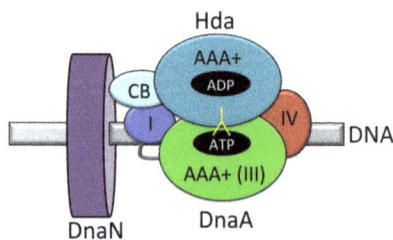

Figure 11. Schematic representation of the interaction of Hda with DnaA and the β-clamp. Hda-ADP (light and dark cyan) makes contacts with domain I (blue) of DnaA–ATP principally through its clamp binding domain (CB), and with domains III (green) and IV (red) of DnaA through its AAA+ domain. An arginine finger from Hda (yellow) projects into the ATP-binding pocket of DnaA and facilitates ATP hydrolysis as part of the regulatory inactivation of DnaA (RIDA). The DNA-bound β-clamp is shown in purple. Figure adapted from [77].

The capacity of Hda to interact with DnaA and DnaN is functionally reminiscent of the *B. subtilis* regulator YabA, which also appears to couple the initiation and elongation steps in DNA replication. However, the mechanism of action of the two proteins is quite different. The proteins are structurally and mechanistically distinct. Hda influences DnaA–ATP hydrolysis in *E. coli*, while YabA has no effect on the hydrolysis of DnaA–ATP in *B. subtilis*. The latter instead appears to influence the oligomerisation of DnaA at the origin.

4.2. IHF and Fis

The DNA-bending proteins integration host factor (IHF) and Fis are thought to play important roles in regulating the binding of DnaA at the origin of replication [159]. Specifically, they have been shown to shape the binding of DnaA to two *cis*-acting regulatory sites on the chromosome, *datA* and DARS [160,161] (see Section 4.3 below). Loss of IHF disrupts synchronous DNA replication. Curiously, Fis has been reported to play both inhibitory and stimulatory roles in DNA replication initiation [162–164].

Both proteins bind to *oriC* and act in an antagonistic manner [159,165]. Binding of IHF to a specific site in the *E. coli* replication origin as shown in Figure 4B promotes binding of DnaA at DnaA-boxes within the origin [166], contributing to DnaA oligomer formation. The binding of IHF induces a bend in the DNA, which is proposed to bring the two adjacent DnaA-boxes into closer proximity, facilitating the extension of the helical DnaA oligomer [65]. Fis has been reported to inhibit DNA unwinding at *oriC* by blocking binding of both DnaA and IHF [159]. Increasing concentrations of DnaA were found to relieve Fis inhibition, and IHF was found to redistribute DnaA molecules at *oriC* [159,166].

4.3. DnaA-Box Sequences: datA and DARS

In stark contrast to the clusters of DnaA-box sequences of *B. subtilis*, which do not play a significant role in replication initiation, *E. coli* possesses three loci with DnaA-box motifs that are used to regulate DNA replication initiation.

One such locus, *datA*, ≈1 kb in length and located at 94.7 min on the *E. coli* chromosome [167], contains five DnaA-boxes with high affinity for DnaA. It acts as a negative regulator of DNA replication initiation; deletion of *datA* or mutations of the DnaA-boxes within *datA* causes over-initiation of DNA replication [167–169]. Binding of IHF to *datA* promotes DnaA-binding and is essential for the regulatory action of *datA* [170]. *datA* had been suggested to act as a sink titrating DnaA away from the replication origin [167–170]. A recent study however, has revealed that *datA* promotes the hydrolysis of DnaA–ATP to DnaA–ADP [161], in a manner that is dependent on both IHF binding to *datA* and the DnaA arginine finger residue (Arg285). This implies that *datA* promotes the formation of a nucleoprotein complex, somewhat reminiscent of that formed at *oriC*, and stimulates the hydrolysis of DnaA–ATP at this site [161]. The binding of IHF to *datA* takes place immediately after initiation, providing a mechanism for the timing of *datA* mediated DnaA–ATP hydrolysis [161].

Two other DnaA-binding loci in *E. coli*, termed DARS1 and DARS2 for DnaA reactivating site 1 and 2, respectively, have been implicated in the reactivation of DnaA by promoting nucleotide exchange, generating DnaA–ATP from DnaA–ADP [171]. Located at 17.5 min and 64 min on the chromosome, respectively, deletion of DARS sequences causes inhibition of DNA replication due to a decrease in the cellular DnaA–ATP concentration [171]. DnaA–ADP molecules have been shown to assemble on DARS1 promoting the regeneration of DnaA–ATP [171]. The simultaneous binding of the DNA bending proteins IHF and Fis to DARS2 has been shown to facilitate DnaA–ATP regeneration in vivo [160], providing a mechanism for the timing of DnaA reactivation. The binding of IHF to DARS2 appears to be cell-cycle regulated and independent of DNA replication, whilst the binding of Fis is linked to growth phase: occurring during exponential growth but not stationary phase [160]. The role of Fis at DARS2 is consistent with a report that Fis is required for the stimulation of replication initiation in rapidly growing cells [164].

The chromosomal positioning of *datA*, DARS1 and, particularly, DARS2, relative to *oriC* has been shown to be important for the proper timing of DNA replication initiation. Translocation of these sites perturbs regulation of initiation [172,173]. However, relocation of *datA* and DARS1 perturbs DNA replication initiation only when they are moved in close proximity to the replication terminus or origin, respectively. In both cases, these effects could be attributed to a gene dosage effect, or decreased/increased proximity to DnaA [172]. The translocation of DARS2 however, has a more significant effect, with relocation of the site proximal to the terminus causing both decreased initiation events, and asynchronous replication. This suggests that the chromosomal location of DARS2 is important for regulating DNA replication synchrony [172,173].

4.4. SeqA

SeqA prevents re-initiation of DNA replication immediately after the previous round of replication has been initiated. It binds to hemimethylated GATC sites [174] in *oriC* and this serves to sequester the origin, preventing DnaA oligomer formation and transcription of the *dnaA* gene by blocking of the *dnaA* promoter [175–178]. The *E. coli* replication origin contains 11 GATC sites which are hemimethylated immediately after DNA replication has been initiated, because the newly synthesized strand is yet to be methylated whilst the parental DNA strand is methylated. SeqA bound to the hemimethylated GATC sites at *oriC* recruits further SeqA proteins. The origin is thus sequestered as long as six or more of the GATC sites are hemimethylated [179]. Sequestration of the origin persists for around a third of the cell cycle, after which time a combination of SeqA dissociation and methylation of the adenosine bases in the GATC sites of the newly synthesized strand by Dam methyltransferase relieves sequestration [177,180,181]. Interestingly, SeqA has also been implicated in faithful chromosome segregation [182].

4.5. DiaA (and HobA)

DiaA is a positive regulator of DNA replication initiation in *E. coli* influencing the frequency and timing of the initiation event [56]. It functions by binding to domain I of DnaA and promoting the oligomerisation of DnaA at the origin [82,183,184]. HobA, an orthologue of DiaA in *H. pylori* [185], is an essential regulator of DNA replication in this organism [55]. DiaA and HobA are tetramers. The structure of a HobA–DnaA domain I complex revealed a 4:4 stoichiometry, with each HobA protomer bound to one DnaA domain I [82]. DiaA has been shown to bind an equivalent site on DnaA [76], although HobA and DiaA are not interchangeable in vivo due to differences in their cognate DnaA domain I sequences [183]. Heterologous complexes can be achieved however with hybrid DnaA molecules. Thus DiaA from *E. coli* can interact with a chimaeric protein resulting from fusion of DnaA domain I from *E. coli* and DnaA domains II-IV from *H. pylori*. This confirms that DiaA and HobA are functional homologs, each promoting DnaA binding at the origin, albeit with different dynamics. HobA accelerates DnaA binding, whilst DiaA decreases the DnaA binding rate [183]. Structural and mutational studies with HobA have led to the suggestion that the tetramers function as molecular scaffolds which promote the formation of DnaA oligomers at *oriC*, however direct experimental evidence is still required [186]. DiaA may play a role in regulating the timing of helicase loading in *E. coli* as both proteins appear to bind to an overlapping site on DnaA, and DiaA has been shown to inhibit helicase loading in vitro [76].

4.6. Lysine Acetylation of DnaA

A recently discovered mechanism for controlling DNA replication initiation in *E. coli* involves the reversible acetylation of lysines within DnaA. Acetylation sites were identified on 13 lysines within natively expressed DnaA, including a key lysine (Lys178) required for the binding of ATP [187]. The acetylation of this residue was growth phase dependent, with peak levels observed in stationary phase. Mutation of Lys178 to Gln or Arg prevented ATP-binding to DnaA, suggesting acetylation of Lys178 would have a similar effect in inactivating the initiator. It is attractive to consider that similar, as yet unidentified, post-translational modifications may exist in other species providing an elegant mechanism for coupling DNA replication initiation to growth phase [187].

5. Regulatory Mechanisms for DNA Replication Regulation: *B. subtilis* vs. *E. coli*

In recent years, opposing themes have emerged in the regulation of DNA replication initiation in *E. coli* and *B. subtilis* (Figure 12). It has long been recognized that the cellular DnaA–ATP concentration plays an important role in the regulation of *E. coli* replication initiation [1,4,188]. Many of the regulatory mechanisms in *E. coli* have been found to influence the adenosine nucleotide bound state of DnaA. Hda, along with the DNA polymerase clamp, DnaN, acts to promote the hydrolysis of DnaA–ATP after initiation [150]. Meanwhile, the *datA*, DARS1 and DARS2 loci influence available DnaA–ATP levels by promoting ATP-hydrolysis (*datA*) [161] or nucleotide exchange from ADP to ATP (DARS) [171]. The DNA-binding proteins Fis and IHF influence DnaA binding at these sites so as to control the cellular DnaA–ATP concentration [159,160]. Finally, lysine acetylation of DnaA coordinated with growth cycle is also believed to affect ATP-binding to DnaA and thus inhibit initiation of DNA replication in later growth phases [187].

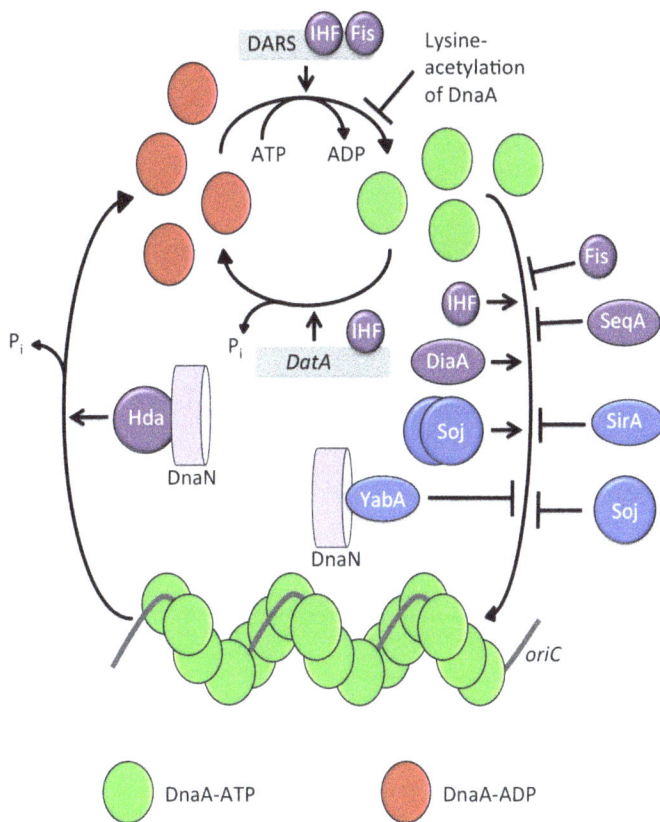

Figure 12. Schematic representation of the mechanisms regulating DNA replication initiation in *E. coli* and *B. subtilis*. Key regulators in *E. coli* influence the adenosine nucleotide bound state of DnaA, whilst those in *B. subtilis* influence the binding of DnaA–ATP to *oriC*. For *E. coli*, the protein regulators are shown in purple, and the DNA binding sites, DatA and DARS, are shown in grey. For *B. subtilis*, protein regulators are shown in blue. Pointed arrows indicate positive effects upon a process, and blunt ended arrows indicate inhibitory effects.

In *B. subtilis*, by contrast, no regulator has been identified which affects the conversion of DnaA–ATP to DnaA–ADP. Instead, replication regulators in *B. subtilis* appear to act directly on the binding of DnaA to *oriC*. YabA is able to inhibit DnaA oligomer formation in vitro [112] and to affect the cooperativity of DnaA–ATP binding at *oriC* [73,122]. Monomeric Soj appears directly to inhibit DnaA oligomer formation on DNA, whilst dimeric Soj seems to be able to promote this oligomerisation [72]. Despite the fact that Soj and YabA interact with the ATPase domain of DnaA, neither protein has an effect on ATP hydrolysis in DnaA–ATP [72,112]. Instead both appear to target the DnaA oligomerisation determinants residing in domain III. Furthermore, DnaA-box clusters with significant roles in the regulation of DNA replication initiation have not been identified in *B. subtilis* [123], in marked contrast to *E. coli*. Together this evidence suggests that the primary mechanisms of DNA replication control are different in *B. subtilis* and *E. coli*. The initiation regulators determine the ligation status (ATP versus ADP) of DnaA in *E. coli* while in *B. subtilis* they act to control the downstream event of DnaA oligomerisation at *oriC*.

Despite this, a number of parallels can be drawn between the regulatory mechanisms in the two species. Both organisms utilize a major regulator during vegetative growth which interacts with both DnaA and DnaN; YabA in *B. subtilis* [110] and Hda in *E. coli* [152]. This may provide the respective species with a mechanism for appropriately timing initiation of replication, since DnaN is a key component of the DNA elongation complex. Both organisms have regulators which are implicated in chromosome segregation, SeqA in *E. coli*, and Soj and SirA during growth and sporulation of *B. subtilis*, respectively. Both organisms appear to utilize a method of origin sequestration to prevent DnaA binding: in *E. coli*, SeqA binds to newly replicated origins, and in *B. subtilis* Spo0A~P is able to bind to the origin, playing an albeit more modest role in inhibiting DNA replication. Furthermore, both organisms have evolved a regulator which targets a structurally equivalent location on DnaA domain I—the sporulation inhibitor of replication in *B. subtilis*, SirA, and the promoter of DNA replication initiation in *E. coli*, DiaA. Thus, these may represent common themes of replication regulation across bacterial species.

Acknowledgments: K.H.J. was the recipient of a DTA Studentship awarded by the Biotechnology and Biological Sciences Research Council, UK. Work on DNA replication proteins at York was supported by the European Integrated Project, BaSysBio LSHG-CT-2006-037469.

Author Contributions: K.H.J. reviewed the literature and prepared the first draft of this review as part of studies towards her PhD which was supervised by A.J.W.

Conflicts of Interest: The authors declare no conflict of interest.

References

1. Kaguni, J.M. Replication initiation at the *Escherichia coli* chromosomal origin. *Curr. Opin. Chem. Biol.* **2011**, *15*, 606–613. [CrossRef] [PubMed]
2. Katayama, T.; Ozaki, S.; Keyamura, K.; Fujimitsu, K. Regulation of the replication cycle: Conserved and diverse regulatory systems for DnaA and *oriC*. *Nat. Rev. Microbiol.* **2010**, *8*, 163–170. [CrossRef] [PubMed]
3. Scholefield, G.; Veening, J.-W.; Murray, H. DnaA and ORC: More than DNA replication initiators. *Trends Cell Biol.* **2011**, *21*, 188–194. [CrossRef] [PubMed]
4. Mott, M.L.; Berger, J.M. DNA replication initiation: Mechanisms and regulation in bacteria. *Nat. Rev. Microbiol.* **2007**, *5*, 343–354. [CrossRef] [PubMed]
5. Leonard, A.C.; Grimwade, J.E. Regulation of DnaA assembly and activity: Taking directions from the genome. *Annu. Rev. Microbiol.* **2011**, *65*, 19–35. [CrossRef] [PubMed]
6. Briggs, G.S.; Smits, W.K.; Soultanas, P. Chromosomal replication initiation machinery of low-G+C-content Firmicutes. *J. Bacteriol.* **2012**, *194*, 5162–5170. [CrossRef] [PubMed]
7. Zorman, S.; Seitz, H.; Sclavi, B.; Strick, T.R. Topological characterization of the DnaA-*oriC* complex using single-molecule nanomanipuation. *Nucleic Acids Res.* **2012**, *40*, 7375–7383. [CrossRef] [PubMed]
8. Erzberger, J.P.; Mott, M.L.; Berger, J.M. Structural basis for ATP-dependent DnaA assembly and replication-origin remodeling. *Nat. Struct. Mol. Biol.* **2006**, *13*, 676–683. [CrossRef] [PubMed]
9. Fuller, R.S.; Funnell, B.E.; Kornberg, A. The dnaA protein complex with the *E. coli* chromosomal replication origin (*oriC*) and other DNA sites. *Cell* **1984**, *38*, 889–900. [CrossRef]

10. Kowalski, D.; Eddy, M.J. The DNA unwinding element: A novel, *cis*-acting component that facilitates opening of the *Escherichia coli* replication origin. *EMBO J.* **1989**, *8*, 4335–4344. [PubMed]

11. Bramhill, D.; Kornberg, A. Duplex opening by dnaA protein at novel sequences in initiation of replication at the origin of the *E. coli* chromosome. *Cell* **1988**, *52*, 743–755. [CrossRef]

12. Marszalek, J.; Kaguni, J.M. DnaA protein directs the binding of DnaB protein in initiation of DNA replication in *Escherichia coli*. *J. Biol. Chem.* **1994**, *269*, 4883–4890. [PubMed]

13. Koboris, J.A.; Kornberg, A. *Escherichia coli dnaC* Gene Product. *Biol. Chem.* **1982**, *257*, 13770–13775.

14. Fang, L.; Davey, M.J.; O'Donnell, M. Replisome assembly at *oriC*, the replication origin of *E. coli*, reveals an explanation for initiation sites outside an origin. *Mol. Cell* **1999**, *4*, 541–553. [CrossRef]

15. Bruand, C.; Ehrlich, S.D.; Jannière, L. Primosome assembly site in *Bacillus subtilis*. *EMBO J.* **1995**, *14*, 2642–2650. [PubMed]

16. Zhang, W.; Carneiro, M.J.V.M.; Turner, I.J.; Allen, S.; Roberts, C.J.; Soultanas, P. The *Bacillus subtilis* DnaD and DnaB proteins exhibit different DNA remodelling activities. *J. Mol. Biol.* **2005**, *351*, 66–75. [CrossRef] [PubMed]

17. Smits, W.K.; Goranov, A.I.; Grossman, A.D. Ordered association of helicase loader proteins with the *Bacillus subtilis* origin of replication in vivo. *Mol. Microbiol.* **2010**, *75*, 452–461. [CrossRef] [PubMed]

18. Bailey, S.; Eliason, W.K.; Steitz, T.A. Structure of hexameric DnaB helicase and its complex with a domain of DnaG primase. *Science* **2007**, *318*, 459–463. [CrossRef] [PubMed]

19. Reyes-Lamothe, R.; Sherratt, D.J.; Leake, M.C. Stoichiometry and architecture of active DNA replication machinery in *Escherichia coli*. *Science* **2010**, *328*, 498–501. [CrossRef] [PubMed]

20. Robinson, A.; Causer, R.J.; Dixon, N.E. Architecture and Conservation of the Bacterial DNA Replication Machinery, an Underexploited Drug Target. *Curr. Drug Targets* **2012**, *13*, 352–372. [CrossRef] [PubMed]

21. Beattie, T.R.; Reyes-Lamothe, R. A Replisome's journey through the bacterial chromosome. *Front. Microbiol.* **2015**, *6*, 1–12. [CrossRef] [PubMed]

22. Voet, D.; Voet, J.G. DNA Replication, Repair, and Recombination. In *Biochemistry*; Wiley: Hoboken, NJ, USA, 2011; pp. 1171–1259.

23. Corn, J.E.; Pelton, J.G.; Berger, J.M. Identification of a DNA primase template tracking site redefines the geometry of primer synthesis. *Nat. Struct. Mol. Biol.* **2008**, *15*, 163–169. [CrossRef] [PubMed]

24. Corn, J.E.; Berger, J.M. Regulation of bacterial priming and daughter strand synthesis through helicase-primase interactions. *Nucleic Acids Res.* **2006**, *34*, 4082–4088. [CrossRef] [PubMed]

25. Corn, J.E.; Pease, P.J.; Hura, G.L.; Berger, J.M. Crosstalk between primase subunits can act to regulate primer synthesis in *trans*. *Mol. Cell* **2005**, *20*, 391–401. [CrossRef] [PubMed]

26. Dervyn, E.; Suski, C.; Daniel, R.; Bruand, C.; Chapuis, J.; Errington, J.; Jannière, L.; Ehrlich, S.D. Two essential DNA polymerases at the bacterial replication fork. *Science* **2001**, *294*, 1716–1719. [CrossRef] [PubMed]

27. Sanders, G.M.; Dallmann, H.G.; McHenry, C.S. Reconstitution of the *B. subtilis* Replisome with 13 Proteins Including Two Distinct Replicases. *Mol. Cell* **2010**, *37*, 273–281. [CrossRef] [PubMed]

28. Hill, T.M.; Henson, J.M.; Kuempel, P.L. The terminus region of the *Escherichia coli* chromosome contains two separate loci that exhibit polar inhibition of replication. *Proc. Natl. Acad. Sci. USA* **1987**, *84*, 1754–1758. [CrossRef] [PubMed]

29. Hidaka, M.; Kobayashi, T.; Takenaka, S.; Takeya, H.; Horiuchi, T. Purification of a DNA replication terminus (*ter*) site-binding protein in *Escherichia coli* and identification of the structural gene. *J. Biol. Chem.* **1989**, *264*, 21031–21037. [PubMed]

30. Neylon, C.; Kralicek, A.V.; Hill, T.M.; Dixon, N.E. Replication Termination in *Escherichia coli*: Structure and Antihelicase Activity of the Tus-*Ter* Complex. *Microbiol. Mol. Biol. Rev.* **2005**, *69*, 501–526. [CrossRef] [PubMed]

31. Kamada, K.; Horiuchi, T.; Ohsumi, K.; Shimamoto, N.; Morikawa, K. Structure of a replication-terminator protein complexed with DNA. *Nature* **1996**, *383*, 598–603. [CrossRef] [PubMed]

32. Mulcair, M.D.; Schaeffer, P.M.; Oakley, A.J.; Cross, H.F.; Neylon, C.; Hill, T.M.; Dixon, N.E. A Molecular Mousetrap Determines Polarity of Termination of DNA Replication in *E. coli*. *Cell* **2006**, *125*, 1309–1319. [CrossRef] [PubMed]

33. Berghuis, B.A.; Dulin, D.; Xu, Z.-Q.; van Laar, T.; Cross, B.; Janissen, R.; Jergic, S.; Dixon, N.E.; Depken, M.; Dekker, N.H. Strand separation establishes a sustained lock at the Tus–Ter replication fork barrier. *Nat. Chem. Biol.* **2015**, *11*, 579–585. [CrossRef] [PubMed]

34. McNicholas, S.; Potterton, E.; Wilson, K.S.; Noble, M.E.M. Presenting your structures: the CCP4mg molecular-graphics software. *Acta. Cryst. D Biol. Crystallogr.* **2011**, *67*, 386–394.
35. Hill, T.M. Arrest of bacterial DNA replication. *Annu. Rev. Microbiol.* **1992**, *46*, 603–633. [CrossRef] [PubMed]
36. Lewis, P.J.; Ralston, G.B.; Christopherson, R.I.; Wake, R.G. Identification of the replication terminator protein binding sites in the terminus region of the *Bacillus subtilis* chromosome and stoichiometry of the binding. *J. Mol. Biol.* **1990**, *214*, 73–84. [CrossRef]
37. Vivian, J.P.; Porter, C.J.; Wilce, J.A.; Wilce, M.C.J. An Asymmetric Structure of the *Bacillus subtilis* Replication Terminator Protein in Complex with DNA. *J. Mol. Biol.* **2007**, *370*, 481–491. [CrossRef] [PubMed]
38. Langley, D.B.; Smith, M.T.; Lewis, P.J.; Wake, R.G. Protein-nucleoside contacts in the interaction between the replication terminator protein of *Bacillus subtilis* and the DNA terminator. *Mol. Microbiol.* **1993**, *10*, 771–779. [CrossRef] [PubMed]
39. Wolanski, M.; Donczew, R.; Zawilak-Pawlik, A.; Zakrzewska-Czerwinska, J. oriC-encoded instructions for the initiation of bacterial chromosome replication. *Front. Microbiol.* **2015**, *5*, 1–14.
40. Leonard, A.C.; Méchali, M. DNA replication origins. *Cold Spring Harb. Perspect. Biol.* **2013**, *5*, a010116. [CrossRef] [PubMed]
41. Ogasawara, N.; Yoshikawa, H. Genes and their organization in the replication origin region of the bacterial chromosome. *Mol. Microbiol.* **1992**, *6*, 629–634. [CrossRef] [PubMed]
42. Moriya, S.; Atlung, T.; Hansen, F.G.; Yoshikawa, H.; Ogasawara, N. Cloning of an autonomously replicating sequence (ars) from the *Bacillus subtilis* chromosome. *Mol. Microbiol.* **1992**, *6*, 309–315. [CrossRef] [PubMed]
43. Donczew, R.; Weigel, C.; Lurz, R.; Zakrzewska-Czerwińska, J.; Zawilak-Pawlik, A. *Helicobacter pylori* oriC-the first bipartite origin of chromosome replication in Gram-negative bacteria. *Nucleic Acids Res.* **2012**, *40*, 9647–9660. [CrossRef] [PubMed]
44. Krause, M.; Rückert, B.; Lurz, R.; Messer, W. Complexes at the replication origin of *Bacillus subtilis* with homologous and heterologous DnaA protein. *J. Mol. Biol.* **1997**, *274*, 365–380. [CrossRef] [PubMed]
45. Messer, W.; Blaesing, F.; Majka, J.; Nardmann, J.; Schaper, S.; Schmidt, A.; Seitz, H.; Speck, C.; Tüngler, D.; Wegrzyn, G.; et al. Functional domains of DnaA proteins. *Biochimie* **1999**, *81*, 819–825. [CrossRef]
46. Sutton, M.D.; Kaguni, J.M. The *Escherichia coli dnaA* gene: Four functional domains. *J. Mol. Biol.* **1997**, *274*, 546–561. [CrossRef] [PubMed]
47. Fukuoka, T.; Moriya, S.; Yoshikawa, H.; Ogasawara, N. Purification and characterization of an initiation protein for chromosomal replication, DnaA, in *Bacillus subtilis*. *J. Biochem.* **1990**, *107*, 732–739. [PubMed]
48. Nishida, S. A Nucleotide Switch in the *Escherichia coli* DnaA Protein Initiates Chromosomal Replication. *J. Biol. Chem.* **2002**, *277*, 14986–14995. [CrossRef] [PubMed]
49. Sekimizu, K.; Bramhill, D.; Kornberg, A. ATP activates dnaA protein in initiating replication of plasmids bearing the origin of the *E. coli* chromosome. *Cell* **1987**, *50*, 259–265. [CrossRef]
50. Roth, A.; Messer, W. The DNA binding domain of the initiator protein DnaA. *EMBO J.* **1995**, *14*, 2106–2111. [PubMed]
51. Fujikawa, N.; Kurumizaka, H.; Nureki, O.; Terada, T. Structural basis of replication origin recognition by the DnaA protein. *Nucleic Acids Res.* **2003**, *31*, 2077–2086. [CrossRef] [PubMed]
52. Kaguni, J.M. DnaA: Controlling the initiation of bacterial DNA replication and more. *Annu. Rev. Microbiol.* **2006**, *60*, 351–375. [CrossRef] [PubMed]
53. Nozaki, S.; Ogawa, T. Determination of the minimum domain II size of *Escherichia coli* DnaA protein essential for cell viability. *Microbiology* **2008**, *154*, 3379–3384. [CrossRef] [PubMed]
54. Rahn-Lee, L.; Merrikh, H.; Grossman, A.D.; Losick, R. The sporulation protein SirA inhibits the binding of DnaA to the origin of replication by contacting a patch of clustered amino acids. *J. Bacteriol.* **2011**, *193*, 1302–1307. [CrossRef] [PubMed]
55. Zawilak-Pawlik, A.; Kois, A.; Stingl, K.; Boneca, I.G.; Skrobuk, P.; Piotr, J.; Lurz, R.; Zakrzewska-Czerwińska, J.; Labigne, A. HobA—A novel protein involved in initiation of chromosomal replication in *Helicobacter pylori*. *Mol. Microbiol.* **2007**, *65*, 979–994. [CrossRef] [PubMed]
56. Ishida, T.; Akimitsu, N.; Kashioka, T.; Hatano, M.; Kubota, T.; Ogata, Y.; Sekimizu, K.; Katayama, T. DiaA, a novel DnaA-binding protein, ensures the timely initiation of *Escherichia coli* chromosome replication. *J. Biol. Chem.* **2004**, *279*, 45546–45555. [CrossRef] [PubMed]
57. Seitz, H.; Weigel, C.; Messer, W. The interaction domains of the DnaA and DnaB replication proteins of *Escherichia coli*. *Mol. Microbiol.* **2000**, *37*, 1270–1279. [CrossRef] [PubMed]

58. Margulies, C.; Kaguni, J.M. Ordered and sequential binding of DnaA protein to *oriC*, the chromosomal origin of *Escherichia coli*. *J. Biol. Chem.* **1996**, *271*, 17035–17040. [CrossRef] [PubMed]

59. Langer, U.; Richter, S.; Roth, A.; Weigel, C.; Messer, W. A comprehensive set of DnaA-box mutations in the replication origin, *oriC*, of *Escherichia coli*. *Mol. Microbiol.* **1996**, *21*, 301–311. [CrossRef] [PubMed]

60. Schaper, S.; Messer, W. Interaction of the initiator protein DnaA of *Escherichia coli* with its DNA target. *J. Biol. Chem.* **1995**, *270*, 17622–17626. [CrossRef] [PubMed]

61. Blaesing, F.; Weigel, C.; Welzeck, M.; Messer, W. Analysis of the DNA-binding domain of *Escherichia coli* DnaA protein. *Mol. Microbiol.* **2000**, *36*, 557–569. [CrossRef] [PubMed]

62. Miller, D.T.; Grimwade, J.E.; Betteridge, T.; Rozgaja, T.; Torgue, J.J.-C.; Leonard, A.C. Bacterial origin recognition complexes direct assembly of higher-order DnaA oligomeric structures. *Proc. Natl. Acad. Sci. USA* **2009**, *106*, 18479–18484. [CrossRef] [PubMed]

63. Rozgaja, T.A.; Grimwade, J.E.; Iqbal, M.; Czerwonka, C.; Vora, M.; Leonard, A.C. Two oppositely oriented arrays of low affinity recognition sites in *oriC* guide progressive binding of DnaA during *E. coli* pre-RC assembly. *Mol. Microbiol.* **2012**, *82*, 475–488. [CrossRef] [PubMed]

64. McGarry, K.C.; Ryan, V.T.; Grimwade, J.E.; Leonard, A.C. Two discriminatory binding sites in the *Escherichia coli* replication origin are required for DNA strand opening by initiator DnaA-ATP. *Proc. Natl. Acad. Sci. USA* **2004**, *101*, 2811–2816. [CrossRef] [PubMed]

65. Kaur, G.; Vora, M.P.; Czerwonka, C.A.; Rozgaja, T.A.; Grimwade, J.E.; Leonard, A.C. Building the bacterial orisome: High affinity DnaA recognition plays a role in setting the conformation of *oriC* DNA. *Mol. Microbiol.* **2014**, *91*, 1148–1163. [CrossRef] [PubMed]

66. Erzberger, J.P.; Pirruccello, M.M.; Berger, J.M. The structure of bacterial DnaA: Implications for general mechanisms underlying DNA replication initiation. *EMBO J.* **2002**, *21*, 4763–4773. [CrossRef] [PubMed]

67. Ogura, T.; Whiteheart, S.W.; Wilkinson, A.J. Conserved arginine residues implicated in ATP hydrolysis, nucleotide-sensing, and inter-subunit interactions in AAA and AAA+ ATPases. *J. Struct. Biol.* **2004**, *146*, 106–112. [CrossRef] [PubMed]

68. Duderstadt, K.E.; Mott, M.L.; Crisona, N.J.; Chuang, K.; Yang, H.; Berger, J.M. Origin remodeling and opening in bacteria rely on distinct assembly states of the DnaA initiator. *J. Biol. Chem.* **2010**, *285*, 28229–28239. [CrossRef] [PubMed]

69. Duderstadt, K.E.; Chuang, K.; Berger, J.M. DNA stretching by bacterial initiators promotes replication origin opening. *Nature* **2011**, *478*, 209–213. [CrossRef] [PubMed]

70. Speck, C.; Messer, W. Mechanism of origin unwinding: Sequential binding of DnaA to double- and single-stranded DNA. *EMBO J.* **2001**, *20*, 1469–1476. [CrossRef] [PubMed]

71. Richardson, T.T.; Harran, O.; Murray, H. The bacterial DnaA-trio replication origin element specifies single-stranded DNA initiator binding. *Nature* **2016**, *534*, 412–416. [CrossRef] [PubMed]

72. Scholefield, G.; Errington, J.; Murray, H. Soj/ParA stalls DNA replication by inhibiting helix formation of the initiator protein DnaA. *EMBO J.* **2012**, *31*, 1–14. [CrossRef] [PubMed]

73. Merrikh, H.; Grossman, A.D. Control of the replication initiator DnaA by an anti-cooperativity factor. *Mol. Microbiol.* **2011**, *82*, 434–446. [CrossRef] [PubMed]

74. Rotoli, S.M.; Biswas-Fiss, E.; Biswas, S.B. Quantitative analysis of the mechanism of DNA binding by *Bacillus* DnaA protein. *Biochimie* **2012**, *94*, 2764–2775. [CrossRef] [PubMed]

75. Sutton, M.D.; Carr, K.M.; Vicente, M.; Kaguni, J.M. *Escherichia coli* DnaA Protein. *J. Biol. Chem.* **1998**, *273*, 34255–34262. [CrossRef] [PubMed]

76. Keyamura, K.; Abe, Y.; Higashi, M.; Ueda, T.; Katayama, T. DiaA dynamics are coupled with changes in initial origin complexes leading to helicase loading. *J. Biol. Chem.* **2009**, *284*, 25038–25050. [CrossRef] [PubMed]

77. Su'etsugu, M.; Harada, Y.; Keyamura, K.; Matsunaga, C.; Kasho, K.; Abe, Y.; Ueda, T.; Katayama, T. The DnaA N-terminal domain interacts with Hda to facilitate replicase clamp-mediated inactivation of DnaA. *Environ. Microbiol.* **2013**, *15*, 3183–3195. [CrossRef] [PubMed]

78. Carr, K.M.; Kaguni, J.M. Stoichiometry of DnaA and DnaB protein in initiation at the *Escherichia coli* chromosomal origin. *J. Biol. Chem.* **2001**, *276*, 44919–44925. [CrossRef] [PubMed]

79. Weigel, C.; Seitz, H. Strand-specific loading of DnaB helicase by DnaA to a substrate mimicking unwound *oriC*. *Mol. Microbiol.* **2002**, *46*, 1149–1156. [CrossRef] [PubMed]

80. Felczak, M.M.; Simmons, L.A.; Kaguni, J.M. An essential tryptophan of *Escherichia coli* DnaA protein functions in oligomerization at the *E. coli* replication origin. *J. Biol. Chem.* **2005**, *280*, 24627–24633. [CrossRef] [PubMed]

81. Simmons, L.A.; Felczak, M.; Kaguni, J.M. DnaA Protein of *Escherichia coli*: Oligomerization at the *E. coli* chromosomal origin is required for initiation and involves specific N-terminal amino acids. *Mol. Microbiol.* **2003**, *49*, 849–858. [CrossRef] [PubMed]

82. Natrajan, G.; Noirot-Gros, M.-F.; Zawilak-Pawlik, A.; Kapp, U.; Terradot, L. The structure of a DnaA/HobA complex from *Helicobacter pylori* provides insight into regulation of DNA replication in bacteria. *Proc. Natl. Acad. Sci. USA* **2009**, *106*, 21115–21120. [CrossRef] [PubMed]

83. Lowery, T.J.; Pelton, J.G.; Chandonia, J.-M.; Kim, R.; Yokota, H.; Wemmer, D.E. NMR structure of the N-terminal domain of the replication initiator protein DnaA. *J. Struct. Funct. Genom.* **2007**, *8*, 11–17. [CrossRef] [PubMed]

84. Abe, Y.; Jo, T.; Matsuda, Y.; Matsunaga, C.; Katayama, T.; Ueda, T. Structure and function of DnaA N-terminal domains: Specific sites and mechanisms in inter-DnaA interaction and in DnaB helicase loading on *oriC*. *J. Biol. Chem.* **2007**, *282*, 17816–17827. [CrossRef] [PubMed]

85. Molt, K.L.; Sutera, V.A.; Moore, K.K.; Lovett, S.T. A role for nonessential domain II of initiator protein, DnaA, in replication control. *Genetics* **2009**, *183*, 39–49. [CrossRef] [PubMed]

86. Arias-Palomo, E.; O'Shea, V.L.; Hood, I.V.; Berger, J.M. The bacterial DnaC helicase loader is a DnaB ring breaker. *Cell* **2013**, *153*, 438–448. [CrossRef] [PubMed]

87. Zhang, W.; Marszalek, J.; Zhang, W.; Hupp, T.R.; Margulies, C.; Carr, K.M.; Cherry, S.; Kaguni, J.M. Domains of DnaA Protein Involved in Interaction with DnaB Protein, and in Unwinding the *Escherichia coli* Chromosomal Origin. *J. Biol. Chem.* **1996**, *271*, 18535–18542.

88. Mott, M.L.; Erzberger, J.P.; Coons, M.M.; Berger, J.M. Structural synergy and molecular crosstalk between bacterial helicase loaders and replication initiators. *Cell* **2008**, *135*, 623–634. [CrossRef] [PubMed]

89. Makowska-Grzyska, M.; Kaguni, J.M. Primase Directs the Release of DnaC from DnaB. *Mol. Cell* **2010**, *37*, 90–101. [CrossRef] [PubMed]

90. Bell, S.P.; Kaguni, J.M. Helicase loading at chromosomal origins of replication. *Cold Spring Harb. Perspect. Biol.* **2013**, *5*, 1–20. [CrossRef] [PubMed]

91. Soultanas, P. Loading mechanisms of ring helicases at replication origins. *Mol. Microbiol.* **2012**, *84*, 6–16. [CrossRef] [PubMed]

92. Velten, M.; McGovern, S.; Marsin, S.; Ehrlich, S.D.; Noirot, P.; Polard, P. A two-protein strategy for the functional loading of a cellular replicative DNA helicase. *Mol. Cell* **2003**, *11*, 1009–1020. [CrossRef]

93. Ioannou, C.; Schaeffer, P.M.; Dixon, N.E.; Soultanas, P. Helicase binding to DnaI exposes a cryptic DNA-binding site during helicase loading in *Bacillus subtilis*. *Nucleic Acids Res.* **2006**, *34*, 5247–5258. [CrossRef] [PubMed]

94. Liu, B.; Eliason, W.K.; Steitz, T.A. Structure of a helicase-helicase loader complex reveals insights into the mechanism of bacterial primosome assembly. *Nat. Commun.* **2013**, *4*, 1–8. [CrossRef] [PubMed]

95. Hoshino, T.; McKenzie, T.; Schmidt, S.; Tanaka, T.; Sueoka, N. Nucleotide sequence of *Bacillus subtilis* dnaB: A gene essential for DNA replication initiation and membrane attachment. *Proc. Natl. Acad. Sci. USA* **1987**, *84*, 653–657. [CrossRef] [PubMed]

96. Marston, F.Y.; Grainger, W.H.; Smits, W.K.; Hopcroft, N.H.; Green, M.; Hounslow, A.M.; Grossman, A.D.; Craven, C.J.; Soultanas, P. When simple sequence comparison fails: The cryptic case of the shared domains of the bacterial replication initiation proteins DnaB and DnaD. *Nucleic Acids Res.* **2010**, *38*, 6930–6942. [CrossRef] [PubMed]

97. Bruand, C.; Farache, M.; McGovern, S.; Ehrlich, S.D.; Polard, P. DnaB, DnaD and DnaI proteins are components of the *Bacillus subtilis* replication restart primosome. *Mol. Microbiol.* **2001**, *42*, 245–255. [CrossRef] [PubMed]

98. Ishigo-oka, D.; Ogasawara, N.; Moriya, S. DnaD Protein of *Bacillus subtilis* Interacts with DnaA, the Initiator Protein of Replication. *J. Bacteriol.* **2001**, *183*, 1–4. [CrossRef] [PubMed]

99. Zhang, W.; Allen, S.; Roberts, C.J.; Soultanas, P. The *Bacillus subtilis* primosomal protein DnaD untwists supercoiled DNA. *J. Bacteriol.* **2006**, *188*, 5487–5493. [CrossRef] [PubMed]

100. Zhang, W.; Machón, C.; Orta, A.; Phillips, N.; Roberts, C.J.; Allen, S.; Soultanas, P. Single-molecule atomic force spectroscopy reveals that DnaD forms scaffolds and enhances duplex melting. *J. Mol. Biol.* **2008**, *377*, 706–714. [CrossRef] [PubMed]

101. Schneider, S.; Zhang, W.; Soultanas, P.; Paoli, M. Structure of the N-Terminal Oligomerization Domain of DnaD Reveals a Unique Tetramerization Motif and Provides Insights into Scaffold Formation. *J. Mol. Biol.* **2008**, *376*, 1237–1250. [CrossRef] [PubMed]

102. Carneiro, M.J.V.M.; Zhang, W.; Ioannou, C.; Scott, D.J.; Allen, S.; Roberts, C.J.; Soultanas, P. The DNA-remodelling activity of DnaD is the sum of oligomerization and DNA-binding activities on separate domains. *Mol. Microbiol.* **2006**, *60*, 917–924. [CrossRef] [PubMed]

103. Bruand, C.; Velten, M.; McGovern, S.; Marsin, S.; Sérèna, C.; Dusko Ehrlich, S.; Polard, P. Functional interplay between the *Bacillus subtilis* DnaD and DnaB proteins essential for initiation and re-initiation of DNA replication. *Mol. Microbiol.* **2005**, *55*, 1138–1150. [CrossRef] [PubMed]

104. Collier, C.; Machón, C.; Briggs, G.S.; Smits, W.K.; Soultanas, P. Untwisting of the DNA helix stimulates the endonuclease activity of *Bacillus subtilis* Nth at AP sites. *Nucleic Acids Res.* **2012**, *40*, 739–750. [CrossRef] [PubMed]

105. Rannou, O.; Le Chatelier, E.; Larson, M.A.; Nouri, H.; Dalmais, B.; Laughton, C.; Jannière, L.; Soultanas, P. Functional interplay of DnaE polymerase, DnaG primase and DnaC helicase within a ternary complex, and primase to polymerase hand-off during lagging strand DNA replication in *Bacillus subtilis*. *Nucleic Acids Res.* **2013**, *41*, 5303–5320. [CrossRef] [PubMed]

106. Rokop, M.E.; Auchtung, J.M.; Grossman, A.D. Control of DNA replication initiation by recruitment of an essential initiation protein to the membrane of *Bacillus subtilis*. *Mol. Microbiol.* **2004**, *52*, 1757–1767. [CrossRef] [PubMed]

107. Grainger, W.H.; Machón, C.; Scott, D.J.; Soultanas, P. DnaB proteolysis in vivo regulates oligomerization and its localization at *oriC* in *Bacillus subtilis*. *Nucleic Acids Res.* **2010**, *38*, 2851–2864. [CrossRef] [PubMed]

108. Hayashi, M.; Ogura, Y.; Harry, E.J.; Ogasawara, N.; Moriya, S. *Bacillus subtilis* YabA is involved in determining the timing and synchrony of replication initiation. *FEMS Microbiol. Lett.* **2005**, *247*, 73–79. [CrossRef] [PubMed]

109. Noirot-Gros, M.-F.; Dervyn, E.; Wu, L.J.; Mervelet, P.; Errington, J.; Ehrlich, S.D.; Noirot, P. An expanded view of bacterial DNA replication. *Proc. Natl. Acad. Sci. USA* **2002**, *99*, 8342–8347. [CrossRef] [PubMed]

110. Noirot-Gros, M.-F.; Velten, M.; Yoshimura, M.; McGovern, S.; Morimoto, T.; Ehrlich, S.D.; Ogasawara, N.; Polard, P.; Noirot, P. Functional dissection of YabA, a negative regulator of DNA replication initiation in *Bacillus subtilis*. *Proc. Natl. Acad. Sci. USA* **2006**, *103*, 2368–2373. [CrossRef] [PubMed]

111. Felicori, L.; Jameson, K.H.; Roblin, P.; Fogg, M.J.; Cherrier, V.; Bazin, A.; Garcia-garcia, T.; Ventroux, M.; Noirot, P.; Wilkinson, A.J.; et al. Tetramerization and interdomain flexibility of the replication initiation controller YabA enables simultaneous binding to multiple partners. *Nucleic Acids Res.* **2016**, *44*, 449–463. [CrossRef] [PubMed]

112. Scholefield, G.; Murray, H. YabA and DnaD inhibit helix assembly of the DNA replication initiation protein DnaA. *Mol. Microbiol.* **2013**, *90*, 147–159. [CrossRef] [PubMed]

113. Cho, E.; Ogasawara, N.; Ishikawa, S. The functional analysis of YabA, which interacts with DnaA and regulates initiation of chromosome replication in *Bacillus subtils*. *Genes Genet. Syst.* **2008**, *83*, 111–125. [CrossRef] [PubMed]

114. Murray, H.; Errington, J. Dynamic control of the DNA replication initiation protein DnaA by Soj/ParA. *Cell* **2008**, *135*, 74–84. [CrossRef] [PubMed]

115. Scholefield, G.; Whiting, R.; Errington, J.; Murray, H. Spo0J regulates the oligomeric state of Soj to trigger its switch from an activator to an inhibitor of DNA replication initiation. *Mol. Microbiol.* **2011**, *79*, 1089–1100. [CrossRef] [PubMed]

116. Leonard, T.A.; Butler, P.J.; Löwe, J. Bacterial chromosome segregation: Structure and DNA binding of the Soj dimer—A conserved biological switch. *EMBO J.* **2005**, *24*, 270–282. [CrossRef] [PubMed]

117. Ringgaard, S.; van Zon, J.; Howard, M.; Gerdes, K. Movement and equipositioning of plasmids by ParA filament disassembly. *Proc. Natl. Acad. Sci. USA* **2009**, *106*, 19369–19374. [CrossRef] [PubMed]

118. Gerdes, K.; Howard, M.; Szardenings, F. Pushing and pulling in prokaryotic DNA segregation. *Cell* **2010**, *141*, 927–942. [CrossRef] [PubMed]

119. Lee, P.S.; Grossman, A.D. The chromosome partitioning proteins Soj (ParA) and Spo0J (ParB) contribute to accurate chromosome partitioning, separation of replicated sister origins, and regulation of replication initiation in *Bacillus subtilis*. *Mol. Microbiol.* **2006**, *60*, 853–869. [CrossRef] [PubMed]

120. Sullivan, N.L.; Marquis, K.A.; Rudner, D.Z. Recruitment of SMC by ParB-parS Organizes the Origin Region and Promotes Efficient Chromosome Segregation. *Cell* **2009**, *137*, 697–707. [CrossRef] [PubMed]

121. Gruber, S.; Errington, J. Recruitment of Condensin to Replication Origin Regions by ParB/SpoOJ Promotes Chromosome Segregation in *B. subtilis*. *Cell* **2009**, *137*, 685–696. [CrossRef] [PubMed]

122. Bonilla, C.Y.; Grossman, A.D. The primosomal protein DnaD inhibits cooperative DNA binding by the replication initiator DnaA in *Bacillus subtilis*. *J. Bacteriol.* **2012**, *194*, 5110–5117. [CrossRef] [PubMed]

123. Okumura, H.; Yoshimura, M.; Ueki, M.; Oshima, T.; Ogasawara, N.; Ishikawa, S. Regulation of chromosomal replication initiation by *oriC*-proximal DnaA-box clusters in *Bacillus subtilis*. *Nucleic Acids Res.* **2012**, *40*, 220–234. [CrossRef] [PubMed]

124. Errington, J. Regulation of endospore formation in *Bacillus subtilis*. *Nat. Rev. Microbiol.* **2003**, *1*, 117–126. [CrossRef] [PubMed]

125. Higgins, D.; Dworkin, J. Recent progress in *Bacillus subtilis* sporulation. *FEMS Microbiol. Rev.* **2012**, *36*, 131–148. [CrossRef] [PubMed]

126. Kay, D.; Warren, S.C. Sporulation in *Bacillus subtilis*. Morphological changes. *Biochem. J.* **1968**, *109*, 819–824. [CrossRef] [PubMed]

127. Jiang, M.; Shao, W.; Perego, M.; Hoch, J.A. Multiple histidine kinases regulate entry into stationary phase and sporulation in *Bacillus subtilis*. *Mol. Microbiol.* **2000**, *38*, 535–542. [CrossRef] [PubMed]

128. Molle, V.; Fujita, M.; Jensen, S.T.; Eichenberger, P.; González-Pastor, J.E.; Liu, J.S.; Losick, R. The SpoOA regulon of *Bacillus subtilis*. *Mol. Microbiol.* **2003**, *50*, 1683–1701. [CrossRef] [PubMed]

129. Veening, J.-W.; Murray, H.; Errington, J. A mechanism for cell cycle regulation of sporulation initiation in *Bacillus subtilis*. *Genes Dev.* **2009**, *23*, 1959–1970. [CrossRef] [PubMed]

130. Mandelstam, J.; Higgs, S.A. Induction of sporulation during synchronized chromosome replication in *Bacillus subtilis*. *J. Bacteriol.* **1974**, *120*, 38–42. [PubMed]

131. Dunn, B.G.; Jeffs, P.; Ma, N.H. The Relationship Between DNA Replication and the Induction of Sporulation in *Bacillus subtilis*. *Microbiology* **1978**, *108*, 189–195. [CrossRef]

132. Narula, J.; Kuchina, A.; Lee, D.D.; Fujita, M.; Süel, G.M.; Igoshin, O.A. Chromosomal Arrangement of Phosphorelay Genes Couples Sporulation and DNA Replication. *Cell* **2015**, *162*, 328–337. [CrossRef] [PubMed]

133. Fujita, M.; González-Pastor, J.E.; Gonza, E.; Losick, R. High- and Low-Threshold Genes in the SpoOA Regulon of *Bacillus subtilis*. *J. Bacteriol.* **2005**, *187*, 1357–1368. [CrossRef] [PubMed]

134. Fujita, M.; Losick, R. Evidence that entry into sporulation in *Bacillus subtilis* is governed by a gradual increase in the level and activity of the master regulator SpoOA. *Genes Dev.* **2005**, *19*, 2236–2244. [CrossRef] [PubMed]

135. Burkholder, W.F.; Kurtser, I.; Grossman, A.D. Replication initiation proteins regulate a developmental checkpoint in *Bacillus subtilis*. *Cell* **2001**, *104*, 269–279. [CrossRef]

136. Bick, M.J.; Lamour, V.; Rajashankar, K.R.; Gordiyenko, Y.; Robinson, C.V.; Darst, S.A. How to Switch Off a Histidine Kinase: Crystal Structure of *Geobacillus stearothermophilus* KinB with the inhibitor Sda. *J. Mol. Biol.* **2009**, *386*, 163–177. [CrossRef] [PubMed]

137. Whitten, A.E.; Jacques, D.A.; Hammouda, B.; Hanley, T.; King, G.F.; Guss, J.M.; Trewhella, J.; Langley, D.B. The Structure of the KinA-Sda Complex Suggests an Allosteric Mechanism of Histidine Kinase Inhibition. *J. Mol. Biol.* **2007**, *368*, 407–420. [CrossRef] [PubMed]

138. Ruvolo, M.V.; Mach, K.E.; Burkholder, W.F. Proteolysis of the replication checkpoint protein Sda is necessary for the efficient initiation of sporulation after transient replication stress in *Bacillus subtilis*. *Mol. Microbiol.* **2006**, *60*, 1490–1508. [CrossRef] [PubMed]

139. Levine, J.H.; Fontes, M.E.; Dworkin, J.; Elowitz, M.B. Pulsed feedback defers cellular differentiation. *PLoS Biol.* **2012**, *10*, e1001252. [CrossRef] [PubMed]

140. Chapman, J.W.; Piggot, P.J. Analysis of the inhibition of sporulation of *Bacillus subtilis* caused by increasing the number of copies of the *spoOF* gene. *J. Gen. Microbiol.* **1987**, *133*, 2079–2088. [CrossRef] [PubMed]

141. Grimshaw, C.E.; Huang, S.; Hanstein, C.G.; Strauch, M.A.; Burbulys, D.; Wang, L.; Hoch, J.A.; Whiteley, J.M. Synergistic kinetic interactions between components of the phosphorelay controlling sporulation in *Bacillus subtilis*. *Biochemistry* **1998**, *37*, 1365–1375. [CrossRef] [PubMed]

142. Rahn-Lee, L.; Gorbatyuk, B.; Skovgaard, O.; Losick, R. The conserved sporulation protein YneE inhibits DNA replication in *Bacillus subtilis*. *J. Bacteriol.* **2009**, *191*, 3736–3739. [CrossRef] [PubMed]

143. Wagner, J.K.; Marquis, K.A.; Rudner, D.Z. SirA enforces diploidy by inhibiting the replication initiator DnaA during spore formation in *Bacillus subtilis*. *Mol. Microbiol.* **2009**, *73*, 963–974. [CrossRef] [PubMed]

144. Jameson, K.H.; Rostami, N.; Fogg, M.J.; Turkenburg, J.P.; Grahl, A.; Murray, H.; Wilkinson, A.J. Structure and interactions of the *Bacillus subtilis* sporulation inhibitor of DNA replication, SirA, with domain I of DnaA. *Mol. Microbiol.* **2014**, *93*, 975–991. [CrossRef] [PubMed]

145. Duan, Y.; Huey, J.D.; Herman, J.K. The DnaA inhibitor SirA acts in the same pathway as Soj (ParA) to facilitate oriC segregation during *Bacillus subtilis* sporulation. *Mol. Microbiol.* **2016**, *102*, 530–544. [CrossRef] [PubMed]

146. Boonstra, M.; de Jong, I.G.; Scholefield, G.; Murray, H.; Kuipers, O.P.; Veening, J.-W.W. Spo0A regulates chromosome copy number during sporulation by directly binding to the origin of replication in *Bacillus subtilis*. *Mol. Microbiol.* **2013**, *87*, 925–938. [CrossRef] [PubMed]

147. Castilla-Llorente, V.; Muñoz-Espín, D.; Villar, L.; Salas, M.; Meijer, W.J.J. Spo0A, the key transcriptional regulator for entrance into sporulation, is an inhibitor of DNA replication. *EMBO J.* **2006**, *25*, 3890–3899. [CrossRef] [PubMed]

148. Camara, J.E.; Breier, A.M.; Brendler, T.; Austin, S.; Cozzarelli, N.R.; Crooke, E. Hda inactivation of DnaA is the predominant mechanism preventing hyperinitiation of *Escherichia coli* DNA replication. *EMBO Rep.* **2005**, *6*, 736–741. [CrossRef] [PubMed]

149. Katayama, T.; Kubota, T.; Kurokawa, K.; Crooke, E.; Sekimizu, K. The initiator function of DnaA protein is negatively regulated by the sliding clamp of the *E. coli* chromosomal replicase. *Cell* **1998**, *94*, 61–71. [CrossRef]

150. Kato, J.; Katayama, T. Hda, a novel DnaA-related protein, regulates the replication cycle in *Escherichia coli*. *EMBO J.* **2001**, *20*, 4253–4262. [CrossRef] [PubMed]

151. Takata, M.; Kubota, T.; Matsuda, Y.; Katayama, T. Molecular mechanism of DNA replication-coupled inactivation of the initiator protein in *Escherichia coli*: Interaction of DnaA with the sliding clamp-loaded DNA and the sliding clamp-Hda complex. *Genes Cells* **2004**, *9*, 509–522.

152. Kurz, M.; Dalrymple, B.; Wijffels, G.; Kongsuwan, K. Interaction of the Sliding Clamp Beta-Subunit and Hda, a DnaA-Related Protein. *J. Bacteriol.* **2004**, *186*, 3508–3515. [CrossRef] [PubMed]

153. Su'etsugu, M.; Shimuta, T.-R.R.; Ishida, T.; Kawakami, H.; Katayama, T. Protein associations in DnaA-ATP hydrolysis mediated by the Hda-replicase clamp complex. *J. Biol. Chem.* **2005**, *280*, 6528–6536. [CrossRef] [PubMed]

154. Fujimitsu, K.; Su'etsugu, M.; Yamaguchi, Y.; Mazda, K.; Fu, N.; Kawakami, H.; Katayama, T. Modes of overinitiation, *dnaA* gene expression, and inhibition of cell division in a novel cold-sensitive *hda* mutant of *Escherichia coli*. *J. Bacteriol.* **2008**, *190*, 5368–5381. [CrossRef] [PubMed]

155. Su'etsugu, M.; Nakamura, K.; Keyamura, K.; Kudo, Y.; Katayama, T. Hda Monomerization by ADP Binding Promotes Replicase Clamp-mediated DnaA-ATP Hydrolysis. *J. Biol. Chem.* **2008**, *283*, 36118–36131. [CrossRef] [PubMed]

156. Xu, Q.; McMullan, D.; Abdubek, P.; Astakhova, T.; Carlton, D.; Chen, C.; Chiu, H.-J.; Clayton, T.; Das, D.; Deller, M.C.; et al. A structural basis for the regulatory inactivation of DnaA. *J. Mol. Biol.* **2009**, *385*, 368–380. [CrossRef] [PubMed]

157. Nakamura, K.; Katayama, T. Novel essential residues of Hda for interaction with DnaA in the regulatory inactivation of DnaA: Unique roles for Hda AAA + Box VI and VII motifs. *Mol. Microbiol.* **2010**, *76*, 302–317. [CrossRef] [PubMed]

158. Keyamura, K.; Katayama, T. DnaA Protein DNA-binding Domain Binds to Hda Protein to Promote Inter-AAA+ Domain Interaction Involved in Regulatory Inactivation of DnaA. *J. Biol. Chem.* **2011**, *286*, 29336–29346. [CrossRef] [PubMed]

159. Ryan, V.T.; Grimwade, J.E.; Camara, J.E.; Crooke, E.; Leonard, A.C. *Escherichia coli* prereplication complex assembly is regulated by dynamic interplay among Fis, IHF and DnaA. *Mol. Microbiol.* **2004**, *51*, 1347–1359. [CrossRef] [PubMed]

160. Kasho, K.; Fujimitsu, K.; Matoba, T.; Oshima, T.; Katayama, T. Timely binding of IHF and Fis to DARS2 regulates ATP-DnaA production and replication initiation. *Nucleic Acids Res.* **2014**, *42*, 13134–13149. [CrossRef] [PubMed]

161. Kasho, K.; Katayama, T. DnaA binding locus datA promotes DnaA-ATP hydrolysis to enable cell cycle-coordinated replication initiation. *Proc. Natl. Acad. Sci. USA* **2012**, *110*, 936–941. [CrossRef] [PubMed]

162. Filutowicz, M.; Ross, W.; Wild, J.; Gourse, R.L. Involvement of Fis protein in replication of the *Escherichia coli* chromosome. *J. Bacteriol.* **1992**, *174*, 398–407. [CrossRef] [PubMed]
163. Pratt, T.S.; Steiner, T.; Feldman, L.S.; Walker, K.A.; Osuna, R. Deletion analysis of the *fis* promoter region in *Escherichia coli*: Antagonistic effects of integration host factor and Fis. *J. Bacteriol.* **1997**, *179*, 6367–6377. [CrossRef] [PubMed]
164. Flåtten, I.; Skarstad, K. The Fis protein has a stimulating role in initiation of replication in *Escherichia coli* in vivo. *PLoS ONE* **2013**, *8*, 1–9. [CrossRef] [PubMed]
165. Cassler, M.R.; Grimwade, J.E.; Leonard, A.C. Cell cycle-specific changes in nucleoprotein complexes at a chromosomal replication origin. *EMBO J.* **1995**, *14*, 5833–5841. [PubMed]
166. Grimwade, J.E.; Ryan, V.T.; Leonard, A.C. IHF redistributes bound initiator protein, DnaA, on supercoiled *oriC* of *Escherichia coli*. *Mol. Microbiol.* **2000**, *35*, 835–844. [CrossRef] [PubMed]
167. Kitagawa, R.; Mitsuki, H.; Okazaki, T.; Ogawa, T. A novel DnaA protein-binding site at 94.7 min on the *Escherichia coli* chromosome. *Mol. Microbiol.* **1996**, *19*, 1137–1147. [CrossRef] [PubMed]
168. Kitagawa, R.; Ozaki, T.; Moriya, S.; Ogawa, T. Negative control of replication initiation by a novel chromosomal locus exhibiting exceptional affinity for *Escherichia coli* DnaA protein. *Genes Dev.* **1998**, *12*, 3032–3043. [CrossRef] [PubMed]
169. Ogawa, T.; Yamada, Y.; Kuroda, T.; Kishi, T.; Moriya, S. The *datA* locus predominantly contributes to the initiator titration mechanism in the control of replication initiation in *Escherichia coli*. *Mol. Microbiol.* **2002**, *44*, 1367–1375. [CrossRef] [PubMed]
170. Nozaki, S.; Yamada, Y.; Ogawa, T. Initiator titration complex formed at *datA* with the aid of IHF regulates replication timing in *Escherichia coli*. *Genes Cells* **2009**, *14*, 329–341. [CrossRef] [PubMed]
171. Fujimitsu, K.; Senriuchi, T.; Katayama, T. Specific genomic sequences of *E. coli* promote replicational initiation by directly reactivating ADP-DnaA. *Genes Dev.* **2009**, *23*, 1221–1233. [CrossRef] [PubMed]
172. Frimodt-Møller, J.; Charbon, G.; Krogfelt, K.A.; Løbner-Olesen, A. DNA Replication Control Is Linked to Genomic Positioning of Control Regions in *Escherichia coli*. *PLoS Genet.* **2016**, *12*, 1–27. [CrossRef] [PubMed]
173. Inoue, Y.; Tanaka, H.; Kasho, K.; Fujimitsu, K.; Oshima, T.; Katayama, T. Chromosomal location of the DnaA-reactivating sequence *DARS2* is important to regulate timely initiation of DNA replication in *Escherichia coli*. *Genes Cells* **2016**, *21*, 1015–1023. [CrossRef] [PubMed]
174. Slater, S.; Wold, S.; Lu, M.; Boye, E.; Skarstad, K.; Kleckner, N. *E. coli* SeqA protein binds *oriC* in two different methyl-modulated reactions appropriate to its roles in DNA replication initiation and origin sequestration. *Cell* **1995**, *82*, 927–936. [CrossRef]
175. Lu, M.; Campbell, J.L.; Boye, E.; Kleckner, N. SeqA: A negative modulator of replication initiation in *E. coli*. *Cell* **1994**, *77*, 413–426. [CrossRef]
176. Von Freiesleben, U.; Rasmussen, K.V.; Schaechter, M. SeqA limits DnaA activity in replication from *oriC* in *Escherichia coli*. *Mol. Microbiol.* **1994**, *14*, 763–772. [PubMed]
177. Campbell, J.L.; Kleckner, N. *E. coli oriC* and the *dnaA* gene promoter are sequestered from dam methyltransferase following the passage of the chromosomal replication fork. *Cell* **1990**, *62*, 967–979. [CrossRef]
178. Nievera, C.; Torgue, J.J.-C.; Grimwade, J.E.; Leonard, A.C. SeqA Blocking of DnaA-*oriC* Interactions Ensures Staged Assembly of the *E. coli* Pre-RC. *Mol. Cell* **2006**, *24*, 581–592. [CrossRef] [PubMed]
179. Han, J.S.; Kang, S.; Lee, H.; Kim, H.K.; Hwang, D.S. Sequential binding of SeqA to paired hemi-methylated GATC sequences mediates formation of higher order complexes. *J. Biol. Chem.* **2003**, *278*, 34983–34989. [CrossRef] [PubMed]
180. Kang, S.; Lee, H.; Han, J.S.; Hwang, D.S. Interaction of SeqA and Dam methylase on the hemimethylated origin of *Escherichia coli* chromosomal DNA replication. *J. Biol. Chem.* **1999**, *274*, 11463–11468. [CrossRef] [PubMed]
181. Waldminghaus, T.; Skarstad, K. The *Escherichia coli* SeqA protein. *Plasmid* **2009**, *61*, 141–150. [CrossRef] [PubMed]
182. Helgesen, E.; Fossum-Raunehaug, S.; Saetre, F.; Schink, K.O.; Skarstad, K. Dynamic *Escherichia coli* SeqA complexes organize the newly replicated DNA at a considerable distance from the replisome. *Nucleic Acids Res.* **2015**, *43*, 2730–2743. [CrossRef] [PubMed]

183. Zawilak-Pawlik, A.; Donczew, R.; Szafrański, S.; Mackiewicz, P.; Terradot, L.; Zakrzewska-Czerwińska, J. DiaA/HobA and DnaA: A pair of proteins co-evolved to cooperate during bacterial orisome assembly. *J. Mol. Biol.* **2011**, *408*, 238–2351. [CrossRef] [PubMed]

184. Keyamura, K.; Fujikawa, N.; Ishida, T.; Ozaki, S.; Su, M.; Fujimitsu, K.; Kagawa, W.; Yokoyama, S.; Kurumizaka, H. The interaction of DiaA and DnaA regulates the replication cycle in *E. coli* by directly promoting ATP—DnaA-specific initiation complexes. *Genes Dev.* **2007**, *21*, 2083–2099. [CrossRef] [PubMed]

185. Natrajan, G.; Hall, D.R.; Thompson, A.C.; Gutsche, I.; Terradot, L. Structural similarity between the DnaA-binding proteins HobA (HP1230) from *Helicobacter pylori* and DiaA from *Escherichia coli*. *Mol. Microbiol.* **2007**, *65*, 995–1005. [CrossRef] [PubMed]

186. Terradot, L.; Zawilak-Pawlik, A. Structural insight into *Helicobacter pylori* DNA replication initiation. *Gut Microbes* **2010**, *1*, 330–334. [CrossRef] [PubMed]

187. Zhang, Q.; Zhou, A.; Li, S.S.; Ni, J.; Tao, J.; Lu, J.; Wan, B.; Li, S.S.; Zhang, J.; Zhao, S.; et al. Reversible lysine acetylation is involved in DNA replication initiation by regulating activities of initiator DnaA in *Escherichia coli*. *Sci. Rep.* **2016**, *6*, 30837. [CrossRef] [PubMed]

188. Leonard, A.C.; Grimwade, J.E. Regulating DnaA complex assembly: It is time to fill the gaps. *Curr. Opin. Microbiol.* **2010**, *13*, 766–772. [CrossRef] [PubMed]

GCAT
TACG
GCAT *genes*

MDPI

Review

Diversity of DNA Replication in the Archaea

Darya Ausiannikava * and Thorsten Allers

School of Life Sciences, University of Nottingham, Nottingham NG7 2UH, UK;
thorsten.allers@nottingham.ac.uk
* Correspondence: darya.ausiannikava@nottingham.ac.uk; Tel.: +44-115-823-0304

Academic Editor: Eishi Noguchi
Received: 29 November 2016; Accepted: 20 January 2017; Published: 31 January 2017

Abstract: DNA replication is arguably the most fundamental biological process. On account of their shared evolutionary ancestry, the replication machinery found in archaea is similar to that found in eukaryotes. DNA replication is initiated at origins and is highly conserved in eukaryotes, but our limited understanding of archaea has uncovered a wide diversity of replication initiation mechanisms. Archaeal origins are sequence-based, as in bacteria, but are bound by initiator proteins that share homology with the eukaryotic origin recognition complex subunit Orc1 and helicase loader Cdc6). Unlike bacteria, archaea may have multiple origins per chromosome and multiple Orc1/Cdc6 initiator proteins. There is no consensus on how these archaeal origins are recognised—some are bound by a single Orc1/Cdc6 protein while others require a multi- Orc1/Cdc6 complex. Many archaeal genomes consist of multiple parts—the main chromosome plus several megaplasmids—and in polyploid species these parts are present in multiple copies. This poses a challenge to the regulation of DNA replication. However, one archaeal species (*Haloferax volcanii*) can survive without replication origins; instead, it uses homologous recombination as an alternative mechanism of initiation. This diversity in DNA replication initiation is all the more remarkable for having been discovered in only three groups of archaea where in vivo studies are possible.

Keywords: DNA replication; replication origin; Orc1/Cdc6; archaea; *Sulfolobus*; *Haloferax*

---

1. Introduction

The principles of DNA replication are common across all three domains of life—bacteria, archaea, and eukaryotes—but there is a fundamental split in terms of the machinery used [1]. The DNA replication proteins found in archaea are homologous to those of eukaryotes, but those encountered in bacteria are quite distinct [1,2]. Nevertheless, phylogenomic studies have shown that the archaeal replication machinery exhibits a striking degree of diversity. In some groups of archaea, components have been lost, while in others, a large number of additional copies have been acquired [2,3]. This is in contrast to eukaryotes where the composition of the replication complex remains constant across the domain [4].

Based on 16S ribosomal RNA (rRNA) sequencing, the archaeal domain was originally divided into two phyla: Crenarchaeota and Euryarchaeota [5]. However, the recent expansion in whole genome sequencing of natural isolates, combined with new statistical models, has challenged the traditional topology of the archaeal tree. It has been proposed that the TACK superphylum (comprising Thaumarchaeota, Aigarchaeota, Crenarchaeota, and Korarchaeota) gave rise to the ancestor of eukaryotes. (Figure 1). It has been suggested [3] that the diversity of replication machinery in the archaeal domain is likely to reflect the evolutionary forces that have fine-tuned their genomes in different environments.

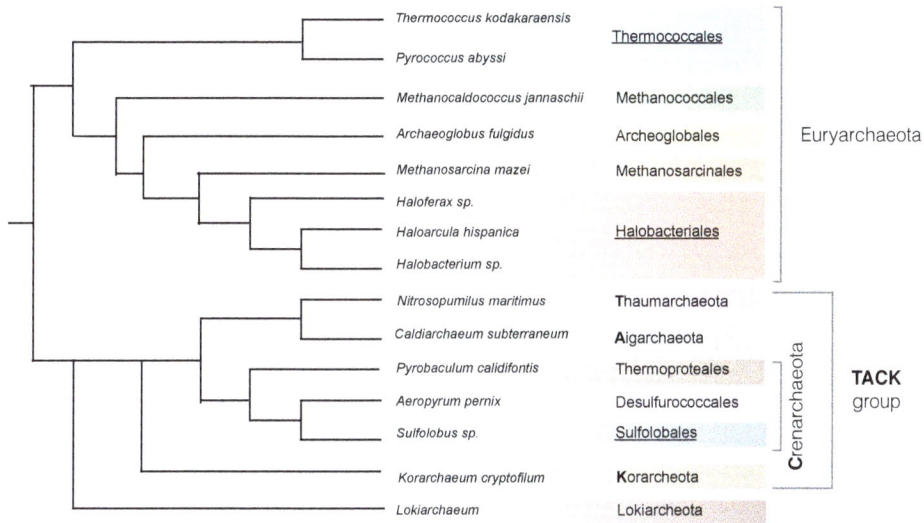

Figure 1. Current view of the archaeal phylogenetic tree. Based on [6,7]. The groups in which in vivo replication initiation studies have been undertaken are underlined.

DNA replication proceeds in three major stages: initiation, DNA synthesis, and termination. Studies of archaeal DNA replication have focused on the biochemical characterization of key enzymes involved in DNA synthesis and, despite the recognized diversity of archaeal domain, have been limited to few species. This is understandable given the interest in exploiting extremophilic enzymes in biotechnology and the difficulty of generating genetic tools for most archaeal species (see Figure 1).

DNA replication initiation is the key regulatory stage for the processes of DNA replication and the cell cycle, and the most powerful methods to study the regulation of DNA replication initiation rely on in vivo genetic analysis. However, these are available for only three groups of archaea: *Sulfolobales, Halobacteriales,* and *Thermococcales.* Here, we review the available knowledge on control of DNA replication initiation in archaea.

2. Machinery for DNA Replication Initiation

2.1. Replication Origins

Similar to the bacterial origins of replication, archaeal replication origins have a clearly defined structure consisting of an AT-rich DNA unwinding element (DUE) flanked by several conserved repeats termed origin recognition boxes (ORBs) that serve as binding sites for the origin recognition protein(s). The number, orientation, sequence, and spacing of ORBs vary among different genera, as reviewed in [8].

The first archaeal replication origin was experimentally identified in the *Pyrococcus* genus and it was shown to have a single origin per chromosome [9]. Since then, experimental studies and in silico predictions have identified several archaeal groups with multiple origins of replication on the same chromosome. For example, *Sulfolobus islandicus* and *Haloferax volcanii* have three replication origins per chromosome [10–13] while *Pyrobaculum calidifontis* has four, the highest number of origins per prokaryotic chromosome identified to date [14]. Interestingly, the number of origins in archaeal genomes does not correlate with genome size (Table 1). It remains an open question what advantages (if any) there are for archaeal cells in having multiple replication origins per chromosome.

Table 1. Chromosome size and number of DNA replication origins in different archaeal species.

	Chromosome Size, kb	Number of Origins per Chromosome
Haloferax mediterranei	2949 *	3 [15]
Haloferax volcanii	2848 *	3 [10]
Haloarcula hispanica	2995 *	2 [16]
Halobacterium sp. strain NRC-1	2014 *	2 [17]
Nitrosopumilus maritimus	1645	1 [14]
Sulfolobus islandicus	2500	3 [18]
Sulfolobus solfataricus	2992	3 [11,13]
Sulfolobus acidocaldarius	2226	3 [11]
Aeropyrum pernix	1670	2 [19]
Pyrobaculum calidifontis	2010	4 [20]
Pyrococcus abyssi	1770	1 [9]
Archaeoglobus fulgidus	2178	1 [21]
Methanococcus jannaschii	1660	1 ** [22]
Methanosarcina mazei	4096	1 ** [23]

* In cases where there are several elements of the genome, only the size of the main chromosome is indicated;
** The number of origins is based on in silico prediction by the Z-curve method and has not been experimentally validated.

2.2. Origin Recognition Proteins

Origins in archaea and bacteria are typically linked to the gene that encodes the replication initiator protein that recognizes the origin. In bacteria, origins are recognized by DnaA-type initiators whereas archaeal origins are recognized by Orc1/Cdc6 proteins that are homologues of the eukaryotic Orc1 origin recognition complex and Cdc6 helicase loader proteins (A confusion in naming of Orc1/Cdc6 proteins exists: in some species they are named Orc1, in others Cdc6; in essence, the same protein has homology to both to Orc1 and Cdc6). In contrast to bacteria, the proteins involved in archaeal origin recognition display a considerable degree of evolutionary flexibility. *Methanococcales* and *Methanopyrales* groups have highly divergent *orc* genes that initially precluded their identification [2], while in *Sulfolobus islandicus,* the third origin of replication *oriC3* is recognized not by the Orc1/Cdc6 protein but instead by WhiP, a distant homologue of Cdt1 [18].

Eukaryotic Orc proteins recognize origins as a preassembled hexameric complex, while bacterial DnaA monomers bind cooperatively to the origin of replication [1]. Most archaea encode at least two Orc1/Cdc6 homologs in their genomes, but the ability of archaeal Orc1/Cdc6 to form homo- or heteromeric complexes for origin recognition in vivo is still unclear and appears to be species-specific. The crystal structure of two Orc1/Cdc6 proteins, Cdc6-1 and Cdc6-3, bound to *Sulfolobus solfataricus* origin *oriC2* was shown to form a heterodimer [24] (Figure 2B). By contrast, the crystal structure of *Aeropyrum pernix* Cdc6-1 bound to the origin *oriC1* indicates binding as a monomer (Figure 2A) [25], while at high concentration Cdc6-1 was shown to form dimers in vitro [25,26]. The second *A. pernix* Cdc6 protein, Cdc6-2, did not bind the origin *oriC1*. Interestingly, none of the two genes for *A. pernix* Cdc6 proteins is located next to the predicted origins [19].

This notable level of diversity exists even among closely related *Sulfolobus* species. *S. solfataricus* has three replication origins (*oriC1*, *oriC2*, *oriC3*) and three Orc1/Cdc6 proteins (Cdc6-1, Cdc6-3, Cdc6-3). Deoxyribonuclease I (DNaseI) footprinting has shown that both Cdc6-1 and Cdc6-2 recognize three sites in *oriC1*, while *oriC2* and *oriC3* are recognized by all three Orc proteins, albeit with different affinities [13,19] (Figure 2B). The solved crystal structure of the Cdc6-1 and Cdc6-3 heterodimer bound to the *oriC2* origin indicates that direct contacts between Cdc6-1 and Cdc6-3 are weak, but they influence one another's DNA binding affinities [24]. It is unclear whether Cdc6-1 and Cdc6-3 recognize the same origin independently or form preassembled complexes. Surprisingly, *oriC1* and *oriC2* origins in two related species, *Sulfolobus islandicus* and *Sulfolobus acidocaldarius,* are only bound by single Orc1-1 and Orc1-3 proteins, respectively (Figure 2C) [18,27]. However, differences in origin binding between the closely-related *Sulfolobus* species may be smaller than these studies imply and could be due to differing experimental techniques.

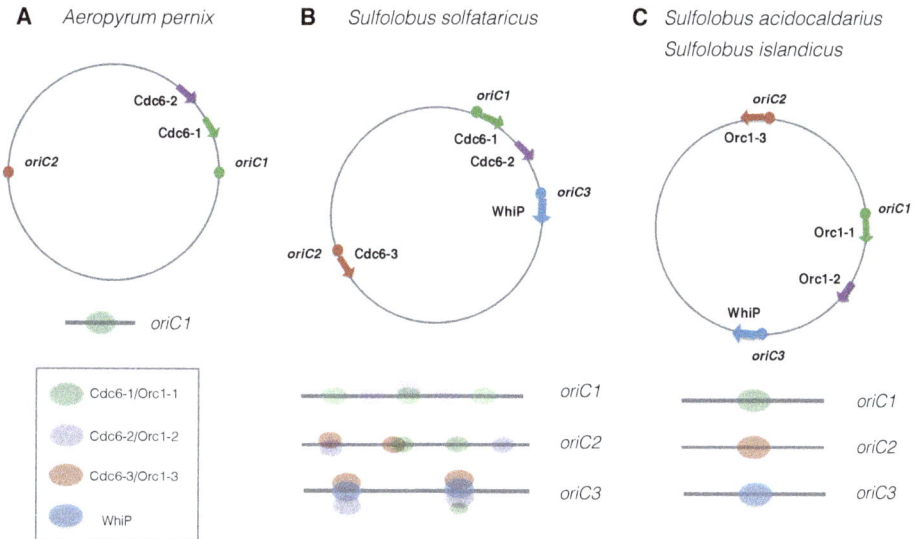

Figure 2. Binding of Orc1/Cdc6 proteins at origins of archaeal chromosomes. (**A**) *Aeropyrum pernix* Cdc6-1 binds to *oriC1* as a monomer; binding to the origin *oriC2* has not been investigated; (**B**) Cdc6-1, Cdc6-2, and Cdc6-3 of *Sulfolobus solfataricus* binds more than one origin each; (**C**) Replication initiation proteins of *Sulfolobus acidocaldarius* bind only one origin each. Similar to *S. acidocaldarius*, initiation proteins in *Sulfolobus islandicus* bind only one origin each.

2.3. Origin Binding and DNA Unwinding

Two crystal structures are available for Orc1/Cdc6 bound to DNA: the Cdc6-1 monomer from *A. pernix* bound to *oriC1* [24] and Cdc6-1/Cdc6-3 heterodimer from *S. solfataricus* bound to *oriC2* [24]. Both structures indicate two general features. Firstly, limited sequence-specific interactions exist between Orc1/Cdc6 and origin DNA (four bases are contacted specifically by *A. pernix* Cdc6-1 and five bases in the case of the *S. solfataricus* Cdc6-1/Cdc6-3 heterodimer). Secondly, Orc1/Cdc6 proteins have bipartite DNA-interaction surfaces: the first one uses a conventional DNA-binding winged-helix domain, while the second involves the AAA+ ATPase domain. This is in contrast to DnaA, where interactions are highly sequence-specific and the ATPase domain is not involved in DNA binding.

Another aspect of Orc1/Cdc6 that differs from DnaA is the formation of higher-order complexes and their effect on DNA unwinding. DnaA binds cooperatively to multiple sites in bacterial origins and there are two reports showing cooperative binding of archaeal Orc1/Cdc6: *Methanothermobacter thermoautotrophicus* Cdc6 [28] and *A. pernix* Cdc6-1 [26]. However, DNA footprinting assays of *Sulfolobus* Orc1/Cdc6 proteins do not support the assembly of a higher-order complex on origin sites [13]. Orc1/Cdc6 have been reported to alter DNA topology in vitro [29,30] and there is one report showing origin unwinding in vitro [29]. In contrast to unwinding by bacterial DnaA, Orc1/Cdc6 were found to act in an ATP-independent manner and did not act at the duplex unwinding element. There is not yet a clear consensus on how origin DNA is unwound by Orc1/Cdc6 proteins [31].

2.4. Multiple Origins on the Chromosome

When multiple origins are found on archaeal chromosomes, have they arisen by duplication or do they have an independent evolutionary history? By comparing two distantly related crenarchaeal species, *Sulfolobus* and *Aeropyrum*, Robinson and Bell demonstrated that multiple origins in both species are likely to have arisen by horizontal gene transfer [19]. The authors proposed that this occurred by integration of extrachromosomal genetic elements into the chromosome, and not by the duplication of existing origins. Similar conclusions have been drawn for the multiple replication origins of haloarchaeal species, which show poor sequence similarities with each other [32,33].

The idea of replicons evolving independently of each other is consistent with in vivo studies from *Sulfolobus islandicus* and *Haloarcula hispanica*. The deletion of a single *orc1/cdc6* gene prevents the origin firing only from the adjacent replication origin but does not affect any other origin. Thus, only an initiation factor genetically linked to the origin is required and sufficient for the replication from that origin; the initiation of the replicons on the same chromosome is independent of each other [16,18]. The fact that *S. solfataricus* origins are bound by several Cdc6 proteins (Figure 2B) points to greater integration among the replicons in this species than in *S. islandicus* (Figure 2C). The exact combination of Orc1/Cdc6 proteins that are necessary and sufficient for origin firing in *S. solfataricus* is unknown.

2.5. Diversity of Functions of Orc1/Cdc6 Proteins in Archaea

The number of *orc1/cdc6* genes present in archaeal genomes is often greater than the number of origins. The extreme situation can be found in the *Halobacteriales* group, where the genome may contain as many as nine *orc1/cdc6* genes on the main chromosome but only three origins, as is the case in *Haloferax volcanii* [10,12,34]. Similar to bacterial DnaA and the eukaryotic ORC complex, archaeal Orc1/Cdc6 proteins are likely to have extended their functions beyond replication initiation. Moreover, some Orc1/Cdc6 proteins may have lost functions connected with replication initiation and have acquired new roles.

A phylogenomic analysis of 140 archaeal genomes found that in each genome, only one or two Orc1/Cdc6 homologs (named core copies) are slow-evolving, while any additional copies (shell copies) are highly divergent [3]. Shell copies of Orc1/Cdc6 might contribute to replication under special circumstances. Thus, when the three main chromosomal origins of *Haloferax mediterranei* are deleted, a dormant origin located next to the shell copy *cdc6H* gene becomes activated [15]. Alternatively, it has been suggested that Orc1/Cdc6 proteins might also work as factors for gene regulation. For example, chromatin immunoprecipitation (ChIP) analysis of *Pyrococcus abyssi* Cdc6 binding indicates that additional regions were bound in addition to *oriC1* [35]. Conversely, the transcription of shell *orc1/cdc6* genes was found to be misregulated when *Halobacteriales* were grown under acidic and alkaline conditions [36].

Intriguingly, when two core copies of *orc1/cdc6* are present in an archaeal genome, only one of these copies is located next to a predicted replication origin; the other *orc1/cdc6* gene is never linked to an origin [3]. The absence of a genetic linkage with origins suggests that the unlinked Orc1/Cdc6 proteins might have acquired functions distinct from replication initiation, for example in the regulation of gene repair, recombination, or replication fork restart. This idea is consistent with the experimental data from *Sulfolobus islandicus*, which has two slow-evolving *orc* copies, *cdc6-1* (adjacent to *oriC1*) and *cdc6-2* (not origin-associated). The deletion of *cdc6-1* inhibits the initiation from *oriC1*, while the deletion of *cdc6-2* does not affect replication initiation from any of the three origins on the chromosome (Figure 2C) [18]. However, in *S. solfataricus* the slowly-evolving Cdc6-2, which is not linked to origins, can bind to the *oriC1*, *oriC2*, and *oriC3* origins both in vivo and in vitro (Figure 2B). This has led to the proposal that Cdc6-2 can negatively regulate replication initiation [13,19].

2.6. Recruitment of a Helicase

The next step of replication initiation after origin recognition is recruitment of a helicase to unwind the DNA duplex. In bacteria, DnaC serves as a DnaB helicase loader, while in eukaryotes the binding of Cdc6 and Cdt1 to the ORC complex helps to recruit the minichromosome maintenance (MCM) helicase and to regulate replication initiation. In eukaryotes, the MCM helicase consists of a heterohexameric complex, whereas most archaeal MCM proteins are homohexamers encoded by a single *mcm* gene [37,38].

Which protein(s) function as a helicase loader in archaea? Archaea do not have a clear homologue of Cdt1 and Orc1/Cdc6 proteins that share homology with both Orc1 and Cdc6. Most archaea have several genes encoding Orc1/Cdc6, therefore, it has been suggested that one of these Orc1/Cdc6 proteins carries out the function of eukaryotic Cdc6 by acting as a helicase loader, while the other Orc1/Cdc6 proteins are responsible for origin recognition. Recent biochemical data support the idea that a single protein can have both Orc1 and Cdc6 features, and in at least two cases, MCM is recruited to origins directly by Orc1/Cdc6 [39,40]. In an in vitro recruitment assay, Cdc6 from *Pyrococcus furiosus* was able to recruit MCM in an *oriC*-dependent manner [40]. In *Sulfolobus islandicus,* the conserved C-terminal winged-helix domain of MCM interacts directly with the ATPase domain of Cdc6-1; this interaction is required for the replication initiation from *oriC1* in vivo [39].

It is likely that there are alternative mechanisms of MCM recruitment in archaea. For example, the MCM-interacting interface appears to be conserved in Cdc6-3, the replication initiator protein in *S. islandicus* that is required for recognition of *oriC2* [39]. By contrast, the third origin of *Sulfolobus islandicus*, *oriC3*, is bound by WhiP, a distant homolog of Cdt1 and not Orc1/Cdc6. It is likely that different interfaces of MCM are involved in its recruitment by WhiP, and that additional partners may play a role in this process [18].

An extreme case in MCM recruitment in archaea is encountered in the *Methanococcales* family. This family has very divergent Orc1/Cdc6s and several copies of MCM encoded in the genome. Although additional copies of *mcm* genes have mostly arisen by the integration of extrachromosomal elements, the mobile elements carrying these *mcm* genes do not appear to have been involved in extensive lateral gene transfer and, thus, may have coevolved with their hosts [41]. Although it is tempting to speculate that under special circumstances (e.g., replication stress) alternative MCM helicases might be recruited to the origins by different Orc1/Cdc6 proteins, the experimental evidence for this is lacking due to difficulties of genetic analysis in *Methanococcales*.

An ancient supergroup of unicellular eukaryotes called Excavates, which is comprised of *Trypanosoma, Giardia,* and *Euglena*, also lacks Cdc6 and Cdt1, and only one Orc-related initiator can be clearly identified by sequence homology [42]. Recently, Orc1/Cdc6-interacting proteins in *Trypanosoma brucei* were shown to act in nuclear DNA replication, and Orc1/Cdc6 was present in a high molecular complex suggesting the presence of a diverged ORC complex [43]. This suggests that a similar situation might exist in archaea, where at least some archaeal Orc1/Cdc6 proteins form complexes with yet-to-be identified proteins, thus increasing the efficiency of replication initiation in vivo.

3. Regulation of DNA Replication Initiation

In eukaryotes, strict regulation of replication initiation is required to ensure one round of chromosome replication per generation. To accomplish this, the cell must ensure that one initiation event occurs per generation per origin, and must prevent a second round of initiation. A regulated cell cycle ensures the temporal separation of DNA replication initiation (from multiple origins) and the onset of cell division, since they occur in wholly distinct phases. This is accomplished by the actions of cyclin-dependent kinases and associated factors.

Bacteria utilise another strategy for DNA replication control. The commitment to replication occurs at a single origin level, and not at a cellular level, and is determined by the concentration of active DnaA and the accessibility of the origin [44]. Thus, initiation of replication in bacteria is growth-dependent, rather than cell cycle-dependent.

Regulation of DNA replication initiation across the archaeal domain is unlikely to be uniform. Firstly, only the Crenarchaeal phylum has haploid species; all Euryarchaeal species that have been examined contain more than one copy of the genome per cell, with the number of copies being variable at different stages of growth. Secondly, species with multiple replication origins per chromosome will need to coordinate their firing. Thirdly, some domains such as *Halobacteriales* have large (up to 0.6 Mb) extrachromosomal megaplasmids that must also be replicated in a cell cycle. These diverse circumstances require a range of mechanisms to regulate DNA replication initiation.

3.1. Cell Cycle Regulation in Haploid Archaea

Limited knowledge of the archaeal cell cycle exists for the most studied archaeal group, the *Sulfolobus* genus, which is a haploid crenarchaeote [45]. Similar to eukaryotes, the *Sulfolobus* cell cycle is divided into pre-replicative G1 phase, S-phase where genome replication happens, post-replicative G2 phase, and M- and D phases when the genome segregation and cell division happen. The longest phase is G2, which takes more than half of the cell cycle. This is in contrast to eukaryotes, where the G2 phase is short.

One method of regulating replication initiation in eukaryotes is cell-cycle specific expression of *cdc6*. The Cdc6 helicase loader is synthesised in late G1 and recruits MCM helicase to the ORC complex in the S-phase. The pattern of Orc binding and expression differs from Cdc6, since the ORC complex is bound to DNA throughout the whole cell cycle. Given that archaeal Orc1/Cdc6 might play the role of both initiator and helicase loader, it would be interesting to know whether the level of its expression is regulated. Again, the expression pattern varies even among closely-related species. In *Sulfolobus solfataricus*, the abundance of three Cdc6 proteins appears to be cell-cycle specific and varies in a cyclin-like fashion. The expression of Cdc6 is increased in or just before the G1 phase, decreased in the S-phase, and is considerably reduced in the non-replicating stationary phase cells [13]. In *S. acidocaldarius*, the expression of Cdc6-1 and Cdc6-3, as well as their binding at origins, remains constant throughout the cell cycle as well as in the stationary phase [27]. A similar case was observed in *Pyrococcus abyssi* where Orc1/Cdc6 remains bound to the replication origin both in the exponential and stationary phases, while MCM is associated with the origin only in the exponential phase [35]. This suggests that there may be additional factors that regulate replication initiation for these species. For example, an additional component of the replication initiation machinery or post-translational modifications.

3.2. Cell Cycle Regulation in Polyploid Archaea

Polyploidy is widespread in the archaeal domain, for example, *Halobacteriales* and *Methanococales* are both highly polyploid [46,47]. Due to their high genome copy number, polyploid species do not have a strict requirement to replicate the genome only once per cell cycle or to evenly distribute the chromosome copies to daughter cells. In fact, it is unclear whether replication of the chromosome copies is synchronous in polyploid archaeal species. Differences in ploidy levels at different stages of growth suggest that cell division and DNA replication are not tightly coupled [46]. Whether DNA replication and cell growth are also uncoupled in archaea, as was recently reported in the polyploid cyanobacterium *Synechococcus* [48], is unknown.

3.3. Regulation of Initiation of Multiple Origins

Having more than one origin per chromosome potentially increases the complexity of regulation of replication initiation; this has been examined in only a few studies. In *Sulfolobus acidocaldarius*, a species with three replication origins per chromosome, there is a close coordination of firing of two origins (*oriC1* and *oriC3*) at the beginning of the S-phase, while the third origin, *oriC2*, is activated slightly later [27]. The mechanisms that ensure simultaneous origin firing are unknown.

In *Haloarcula hispanica*, a halophile with two origins per chromosome, the sequences located next to the origins appear to influence the activity of origin firing: *oriC1* has a G-rich inverted repeat that serves as an enhancer, while *oriC2* is negatively regulated by an ORB-rich region [16]. The stoichiometry between different origins might be important. *Haloarcula hispanica* wild-type cells fail to replicate a plasmid bearing an additional copy of the *oriC2* origin, while the cells lacking *oriC2* on the chromosome tolerate the plasmid-borne *oriC2* origin. This suggests that the Orc1/Cdc6 that binds to *oriC2* may be rate-limiting.

3.4. Regulation of Replication of Multiple Chromosomal Elements

The genomes of *Halobacteriales* consist of several parts, the main chromosome and several large extrachromosomal DNA species named megaplasmids or minichromosomes. The megaplasmids tend to have Orc1/Cdc6-based replication initiators of their own. Because the chromosome and most megaplasmids are present at a similar copy number, it is likely that for some megaplasmids there is coordination of their replication initiation with the main chromosome [10,46]. However, pHV1 (a megaplasmid found in *Haloferax volcanii*) was found to have a copy number different from that of the main chromosome, indicating that it has inputs from alternative regulation circuits [46].

4. Alternative Mechanisms of Replication Initiation

Genetic experiments where *orc1/cdc6* genes and origins have been deleted suggest that replication initiation is quite flexible in archaea. The deletion of a single *orc1/cdc6* gene (thus inactivating the adjacent origin) in *Sulfolobus islandicus* does not affect cell growth, while the inactivation of two out of three *orc1/cdc6* genes leads only to a moderate growth defect. However, the deletion of all three *orc1/cdc6* genes is impossible [18] (Figure 3).

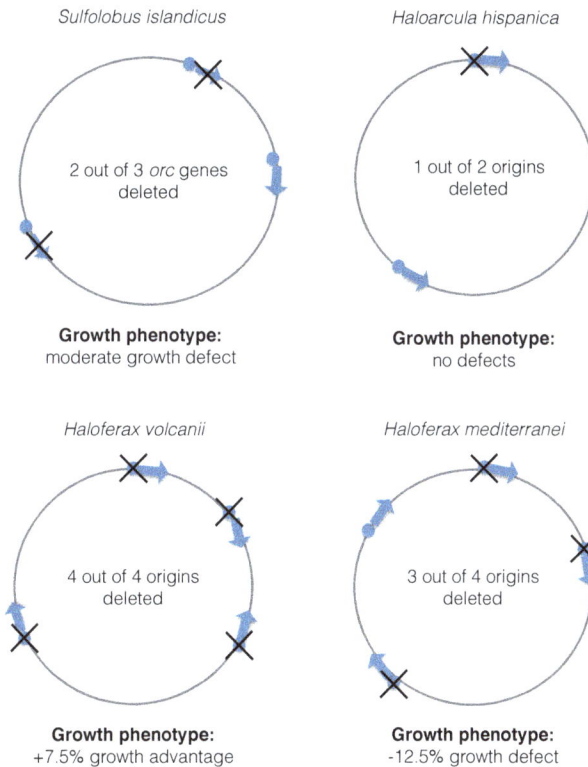

Figure 3. Serial deletion of *orc* genes or origins in different archaeal species. The highest number of *orc*/origin deletions possible in one strain is shown. Chromosomes are not drawn to scale.

The consequences of deleting multiple origins or *orc1/cdc6* genes has also been examined in four halobacterial species: *Haloferax mediterranei, Haloferax volcanii, Haloarcula hispanica,* and *Halobacterium* NRC-1 [10,15,17,32] (Figure 3). Seven out of ten *orc1/cdc6* genes can be deleted simultaneously in *Halobacterium* NRC-1 [17]. In *Haloarcula hispanica,* five out of six *orc1/cdc6* genes located on the main chromosome and three out of four *orc1/cdc6* genes on the megaplasmid can be also deleted at the same time [32]. Similar to *Sulfolobus islandicus,* the deletion of one of the two origins on the main chromosome of *Haloarcula hispanica* did not lead to any growth defects under normal conditions [32]. In *Haloferax mediterranei,* it was possible to delete all three replication origins on the main chromosome, and growth of the strain lacking *oriC1, oriC2* and *oriC3* is 12.4% slower than the wild type [15]. However, a dormant origin, named *oriC4,* became activated in the triple origin deletion strain. As the growth defect of a triple-deleted strain suggests, this dormant origin is not able to restore growth to wild-type levels. Similar to *Sulfolobus islandicus,* the generation of a quadruple Δ*oriC* mutant was found to be impossible [15]. These studies indicate that the loss of a single *orc1/cdc6* gene or origin does not affect growth, while the loss of multiple *orc1/cdc6* genes or origins leads to slower growth, and it is impossible to delete all *orc1/cdc6* genes and/or origins.

However, *Haloferax volcanii* is a notable exception in this regard: the deletion of two or more origins does not result in growth defects and the deletion of all origins leads to 7.5% faster growth than wild type; however, unlike *Haloferax mediterranei,* there is no activation of dormant origins [10]. This indicates that an alternative, highly efficient mechanism for replication initiation exists in *Haloferax volcanii.* Given the common evolutionary history of *Halobacteriales,* it is likely that the core machinery for origin-independent replication exists in all species, but that *Haloferax volcanii* has lost an inhibitory component that prevents this mode of replication. Alternatively, it might have acquired an activating component that promotes origin-independent replication. Indeed, horizontal gene transfer is highly prevalent in *Halobacteriales,* as evident by a large number of gene duplications in the genome. Low species barriers exist in halophilic archaea for gene transfer and the exchange of large chromosomal fragments between *Haloferax volcanii* and *Haloferax mediterranei* has been detected in vivo [49]. Interestingly, the dormant origin that becomes activated upon deletion of three chromosomal origins in *Haloferax mediterranei* is "foreign" to its genome—its chromosomal context indicates that it was acquired during a recent lateral gene transfer event [15]. Furthermore, it is not found in *Haloferax volcanii,* which explains why it is not activated in an origin-less *Haloferax volcanii* mutant.

Some viruses, such as bacteriophage T4, use recombination-dependent DNA replication initiation at certain life stages, where the invading 3' DNA end of a displacement loop (D-loop) recombination intermediate is used as a primer for leading strand DNA synthesis (Figure 4). In contrast, the nuclear genomes of eukaryotes are replicated from internal origins using the replication-fork model. The case of plastids (mitochondria, chloroplasts, and kinetoplastids in *Trypanosoma*) is often overlooked. Replication is assumed to occur using the single-strand displacement model, and similar to euryarchaeal genomes, plastids contain many copies of their respective genomes. The best studied example is mitochondrial DNA replication, which begins at a site of gene transcription and proceeds unidirectionally by displacing one of the template strands as single-stranded DNA. Thus, a triple-stranded D-loop replication intermediate is formed [50].

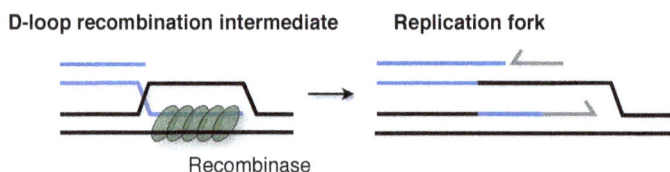

D-loop recombination intermediate   **Replication fork**

Recombinase

Figure 4. Recombination-dependent replication initiation. The invading 3' DNA end of a displacement loop (D-loop) recombination intermediate is used as a primer for leading strand DNA synthesis. Formation of a D-loop requires a RecA-family recombinase.

How is recombination-dependent replication of T4 phage and single-strand displacement replication of plastids related to the origin-independent replication seen in *Haloferax volcanii*? Given that the *Haloferax volcanii* strain without origins has an absolute requirement for the recombinase RadA, it is likely that this model of replication involves D-loop intermediates that are formed by homologous recombination.

5. Perspectives and Open Questions

5.1. Tools to Control Replication Initiation in Archaeal Cells

Regulation of the cell cycle in eukaryotes is dependent on post-translational modifications of proteins. Archaea have eukaryotic-like phosphatases and kinases that may potentially phosphorylate serine, threonine, and tyrosine residues, as reviewed in [51,52]. Phosphorylation of the *Haloferax volcanii* Orc1 protein was detected by shotgun proteomic approaches [53]. The ubiquitin family of protein modification features prominently in the control of eukaryotic DNA replication [54], and ubiquitin-like small archaeal modifier proteins have been discovered in archaea [55].

Similar to eukaryotes, GTPases could be involved in mechanisms of DNA replication and repair in archaea. In several cases, genes for GTPases are located in the genomic neighbourhood of replication genes, and in *Pyrococcus abyssi,* a GTPase has been found in association with the RFC (replication factor C) clamp loader [56].

5.2. Spatial Organisation of Genome and Replication

In both bacteria and eukaryotes, it is well known that the three-dimensional organisation of the genome inside the cell is an important determinant in the regulation of replication. Different chromatin proteins have been described in archaea, with Alba and histone proteins being the most widespread; Alba proteins are characteristic for Crenarchaeota, while histones are found in Euryarchaeota [57]. Could it be that archaeal chromatin provides a barrier for replication fork progression, and if so, what role does it have in replication regulation? Most studies tackling this question have been focused on the Alba protein, which has been shown to exist in acetylated and non-acetylated forms; the deacetylated form represses transcription in vitro [58], and the acetylated form of Alba alleviates repression of MCM in vitro [59]. However, a direct role in DNA replication has yet to be determined.

The first attempts to correlate spatial organisation and replication have been made by Gristwood et al. [60], who used a nucleoside analogue incorporation assay to observe the sub-cellular localisation of *Sulfolobus* DNA replication. Replisomes were located at the periphery, with the three origin loci being separated in space. This suggests that replication initiation at the three origins may be regulated semi-independently.

6. Conclusions

The archaeal domain is the most underexplored branch of the tree of life, not only in terms of DNA replication control. Nevertheless, there are clear indications that archaea exhibit unprecedented diversity in their cellular mechanisms, while at the same time they serve as a simplified model to study many eukaryotic processes. For example, archaea with several origins per chromosome provide an excellent model for studying the coordination of replication initiation. The control of replication of polyploid archaea may give insights into DNA replication in cancer cells with multiple copies of the genome. Cell cycle studies in *Sulfolobus* can trace the development of a sophisticated cell cycle in eukaryotes. Similar arguments can be applied for unravelling connections between chromatin organisation and replication in archaea. The diversity of archaeal DNA replication resembles a melting pot of mechanisms, from which the refined system that is common to eukaryotes has emerged.

Acknowledgments: Work in T.A's laboratory is funded by the Biotechnology and Biological Sciences Research Council (BBSRC) (BB/M001393/1), this includes funds for covering the costs to publish in open access. We thank Tom Williams for assistance with Figure 1 and Hannah Marriott for helpful comments on the manuscript.

Author Contributions: D.A. and T.A. wrote the manuscript.

Conflicts of Interest: The authors declare no conflict of interest. The funding sponsors had no role in the writing of the manuscript.

References

1. O'Donnell, M.; Langston, L.; Stillman, B. Principles and concepts of DNA replication in bacteria, archaea, and eukarya. *Cold Spring Harb. Perspect. Biol.* **2013**, *5*, a010108.
2. Makarova, K.S.; Koonin, E.V. Archaeology of eukaryotic DNA replication. *Cold Spring Harb. Perspect. Biol.* **2013**, *5*, a012963. [CrossRef] [PubMed]
3. Raymann, K.; Forterre, P.; Brochier-Armanet, C.; Gribaldo, S. Global phylogenomic analysis disentangles the complex evolutionary history of DNA replication in archaea. *Genome Biol. Evol.* **2014**, *6*, 192–212. [CrossRef] [PubMed]
4. Aves, S.J.; Liu, Y.; Richards, T.A. Evolutionary diversification of eukaryotic DNA replication machinery. *Subcell. Biochem.* **2012**, *62*, 19–35. [PubMed]
5. Woese, C.R.; Kandler, O.; Wheelis, M.L. Towards a natural system of organisms: Proposal for the domains Archaea, Bacteria, and Eucarya. *Proc. Natl. Acad. Sci. USA* **1990**, *87*, 4576–4579. [CrossRef] [PubMed]
6. Petitjean, C.; Deschamps, P.; Lopez-Garcia, P.; Moreira, D. Rooting the domain archaea by phylogenomic analysis supports the foundation of the new kingdom Proteoarchaeota. *Genome Biol. Evol.* **2014**, *7*, 191–204. [CrossRef] [PubMed]
7. Williams, T.A.; University of Bristol, Bristol, UK. Personal communication, 2016.
8. Wu, Z.; Liu, J.; Yang, H.; Xiang, H. DNA replication origins in archaea. *Front. Microbiol.* **2014**, *5*, 179. [CrossRef] [PubMed]
9. Myllykallio, H.; Lopez, P.; Lopez-Garcia, P.; Heilig, R.; Saurin, W.; Zivanovic, Y.; Philippe, H.; Forterre, P. Bacterial mode of replication with eukaryotic-like machinery in a hyperthermophilic archaeon. *Science* **2000**, *288*, 2212–2215. [CrossRef] [PubMed]
10. Hawkins, M.; Malla, S.; Blythe, M.J.; Nieduszynski, C.A.; Allers, T. Accelerated growth in the absence of DNA replication origins. *Nature* **2013**, *503*, 544–547. [CrossRef] [PubMed]
11. Lundgren, M.; Andersson, A.; Chen, L.; Nilsson, P.; Bernander, R. Three replication origins in *Sulfolobus* species: Synchronous initiation of chromosome replication and asynchronous termination. *Proc. Natl. Acad. Sci. USA* **2004**, *101*, 7046–7051. [CrossRef] [PubMed]
12. Norais, C.; Hawkins, M.; Hartman, A.L.; Eisen, J.A.; Myllykallio, H.; Allers, T. Genetic and physical mapping of DNA replication origins in *Haloferax volcanii*. *PLoS Genet.* **2007**, *3*, e77. [CrossRef] [PubMed]
13. Robinson, N.P.; Dionne, I.; Lundgren, M.; Marsh, V.L.; Bernander, R.; Bell, S.D. Identification of two origins of replication in the single chromosome of the archaeon *Sulfolobus solfataricus*. *Cell* **2004**, *116*, 25–38. [CrossRef]
14. Pelve, E.A.; Martens-Habbena, W.; Stahl, D.A.; Bernander, R. Mapping of active replication origins in vivo in thaum- and euryarchaeal replicons. *Mol. Microbiol.* **2013**, *90*, 538–550. [CrossRef] [PubMed]
15. Yang, H.; Wu, Z.; Liu, J.; Liu, X.; Wang, L.; Cai, S.; Xiang, H. Activation of a dormant replication origin is essential for *Haloferax mediterranei* lacking the primary origins. *Nat. Commun.* **2015**, *6*, 8321. [CrossRef] [PubMed]
16. Wu, Z.; Liu, J.; Yang, H.; Liu, H.; Xiang, H. Multiple replication origins with diverse control mechanisms in *Haloarcula hispanica*. *Nucleic Acids Res.* **2014**, *42*, 2282–2294. [CrossRef] [PubMed]
17. Coker, J.A.; DasSarma, P.; Capes, M.; Wallace, T.; McGarrity, K.; Gessler, R.; Liu, J.; Xiang, H.; Tatusov, R.; Berquist, B.R.; et al. Multiple replication origins of *Halobacterium* sp. strain NRC-1: Properties of the conserved *orc7*-dependent *oriC1*. *J. Bacteriol.* **2009**, *191*, 5253–5261. [PubMed]
18. Samson, R.Y.; Xu, Y.; Gadelha, C.; Stone, T.A.; Faqiri, J.N.; Li, D.; Qin, N.; Pu, F.; Liang, Y.X.; She, Q.; et al. Specificity and function of archaeal DNA replication initiator proteins. *Cell Rep.* **2013**, *3*, 485–496. [CrossRef] [PubMed]
19. Robinson, N.P.; Bell, S.D. Extrachromosomal element capture and the evolution of multiple replication origins in archaeal chromosomes. *Proc. Natl. Acad. Sci. USA* **2007**, *104*, 5806–5811. [CrossRef] [PubMed]
20. Pelve, E.A.; Lindas, A.C.; Knoppel, A.; Mira, A.; Bernander, R. Four chromosome replication origins in the archaeon *Pyrobaculum calidifontis*. *Mol. Microbiol.* **2012**, *85*, 986–995. [CrossRef] [PubMed]

21. Maisnier-Patin, S.; Malandrin, L.; Birkeland, N.K.; Bernander, R. Chromosome replication patterns in the hyperthermophilic euryarchaea *Archaeoglobus fulgidus* and *Methanocaldococcus* (*Methanococcus*) *jannaschii*. *Mol. Microbiol.* **2002**, *45*, 1443–1450. [CrossRef] [PubMed]

22. Zhang, R.; Zhang, C.T. Identification of replication origins in the genome of the methanogenic archaeon, *Methanocaldococcus jannaschii*. *Extremophiles* **2004**, *8*, 253–258. [CrossRef] [PubMed]

23. Zhang, R.; Zhang, C.T. Single replication origin of the archaeon *Methanosarcina mazei* revealed by the Z curve method. *Biochem. Biophys. Res. Commun.* **2002**, *297*, 396–400. [CrossRef]

24. Dueber, E.L.; Corn, J.E.; Bell, S.D.; Berger, J.M. Replication origin recognition and deformation by a heterodimeric archaeal Orc1 complex. *Science* **2007**, *317*, 1210–1213. [CrossRef] [PubMed]

25. Gaudier, M.; Schuwirth, B.S.; Westcott, S.L.; Wigley, D.B. Structural basis of DNA replication origin recognition by an ORC protein. *Science* **2007**, *317*, 1213–1216. [CrossRef] [PubMed]

26. Grainge, I.; Gaudier, M.; Schuwirth, B.S.; Westcott, S.L.; Sandall, J.; Atanassova, N.; Wigley, D.B. Biochemical analysis of a DNA replication origin in the archaeon *Aeropyrum pernix*. *J. Mol. Biol.* **2006**, *363*, 355–369. [CrossRef] [PubMed]

27. Duggin, I.G.; McCallum, S.A.; Bell, S.D. Chromosome replication dynamics in the archaeon *Sulfolobus acidocaldarius*. *Proc. Natl. Acad. Sci. USA* **2008**, *105*, 16737–16742. [CrossRef] [PubMed]

28. Capaldi, S.A.; Berger, J.M. Biochemical characterization of Cdc6/Orc1 binding to the replication origin of the euryarchaeon *Methanothermobacter thermoautotrophicus*. *Nucleic Acids Res.* **2004**, *32*, 4821–4832. [CrossRef] [PubMed]

29. Matsunaga, F.; Takemura, K.; Akita, M.; Adachi, A.; Yamagami, T.; Ishino, Y. Localized melting of duplex DNA by Cdc6/Orc1 at the DNA replication origin in the hyperthermophilic archaeon *Pyrococcus furiosus*. *Extremophiles* **2010**, *14*, 21–31. [CrossRef] [PubMed]

30. Dueber, E.C.; Costa, A.; Corn, J.E.; Bell, S.D.; Berger, J.M. Molecular determinants of origin discrimination by Orc1 initiators in archaea. *Nucleic Acids Res.* **2011**, *39*, 3621–3631. [CrossRef] [PubMed]

31. Bell, S.D. Archaeal *orc1/cdc6* proteins. *Subcell. Biochem.* **2012**, *62*, 59–69. [PubMed]

32. Wu, Z.; Liu, H.; Liu, J.; Liu, X.; Xiang, H. Diversity and evolution of multiple *orc/cdc6*-adjacent replication origins in haloarchaea. *BMC Genom.* **2012**, *13*, 478. [CrossRef] [PubMed]

33. Wu, Z.; Yang, H.; Liu, J.; Wang, L.; Xiang, H. Association between the dynamics of multiple replication origins and the evolution of multireplicon genome architecture in haloarchaea. *Genome Biol. Evol.* **2014**, *6*, 2799–2810. [CrossRef] [PubMed]

34. Hartman, A.L.; Norais, C.; Badger, J.H.; Delmas, S.; Haldenby, S.; Madupu, R.; Robinson, J.; Khouri, H.; Ren, Q.; Lowe, T.M.; et al. The complete genome sequence of *Haloferax volcanii* DS2, a model archaeon. *PLoS ONE* **2010**, *5*, e9605. [CrossRef] [PubMed]

35. Matsunaga, F.; Glatigny, A.; Mucchielli-Giorgi, M.H.; Agier, N.; Delacroix, H.; Marisa, L.; Durosay, P.; Ishino, Y.; Aggerbeck, L.; Forterre, P. Genomewide and biochemical analyses of DNA-binding activity of Cdc6/Orc1 and Mcm proteins in *Pyrococcus* sp. *Nucleic Acids Res.* **2007**, *35*, 3214–3222. [CrossRef] [PubMed]

36. Moran-Reyna, A.; Coker, J.A. The effects of extremes of pH on the growth and transcriptomic profiles of three haloarchaea. *F1000Research* **2014**, *3*, 168. [CrossRef] [PubMed]

37. Chong, J.P.; Hayashi, M.K.; Simon, M.N.; Xu, R.M.; Stillman, B. A double-hexamer archaeal minichromosome maintenance protein is an ATP-dependent DNA helicase. *Proc. Natl. Acad. Sci. USA* **2000**, *97*, 1530–1535. [CrossRef] [PubMed]

38. Grainge, I.; Scaife, S.; Wigley, D.B. Biochemical analysis of components of the pre-replication complex of *Archaeoglobus fulgidus*. *Nucleic Acids Res.* **2003**, *31*, 4888–4898. [CrossRef] [PubMed]

39. Samson, R.Y.; Abeyrathne, P.D.; Bell, S.D. Mechanism of Archaeal MCM Helicase Recruitment to DNA Replication Origins. *Mol. Cell* **2016**, *61*, 287–296. [CrossRef] [PubMed]

40. Akita, M.; Adachi, A.; Takemura, K.; Yamagami, T.; Matsunaga, F.; Ishino, Y. Cdc6/Orc1 from *Pyrococcus furiosus* may act as the origin recognition protein and Mcm helicase recruiter. *Genes Cells* **2010**, *15*, 537–552. [PubMed]

41. Krupovic, M.; Gribaldo, S.; Bamford, D.H.; Forterre, P. The evolutionary history of archaeal MCM helicases: A case study of vertical evolution combined with hitchhiking of mobile genetic elements. *Mol. Biol. Evol.* **2010**, *27*, 2716–2732. [CrossRef] [PubMed]

42. Li, Z. Regulation of the cell division cycle in *Trypanosoma brucei*. *Eukaryot. Cell* **2012**, *11*, 1180–1190. [CrossRef] [PubMed]

43. Marques, C.A.; Tiengwe, C.; Lemgruber, L.; Damasceno, J.D.; Scott, A.; Paape, D.; Marcello, L.; McCulloch, R. Diverged composition and regulation of the *Trypanosoma brucei* origin recognition complex that mediates DNA replication initiation. *Nucleic Acids Res.* **2016**, *44*, 4763–4784. [CrossRef] [PubMed]

44. Marczynski, G.T.; Rolain, T.; Taylor, J.A. Redefining bacterial origins of replication as centralized information processors. *Front. Microbiol.* **2015**, *6*, 610. [CrossRef] [PubMed]

45. Lindas, A.C.; Bernander, R. The cell cycle of archaea. *Nat. Rev. Microbiol.* **2013**, *11*, 627–638. [CrossRef] [PubMed]

46. Breuert, S.; Allers, T.; Spohn, G.; Soppa, J. Regulated polyploidy in halophilic archaea. *PLoS ONE* **2006**, *1*, e92. [CrossRef] [PubMed]

47. Hildenbrand, C.; Stock, T.; Lange, C.; Rother, M.; Soppa, J. Genome copy numbers and gene conversion in methanogenic archaea. *J. Bacteriol.* **2011**, *193*, 734–743. [CrossRef] [PubMed]

48. Watanabe, S.; Ohbayashi, R.; Kanesaki, Y.; Saito, N.; Chibazakura, T.; Soga, T.; Yoshikawa, H. Intensive DNA Replication and Metabolism during the Lag Phase in Cyanobacteria. *PLoS ONE* **2015**, *10*, e0136800. [CrossRef] [PubMed]

49. Naor, A.; Lapierre, P.; Mevarech, M.; Papke, R.T.; Gophna, U. Low species barriers in halophilic archaea and the formation of recombinant hybrids. *Curr. Biol.* **2012**, *22*, 1444–1448. [CrossRef] [PubMed]

50. Holt, I.J.; Reyes, A. Human mitochondrial DNA replication. *Cold Spring Harb. Perspect. Biol.* **2012**, *4*, a012971. [CrossRef] [PubMed]

51. Eichler, J.; Maupin-Furlow, J. Post-translation modification in Archaea: Lessons from *Haloferax volcanii* and other haloarchaea. *FEMS Microbiol. Rev.* **2013**, *37*, 583–606. [CrossRef] [PubMed]

52. Kennelly, P.J. Protein Ser/Thr/Tyr phosphorylation in the Archaea. *J. Biol. Chem.* **2014**, *289*, 9480–9487. [CrossRef] [PubMed]

53. Kirkland, P.A.; Gil, M.A.; Karadzic, I.M.; Maupin-Furlow, J.A. Genetic and proteomic analyses of a proteasome-activating nucleotidase A mutant of the haloarchaeon *Haloferax volcanii*. *J. Bacteriol.* **2008**, *190*, 193–205. [CrossRef] [PubMed]

54. Garcia-Rodriguez, N.; Wong, R.P.; Ulrich, H.D. Functions of Ubiquitin and SUMO in DNA Replication and Replication Stress. *Front. Genet.* **2016**, *7*, 87. [CrossRef] [PubMed]

55. Humbard, M.A.; Miranda, H.V.; Lim, J.M.; Krause, D.J.; Pritz, J.R.; Zhou, G.; Chen, S.; Wells, L.; Maupin-Furlow, J.A. Ubiquitin-like small archaeal modifier proteins (SAMPs) in *Haloferax volcanii*. *Nature* **2010**, *463*, 54–60. [CrossRef] [PubMed]

56. Gras, S.; Chaumont, V.; Fernandez, B.; Carpentier, P.; Charrier-Savournin, F.; Schmitt, S.; Pineau, C.; Flament, D.; Hecker, A.; Forterre, P.; et al. Structural insights into a new homodimeric self-activated GTPase family. *EMBO Rep.* **2007**, *8*, 569–575. [CrossRef] [PubMed]

57. Peeters, E.; Driessen, R.P.; Werner, F.; Dame, R.T. The interplay between nucleoid organization and transcription in archaeal genomes. *Nat. Rev. Microbiol.* **2015**, *13*, 333–341. [CrossRef] [PubMed]

58. Bell, S.D.; Botting, C.H.; Wardleworth, B.N.; Jackson, S.P.; White, M.F. The interaction of Alba, a conserved archaeal chromatin protein, with Sir2 and its regulation by acetylation. *Science* **2002**, *296*, 148–151. [CrossRef] [PubMed]

59. Marsh, V.L.; McGeoch, A.T.; Bell, S.D. Influence of chromatin and single strand binding proteins on the activity of an archaeal MCM. *J. Mol. Biol.* **2006**, *357*, 1345–1350. [CrossRef] [PubMed]

60. Gristwood, T.; Duggin, I.G.; Wagner, M.; Albers, S.V.; Bell, S.D. The sub-cellular localization of *Sulfolobus* DNA replication. *Nucleic Acids Res.* **2012**, *40*, 5487–5496. [CrossRef] [PubMed]

![genes logo](GCAT TACG GCAT) *genes*

MDPI

Review

Non-Canonical Replication Initiation: You're Fired!

Bazilė Ravoitytė [1] and Ralf Erik Wellinger [2,*]

[1]   Nature Research Centre, Akademijos g. 2, LT-08412 Vilnius, Lithuania; bazilerav@gmail.com
[2]   CABIMER-Universidad de Sevilla, Avd Americo Vespucio sn, 41092 Sevilla, Spain
*   Correspondence: ralf.wellinger@cabimer.es; Tel.: +34-954-467-968

Academic Editor: Eishi Noguchi
Received: 10 November 2016; Accepted: 19 January 2017; Published: 27 January 2017

Abstract: The division of prokaryotic and eukaryotic cells produces two cells that inherit a perfect copy of the genetic material originally derived from the mother cell. The initiation of canonical DNA replication must be coordinated to the cell cycle to ensure the accuracy of genome duplication. Controlled replication initiation depends on a complex interplay of *cis*-acting DNA sequences, the so-called origins of replication (*ori*), with *trans*-acting factors involved in the onset of DNA synthesis. The interplay of *cis*-acting elements and *trans*-acting factors ensures that cells initiate replication at sequence-specific sites only once, and in a timely order, to avoid chromosomal endoreplication. However, chromosome breakage and excessive RNA:DNA hybrid formation can cause break-induced (BIR) or transcription-initiated replication (TIR), respectively. These non-canonical replication events are expected to affect eukaryotic genome function and maintenance, and could be important for genome evolution and disease development. In this review, we describe the difference between canonical and non-canonical DNA replication, and focus on mechanistic differences and common features between BIR and TIR. Finally, we discuss open issues on the factors and molecular mechanisms involved in TIR.

Keywords: replication control; RNA:DNA hybrid; transcription-initiated replication

1. Origin-Dependent Replication

1.1. Chromosomal DNA Replication Initiation in Escherichia coli and Saccharomyces cerevisiae

Replication initiation at a single origin (*ori*) in the bacteria *Escherichia coli* has been the first, and until present, best-described mechanism of a classical replication initiation (see Figure 1; for reviews, see References [1–5]). Within the circular *E. coli* chromosome [6], a single origin called *oriC* provides a platform for protein recognition, local double-stranded DNA (dsDNA) opening, and access of the replication machinery [1]. *OriC* contains multiple repeats of the DnaA-box consensus sequence, and an AT-rich DNA-unwinding element (DUE) adjacent to the DnaA box [7] for the ATP-driven binding of the initiator protein DnaA [1]. *OriC* activation is coupled with bacterial growth rate [8], to efficiently initiate replication at the appropriate time and to avoid replication initiation at particular origins more than once [9–13]. DnaA binds to *oriC* and facilitates binding of the helicase loader-helicase DnaC–DnaB complex to form the pre-priming complex [4,14]. The DnaB helicase then stably interacts with the DnaG primase until RNA primer synthesis is accomplished [15]. Probably, RNA primer synthesis induces conformational changes that release DnaB from DnaG, because primer synthesis is coordinated with or followed by translocation of DnaB to the junction of the replication fork (reviewed in [16]). Subsequently, primer elongation by the DNA polymerase III (DNA Pol III) holoenzyme marks the switch from replication initiation to elongation [17,18]. In contrast to the single origin found in *E. coli*, the budding yeast *Saccharomyces cerevisiae* contains about 400 replication origins. The number of origins per genome is related to the genome size, explaining why eukaryotic genomes require more replication origins for their timely genome duplication [19]. Yeast continues to be one of the most advantageous model systems to study the basis of eukaryotic replication, but in contrast to prokaryotic cells, yeast chromosomes are packaged into nucleosomes. Dependent on their activation timing, replication origins can be separated into early and late replicating origins ([20–22], reviewed in [23]). In general, origin-dependent replication initiation requires the following conditions to be fulfilled: recognition of origins, pre-replicative complex (pre-RC) assembly during G1 phase (origin-licensing), and activation of the pre-RC at G1/S-phase (origin-firing; see Figure 1 and Table 1). *S. cerevisiae* origins are defined by a specific consensus sequence, known as autonomously replicating sequence (ARS) [24–26]. The AT-rich ARS consensus sequence (ACS) itself is not sufficient for replication initiation [27] but is required for the loading of the pre-RC during G1 phase ([28,29]). The pre-RC is composed of the origin recognition complex proteins Orc1–6 (ORC), Cdc6, Cdt1, and an inactive form of the replicative helicase Mcm2–7 complex ([30–32], reviewed in [33]). At G1/S-phase, the Dbf4-dependent kinase (DDK) and S-phase-dependent cyclin-dependent kinases (S-CDKs) phosphorylate Mcm4, Sld2, and Sld3 ([34,35]), prior to the stepwise recruitment of replication factors Cdc45/Sld3/Sld7 and Sld2/Dpb11/Mcm10/GINS/DNA Pol-ε ([36–39], see [40] for a review). Building up of the active Cdc45/Mcm2–7/GINS (CMG) helicase complex completes the replisome formation [41] and, consequently, DNA synthesis by the DNA Pol-α-primase complex is initiated [42]. Replication initiation is completed by the loading of the proliferating cell nuclear antigen (PCNA) onto the DNA Pol-α synthesized primer to switch to processive DNA synthesis by DNA Pol-ε and Pol-δ (see [43]).

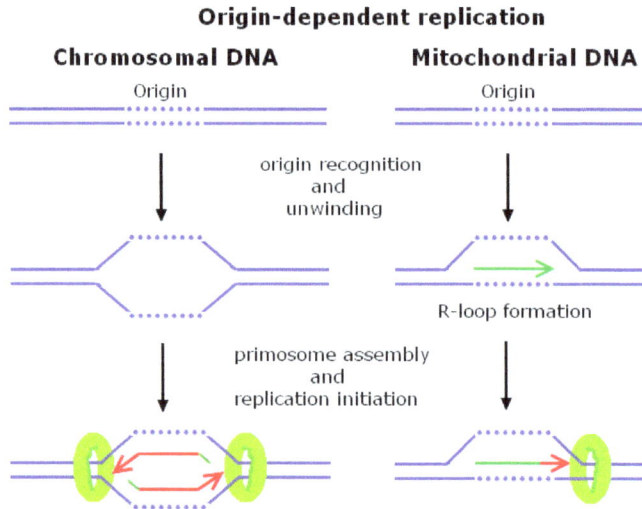

Figure 1. Schematic outline of origin-dependent initiation of chromosomal and mitochondrial DNA replication. *cis*-acting origin DNA sequences (dotted lines), RNA (green), newly synthesized DNA (red), and helicases (green circle) are indicated. Note that chromosomal origin unwinding is driven by protein–DNA interactions, while transcription-dependent R-loop formation is a key step in mitochondrial origin-unwinding. See text for more details.

Yeast has developed sophisticated mechanisms to avoid endoreplication events caused by replication re-initiation of already replicated origins. B-type CDKs prevent re-initiation through multiple overlapping mechanisms, including phosphorylation of ORC factors [44], nuclear exclusion of the Mcm2–7 complex and Cdc6 [45,46], transcriptional downregulation, polyubiquitination, and degradation of phosphorylated Cdc6 ([47–49]). Under certain conditions, traces of non-phosphorylatable Cdc6 [50] or mutations in components of the pre-replicative complex (origin recognition complex, Cdc6, and MCM proteins are sufficient to re-initiate DNA replication in G2/M cells. In the latter case, a Mec1 and

Mre11-Rad50-Xrs2 (MRX) complex-dependent DNA damage signaling pathway is activated to restrain the extent of re-replication and to promote survival when origin-localized replication control pathways are abrogated [51]. Genome-wide analysis suggests that replication re-initiation in G2/M phase primarily occurs at a subset of both active and latent origins, but is independent of chromosomal determinants that specify the use and timing of these origins in S phase [52]. Moreover, the frequency and locations of re-replication events differ from the S to the G2/M phase, illustrating the dynamic nature of DNA replication controls [52]. Additional mechanisms may exist to prevent chromosomal re-replication in metazoans [53]. Interestingly, a recent study identified 42 uncharacterized human genes that are required to prevent either DNA re-replication or unscheduled endoreplication [54].

1.2. Mitochondrial DNA Replication Initiation

The variation in mitochondrial DNA (mtDNA) copy number reflects the fact that its replication cycle is not coupled with S phase-restricted, chromosomal DNA replication. Replication of mtDNA is connected with mtDNA transcription through the formation of a RNA:DNA hybrid that has been first detected by electron microscopy as a short three-stranded DNA region [55]. During transcription, the nascent transcript behind an elongating RNA polymerase (RNAP) can invade the double stranded DNA duplex and hybridize with the complementary DNA template strand. The formation of an RNA:DNA hybrid, opposite to an unpaired non-template DNA strand, results in a so-called R-loop structure (for a review see [56]). RNA:DNA hybrids are also the onset of Okazaki fragments, which serve as primers during DNA lagging-strand replication (for a review see [57]; see Figure 1 and Table 1). In the case of mtDNA replication, an R-loop is required for replication priming [58] at the mtDNA heavy-strand replication origin (*OriH*) and light-strand replication origin (*OriL*) [59]. *OriH* and *OriL* consist of a promoter and downstream conserved sequences with a high GC content, and are conserved from *S. cerevisiae* to humans [60]. Budding yeast contains about eight *OriH*-like regions (*ori1–8*; [60]) of which *ori1–3* and *ori5* represent bona fide origins of replication (see [61,62]). The *OriH* region of many organisms includes three conserved sequence blocks called *CSB1*, *CSB2*, and *CSB3* [58], and transition from RNA to DNA synthesis is thought to happen at *CSB2* [63]. Yeast mitochondrial RNA polymerase Rpo41, the helicase Irc3, and the single-stranded DNA (ssDNA)-binding protein Rim1 are the main factors involved in DNA strand separation during mtDNA replication [64–66]. After processing by RNase H1, the RNA molecule is used as a primer for DNA synthesis by the *MIP1* encoded mitochondrial DNA polymerase γ (DNA Pol-γ) in budding yeasts [59]. Interestingly, in the absence of RNase H1, primer retention at *OriL* provides an obstacle for DNA Pol-γ [67], leading to mtDNA depletion and embryonic lethality in mice [68].

Apart from DNA Pol-γ, in metazoans the replicative mtDNA helicase Twinkle and the mitochondrial single-stranded DNA-binding protein (mtSSB) play key roles mtDNA replication fork progression (reviewed in [69,70]). The mechanism of mtDNA replication is not fully understood, and various possible mechanisms have been proposed ([71], reviewed in [72]). Currently, there are three main models of mtDNA replication. One is the initial "strand-displacement model", proposing that leading strand DNA synthesis begins at a specific site and advances approximately two-thirds of the way around the molecule before DNA synthesis is initiated on the lagging strand [73]. A second "strand-coupled model" refers to a strand-asynchronous, unidirectional replication mode [74]. A third "RITOLS model" (RNA incorporation throughout the lagging strand) proposes that replication initiates in the major noncoding region at *OriH*, while *OriL* is a major initiation site of lagging-strand DNA synthesis but the lagging strand is laid down initially as RNA [75]. The idea of transcription-dependent mtDNA replication initiation has been unanimously accepted. However, by taking advantage of mutants devoid of the mitochondrial RNA polymerase Rpo41, Fangman et. al. suggested that replication priming by transcription is not the only mechanism for mtDNA replication initiation in yeast [76–78]. Alternatively, the mitochondrial *ori5* has been shown to initiate mtDNA amplification by a rolling circle mechanism [79]. These kinds of replication events are linked to increased mtDNA damage and breaks by oxidative stress, and can be modulated by nuclease and recombinase activities carried out by Din7 and Mhr1, respectively [80].

Table 1. Factors required for origin-dependent DNA replication initiation in *Escherichia coli* and *Saccharomyces cerevisiae*.

Origin-Dependent Replication	E. coli	S. cerevisiae		
	Chromosomal DNA Replication	Chromosomal DNA Replication	Mitochondrial DNA Replication
Origin	OriC	ARS	OriH, OriL
DNA unwinding	DnaA, DnaB, DnaC, SSB	Cdc45, GINS, Mcm2–7, Mcm10, RPA	Rpo41, Irc3, Rim1
Replication priming/elongation	DnaG, DNA Pol III	DNA Pol-α-primase, DNA Pol-ε and Pol-δ	Rpo41, DNA Pol-γ

SSB: single-stranded DNA-binding protein; DNA Pol: DNA polymerase; RPA: replication protein A; ARS: autonomously replicating sequence.

Collectively, these findings demonstrate that mtDNA replication initiation is capable of adapting to stress situations, and that the stress-dependent, mitochondrial import of nuclear-encoded proteins such as Din7 and Mhr1 could provide another layer of mtDNA replication control. Interestingly, all other proteins involved in replication initiation are nuclear-encoded, and some genes, such as *RNH1*, encode both nuclear and mitochondrial protein isoforms [81]. It will be exciting to see if new players in mtDNA replication initiation may appear in response to different endogenous or exogenous stimuli. To date, little is known about how nuclear and mitochondrial replication checkpoints are interconnected, and how they control mtDNA replication initiation. Interestingly, a recent study showed that the DNA damage response protein kinase Rad53 (hChk2) is essential for an mtDNA inheritance checkpoint [82]. In mtDNA-depleted rho° cells, the DNA helicase Pif1 (petite integration frequency 1) undergoes Rad53-dependent phosphorylation. Pif1 is a highly conservative helicase localized to both nucleus and mitochondria in yeast and human cells [83] and promotes DNA replication through interaction with G-quadruplex DNA sequences ([84], reviewed in [85]). Thus, loss of mtDNA activates a nuclear checkpoint kinase that inhibits G1- to S-phase progression [82]. Pif1 is only one example of nuclear DNA helicases to protect mtDNA but, notably [86], it also has an essential role in recombination-dependent replication (as discussed subsequently). Future research may lead to the identification of other factors involved in the crosstalk between nuclear and mitochondrial genome duplication, and even improve our understanding of how the control of mitochondrial replication initiation is related to genome stability, aging, and mitochondrial diseases.

2. Origin-Independent Replication

2.1. Break-Induced Replication

A classic example of the initiation of origin-independent DNA replication events is recombination-dependent DNA replication, often called break-induced replication (BIR; see Figure 2 and Table 2, and [87] for a review). Kogoma and colleagues originally designated BIR in bacteria as DNA damage-inducible DNA replication, termed inducible stable DNA replication ((iSDR) [88,89], and reviewed in [90]). Double-strand end repair is initiated by break recognition and loading of the RecBCD helicase/nuclease complex. DNA unwinding by RecBCD leads to subsequent binding of RecA to ssDNA. Then, the strand exchange reaction between two recombining DNA double helices was proposed to as the mechanism by which DNA replication is primed [91,92]. DnaA is essential for helicase loading at *oriC*, whereas PriA, PriB, PriC, and DnaT appear to load DnaB into the forming replisome to promote replication fork assembly at a recombinational D-loop structure ([93], see [94] for a review). Finally, the branch migration and Holliday-junction resolving activities of the RuvABC complex are involved in the resolution of converging replication intermediates generated during iSDR [95].

BIR was later found to occur in yeast upon transformation of yeast with linearized DNA fragments [96,97]. BIR turned out to promote DNA replication restart at broken replication forks and telomeres ([98,99], and reviewed in [87,100,101]) being an error-prone recombination-dependent DNA repair process that occurs in G2/M when only one end of a double-strand break (DSB) is available for recombination [102]. BIR can be Rad51-dependent or independent [102,103]. Rad51 is homologous to the bacterial ssDNA-binding protein RecA, and mainly involved in the search for homology and strand-pairing stages of homologous recombination [104]. Rad51-independent BIR at a one-ended break can occur when long-range strand invasion is not required. It primarily operates during intramolecular recombination; however, intermolecular events mostly rely on Rad51-dependent strand invasion [98,105]. More than 95% of BIR events in *S. cerevisiae* are reported to be Rad51-dependent and do not require either Rad50 or Rad59 [98,106], thus we discuss the Rad51-dependent pathway in more detail. During Rad51-dependent BIR, a DSB end is resected to produce a 3'-ended single-stranded DNA tail, subsequently coated by Rad51 nucleoprotein filaments [102]. This Rad51 filament then invades a homologous sequence and a D-loop is created, followed by an extension of the invading strand by new DNA synthesis using the paired homologous sequence as a template [107]. BIR is known to be a multistep process in which strand invasion occurs rapidly; by contrast, new DNA synthesis does not initiate until 3–4 h after strand invasion [99,102,108]. Once initiated, DNA synthesis may be very processive and continue to the end of the donor chromosome (reviewed in [109]).

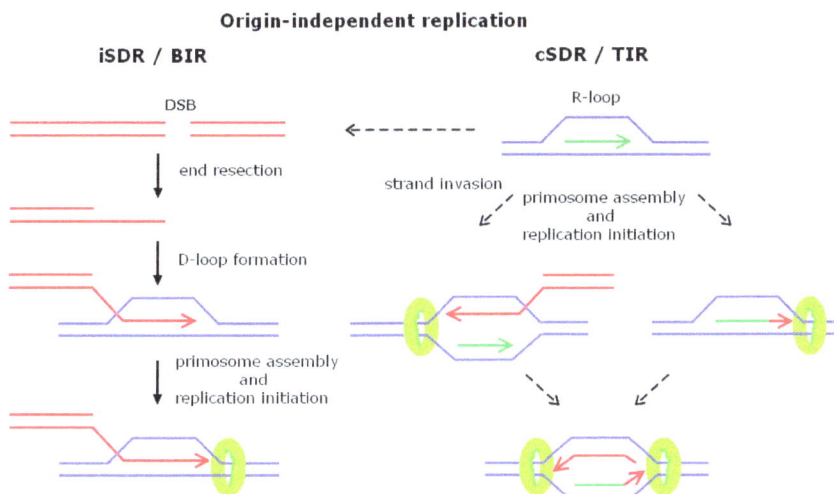

Figure 2. Schematic representation of possible mechanism involved in origin-independent replication initiation by inducible stable DNA replication/break-induced replication (iSDR/BIR) or constitutive stable DNA replication/transcription-initiated replication (cSDR/TIR). Invading and newly synthesized DNA (red), RNA (green), and helicases (green circle) are indicated. Dashed arrows indicate putative scenarios for TIR-dependent replication initiation. Note that none of these scenarios have been experimentally verified. See text for more details. DSB: double-strand break.

Yeast proteins taking part in BIR also play a role in recombination. Recombination proteins Rad51, Rad52, Rad54, Rad55, and Rad57 initiate BIR by promoting strand invasion and D-loop formation [88,98]. BIR requires leading- and lagging-strand DNA synthesis and all essential DNA replication factors, including Pol-α-primase,Cdc7,Cdt1, Mcm10, Ctf4 and CMG helicase complex (except Cdc6 and ORC proteins), specific for pre-RC assembly and specifically needed for origin-dependent

DNA replication [99,110]. It still remains to be determined how MCMs are recruited to the D-loop, but it is important to note that BIR occurs at the G2/M phase and normally depends on the Pif1 helicase. BIR may initiate in the absence of Pif1, but Pif1 appears to be required for long-range synthesis during BIR that proceeds by asynchronous synthesis of leading and lagging strands and leads to conservative inheritance of the new genetic material [111,112]. Analysis of BIR-dependent replication intermediates by 2D-agarose gels [113] revealed bubble arc-like migrating structures suggesting the accumulation of ssDNA at unrepaired DNA lesions within the template strand [112,114]. Investigation of BIR in yeast diploid cells led to observation of frequent switches of BIR between two homologous DNA templates, leading to the proposal that BIR is initiated via an unstable replication fork [115]. It was proposed that BIR could occur by several rounds of strand invasion, even at dispersed repeated sequences [115], leading to chromosome rearrangements [116]. However, the specific mechanisms of multiple strand invasions, D-loop displacement, and transition to a stable replication fork remain unknown.

Pol32, a nonessential subunit of Pol-δ, is another key player in BIR [111]. Pol32's role in BIR is not unequivocally clear, but it has been reported to be essential for Rad51-dependent BIR [99] and required for replication fork processivity [111]. Interestingly, it has been recently shown that theMus81 endonuclease is required to limit BIR-associated template switching during Pol32-dependent DNA synthesis [117]. The involvement of structure-specific nucleases in BIR, such as Mus81-Mms4, Slx1-Slx4, and Yen1, suggests that these nucleases are needed for the processing or resolution of various types of BIR-dependent replication intermediates [118].

Table 2. Factors required for origin-independent DNA replication by iSDR/BIR or cSDR/TIR.

	Function	iSDR	cSDR
	End processing	RecBCD	RecBCD
	Strand invasion	RecA	RecA
E. coli		DnaBC, PriAB	DnaBC, PriAB
	DNA unwinding	RecG	?
		DnaT	?
	Replication	DnaG,	DnaG,
	priming/elongation	DNA Pol III	DNA Pol I/Pol III
	Resolution	RuvABC	?

	Function	BIR	TIR
	End processing	MRX (Mre11-Rad50-Xrs2)	?
S. cerevisiae	Strand invasion	Rad51*, Rad52, Rad54, Rad55, Rad57	?
	DNA unwinding	Cdc45-MCM-GINS, DDK, Mcm10, Ctf4, RPA, Pif1	RNA:DNA hybrid
	Replication priming/elongation	Pol-α-primase, Pol-δ, Pol32*	?
	Resolution	Mus81-MMS4, Slx1–Slx4, Yen1	?

Note that BIR can be Rad51 and/or Pol32 independent (*). MCM: minichromosome maintenance complex; DDK: Dbf4-dependent kinase; Pif1: petite integration frequency 1.

The establishment of a replication fork appears to be the slowest step in BIR. In bacteria, the normal initiation role of the DnaA and DnaC proteins in loading DnaB helicase at origins is replaced by the PriA complex (reviewed in [119,120]). PriA is implicated in loading DnaB onto replication fork structures other than replisomes, thus making PriA indispensable for the completion of any replication fork repair [121]. There is no obvious PriA homologue in eukaryotes, but it has been speculated that such a protein must exist. In yeast, the DnaB helicase function is provided by the Mcm2–7complex, which is conserved in all eukaryotes. The Cdc7–Dbf4 protein kinase promotes assembly of a stable Cdc45–MCM complex exclusively on chromatin in S phase [37], and, interestingly, BIR also requires the cell cycle-dependent kinase Cdc7 to initiate BIR [110]. As Rad51-dependent

BIR occurs efficiently in G2-arrested yeast cells [102], either a subset of replication-competent MCM helicases remain bound to already replicated DNA, or DNA damage signaling leads to MCM-complex loading and Cdc7-dependent BIR activation in G2 phase. Recent studies show that SUMOylation and polyubiquitylation of MCM proteins have a role in replication initiation and termination, respectively [122–124]. It still remains to be determined if these post-translational MCM modifications affect BIR and if other helicases can drive BIR in the absence of MCM proteins. Pif1 may do so, as it already has a known role in BIR [111]. Pif1 is phosphorylated in response to DNA breaks by the Mec1/Rad53 DNA damage pathway in order to block the activity of telomerase at DNA breaks but not at chromosome ends [125], and its phosphorylation is required for BIR-mediated telomere replication in yeast [126]. Although this is pure speculation, it is conceivable that Pif1 might also be prone to Cdc7-dependent phosphorylation in order to fulfill its function in recombination-coupled DNA synthesis.

2.2. Transcription-Initiated Replication

R-loops have been shown to have roles in T4 bacteriophage, *E. coli* ColE1 plasmid, and mtDNA replication as well as B-cell immunoglobulin class switch recombination. R-loops are abundant structures, however, unscheduled R-loop formation challenges genome dynamics and function [127,128], and is related to neurological diseases and cancer (reviewed in [129–133]).

The role of R-loops in replication initiation was first demonstrated in *E. coli* ColE1 plasmid [134–136] and bacteriophage T4 replication (reviewed in [137]). Another legacy of Tokio Kogoma and colleagues was the discovery of *oriC*-independent DNA replication events ([138–140], reviewed in [90]). This type of replication was named constitutive stable DNA replication (cSDR) and, surprisingly, *E. coli* cells can stay alive exclusively on these origin-independent initiation events. One mutation that conferred this phenotype was found to inactivate the *rnhA* gene encoding RNase H1, an RNase specific to RNA in the RNA:DNA hybrid form [141,142]. cSDR was thought to originate from chromosomal sites named *oriK*, and only recently have specific candidate locations for *oriK* been mapped [143]. Moreover, it has been shown that origin-independent DNA synthesis arises in *E. coli* cells lacking the RecG helicase and results in chromosome duplication [144]. In contrast to RNase H1, RecG deals with replication fork fusion intermediates [145,146]; hence, origin-independent synthesis is initiated in different ways, but in both cases a fraction of forks will proceed in an orientation opposite to normal [144]. Drolet et al. [147] provided first evidence that R-loops can accumulate incells lacking *topA*, which encodes a type 1A topoisomerase that relieves negative supercoiling behind the RNAP, by showing that overexpression of *rnhA* partially compensates for the lack of *topA*. Notably, *E. coli* possesses two type 1A enzymes, Top1 (*topA*-encoded) and Top3 (*topB*-encoded), but only cells lacking Top1 are prone to cSDR [148]. Apart from transcription, cSDR requires RecA, and the primosome-complex including PriA, PriB, DnaT, and DNA Pol I [90,149,150]. RecA may also participate in cSDR by binding to ssDNA to stabilize an R-loop, or facilitate an inverse strand exchange reaction performed by RecA ([151,152], see Figure 2). In cSDR, DNA Pol I is thought to extend the RNA of the R-loop and to provide a substrate for PriA binding, as well as DnaB and DNA Pol III loading [90]. Interestingly, cSDR uses the same replicative helicase (DnaB) and replisome components (DNA Pol III) to initiate replication from *oriC*, but uses the PriA-dependent primosome for replicative helicase loading [90], as is the case for replication restart of disassembled replisomes [94]. Improperly regulated DNA replication may lead to various consequences related to genome instability. Interestingly, evidence that R-loop-dependent replication leads to DNA breakage and genome instability in non-growing *E. coli* cells has been presented [153], and mutations reducing replication from R-loops suppress the defects of growth, chromosome segregation, and DNA supercoiling in cells lacking Top1 and RNase H1 activity [154].

Transcription-linked replication initiation in eukaryotic cells was thought to be an exclusive feature of mtDNA replication. Yet, some highly transcribed DNA regions, such as RNAPI-transcribed ribosomal DNA (rDNA) or RNAP III-transcribed genes, were shown to be hot spots for R-loop formation in yeast mutants lacking RNases H [155,156]. In addition, mutants lacking an RNA/DNA helicase Sen1 [157,158] or the yeast Pab1-binding protein Pbp1 (hAtaxin-2) had been found to increase R-loop formation [159]. The absence of RNase H and Top1 activities causes synthetic lethality in yeast, suggesting that persistent R-loop formation could constrain cell viability [160,161]. Accordingly, persistent R-loop formation could be induced by treatment of RNase H mutants with the Top1 inhibitor camptothecin (CPT) leading to the detection of unscheduled transcription-initiated replication (TIR) events in yeast ([161], see Figure 2). TIR initiation intermediates were observed within the rDNA region, but were not linked to a defined replication origin; moreover, they were observed in the late S/G2 phase of the cell cycle, when replication termination and completion was expected to take place [161]. TIR was RNAPI transcription-dependent and led to replication fork pausing sites at sites of protein–DNA interaction. Taken together, these results suggest that R-loops could mediate origin-independent replication initiation events that constitute a non-canonical replisome, lacking the factors required to bypass replication constrains.

The factors and mechanisms participating in transcription-initiated replication events still remain to be elucidated. Various nonexclusive mechanisms could cooperate to trigger TIR events (summarized in Figure 2). These include strand invasion-dependent replication events that might be stimulated by the presence of single-stranded DNA within R-loops. In the absence of RNase H and Top1 activities, the rDNA locus turns into a hotspot for DSBs [161], thus it is conceivable that these DSBs drive recombination-dependent replication such as BIR. Other possibilities include that R-loops cause replication fork collapse and TIR is the result of replication restart of a replisome–RNAP complex [162,163]. An interesting possibility would be de novo replisome assembly at an R-loop. The RNA present within the R-loop could prime leading-strand synthesis and provoke assembly of replication-competent replicases at S/G2 phase [164]. Apparently, ssDNA opposite an RNA:DNA hybrid could activate Mec1-mediated checkpoint activation and binding of the replication protein A (RPA) complex, which has been shown to be involved in replication initiation as well as DNA repair by interacting with both the DNA Pol-α-primase complex and with DNA Pol-δ [164,165]. An R-loop may promote DNA replication restart by Pol-α-driven DNA synthesis, since the essential DNA Pol-α-primase subunit Pol12 remains active and phosphorylated in S/G2 and is inactivated while cells exit mitosis [44,161,166]. Moreover, a recent work by Symington and coworkers suggests that BIR occurs by a conservative mode of DNA synthesis [107]. Thus, it will be interesting to determine whether the same is true for TIR, or if TIR pursues a semiconservative replication mode. It is striking that in *E. coli*, many factors involved in iSDR are also needed for cSDR. These findings suggest that in yeast, many factors involved in BIR might be required for TIR. These factors include proteins involved in homologous recombination, DNA end-processing, helicases, primases, DNA polymerases, and, finally, structure-specific endonucleases (as listed in Table 2). Nevertheless, genetic interactions in yeast cells between RNase H deficiencies and proteins involved in BIR still remain to be determined.

Yet-to-be determined questions include whether TIR is limited to rDNA, and whether TIR can be observed in other RNA/DNA helicases mutants, including Sen1 [156–158] or the yeast ataxin-2 protein Pbp1 [159]. Recently, it has been shown that replication initiates, albeit very infrequently, within the telomeric repeats [167]. A long noncoding telomeric repeat-containing RNA (TERRA) has been implicated in telomere maintenance during replicative senescence and cancer [168,169]. TERRA accumulates specifically at short telomeres and may promote replication-fork restarting by recruiting homology-directed repair (HDR) mediators or even by directly priming replication in an origin-independent manner [167], similar to what was reported by Stuckey et al. [161]. This proposal might be supported by the fact that the cell cycle regulation of TERRA becomes perturbed at telomeres that are maintained by HDR, and that TERRA remains telomere-associated at G2/M in cells that use the alternative lengthening of telomeres (ALT) mechanism [170]. Interestingly, loss of ATP-dependent helicase ATRX that is frequently mutated in ALT-positive cancers, leads to persistent association of RPA with telomeres after DNA replication [170]. ATRX is involved in establishing transcriptionally silenced heterochromatin, and one hypothesis is that ATRX helicase and ATPase activity resolves

G4 DNA secondary structures formed opposite of a TERRA-containing R-loop ([169,171], reviewed in [167]).

3. Conclusions

Since the detection of recombination-dependent replication of the *E. coli* chromosome by Lark and Kogoma about 50 years ago [172], we have learned a lot about mechanisms that can lead to non-canonical replication initiation in prokaryotic and eukaryotic cells. It is generally accepted that recombination serves to rescue broken chromosomes and stalled replication forks, however, we are far away from the complete picture on how cells manage to bypass the need for origin-dependent replication initiation. The mechanistic models and enzymatic steps leading to iSDR and cSDR in *E. coli* can be considered as a blueprint for BIR and TIR events in eukaryotic systems. Interestingly, all known features of BIR and TIR can participate in mtDNA replication events. Nevertheless, an important difference is noted by the fact that nuclear BIR and TIR events happen in a chromatin context with eukaryotic replication, starting with nucleosome packaging.

Many aspects of non-canonical DNA replication in eukaryotes still remain unknown and deserve to be addressed in the future; in particular, the factors driving replication fork progression and the mode of TIR-dependent DNA synthesis need to be characterized. Special attention should be given to the identification of key replication factors involved in TIR, such as DNA polymerases and helicases, but also to otherwise auxiliary replication proteins such as Pol32. R-loops are essential for the onset of TIR, and this might not be the only difference between TIR and BIR events. As outlined in Figure 2, the question remains if TIR is driven by strand invasion of the R-loop. TIR has been characterized only in repetitive ribosomal DNA sequences, raising the question of whether it is sister-chromatid-dependent, or if it uses non-sister chromatids as a template for DNA synthesis. In either case, strand invasion could be Rad51-dependent or independent. However, the role of Rad51 in TIR still needs to be determined. Genetic screens might help to shed light on factors required for TIR initiation and provide more insight to the differences between TIR and BIR.

The other model proposed in Figure 2 includes de novo assembly of a replication fork at an R-loop. In this case, which replication factors would be assembled at an R-loop, and would this kind of non-canonical replication restart be S phase-dependent? Would conservative or semiconservative replication account for the newly synthesized DNA? Could an R-loop even contribute to the activation of less defined replication origins in higher eukaryotes? Unrevealed functions of R-loops in higher eukaryotes may include a role the epigenetic regulation of origin-dependent replication initiation [173,174]. Interestingly, a nuclease-resistant G-quadruplex hybrid structure involving both RNA and DNA is present at the mtDNA replication initiation site [65]. G-rich RNA mediates Epstein-Barr virus nuclear antigen 1 EBNA1 and ORC interaction [175], thus it is conceivable that that transcription-related RNA structures might replace the need for specific origin-recognition sequences. By using a high-resolution PCR strategy to localize replication origins directly on total unfractionated human DNA, over-replicated regions were found to overlap with transcription initiation sites of CpG island promoters [176] and, recently, active transcription was proposed to be a driving force for the human parasite *Leishmania major* spatial and the temporal program of DNA replication [177]. Last but not least, TIR could be considered as an ancient mechanism to promote gene amplification events linked to nuclear differentiation and evolution. In order to resolve these questions, future studies should include higher eukaryotic model systems to see if TIR has a role in genome stability connected to various human diseases, including cancer.

Acknowledgments: This work was supported in part by the Spanish Ministry of Science and Innovation (MINECO; BFU2015-69183-P) and the European Union (FEDER; R.E.W.) and an ERASMUS+ fellowship of the European Union (B.R.). We would like to thank Hélène Gaillard and Daniel Fitzgerald for critical reading of the manuscript, and Zoe Cooper for style correction.

Author Contributions: R.E.W. organized and wrote the paper. B.R. contributed to writing the paper and designing the figures.

Conflicts of Interest: The authors declare no conflict of interest. The founding sponsors had no role in the design of the study; in the collection, analyses, or interpretation of data; in the writing of the manuscript, and in the decision to publish the results.

References

1. Fuller, R.S.; Funnell, B.E.; Kornberg, A. The DnaA protein complex with the *E. coli* chromosomal replication origin (*oriC*) and other DNA sites. *Cell* **1984**, *38*, 889–900. [CrossRef]
2. Hwang, D.S.; Kornberg, A. Opening of the replication origin of *Escherichia coli* by DnaA protein with protein HU or IHF. *J. Biol. Chem.* **1992**, *267*, 23083–23086. [PubMed]
3. Boye, E.; Lobner-Olesen, A.; Skarstad, K. Limiting DNA replication to once and only once. *EMBO Rep.* **2000**, *1*, 479–483. [CrossRef] [PubMed]
4. Mott, M.L.; Berger, J.M. DNA replication initiation: Mechanisms and regulation in bacteria. *Nat. Rev. Microbiol.* **2007**, *5*, 343–354. [CrossRef] [PubMed]
5. Wolanski, M.; Donczew, R.; Zawilak-Pawlik, A.; Zakrzewska-Czerwinska, J. OriC-encoded instructions for the initiation of bacterial chromosome replication. *Front. Microbiol.* **2014**, *5*, 735. [PubMed]
6. Kohara, Y.; Akiyama, K.; Isono, K. The physical map of the whole *E. coli* chromosome: Application of a new strategy for rapid analysis and sorting of a large genomic library. *Cell* **1987**, *50*, 495–508. [CrossRef]
7. Speck, C.; Messer, W. Mechanism of origin unwinding: Sequential binding of DnaA to double- and single-stranded DNA. *EMBO J.* **2001**, *20*, 1469–1476. [CrossRef] [PubMed]
8. Wold, S.; Skarstad, K.; Steen, H.B.; Stokke, T.; Boye, E. The initiation mass for DNA replication in *Escherichia coli* K-12 is dependent on growth rate. *EMBO J.* **1994**, *13*, 2097–2102. [PubMed]
9. Yamaki, H.; Ohtsubo, E.; Nagai, K.; Maeda, Y. The *oriC* unwinding by dam methylation in *Escherichia coli*. *Nucleic Acids Res.* **1988**, *16*, 5067–5073. [CrossRef] [PubMed]
10. Campbell, J.L.; Kleckner, N. *E. coli oriC* and the *dnaA* gene promoter are sequestered from dam methyltransferase following the passage of the chromosomal replication fork. *Cell* **1990**, *62*, 967–979.
11. Boye, E.; Stokke, T.; Kleckner, N.; Skarstad, K. Coordinating DNA replication initiation with cell growth: Differential roles for DnaA and SeqA proteins. *Proc. Natl. Acad. Sci. USA* **1996**, *93*, 12206–12211.
12. Torheim, N.K.; Boye, E.; Løbner-Olesen, A.; Stokke, T.; Skarstad, K. The *Escherichia coli* SeqA protein destabilizes mutant DnaA204 protein. *Mol. Microbiol.* **2000**, *37*, 629–638. [CrossRef] [PubMed]
13. Fujimitsu, K.; Senriuchi, T.; Katayama, T. Specific genomic sequences of *E. coli* promote replicational initiation by directly reactivating ADP-DnaA. *Genes Dev.* **2009**, *23*, 1221–1233. [CrossRef] [PubMed]
14. Bramhill, D.; Kornberg, A. Duplex opening by DnaA protein at novel sequences in initiation of replication at the origin of the *E. coli* chromosome. *Cell* **1988**, *52*, 743–755. [CrossRef]
15. Chang, P.; Marians, K.J. Identification of a region of *Escherichia coli* DnaB required for functional interaction with DnaG at the replication fork. *J. Biol. Chem.* **2000**, *275*, 26187–26195. [CrossRef] [PubMed]
16. Chodavarapu, S.; Kaguni, J.M. Replication initiation in bacteria. In *The Enzymes*; Academic Press: New York, NY, USA, 2016; Volume 39, Chapter 1, pp. 1–30.
17. O'Donnell, M.E.; Kornberg, A. Complete replication of templates by *Escherichia coli* DNA polymerase III holoenzyme. *J. Biol. Chem.* **1985**, *260*, 12884–12889. [PubMed]
18. Lewis, J.S.; Jergic, S.; Dixon, N.E. The *E. coli* DNA replication fork. In *The Enzymes*; Academic Press: New York, NY, USA, 2016; Volume 39, Chapter 2, pp. 31–88.
19. Gilbert, D.M. Replication origins in yeast versus metazoa: Separation of the haves and the have nots. *Curr. Opin. Genet. Dev.* **1998**, *8*, 194–199. [CrossRef]
20. Yamashita, M.; Hori, Y.; Shinomiya, T.; Obuse, C.; Tsurimoto, T.; Yoshikawa, H.; Shirahige, K. The efficiency and timing of initiation of replication of multiple replicons of *Saccharomyces cerevisiae* chromosome VI. *Genes Cells* **1997**, *2*, 655–665. [CrossRef] [PubMed]
21. Das, S.P.; Borrman, T.; Liu, V.W.; Yang, S.C.; Bechhoefer, J.; Rhind, N. Replication timing is regulated by the number of MCMs loaded at origins. *Genome Res.* **2015**, *25*, 1886–1892. [CrossRef] [PubMed]
22. Peace, J.M.; Ter-Zakarian, A.; Aparicio, O.M. Rif1 regulates initiation timing of late replication origins throughout the *S. cerevisiae* genome. *PLoS ONE* **2014**, *9*, e98501. [CrossRef] [PubMed]
23. Goren, A.; Cedar, H. Replicating by the clock. *Nat. Rev. Mol. Cell Biol.* **2003**, *4*, 25–32. [CrossRef] [PubMed]

24. Stinchcomb, D.T.; Struhl, K.; Davis, R.W. Isolation and characterisation of a yeast chromosomal replicator. *Nature* **1979**, *282*, 39–43. [CrossRef] [PubMed]

25. Bell, S.P.; Stillman, B. ATP-dependent recognition of eukaryotic origins of DNA replication by a multiprotein complex. *Nature* **1992**, *357*, 128–134. [CrossRef] [PubMed]

26. Marahrens, Y.; Stillman, B. A yeast chromosomal origin of DNA replication defined by multiple functional elements. *Science* **1992**, *255*, 817–823. [CrossRef] [PubMed]

27. Nieduszynski, C.A.; Knox, Y.; Donaldson, A.D. Genome-wide identification of replication origins in yeast by comparative genomics. *Genes Dev.* **2006**, *20*, 1874–1879. [CrossRef] [PubMed]

28. Rao, H.; Stillman, B. The origin recognition complex interacts with a bipartite DNA binding site within yeast replicators. *Proc. Natl. Acad. Sci. USA* **1995**, *92*, 2224–2228. [CrossRef] [PubMed]

29. Speck, C.; Chen, Z.; Li, H.; Stillman, B. ATPase-dependent cooperative binding of ORC and Cdc6 to origin DNA. *Nat. Struct. Mol. Biol.* **2005**, *12*, 965–971. [CrossRef] [PubMed]

30. Aparicio, O.M.; Weinstein, D.M.; Bell, S.P. Components and dynamics of DNA replication complexes in *S. cerevisiae*: Redistribution of MCM proteins and Cdc45p during S phase. *Cell* **1997**, *91*, 59–69. [CrossRef]

31. Kawasaki, Y.; Kim, H.D.; Kojima, A.; Seki, T.; Sugino, A. Reconstitution of *Saccharomyces cerevisiae* prereplicative complex assembly in vitro. *Genes Cells* **2006**, *11*, 745–756. [CrossRef] [PubMed]

32. Tanaka, S.; Diffley, J.F. Interdependent nuclear accumulation of budding yeast Cdt1 and Mcm2–7 during G1 phase. *Nat. Cell Biol.* **2002**, *4*, 198–207. [CrossRef] [PubMed]

33. Chesnokov, I.N. Multiple functions of the origin recognition complex. *Int. Rev. Cytol.* **2007**, *256*, 69–109.

34. Tanaka, S.; Umemori, T.; Hirai, K.; Muramatsu, S.; Kamimura, Y.; Araki, H. CDK-dependent phosphorylation of Sld2 and Sld3 initiates DNA replication in budding yeast. *Nature* **2007**, *445*, 328–332. [CrossRef] [PubMed]

35. Sheu, Y.J.; Stillman, B. Cdc7-Dbf4 phosphorylates MCM proteins via a docking site-mediated mechanism to promote S phase progression. *Mol. Cell* **2006**, *24*, 101–113. [CrossRef] [PubMed]

36. Zou, L.; Stillman, B. Formation of a preinitiation complex by S-phase cyclin CDK-dependent loading of Cdc45p onto chromatin. *Science* **1998**, *280*, 593–596. [CrossRef] [PubMed]

37. Muramatsu, S.; Hirai, K.; Tak, Y.S.; Kamimura, Y.; Araki, H. CDK-dependent complex formation between replication proteins Dpb11, Sld2, Polε, and GINS in budding yeast. *Genes Dev.* **2010**, *24*, 602–612.

38. Homesley, L.; Lei, M.; Kawasaki, Y.; Sawyer, S.; Christensen, T.; Tye, B.K. Mcm10 and the Mcm2–7 complex interact to initiate DNA synthesis and to release replication factors from origins. *Genes Dev.* **2000**, *14*, 913–926.

39. Gambus, A.; Jones, R.C.; Sanchez-Diaz, A.; Kanemaki, M.; van Deursen, F.; Edmondson, R.D.; Labib, K. GINS maintains association of Cdc45 with MCM in replisome progression complexes at eukaryotic DNA replication forks. *Nat. Cell Biol.* **2006**, *8*, 358–366. [CrossRef] [PubMed]

40. Tanaka, S.; Araki, H. Helicase activation and establishment of replication forks at chromosomal origins of replication. *Cold Spring Harb. Perspect. Biol.* **2013**, *5*, a010371. [CrossRef] [PubMed]

41. Moyer, S.E.; Lewis, P.W.; Botchan, M.R. Isolation of the Cdc45/Mcm2–7/GINS (CMG) complex, a candidate for the eukaryotic DNA replication fork helicase. *Proc. Natl. Acad. Sci. USA* **2006**, *103*, 10236–10241.

42. Foiani, M.; Marini, F.; Gamba, D.; Lucchini, G.; Plevani, P. The B subunit of the DNA polymerase alpha-primase complex in *Saccharomyces cerevisiae* executes an essential function at the initial stage of DNA replication. *Mol. Cell. Biol.* **1994**, *14*, 923–933. [CrossRef] [PubMed]

43. Garg, P.; Burgers, P.M. DNA polymerases that propagate the eukaryotic DNA replication fork. *Crit. Rev. Biochem. Mol. Biol.* **2005**, *40*, 115–128. [CrossRef] [PubMed]

44. Weinreich, M.; Liang, C.; Chen, H.H.; Stillman, B. Binding of cyclin-dependent kinases to ORC and Cdc6p regulates the chromosome replication cycle. *Proc. Natl. Acad. Sci. USA* **2001**, *98*, 11211–11217.

45. Labib, K.; Diffley, J.F.; Kearsey, S.E. G1-phase and B-type cyclins exclude the DNA-replication factor Mcm4 from the nucleus. *Nat. Cell Biol.* **1999**, *1*, 415–422. [PubMed]

46. Nguyen, V.Q.; Co, C.; Li, J.J. Cyclin-dependent kinases prevent DNA re-replication through multiple mechanisms. *Nature* **2001**, *411*, 1068–1073. [CrossRef] [PubMed]

47. Moll, T.; Tebb, G.; Surana, U.; Robitsch, H.; Nasmyth, K. The role of phosphorylation and the CDC28 protein kinase in cell cycle-regulated nuclear import of the *S. cerevisiae* transcription factor SWI5. *Cell* **1991**, *66*, 743–758. [CrossRef]

48. Drury, L.S.; Perkins, G.; Diffley, J.F. The Cdc4/34/53 pathway targets Cdc6p for proteolysis in budding yeast. *EMBO J.* **1997**, *16*, 5966–5976. [CrossRef] [PubMed]

49. Drury, L.S.; Perkins, G.; Diffley, J.F. The cyclin-dependent kinase Cdc28p regulates distinct modes of Cdc6p proteolysis during the budding yeast cell cycle. *Curr. Biol.* **2000**, *10*, 231–240. [CrossRef]

50. Honey, S.; Futcher, B. Roles of the CDK Phosphorylation Sites of Yeast Cdc6 in Chromatin Binding and Rereplication. *Mol. Biol. Cell* **2007**, *18*, 1324–1336. [CrossRef] [PubMed]

51. Archambault, V.; Ikui, A.E.; Drapkin, B.J.; Cross, F.R. Disruption of mechanisms that prevent rereplication triggers a DNA damage response. *Mol. Cell. Biol.* **2005**, *25*, 6707–6721. [CrossRef] [PubMed]

52. Green, B.M.; Morreale, R.J.; Ozaydin, B.; Derisi, J.L.; Li, J.J. Genome-wide mapping of DNA synthesis in *Saccharomyces cerevisiae* reveals that mechanisms preventing reinitiation of DNA replication are not redundant. *Mol. Biol. Cell* **2006**, *17*, 2401–2414. [CrossRef] [PubMed]

53. Blow, J.J.; Dutta, A. Preventing re-replication of chromosomal DNA. *Nat. Rev. Mol. Cell Biol.* **2005**, *6*, 476–486. [CrossRef] [PubMed]

54. Vassilev, A.; Lee, C.Y.; Vassilev, B.; Zhu, W.; Ormanoglu, P.; Martin, S.E.; DePamphilis, M.L. Identification of genes that are essential to restrict genome duplication to once per cell division. *Oncotarget* **2016**, *7*, 34956–34976. [CrossRef] [PubMed]

55. Kasamatsu, H.; Robberson, D.L.; Vinograd, J. A novel closed-circular mitochondrial DNA with properties of a replicating intermediate. *Proc. Natl. Acad. Sci. USA* **1971**, *68*, 2252–2257. [CrossRef] [PubMed]

56. Aguilera, A.; Garcia-Muse, T. R loops: From transcription byproducts to threats to genome stability. *Mol. Cell* **2012**, *46*, 115–124. [CrossRef] [PubMed]

57. Lujan, S.A.; Williams, J.S.; Kunkel, T.A. DNA polymerases divide the labor of genome replication. *Trends Cell Biol.* **2016**, *26*, 640–654. [CrossRef] [PubMed]

58. Xu, B.; Clayton, D.A. RNA-DNA hybrid formation at the human mitochondrial heavy-strand origin ceases at replication start sites: An implication for RNA-DNA hybrids serving as primers. *EMBO J.* **1996**, *15*, 3135–3143. [PubMed]

59. Nicholls, T.J.; Minczuk, M. In D-loop: 40 years of mitochondrial 7S DNA. *Exp. Gerontol.* **2014**, *56*, 175–181. [CrossRef] [PubMed]

60. Baldacci, G.; Cherif-Zahar, B.; Bernardi, G. The initiation of DNA replication in the mitochondrial genome of yeast. *EMBO J.* **1984**, *3*, 2115–2120. [PubMed]

61. Shadel, G.S. Yeast as a model for human mtDNA replication. *Am. J. Hum. Genet.* **1999**, *65*, 1230–1237. [CrossRef] [PubMed]

62. Williamson, D. The curious history of yeast mitochondrial DNA. *Nat. Rev. Genet.* **2002**, *3*, 475–481. [PubMed]

63. Wanrooij, P.H.; Uhler, J.P.; Shi, Y.; Westerlund, F.; Falkenberg, M.; Gustafsson, C.M. A hybrid G-quadruplex structure formed between RNA and DNA explains the extraordinary stability of the mitochondrial R-loop. *Nucleic Acids Res.* **2012**, *40*, 10334–10344. [CrossRef] [PubMed]

64. Sanchez-Sandoval, E.; Diaz-Quezada, C.; Velazquez, G.; Arroyo-Navarro, L.F.; Almanza-Martinez, N.; Trasvina-Arenas, C.H.; Brieba, L.G. Yeast mitochondrial RNA polymerase primes mitochondrial DNA polymerase at origins of replication and promoter sequences. *Mitochondrion* **2015**, *24*, 22–31. [CrossRef] [PubMed]

65. Sedman, T.; Gaidutsik, I.; Villemson, K.; Hou, Y.; Sedman, J. Double-stranded DNA-dependent ATPase Irc3p is directly involved in mitochondrial genome maintenance. *Nucleic Acids Res.* **2014**, *42*, 13214–13227. [CrossRef] [PubMed]

66. Van Dyck, E.; Foury, F.; Stillman, B.; Brill, S.J. A single-stranded DNA binding protein required for mitochondrial DNA replication in *S. cerevisiae* is homologous to *E. coli* SSB. *EMBO J.* **1992**, *11*, 3421–3430. [PubMed]

67. Holmes, J.B.; Akman, G.; Wood, S.R.; Sakhuja, K.; Cerritelli, S.M.; Moss, C.; Bowmaker, M.R.; Jacobs, H.T.; Crouch, R.J.; Holt, I.J. Primer retention owing to the absence of RNase H1 is catastrophic for mitochondrial DNA replication. *Proc. Natl. Acad. Sci. USA* **2015**, *112*, 9334–9339. [CrossRef] [PubMed]

68. Cerritelli, S.M.; Frolova, E.G.; Feng, C.; Grinberg, A.; Love, P.E.; Crouch, R.J. Failure to produce mitochondrial DNA results in embryonic lethality in *Rnaseh1* null mice. *Mol. Cell* **2003**, *11*, 807–815. [CrossRef]

69. Holt, I.J. Mitochondrial DNA replication and repair: All a flap. *Trends Biochem. Sci.* **2009**, *34*, 358–365. [CrossRef] [PubMed]

70. Ciesielski, G.L.; Oliveira, M.T.; Kaguni, L.S. Animal mitochondrial DNA replication. In *The Enzymes*; Academic Press: New York, NY, USA, 2016; Volume 39, Chapter 8, pp. 255–292.

71. Gerhold, J.M.; Aun, A.; Sedman, T.; Joers, P.; Sedman, J. Strand invasion structures in the inverted repeat of *Candida albicans* mitochondrial DNA reveal a role for homologous recombination in replication. *Mol. Cell* **2010**, *39*, 851–861. [CrossRef] [PubMed]

72. Gustafsson, C.M.; Falkenberg, M.; Larsson, N.G. Maintenance and expression of mammalian mitochondrial DNA. *Annu. Rev. Biochem.* **2016**, *85*, 133–160. [CrossRef] [PubMed]

73. Clayton, D.A. Replication of animal mitochondrial DNA. *Cell* **1982**, *28*, 693–705. [CrossRef]

74. Holt, I.J.; Lorimer, H.E.; Jacobs, H.T. Coupled leading- and lagging-strand synthesis of mammalian mitochondrial DNA. *Cell* **2000**, *100*, 515–524. [CrossRef]

75. Yasukawa, T.; Reyes, A.; Cluett, T.J.; Yang, M.Y.; Bowmaker, M.; Jacobs, H.T.; Holt, I.J. Replication of vertebrate mitochondrial DNA entails transient ribonucleotide incorporation throughout the lagging strand. *EMBO J.* **2006**, *25*, 5358–5371. [CrossRef] [PubMed]

76. Fangman, W.L.; Henly, J.W.; Churchill, G.; Brewer, B.J. Stable maintenance of a 35-base-pair yeast mitochondrial genome. *Mol. Cell. Biol.* **1989**, *9*, 1917–1921. [CrossRef] [PubMed]

77. Fangman, W.L.; Henly, J.W.; Brewer, B.J. *RPO41*-independent maintenance of [*rho-*] mitochondrial DNA in *Saccharomyces cerevisiae*. *Mol. Cell. Biol.* **1990**, *10*, 10–15. [CrossRef] [PubMed]

78. Lorimer, H.E.; Brewer, B.J.; Fangman, W.L. A test of the transcription model for biased inheritance of yeast mitochondrial DNA. *Mol. Cell. Biol.* **1995**, *15*, 4803–4809. [CrossRef] [PubMed]

79. Ling, F.; Hori, A.; Shibata, T. DNA recombination-initiation plays a role in the extremely biased inheritance of yeast [*rho-*] mitochondrial DNA that contains the replication origin *ori5*. *Mol. Cell. Biol.* **2007**, *27*, 1133–1145. [CrossRef] [PubMed]

80. Ling, F.; Hori, A.; Yoshitani, A.; Niu, R.; Yoshida, M.; Shibata, T. Din7 and Mhr1 expression levels regulate double-strand-break-induced replication and recombination of mtDNA at *ori5* in yeast. *Nucleic Acids Res.* **2013**, *41*, 5799–5816. [CrossRef] [PubMed]

81. Engel, M.L.; Hines, J.C.; Ray, D.S. The *Crithidia fasciculata RNH1* gene encodes both nuclear and mitochondrial isoforms of RNase H. *Nucleic Acids Res.* **2001**, *29*, 725–731. [CrossRef] [PubMed]

82. Crider, D.G.; Garcia-Rodriguez, L.J.; Srivastava, P.; Peraza-Reyes, L.; Upadhyaya, K.; Boldogh, I.R.; Pon, L.A. Rad53 is essential for a mitochondrial DNA inheritance checkpoint regulating G1 to S progression. *J. Cell Biol.* **2012**, *198*, 793–798. [CrossRef] [PubMed]

83. Futami, K.; Shimamoto, A.; Furuichi, Y. Mitochondrial and nuclear localization of human Pif1 helicase. *Biol. Pharm. Bull.* **2007**, *30*, 1685–1692. [CrossRef] [PubMed]

84. Sanders, C.M. Human Pif1 helicase is a G-quadruplex DNA-binding protein with G-quadruplex DNA-unwinding activity. *Biochem. J.* **2010**, *430*, 119–128. [CrossRef] [PubMed]

85. Bochman, M.L.; Sabouri, N.; Zakian, V.A. Unwinding the functions of the Pif1 family helicases. *DNA Repair (Amst)* **2010**, *9*, 237–249. [CrossRef] [PubMed]

86. Ding, L.; Liu, Y. Borrowing nuclear DNA helicases to protect mitochondrial DNA. *Int. J. Mol. Sci.* **2015**, *16*, 10870–10887. [CrossRef] [PubMed]

87. Anand, R.P.; Lovett, S.T.; Haber, J.E. Break-induced DNA replication. *Cold Spring Harb. Perspect. Biol.* **2013**, *5*, a010397. [CrossRef] [PubMed]

88. Asai, T.; Sommer, S.; Bailone, A.; Kogoma, T. Homologous recombination-dependent initiation of DNA replication from DNA damage-inducible origins in *Escherichia coli*. *EMBO J.* **1993**, *12*, 3287–3295. [PubMed]

89. Asai, T.; Bates, D.B.; Kogoma, T. DNA replication triggered by double-stranded breaks in *E. coli*: Dependence on homologous recombination functions. *Cell* **1994**, *78*, 1051–1061. [CrossRef]

90. Kogoma, T. Stable DNA replication: Interplay between DNA replication, homologous recombination, and transcription. *Microbiol. Mol. Biol. Rev.* **1997**, *61*, 212–238. [PubMed]

91. Kuzminov, A.; Stahl, F.W. Double-strand end repair via the RecBC pathway in *Escherichia coli* primes DNA replication. *Genes Dev.* **1999**, *13*, 345–356. [CrossRef] [PubMed]

92. Magee, T.R.; Kogoma, T. Requirement of RecBC enzyme and an elevated level of activated RecA for induced stable DNA replication in *Escherichia coli*. *J. Bacteriol.* **1990**, *172*, 1834–1839. [CrossRef] [PubMed]

93. Liu, J.; Xu, L.; Sandler, S.J.; Marians, K.J. Replication fork assembly at recombination intermediates is required for bacterial growth. *Proc. Natl. Acad. Sci. USA* **1999**, *96*, 3552–3555. [CrossRef] [PubMed]

94. Gabbai, C.B.; Marians, K.J. Recruitment to stalled replication forks of the PriA DNA helicase and replisome-loading activities is essential for survival. *DNA Repair (Amst)* **2010**, *9*, 202–209. [CrossRef] [PubMed]

95. Asai, T.; Imai, M.; Kogoma, T. DNA damage-inducible replication of the *Escherichia coli* chromosome is initiated at separable sites within the minimal *oriC*. *J. Mol. Biol.* **1994**, *235*, 1459–1469. [CrossRef] [PubMed]

96. Voelkel-Meiman, K.; Roeder, G.S. Gene conversion tracts stimulated by *HOT1*-promoted transcription are long and continuous. *Genetics* **1990**, *126*, 851–867. [PubMed]

97. Morrow, D.M.; Connelly, C.; Hieter, P. "Break copy" duplication: A model for chromosome fragment formation in *Saccharomyces cerevisiae*. *Genetics* **1997**, *147*, 371–382. [PubMed]

98. Davis, A.P.; Symington, L.S. *RAD51*-dependent break-induced replication in yeast. *Mol. Cell. Biol.* **2004**, *24*, 2344–2351. [CrossRef] [PubMed]

99. Lydeard, J.R.; Jain, S.; Yamaguchi, M.; Haber, J.E. Break-induced replication and telomerase-independent telomere maintenance require Pol32. *Nature* **2007**, *448*, 820–823. [CrossRef] [PubMed]

100. Kraus, E.; Leung, W.Y.; Haber, J.E. Break-induced replication: A review and an example in budding yeast. *Proc. Natl. Acad. Sci. USA* **2001**, *98*, 8255–8262. [CrossRef] [PubMed]

101. Malkova, A.; Ira, G. Break-induced replication: Functions and molecular mechanism. *Curr. Opin. Genet. Dev.* **2013**, *23*, 271–279. [CrossRef] [PubMed]

102. Malkova, A.; Naylor, M.L.; Yamaguchi, M.; Ira, G.; Haber, J.E. *RAD51*-dependent break-induced replication differs in kinetics and checkpoint responses from *RAD51*-mediated gene conversion. *Mol. Cell. Biol.* **2005**, *25*, 933–944. [CrossRef] [PubMed]

103. Malkova, A.; Signon, L.; Schaefer, C.B.; Naylor, M.L.; Theis, J.F.; Newlon, C.S.; Haber, J.E. *RAD51*-independent break-induced replication to repair a broken chromosome depends on a distant enhancer site. *Genes Dev.* **2001**, *15*, 1055–1060. [CrossRef] [PubMed]

104. Godin, S.K.; Sullivan, M.R.; Bernstein, K.A. Novel insights into *RAD51* activity and regulation during homologous recombination and DNA replication. *Biochem. Cell Biol.* **2016**, *94*, 407–418. . [CrossRef]

105. Ira, G.; Haber, J.E. Characterization of *RAD51*-independent break-induced replication that acts preferentially with short homologous sequences. *Mol. Cell. Biol.* **2002**, *22*, 6384–6392. [CrossRef] [PubMed]

106. Verma, P.; Greenberg, R.A. Noncanonical views of homology-directed DNA repair. *Genes Dev.* **2016**, *30*, 1138–1154. [CrossRef] [PubMed]

107. Donnianni, R.A.; Symington, L.S. Break-induced replication occurs by conservative DNA synthesis. *Proc. Natl. Acad. Sci. USA* **2013**, *110*, 13475–13480. [CrossRef] [PubMed]

108. Jain, S.; Sugawara, N.; Lydeard, J.; Vaze, M.; Tanguy Le Gac, N.; Haber, J.E. A recombination execution checkpoint regulates the choice of homologous recombination pathway during DNA double-strand break repair. *Genes Dev.* **2009**, *23*, 291–303. [CrossRef] [PubMed]

109. McEachern, M.J.; Haber, J.E. Break-induced replication and recombinational telomere elongation in yeast. *Annu. Rev. Biochem.* **2006**, *75*, 111–135. [CrossRef] [PubMed]

110. Lydeard, J.R.; Lipkin-Moore, Z.; Sheu, Y.J.; Stillman, B.; Burgers, P.M.; Haber, J.E. Break-induced replication requires all essential DNA replication factors except those specific for pre-RC assembly. *Genes Dev.* **2010**, *24*, 1133–1144. [CrossRef] [PubMed]

111. Wilson, M.A.; Kwon, Y.; Xu, Y.; Chung, W.H.; Chi, P.; Niu, H.; Mayle, R.; Chen, X.; Malkova, A.; Sung, P.; et al. Pif1 helicase and Polδ promote recombination-coupled DNA synthesis via bubble migration. *Nature* **2013**, *502*, 393–396. [CrossRef] [PubMed]

112. Saini, N.; Ramakrishnan, S.; Elango, R.; Ayyar, S.; Zhang, Y.; Deem, A.; Ira, G.; Haber, J.E.; Lobachev, K.S.; Malkova, A. Migrating bubble during break-induced replication drives conservative DNA synthesis. *Nature* **2013**, *502*, 389–392. [CrossRef] [PubMed]

113. Fangman, W.L.; Brewer, B.J. Activation of replication origins within yeast chromosomes. *Annu. Rev. Cell Biol.* **1991**, *7*, 375–402. [CrossRef] [PubMed]

114. Yang, Y.; Sterling, J.; Storici, F.; Resnick, M.A.; Gordenin, D.A. Hypermutability of damaged single-strand DNA formed at double-strand breaks and uncapped telomeres in yeast *Saccharomyces cerevisiae*. *PLoS Genet.* **2008**, *4*, e1000264. [CrossRef] [PubMed]

115. Smith, C.E.; Llorente, B.; Symington, L.S. Template switching during break-induced replication. *Nature* **2007**, *447*, 102–105. [CrossRef] [PubMed]

116. Llorente, B.; Smith, C.E.; Symington, L.S. Break-induced replication: What is it and what is it for? *Cell Cycle* **2008**, *7*, 859–864. [CrossRef] [PubMed]

117. Mayle, R.; Campbell, I.M.; Beck, C.R.; Yu, Y.; Wilson, M.; Shaw, C.A.; Bjergbaek, L.; Lupski, J.R.; Ira, G. DNA REPAIR. Mus81 and converging forks limit the mutagenicity of replication fork breakage. *Science* **2015**, *349*, 742–747. [CrossRef] [PubMed]

118. Pardo, B.; Aguilera, A. Complex chromosomal rearrangements mediated by break-induced replication involve structure-selective endonucleases. *PLoS Genet.* **2012**, *8*, e1002979. [CrossRef] [PubMed]

119. Lovett, S.T. Connecting replication and recombination. *Mol. Cell* **2003**, *11*, 554–556. [CrossRef]

120. Heller, R.C.; Marians, K.J. Replisome assembly and the direct restart of stalled replication forks. *Nat. Rev. Mol. Cell Biol.* **2006**, *7*, 932–943. [CrossRef] [PubMed]

121. Sandler, S.J.; Marians, K.J. Role of PriA in replication fork reactivation in *Escherichia coli. J. Bacteriol.* **2000**, *182*, 9–13. [CrossRef] [PubMed]

122. Wei, L.; Zhao, X. A new MCM modification cycle regulates DNA replication initiation. *Nat. Struct. Mol. Biol.* **2016**, *23*, 209–216. [CrossRef] [PubMed]

123. Maric, M.; Maculins, T.; De Piccoli, G.; Labib, K. Cdc48 and a ubiquitin ligase drive disassembly of the CMG helicase at the end of DNA replication. *Science* **2014**, *346*, 1253596. [CrossRef] [PubMed]

124. Moreno, S.P.; Bailey, R.; Campion, N.; Herron, S.; Gambus, A. Polyubiquitylation drives replisome disassembly at the termination of DNA replication. *Science* **2014**, *346*, 477–481. [CrossRef] [PubMed]

125. Makovets, S.; Blackburn, E.H. DNA damage signalling prevents deleterious telomere addition at DNA breaks. *Nat. Cell Biol.* **2009**, *11*, 1383–1386. [CrossRef] [PubMed]

126. Vasianovich, Y.; Harrington, L.A.; Makovets, S. Break-induced replication requires DNA damage-induced phosphorylation of Pif1 and leads to telomere lengthening. *PLoS Genet.* **2014**, *10*, e1004679. [CrossRef] [PubMed]

127. Aguilera, A.; Garcia-Muse, T. Causes of genome instability. *Annu. Rev. Genet.* **2013**, *47*, 1–32. [CrossRef] [PubMed]

128. Costantino, L.; Koshland, D. The Yin and Yang of R-loop biology. *Curr. Opin. Cell Biol.* **2015**, *34*, 39–45. [CrossRef] [PubMed]

129. Hamperl, S.; Cimprich, K.A. The contribution of co-transcriptional RNA:DNA hybrid structures to DNA damage and genome instability. *DNA Repair (Amst)* **2014**, *19*, 84–94. [CrossRef] [PubMed]

130. Groh, M.; Gromak, N. Out of balance: R-loops in human disease. *PLoS Genet.* **2014**, *10*, e1004630. [CrossRef] [PubMed]

131. Santos-Pereira, J.M.; Aguilera, A. R loops: New modulators of genome dynamics and function. *Nat. Rev. Genet.* **2015**, *16*, 583–597. [CrossRef] [PubMed]

132. Skourti-Stathaki, K.; Kamieniarz-Gdula, K.; Proudfoot, N.J. R-loops induce repressive chromatin marks over mammalian gene terminators. *Nature* **2014**, *516*, 436–439. [CrossRef] [PubMed]

133. Gaillard, H.; Aguilera, A. Transcription as a threat to genome integrity. *Annu. Rev. Biochem.* **2016**, *85*, 291–317. [CrossRef] [PubMed]

134. Dasgupta, S.; Masukata, H.; Tomizawa, J. Multiple mechanisms for initiation of ColE1 DNA replication: DNA synthesis in the presence and absence of ribonuclease H. *Cell* **1987**, *51*, 1113–1122. [CrossRef]

135. Masukata, H.; Dasgupta, S.; Tomizawa, J. Transcriptional activation of ColE1 DNA synthesis by displacement of the nontranscribed strand. *Cell* **1987**, *51*, 1123–1130. [CrossRef]

136. Marians, K.J. Prokaryotic DNA replication. *Annu. Rev. Biochem.* **1992**, *61*, 673–719. [CrossRef] [PubMed]

137. Mosig, G. The essential role of recombination in phage T4 growth. *Annu. Rev. Genet.* **1987**, *21*, 347–371. [CrossRef] [PubMed]

138. de Massy, B.; Fayet, O.; Kogoma, T. Multiple origin usage for DNA replication in *sdrA(rnh)* mutants of *Escherichia coli* K-12. Initiation in the absence of oriC. *J. Mol. Biol.* **1984**, *178*, 227–236. [CrossRef]

139. Kogoma, T. Absence of RNase H allows replication of pBR322 in *Escherichia coli* mutants lacking DNA polymerase I. *Proc. Natl. Acad. Sci. USA* **1984**, *81*, 7845–7849. [CrossRef] [PubMed]

140. Von Meyenburg, K.; Boye, E.; Skarstad, K.; Koppes, L.; Kogoma, T. Mode of initiation of constitutive stable DNA replication in RNase H-defective mutants of *Escherichia coli* K-12. *J. Bacteriol.* **1987**, *169*, 2650–2658. [CrossRef] [PubMed]

141. Horiuchi, T.; Maki, H.; Sekiguchi, M. RNase H-defective mutants of *Escherichia coli*: A possible discriminatory role of RNase H in initiation of DNA replication. *Mol. Gen. Genet.* **1984**, *195*, 17–22. [CrossRef] [PubMed]

142. Ogawa, T.; Pickett, G.G.; Kogoma, T.; Kornberg, A. RNase H confers specificity in the DnaA-dependent initiation of replication at the unique origin of the *Escherichia coli* chromosome in vivo and in vitro. *Proc. Natl. Acad. Sci. USA* **1984**, *81*, 1040–1044. [CrossRef] [PubMed]

143. Maduike, N.Z.; Tehranchi, A.K.; Wang, J.D.; Kreuzer, K.N. Replication of the *Escherichia coli* chromosome in RNase HI-deficient cells: Multiple initiation regions and fork dynamics. *Mol. Microbiol.* **2014**, *91*, 39–56. [CrossRef] [PubMed]

144. Dimude, J.U.; Stockum, A.; Midgley-Smith, S.L.; Upton, A.L.; Foster, H.A.; Khan, A.; Saunders, N.J.; Retkute, R.; Rudolph, C.J. The consequences of replicating in the wrong orientation: Bacterial chromosome duplication without an active replication origin. *MBio* **2015**, *6*, e01294-15. [CrossRef] [PubMed]

145. Hong, X.; Cadwell, G.W.; Kogoma, T. *Escherichia coli* RecG and RecA proteins in R-loop formation. *EMBO J.* **1995**, *14*, 2385–2392. [PubMed]

146. Hong, X.; Cadwell, G.W.; Kogoma, T. Activation of stable DNA replication in rapidly growing *Escherichia coli* at the time of entry to stationary phase. *Mol. Microbiol.* **1996**, *21*, 953–961. [CrossRef] [PubMed]

147. Drolet, M.; Phoenix, P.; Menzel, R.; Masse, E.; Liu, L.F.; Crouch, R.J. Overexpression of RNase H partially complements the growth defect of an *Escherichia coli* delta *topA* mutant: R-loop formation is a major problem in the absence of DNA topoisomerase I. *Proc. Natl. Acad. Sci. USA* **1995**, *92*, 3526–3530. [CrossRef] [PubMed]

148. Martel, M.; Balleydier, A.; Sauriol, A.; Drolet, M. Constitutive stable DNA replication in *Escherichia coli* cells lacking type 1A topoisomerase activity. *DNA Repair (Amst)* **2015**, *35*, 37–47. [CrossRef] [PubMed]

149. Sandler, S.J. Requirements for replication restart proteins during constitutive stable DNA replication in *Escherichia coli* K-12. *Genetics* **2005**, *169*, 1799–1806. [CrossRef] [PubMed]

150. Usongo, V.; Drolet, M. Roles of type 1A topoisomerases in genome maintenance in *Escherichia coli*. *PLoS Genet.* **2014**, *10*, e1004543. [CrossRef] [PubMed]

151. Kasahara, M.; Clikeman, J.A.; Bates, D.B.; Kogoma, T. RecA protein-dependent R-loop formation in vitro. *Genes Dev.* **2000**, *14*, 360–365. [PubMed]

152. Zaitsev, E.N.; Kowalczykowski, S.C. A novel pairing process promoted by *Escherichia coli* RecA protein: Inverse DNA and RNA strand exchange. *Genes Dev.* **2000**, *14*, 740–749. [PubMed]

153. Wimberly, H.; Shee, C.; Thornton, P.C.; Sivaramakrishnan, P.; Rosenberg, S.M.; Hastings, P.J. R-loops and nicks initiate DNA breakage and genome instability in non-growing *Escherichia coli*. *Nat. Commun.* **2013**, *4*, 2115. [CrossRef] [PubMed]

154. Usongo, V.; Martel, M.; Balleydier, A.; Drolet, M. Mutations reducing replication from R-loops suppress the defects of growth, chromosome segregation and DNA supercoiling in cells lacking topoisomerase I and RNase HI activity. *DNA Repair (Amst)* **2016**, *40*, 1–17. [CrossRef] [PubMed]

155. El Hage, A.; Webb, S.; Kerr, A.; Tollervey, D. Genome-wide distribution of RNA-DNA hybrids identifies RNase H targets in tRNA genes, retrotransposons and mitochondria. *PLoS Genet.* **2014**, *10*, e1004716. [CrossRef] [PubMed]

156. Chan, Y.A.; Aristizabal, M.J.; Lu, P.Y.; Luo, Z.; Hamza, A.; Kobor, M.S.; Stirling, P.C.; Hieter, P. Genome-wide profiling of yeast DNA:RNA hybrid prone sites with DRIP-chip. *PLoS Genet.* **2014**, *10*, e1004288. [CrossRef] [PubMed]

157. Mischo, H.E.; Gomez-Gonzalez, B.; Grzechnik, P.; Rondon, A.G.; Wei, W.; Steinmetz, L.; Aguilera, A.; Proudfoot, N.J. Yeast Sen1 helicase protects the genome from transcription-associated instability. *Mol. Cell* **2011**, *41*, 21–32. [CrossRef] [PubMed]

158. Alzu, A.; Bermejo, R.; Begnis, M.; Lucca, C.; Piccini, D.; Carotenuto, W.; Saponaro, M.; Brambati, A.; Cocito, A.; Foiani, M.; et al. Senataxin associates with replication forks to protect fork integrity across RNA-polymerase-II-transcribed genes. *Cell* **2012**, *151*, 835–846. [CrossRef] [PubMed]

159. Salvi, J.S.; Chan, J.N.; Szafranski, K.; Liu, T.T.; Wu, J.D.; Olsen, J.B.; Khanam, N.; Poon, B.P.; Emili, A.; Mekhail, K. Roles for Pbp1 and caloric restriction in genome and lifespan maintenance via suppression of RNA-DNA hybrids. *Dev. Cell* **2014**, *30*, 177–191. [CrossRef] [PubMed]

160. El Hage, A.; French, S.L.; Beyer, A.L.; Tollervey, D. Loss of Topoisomerase I leads to R-loop-mediated transcriptional blocks during ribosomal RNA synthesis. *Genes Dev.* **2010**, *24*, 1546–1558. [CrossRef] [PubMed]

161. Stuckey, R.; Garcia-Rodriguez, N.; Aguilera, A.; Wellinger, R.E. Role for RNA:DNA hybrids in origin-independent replication priming in a eukaryotic system. *Proc. Natl. Acad. Sci. USA* **2015**, *112*, 5779–5784. [CrossRef] [PubMed]

162. Pomerantz, R.T.; O'Donnell, M. The replisome uses mRNA as a primer after colliding with RNA polymerase. *Nature* **2008**, *456*, 762–766. [CrossRef] [PubMed]

163. Pomerantz, R.T.; O'Donnell, M. Direct restart of a replication fork stalled by a head-on RNA polymerase. *Science* **2010**, *327*, 590–592. [CrossRef] [PubMed]

164. Longhese, M.P.; Plevani, P.; Lucchini, G. Replication factor A is required *in vivo* for DNA replication, repair, and recombination. *Mol. Cell. Biol.* **1994**, *14*, 7884–7890. [CrossRef] [PubMed]

165. Bartrand, A.J.; Iyasu, D.; Brush, G.S. DNA stimulates Mec1-mediated phosphorylation of replication protein A. *J. Biol. Chem.* **2004**, *279*, 26762–26767. [CrossRef] [PubMed]

166. Foiani, M.; Liberi, G.; Lucchini, G.; Plevani, P. Cell cycle-dependent phosphorylation and dephosphorylation of the yeast DNA polymerase alpha-primase B subunit. *Mol. Cell. Biol.* **1995**, *15*, 883–891. [CrossRef] [PubMed]

167. Rippe, K.; Luke, B. TERRA and the state of the telomere. *Nat. Struct. Mol. Biol.* **2015**, *22*, 853–858. [CrossRef] [PubMed]

168. Balk, B.; Maicher, A.; Dees, M.; Klermund, J.; Luke-Glaser, S.; Bender, K.; Luke, B. Telomeric RNA-DNA hybrids affect telomere-length dynamics and senescence. *Nat. Struct. Mol. Biol.* **2013**, *20*, 1199–1205. [CrossRef] [PubMed]

169. Arora, R.; Lee, Y.; Wischnewski, H.; Brun, C.M.; Schwarz, T.; Azzalin, C.M. RNase H1 regulates TERRA-telomeric DNA hybrids and telomere maintenance in ALT tumour cells. *Nat. Commun.* **2014**, *5*, 5220. [CrossRef] [PubMed]

170. Flynn, R.L.; Cox, K.E.; Jeitany, M.; Wakimoto, H.; Bryll, A.R.; Ganem, N.J.; Bersani, F.; Pineda, J.R.; Suva, M.L.; Benes, C.H.; et al. Alternative lengthening of telomeres renders cancer cells hypersensitive to ATR inhibitors. *Science* **2015**, *347*, 273–277. [CrossRef] [PubMed]

171. Law, M.J.; Lower, K.M.; Voon, H.P.; Hughes, J.R.; Garrick, D.; Viprakasit, V.; Mitson, M.; De Gobbi, M.; Marra, M.; Morris, A.; et al. ATR-X syndrome protein targets tandem repeats and influences allele-specific expression in a size-dependent manner. *Cell* **2010**, *143*, 367–378. [CrossRef] [PubMed]

172. Kogoma, T.; Lark, K.G. Characterization of the replication of *Escherichia coli* DNA in the absence of protein synthesis: Stable DNA replication. *J. Mol. Biol.* **1975**, *94*, 243–256. [CrossRef]

173. Lombrana, R.; Almeida, R.; Alvarez, A.; Gomez, M. R-loops and initiation of DNA replication in human cells: A missing link? *Front. Genet.* **2015**, *6*, 158. [CrossRef] [PubMed]

174. Leonard, A.C.; Mechali, M. DNA replication origins. *Cold Spring Harb. Perspect. Biol.* **2013**, *5*, a010116. [CrossRef] [PubMed]

175. Norseen, J.; Thomae, A.; Sridharan, V.; Aiyar, A.; Schepers, A.; Lieberman, P.M. RNA-dependent recruitment of the origin recognition complex. *EMBO J.* **2008**, *27*, 3024–3035. [CrossRef] [PubMed]

176. Gomez, M.; Antequera, F. Overreplication of short DNA regions during S phase in human cells. *Genes Dev.* **2008**, *22*, 375–385. [CrossRef] [PubMed]

177. Lombrana, R.; Alvarez, A.; Fernandez-Justel, J.M.; Almeida, R.; Poza-Carrion, C.; Gomes, F.; Calzada, A.; Requena, J.M.; Gomez, M. Transcriptionally driven DNA replication program of the human parasite *Leishmania major*. *Cell Rep.* **2016**, *16*, 1774–1786. [CrossRef] [PubMed]

![genes logo](GCAT TACG GCAT *genes*)

MDPI

Review

Maintenance of Genome Integrity: How Mammalian Cells Orchestrate Genome Duplication by Coordinating Replicative and Specialized DNA Polymerases

Ryan Barnes [1] and Kristin Eckert [2,*]

[1] Biomedical Sciences Graduate Program, Pennsylvania State University College of Medicine, Hershey, PA 17033, USA; rbarnes1@hmc.psu.edu
[2] Departments of Pathology and Biochemistry & Molecular Biology, The Jake Gittlen Laboratories for Cancer Research, Pennsylvania State University College of Medicine, Hershey, PA 17033, USA
*   Correspondence: kae4@psu.edu; Tel.: +1-717-531-4065

Academic Editor: Eishi Noguchi
Received: 11 November 2016; Accepted: 27 December 2016; Published: 6 January 2017

Abstract: Precise duplication of the human genome is challenging due to both its size and sequence complexity. DNA polymerase errors made during replication, repair or recombination are central to creating mutations that drive cancer and aging. Here, we address the regulation of human DNA polymerases, specifically how human cells orchestrate DNA polymerases in the face of stress to complete replication and maintain genome stability. DNA polymerases of the B-family are uniquely adept at accurate genome replication, but there are numerous situations in which one or more additional DNA polymerases are required to complete genome replication. Polymerases of the Y-family have been extensively studied in the bypass of DNA lesions; however, recent research has revealed that these polymerases play important roles in normal human physiology. Replication stress is widely cited as contributing to genome instability, and is caused by conditions leading to slowed or stalled DNA replication. Common Fragile Sites epitomize "difficult to replicate" genome regions that are particularly vulnerable to replication stress, and are associated with DNA breakage and structural variation. In this review, we summarize the roles of both the replicative and Y-family polymerases in human cells, and focus on how these activities are regulated during normal and perturbed genome replication.

Keywords: translesion synthesis; replication stress; transcriptional regulation; polymerase interactions; polymerase domains; polymerase modifications

---

1. Introduction

Human cells encode 15 distinct nuclear DNA polymerases with widely varying enzymatic properties and accuracies, reflecting the need for biochemical flexibility during genome maintenance. DNA polymerases of the Y-family are characterized by their unique ability to efficiently replicate non-B DNA structures, as well as numerous DNA lesions formed by endogenous cellular processes and exposure to exogenous agents (reviewed in [1,2]). The need for this enzymatic flexibility during DNA replication is reflected in the conservation of the Y-family from *E. coli* and yeast, to rodents and mammals. DNA polymerase errors during DNA synthesis pathways associated with replication, repair, and recombination can cause mutations that drive cancer and aging. Y-family polymerases, although essential, have higher error rates than replicative polymerases. While the biochemistry of DNA lesion bypass or translesion synthesis (TLS) by Y-family polymerases has been extensively studied (reviewed in [3]), the regulation of these polymerases is often viewed in that narrow context, and how mammalian cells orchestrate DNA polymerase activities to maintain genome stability is an open question. In this review, we summarize the factors regulating both the expression and activity of the Y-family polymerases, focusing primarily on mammalian cells, and compare such regulation to the major replicative polymerases of the B-family.

2. Overview of Polymerase Functions

Currently known functions of the mammalian DNA polymerases to be discussed in this review, as well as gene and protein nomenclature, are summarized in Table 1. Replication of the human genome is carried out primarily by the replicative B-family polymerases (pols) α, δ, and ε [4]. The coordinated activities of several DNA polymerases are required for DNA repair pathways including base excision repair (BER), nucleotide excision repair (NER), mismatch repair (MMR), double-strand break repair (DSBR), and homologous recombination (HR) (Table 1). In response to replication stress, the ATR-mediated intra-S phase checkpoint coordinates DNA replication, repair and recombination processes at stalled replication forks [5]. Polymerases required to activate the ATR checkpoint include Pol α, Pol κ, Pol δ and Pol ε (Table 1). The replisome is a highly dynamic structure, and current models to explain resolution of stalled replication forks specialized polymerases (Y-family and Pol ζ) include performing DNA synthesis at the fork when replicative polymerases (B-family) are inhibited, or post-replicative gap-filling synthesis behind the replication fork [6]. Repetitive sequences make up ~67% of the human genome [7], and are enriched within rare and common fragile sites (CFS), chromosomal regions susceptible to breakage, particularly under replication stress [8]. Our laboratory has proposed that the presence of multiple DNA polymerases with complementary biochemical activities and accuracies reflects the complexity of completing DNA replication in genomes with a high density of repetitive DNA sequences [9]. We demonstrated biochemically that microsatellite sequences and high flexibility AT-rich repeats are particularly inhibitory to replicative DNA α and δ polymerase elongation [9–11]. We also made the novel discovery that Pols η and κ efficiently replicate through repetitive DNA sequences [11,12]. Loss of either Pol η or Pol ε increases CFS breakage [13,14], underscoring the importance of these enzymes in maintaining genome integrity. While classified as a Y-family polymerase gene, REV1's catalytic terminal transferase activity is overwhelmingly dispensable [15]. Instead, REV1's crucial function is to serve as a scaffolding protein and assist the function of other polymerases.

Although they have dramatically different biochemical capacities (i.e., DNA synthesis efficiency and fidelity using defined DNA substrates), the B- and Y-family DNA polymerases adopt the conserved "right hand" structure found in polymerases of all forms of life. B-family polymerases utilize the thumb domain to make extensive contact with both the primer and template DNA; the palm domain contains catalytic residues and coordinates the Mg^{2+} ions; and the finger domain makes extensive conformational swings to coordinate the incoming deoxynucleotide (dNTP) with the template. While the Y-family enzymes share these general features and similar overall structure, they have several key differences. Foremost, all of the Y-family polymerases have a little finger domain which compensates for diminished interaction of the thumb domain with the template. Compared to B-family polymerases, the finger domains are also smaller and more rigid, making little-to-no movement when interacting with a dNTP, leaving the active site largely solvent exposed [16]. This finding has led to the proposal that in contrast to the tight complex between polymerase, DNA, and dNTP made by the fingers of B-family polymerases, Y-family polymerases have a preformed active site, explaining their low fidelity and catalytic efficiency [17].

Moreover, the little finger domain is believed to ascribe unique biochemical functions to certain polymerases. For Pol η, this domain was shown to interact tightly with the catalytic core and act as a molecular splint, forcing DNA to adopt a B-form structure in the active site [18]. This ability may explain Pol η's ability to both accurately replicate certain DNA lesions, as well as repetitive DNA [2,17,19]. Pol κ also contains an N-terminal clasp domain which allows it to encircle DNA, linking the little finger and thumb, while also interacting with the primer [20]. In contrast to Pol η,

Pol κ has a large gap between the little finger and thumb domain which may accommodate bulky minor groove lesions [21]. Indeed, recent structural work has shown that the bulky benzo[a]pyrene adduct is easily accommodated by the Pol κ active site, as the adducted DNA remained in B-form, displaying little difference to normal DNA [22]. This finding suggests that the function of specialized polymerases to replicate non-B DNA may be a result of their ability to force DNA into a B-form.

Table 1. Known functions of mammalian replicative and specialized polymerases.

Polymerase	Gene/ Subunit	Cellular Functions	References
Replicative	**(B-family)**		
Alpha (α)	POLA1-A2 p180, p70	Replication: initiator DNA synthesis Checkpoint signaling	[23] [24,25]
Delta (δ)	POLD1-D4 p125, p50, p66, p12	Replication: Lagging strand; late S/G2 DNA repair synthesis (BER, NER, MMR) Checkpoint Signaling	[4,26] [27–29] [24]
Epsilon (ε)	POLE1-E4 p261, p59, p17, p12	Replication: Leading strand DNA repair synthesis (NER) Checkpoint signaling	[4,26] [30] [24]
Specialized	**(B-family)**		
Zeta (ζ)	REV3L REV7	Translesion synthesis Common Fragile Site stability	[1,13]
Specialized	**(Y-family)**		
Eta (η)	POLH	Translesion synthesis Common Fragile Site stability DNA repair synthesis (MMR; HR) Somatic Hypermutation	[3] [14] [31,32] [33]
Kappa (κ)	POLK	Translesion synthesis G4 and Microsatellite DNA synthesis DNA repair synthesis (NER, DSBR) ATR signaling	[3] [34] [30,35,36] [37]
Iota (ι)	POLI	Translesion synthesis Somatic Hypermutation	[3] [38]
Rev1	REV1	Translesion synthesis	[3]

3. DNA Polymerase Expression during an Unperturbed Mitotic Cell Cycle

All four Y-family polymerases are expressed throughout the adult organism in mice and humans. Comparatively, the expression of *POLH*, *POLK*, and *REV1* genes is high in testis and ovaries, moderate in tissues such as kidney, liver, and spleen, and low in slow proliferating tissues, such as skeletal muscle and brain [39–43]. The *POLI* gene is expressed highly in testis and ovaries and present in other adult human tissues, but at low levels [44,45]. The Y-family polymerase proteins are expressed at very low levels, with as few as 60,000 molecules of Pol η and REV1 estimated in unperturbed human cells [46]. For comparison, each human cell is estimated to have ~3 million molecules of Pol ε and 500,000 molecules of Pol δ, based on the abundance of the catalytic subunits [47]. Additionally, unlike replicative polymerases and PCNA which increase transcript and protein just before S-phase, Y-family polymerases either do not change expression during the cell cycle (*POLI* and *REV1*) or increase only in G2/M (*POLH*) [48].

3.1. Transcriptional Regulation

3.1.1. Sp1

The Sp1 transcription factor regulates numerous genes in processes such as apoptosis, cell growth, and the immune response [49]. Sp1 regulation of basal transcription has been functionally characterized for several mammalian replicative polymerase genes (*POLA1*, *POLD1*, *POLE2*) [50–52], and both direct and indirect evidence suggests that Sp1 regulation extends to all four Y-family polymerase genes (Figure 1A).

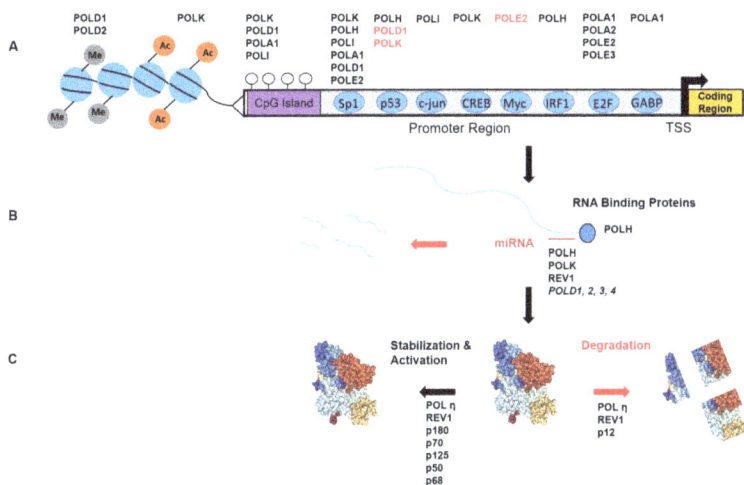

Figure 1. Overview of DNA polymerase regulation. (**A**) Transcriptional regulation of the B and Y-family polymerases genes (Top, see Table 1), as controlled by histone modifications (grey and orange circles), CpG methylation (open circles), and transcription factors (blue circles). Genes in red are negatively regulated by the factor below. TSS = transcription start site; (**B**) Post-transcriptional regulation: Polymerase mRNA stability is controlled by mRNA binding proteins and microRNA binding at the 3′ UTR; (**C**) Post-translational regulation: Polymerase proteins can be stabilized and functionally activated by various modifications, or prompted for degradation (red). See text for details.

Early work characterizing the *POLK* promoter showed the presence of both *cis* repressive (−1413/−395) and activating elements (−395/−83) [53]. Mutation of a CREB binding element or an Sp1 site (−180 and −78 respectively) reduced *POLK* promoter activity, as measured using luciferase reporter constructs (pGL3-Basic). Indeed, these proteins were shown to bind their cognate sequences in vitro by mobility shift assays, and over-expression of CREB, Sp1, or Sp3 enhanced luciferase expression via the *POLK* promoter. The *POLK* gene also harbors an Sp1 motif at position +60, and this upstream site was confirmed to positively regulate *POLK* [54]. However, several other putative transcription factor binding sites were shown not to affect *POLK* promoter activity, including SMAD and NFκB.

The *POLI* promoter contains functional Sp1 binding motifs [55], and an early study documented the control of *POLI* expression by Sp1 [56]. Using chromatin immunoprecipitation (ChIP), Sp1 was confirmed to bind within the *POLI* promoter, and overexpression of Sp1 enhanced luciferase expression driven by the *POLI* promoter. Interestingly, overexpression of Sp1, but not Oct-1, increased *POLI* mRNA, despite the presence of predicted Oct-1 sites.

The *POLH* gene promoter also contains putative Sp1 motifs, and deletion of Sp1 motifs reduced luciferase expression to levels comparable to the empty pGL3 vector [57]. Additionally, publicly available ChIP data (UCSC Genome Browser, [58]) shows an Sp1 signal in the *POLH* promoter, at the consensus sequence. To our knowledge, there are no reports characterizing the human *REV1* promoter. However, the UCSC Genome Browser also shows an Sp1 ChIP signal at the 3′ of the *REV1* gene in intron 17. Although this peak lacks the consensus motif, a possible binding site can be found in intron 22 (5′-AGGGCGGATC-3′) and several 5′-GGGCGG-3′ motifs are present in the promoter region.

3.1.2. p53

The p53 transcription factor was first reported a decade ago to positively regulate the *POLH* [59]. Overexpression of *TP53* increased Pol η mRNA levels and enhanced luciferase activity in a reporter assay. Importantly, overexpression of a mutant *TP53* (R175H) was unable to enhance luciferase expression. Unpublished data from our laboratory and other studies have confirmed higher expression levels of human *POLH* in *TP53* proficient cells compared to deficient cells [60]. *POLH* gene regulation by p53 is conserved in murine cells [61].

In contrast, the effect of p53 on *POLK* expression appears to have diverged between humans and mice. While murine *POLK* expression is enhanced by p53 in the absence of DNA damage, human *POLK* expression is either unaffected [39], or negatively regulated [62]. In the latter study, luciferase constructs containing the human *POLK* promoter were inhibited when *TP53* was transiently overexpressed, compared to controls, and this inhibition was dependent on p53 DNA binding activity [62]. Consistently, using the same human constructs in mouse cells, *POLK* promoter-dependent luciferase activity was increased in *TP53* null cells, compared to wild-type. Similar to *POLK*, *POLD1* gene transcription is repressed by p53 binding to the core promoter, in a mechanism that excludes Sp1 binding [63,64].

3.1.3. E2F

Replicative polymerase gene transcription is increased upon mitogen-stimulated entry into the mitotic cell cycle after serum deprivation (G0) growth arrest. This response has been functionally characterized for several genes (*POLA1*, *POLA2*, *POLD1*, *POLE2*, and *POLE3*), and is dependent upon E2F transcription factor binding [50–52,65,66].

3.1.4. Epigenetic Regulation

Very little is known regarding the epigenetic regulation of polymerase genes. The promoters of both *POLI* and *POLK* are unmethylated and treatment of cells with 5-azacytidine did not alter expression [53,55]. In these same studies, treatment with Trichostatin A and did not change the expression of *POLI*, but *POLK* expression was increased ~five-fold, suggesting histone acetylation status is used to regulate *POLK* expression. *POLA1* and *POLD1* gene expression is also unresponsive to 5-azacytidine and Trichostatin A suggesting an absence of repressive epigenetic modification at their promoters [53]. However, the PRMT7 histone methyltransferase is a negative regulator for *POLD1* and *POLD2* gene expression [67].

3.2. Post-Transcriptional Regulation

An important point for regulating gene expression is at the level of the mRNA half-life. The stability of *POLH* mRNA is enhanced by binding of PRCB1 (or hnRNP E1) to an AU-rich element within the 3′ UTR [68] (Figure 1B). Knock-down of this protein reduces Pol η protein levels via a reduction in *POLH* mRNA half-life.

Overexpression of miR-155 causes down-regulation of all four *POLD* genes [69], but it is unclear whether miR-155 regulates Pol δ expression by directly binding to *POLD* gene transcripts. However, micoRNAs have been shown to regulate most of the Y-Family polymerases. miR-96 negatively regulates *REV1* in human cells by interacting with a predicated binding site in the 3′ UTR [70]. miR-20b is predicated to bind the 3′ UTR of both *POLH* and *POLK* transcripts, and the miR-20b binding site was confirmed to be functional for the *POLK* 3′UTR. Overexpressing a miR-20b mimic reduces, while a miR-20b inhibitor elevates, Pol κ protein levels [71]. In a separate report, the downregulation of miR-93 expression in ovarian cancer cells caused an increase in Pol η levels. This negative regulation was validated using both a miR-93 mimic and an inhibitor [72]. In contrast to the study by Guo et al. [72], *POLK* transcript was not affected even though miR-20b was downregulated in the ovarian cancer cells, and a miR-20b mimic did not alter Pol η expression [72].

3.3. Post-Translational Modifications-Functional

Phosphorylation is an important mechanism regulating replicative polymerases (Figure 1C). Pol α-primase holoenzyme activity is regulated by cyclin-dependent kinases (CDKs) in a cell cycle-dependent manner [73]. The p180 catalytic subunit is a phosphoprotein that becomes hyperphosphorylated in G_2/M phase, while the regulatory p70 subunit is phosphorylated only in G_2/M [74]. Pol α phosphorylation results in lowered single-stranded DNA binding affinity, lowered DNA synthesis activity, and an inhibition of DNA replication [73,74].

The mammalian Pol δ holoenzyme consists of catalytic p125 (POLD1), regulatory p50 (POLD2), regulatory p68 (POLD3) and p12 (POLD4) subunits [75]. The Pol δ holoenzyme is phosphorylated in a cell cycle-dependent manner (see [76] for review). The catalytic p125 subunit is phosphorylated primarily during S-phase [77]. The regulatory B subunit (p50) is phosphorylated in vivo, and is an in vitro substrate of the Cyclin A-CDK2 cell cycle-dependent kinase [78]. The regulatory C subunit (p68) can be phosphorylated by G1/S phase and S-phase cyclin-dependent kinases in vitro, and PCNA interferes with this phosphorylation [79]. Phosphorylation of p68 coincides with Pol δ association with chromatin at the start of S-phase [80]. The regulatory p68 subunit also contains a phosphorylation site for Protein Kinase A, and phosphomimetic mutation of this residue decreases Pol δ affinity for PCNA and processivity [81]. In addition, mammalian p125, p68 and p12 subunits can be phosphorylated by Casein Kinase 2 in vitro, and subsequently dephosphorylated by protein phosphatase-1 [82], suggesting an additional regulatory circuit for regulation. Thus, phosphorylation may serve to regulate Pol δ activity by controlling its interaction with DNA and/or auxiliary proteins during replication.

3.4. Post-Translational Modifications-Degradation

While the mammalian Pol δ holoenzyme is a heterotetrameric protein (Pol δ4), the Pol δ holoenzyme found in budding yeast, *Saccharomyces cerevisiae*, lacks the small subunit, and exists only in the three subunits assembly (Pol δ3). The human p12 subunit interacts with the p125 and p50 subunits, increasing stability of the Pol δ holoenzyme and increasing PCNA-dependent DNA synthesis activity [78]. During an unperturbed mitotic cell cycle, p12 levels fall during G1 phase, preceding the initiation of DNA synthesis, and rapidly rise again upon completion of DNA synthesis and transition to the G2/M phase [83] (Figure 1C). During S- phase, the majority of Pol δ activity is attributed to the Pol δ3 form [83,84]. This partial degradation of p12 occurs via a PCNA interacting peptide (PIP) degron sequence and is controlled by the CRL4^{Cdt2} E3 ligase [85]. CRL4^{Cdt2} recognizes substrates bound to chromatin-loaded PCNA and is a key regulator of replication [86]. Another CRL4^{Cdt2} substrate, the p21 protein, directly interacts with the PolD2/p50 subunit, and p21 and p12 are coordinately degraded in S-phase. The biochemical properties of the human Pol δ3 and Pol δ4 forms differ, with the Pol δ3 form being more adapted for completion of Okazaki fragment processing and DNA repair synthesis [87].

4. DNA Polymerase Expression under Stress

4.1. Transcriptional Regulation

The cellular response to ultraviolet (UV) radiation in human cells involves upregulation of Pol ι expression [88] that is dependent on ATR activation of c-jun [89] (Figure 1A). This is surprising, considering Pol ι over-expression is unable to rescue the UV sensitivity of patient derived, Pol η-deficient cells [90]. In mice, UV induces Pol κ expression in a p53-dependent manner, whereas in human cells, UV induces either no change (p53 positive cells) or a reduction (p53 deficient cells) in

POLK expression [39]. Notably, *POLK* was upregulated following UV in patient derived Pol η-deficient cells [91], suggesting a regulatory adaptation to loss of Pol η. Surprisingly, despite its function as an accurate and efficient TLS polymerase for UV induced pyrimidine dimers, the levels of Pol η actually decrease following UV irradiation in human and murine cells [42,92] (see Section 4.3).

Pol η levels are induced following treatments that create double strand breaks. Exposure to both camptothecin (CPT) or ionizing radiation (IR) induces transactivation of *POLH* in human cells in a p53-dependent manner [59]. Studies examining the relationship between p53 and Pol κ have produced interesting results. Murine cells treated with doxorubicin, which can cause strand breaks, causes *POLK* upregulation in a p53 dependent manner, whereas similar treatment of human cells caused either a reduction (p53 deficient) or no change (p53 wild-type) in *POLK* expression [39]. These findings are consistent with the basal levels of *POLK* mentioned above, and provide further support that p53's role in Y-family polymerase regulation has diverged between rodents and primates.

Notably, *POLH*, *POLI*, and *POLK* gene expression are all induced by alkylation damage. *N*-methyl-*N'*-nitro-*N*-nitrosoguanidine treatment of human cells induces *POLH* expression in a pathway dependent on interferon regulatory factor 1 (IRF1), and *POLI* gene expression in an Sp1 dependent manner [56,57]. Temozolomide, an alkylating drug used in chemotherapy, upregulates the expression of Pol κ at both the mRNA and protein levels [93].

4.2. Post-Translational Modifications—Functional

4.2.1. Phosphorylation

Early work showed that following UV radiation, human Pol η was phosphorylated and its foci formation was reduced in ATR-depleted cells [94]. Later work demonstrated that Pol η is directly phosphorylated at Ser 601 by ATR in vitro and in cells following UV, hydroxyurea, cisplatin, and CPT treatment (Figure 1C). This phosphorylation is dependent on Pol η's interaction with Rad18 and Pol η's ubiquitin binding zinc finger (UBZ) domain but independent of Rad18 catalytic activity and PCNA ubiquitination [95]. Importantly, following UV treatment, loss of Ser601 phosphorylation does not impact Pol η chromatin localization or foci formation, but does reduce cell survival. A recent report using LC-MS/MS discovered that both Ser601 and Ser687 are phosphorylated in untreated human cells [96]. The Ser687 phosphorylation is induced following UV by Cyclin A2- CDK2 [96]. Again, loss of this Ser687 phosphorylation does not impact Pol η nuclear localization but does reduce cell viability following UV treatment. Interestingly, a phospho-mimetic mutant (S687D) had reduced PCNA interaction, but no defect in a cellular TLS assay compared to wild-type cells. These findings together suggest that despite reduced PCNA interaction, Pol η phosphorylation promotes its activity. Further biochemical studies are required to determine if phosphorylation impacts Pol η activity per se.

4.2.2. Ubiquitination

Multiple reports have shown that Pol η is mono-ubiquitinated at Lys682, under normal conditions [96,97]. The nuclear localization sequence of Pol η, including Lys682, Lys686, Lys694, and Lys709, is ubiquitinated in cells by the PIRH2 E3 ligase [98], and these modifications act as a surface for PCNA interaction along with the PIP box. In response to UV, ubiquitinated Pol η disappears, suggesting that removal of the modification is required for function [97]. Consistent with this, a ubiquitin-Pol η chimera has reduced nuclear foci and compromised PCNA interaction, due to intramolecular interaction between ubiquitin and the UBZ domain of Pol η [97,98]. However, this mutant has only slightly reduced clonogenic survival in comparison to wild-type cells [97,98]. These studies suggest that while mono-ubiquitination is a bona-fide mechanism for regulating Pol η/PCNA interaction and foci formation, this modification does not dramatically reduce its ability to prevent UV sensitivity, especially in comparison to Pol η-deficient cells.

In contrast to Pol η, there is little experimental evidence concerning functional consequences of post-translational modifications of the other Y-family polymerases, although all three are ubiquitinated in a UBZ/UBD dependent fashion [99–103]. The E3 ligase TRIP has been shown to interact with Pol κ [104]. Polymerase δ subunits p68 and p12 are modified by ubiquitination, and the p68 subunit is primarily mono-ubiquitinated [105]. The p68 subunit also is SUMOylated by SUMO3 in unperturbed cells [105] and by SUMO2 in cells under replication stress [106]. The functional consequences of the p68 SUMOylation have not been determined, although the lysine residues modified by SUMO3 lie outside of the known p50 and PCNA binding domains.

4.3. Post-Translational Modifications—Degradation and Stability

Human Pol η is degraded following UV treatment by the proteasome. MDM2 negatively regulates Pol η following UV via poly-ubiquitination [107], while, USP7 acts as a de-ubiquitinase, preventing the poly-ubiquitination of Pol η and its degradation [108]. USP7 also negatively regulates the stability of MDM2. Thus, MDM2 and USP7 regulate Pol η levels in an inverse manner. Following UV treatment, Pol η is also targeted for degradation in a ubiquitin independent fashion by PIRH2 [92]. TRIP is also known to poly-ubiquitinate human Pol η, but the consequences of this on stability have not been examined [104]. In *C. elegans*, degradation of Pol η is prevented by GEI-17 [109]. GEI-17 (PIAS homolog) SUMOylates Pol η to halt CRL4-Cdt2 mediated degradation following UV and methyl methanesulfonate (MMS). An epistatic relationship between *C. elegans* Pol η and GEI-17 following UV confirms that the two proteins act in the same pathway [110]. Recently, the SUMOylation of human Pol η was shown to occur following UV and replication stress, although its functional consequence is unclear [111]. PIAS1 acts as the SUMO ligase for human Pol η and its function depends on Rad18. *POLK* was also implicated to act in the GEI-17 pathway in the *C. elegans* study, and identified as a SUMOylated peptide in a human cell proteomic screen [112], but there have been no reports validating human Pol κ SUMOylation. Finally, REV1 is SUMOylated by PIAS in a cell model of doxorubicin sensitization by starvation [113]. This modification promotes REV1 stability, and was also demonstrated in H_2O_2 treated cells. Since PIAS is the human homolog of *C. elegans* GEI-17, the data to date suggest that SUMOylation is a conserved mechanism for regulating Y-family polymerases.

Degradation of the human Pol δ p12 subunit (Section II D) also has been studied during the DNA damage response (reviewed in [76]). Upon treatment of cells with various DNA damage-inducing agents, including UV, MMS, hydroxyurea, aphidicolin, and IR, the p12 subunit undergoes complete ubiquitylation-dependent degradation, to form Pol δ3. This process requires both the CRL4^{Cdt2} and the RNF8 E3 ligases [84,114]. Under conditions of low UV doses, Pol δ3 formation is dependent on activation of ATR [115]. After UV irradiation, ~70% of cells with cyclobutane pyrimidine dimer foci co-localize with the Pol δ3 form exclusively, consistent with a role for Pol δ3 in NER re-synthesis [116]. Interestingly, Pol δ3 displays increased exonuclease partitioning and decreased potential for bypass of various DNA lesions [76]. These findings led to a model in which Pol δ3 may slow fork replication progression at sites of DNA damage, allowing for switching to a specialized DNA polymerase [76].

Both Pol η and REV1 interact with Hsp90, suggesting that proper folding of polymerases is required for optimal function [117,118]. Inhibition of this interaction reduced UV induced foci formation of both polymerases and, in some cell lines, reduced protein stability. Inhibition of Hsp90 also reduced the interaction of these polymerases with PCNA after UV and was epistatic to knock-down of either protein in UV induced mutagenesis.

5. Orchestration of DNA Polymerases

Maintenance of genome stability requires the formation of distinct replication and repair complexes that include both replicative and specialized DNA polymerases. A summary of known protein interactions of the Y-family polymerases is given in Figure 2. The regulation of Y-family polymerase complex formation is fairly well studied, and cross-talk between the B- and Y-family polymerases also occurs through several mechanisms. Human Pol η directly interacts with the p50 subunit of Pol δ, but Pol κ and ι do not [119]. This interaction occurs via an FF motif (F1) and is required for optimal UV survival, but not Pol η foci formation. Mutation of this site also reduced the cellular interaction between Pol η and p50, as well as Pol η and PCNA. The F1 motif was later described as PIP3, and while still putative, may explain the latter phenotype [120].

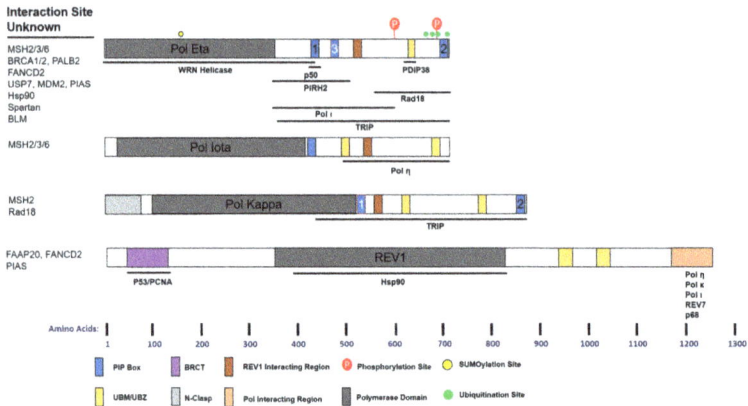

Figure 2. Schematic of Y-Family Polymerase Domains and Interaction Sites. Functional domains that have been experimentally validated are indicated and drawn to scale along the length of the protein. PIP (PCNA Interacting Peptide) boxes with red highlight are putative. Below each cartoon are the known sites of interaction between the polymerase and the indicated protein. Proteins whose interaction has been suggested but the precise site is unknown are listed to the left. See text for details.

Human REV1 harbors a C-terminal domain capable of interacting with the other Y-family polymerases, REV7, and the p68 subunit of Pol δ [121] (Figure 2). Interaction with REV1 promotes Pols η, κ, and ι UV-induced foci formation and UV lesion TLS [122]. Structural studies suggest that REV1 may orchestrate multiple polymerases simultaneously [123,124]. This is of particular interest as REV1 interacts with p68 in a complex similar to Pol η or κ [125]. Moreover, this may be a function unique to higher organisms, as binding of REV1 to Y-family polymerases is found in humans, mice, and flies, but not worms and yeast [126]. Further studies are required to assess the cellular consequences of these interactions.

PDIP38 also provides a link between Pol δ and specialized polymerases, binding both p50 and Pol η [127,128]. This protein interacts with Pol η in an area overlapping the UBZ, but independent of ubiquitin binding. Loss of PDIP38 does not reduce Pol η foci following UV, but does impair viability to the same extent as Pol η knock-down. Moreover, PDIP38 interacts with REV1, but not Pols κ and ι. Combined with the above report, it seems PDIP38 may facilitate the interaction between Pol δ and η and thereby promote TLS following UV. Several questions remain, however, as to the biochemical consequences of these interactions.

Spartan (C1orf124), a ubiquitin binding protein, was recently identified as a regulator of TLS. Spartan interacts directly with the p68 subunit of Pol δ through its zinc metalloprotease domain [129]. Spartan interaction with p68 is lost following UV, and Spartan interacts strongly with Pol η instead. Depletion of Spartan increases the interaction of p68 with REV1, and reduces UV induced Pol η foci, suggesting that Spartan positively regulates TLS [130,131]. Although controversial, Spartan is proposed to positively regulate TLS by promoting Rad18 activity, PCNA ubiquitination, and inhibiting USP1 [130,132,133]. However, it has also been inferred from mutagenesis studies that Spartan negatively regulates TLS [129,133,134]. In these reports, knock-down of Spartan elevates UV mutagenesis. However, these studies were conducted in *POLH* and NER proficient cells, which actually suggests that Spartan either suppresses erroneous repair processes, or promotes error-free pathways. In agreement with this, knock-down of REV1, which promotes error-prone TLS, eliminates the increased mutagenesis in Spartan depleted cells [129]. Finally, Spartan depletion sensitizes cells to

UV, again suggesting that it promotes TLS activity [130,131]. Future studies examining the epistasis between Spartan and accurate TLS mechanisms, as well as the functional consequences of Spartan interactions, are required.

There is growing evidence that the Fanconi anemia (FA) pathway, traditionally known for inter-strand crosslink repair, also regulates the Y-family polymerases. FANCD2 interacts with Pol η following UV and this is dependent on FANCD2 ubiquitination [135]. Pol η/FANCD2 interaction precedes Pol η/PCNA interaction following UV, suggesting FANCD2 is involved in the early regulation of Pol η. Recently, FANCD2 was shown to regulate chromatin localization of Pol η, but not Pol κ, following hydroxyurea [136]. REV1 also interacts with FANCD2, and loss of FANCD2 impairs REV1 accumulation at sites of laser microirradiation [137]. Interestingly, knock-down of *REV1* or *POLH* reduces FANCD2 foci following UV, consistent with co-regulation of FA proteins and Y-family polymerases. The FA core complex (A, B, C, E, G, F, M, and L) also regulates REV1 and Pol η. Following UV, FANCA and FANCG deficient cells display reduced REV1 foci compared to proficient cells [138]. Moreover, REV1 directly interacts with FAAP20, which stabilizes the FA core and promotes REV1 foci following UV. Additionally, the FA core may influence the expression of Pol η [139].

BRCA1 and BRCA2 have been extensively studied due to their roles in HR and dysregulation in breast cancers and are thought of as members of the FA pathway. In addition to lesion bypass, there is accumulating evidence that Y-family polymerases play an important role in the response to DSBs and are regulated in accordingly. BRCA1 was first reported to interact with REV1 and Pol η and its knock-down reduced their foci formation following UV [140]. BRCA2 as well as PALB2 were shown to interact with Pol η and co-localize following UV and replication stress, and to a lesser extent following IR [141]. Interaction with BRAC2/PALB2 was required for optimal Pol η foci formation following hydroxyurea. Both BRCA2 and PALB2, but not BRCA1, stimulated Pol η's DNA synthesis activity using a model recombination (D-loop) substrate in vitro.

6. Summary and Model for Orchestration

Our knowledge of how mammalian cells regulate the levels and activities of replicative and specialized polymerases to maintain genome integrity is in a state of infancy. Such regulation is quite intricate, and occurs at the transcriptional, post-transcriptional and translational levels (Figure 1). UV irradiation is the most well characterized model for polymerase orchestration (Figure 3). Following UV irradiation, Pol κ synthesis promotes ATR activation which, in turn, enhances Pol ι expression and phosphorylates Pol η [37,89,95]. This is likely concomitant with Pol η deubiquitination by USP7 and PIAS1/Rad18 dependent SUMOylation [108,111] (Figure 3A).

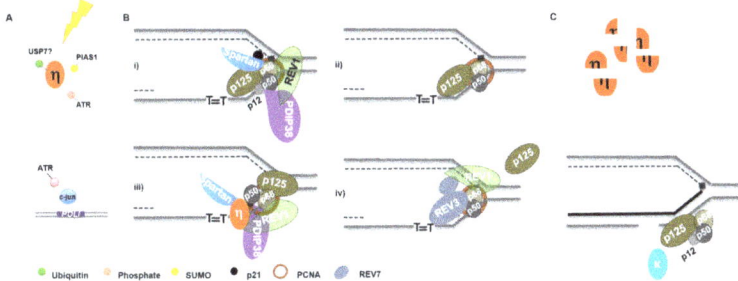

Figure 3. Orchestration of DNA Polymerases Following UV Irradiation: (**A**) following UV, Pol η is deubiquitinated, phosphorylated, and SUMOylated while *POLI* gene expression is induced; (**B**) thymine dimers stall replication forks requiring exchange, or altered polymerase activity (see text for details); and (**C**) following lesion bypass, Pol η is degraded by the proteasome while Pol κ and Pol δ (as well as Pol ε) filling in gaps generated by lesion incision by NER.

Recruitment of Pol η to sites of UV lesions is facilitated by numerous factors, as discussed above. REV1 may coordinate an exchange of DNA synthesis activity with Pol δ by interacting with p68 [125], with assistance by PDIP38/p50 and Spartan/p68 interactions (Figure 3Bi) [119,127,129,131]. Replicative Pol δ subunit composition is altered with the concomitant degradation of p12 and p21 (Figure 3Bii), which may aid polymerase orchestration at replication forks [85,142]. Pol η engagement at the fork may be facilitated by numerous protein-protein interactions (Figure 3Biii). Additionally, Pol ζ may be recruited to exchange subunits with Pol δ (p68 and p50) and assist in lesion bypass, in a manner likely facilitated by REV1 [143,144] (Figure 3Biv). Following lesion bypass, Pol η is degraded, while Pol κ and δ can participate in gap filling following lesion excision by NER (Figure 3C) [30,107]. The timing of PCNA mono-ubiquitination following UV irradiation and the nature of its function vis-à-vis polymerase orchestration is controversial [135,145–147]. Considering Pol η facilitates Rad18 recruitment to chromatin and it and Pol κ promote PCNA mono-ubiquitination, this modification may actually occur following specialized polymerase synthesis [60,120].

Much work remains to be done characterizing the orchestration of DNA polymerases during the replication stress response. However, research over the past decade clearly has shown this to be a more complex process than a single post-translational modification of PCNA. Numerous proteins converge on the Y-family polymerases to facilitate their recruitment to (presumably) stalled replication forks, including the Fanconi Anemia and BRCA pathways [136,141]. Interestingly, these pathways also promote repair of collapsed forks through DSBR and HR, and Y-family polymerases perform synthesis during these repair processes [31,36,148]. Therefore, recruitment of Y-family polymerases during replication stress may serve as an attempt to both relieve stalled replication and repair collapsed forks simultaneously. While such widespread utilization of error-prone polymerases may seem counter-intuitive, Pols η, κ, and ι interact with MMR proteins, opening the formal possibility that errors in the DNA synthesis products of these enzymes may be removed to maintain overall genome stability [32,36,149,150].

Acknowledgments: Research in our laboratory was supported by the Donald B. and Dorothy L. Stabler Foundation and generous donations to the Jake Gittlen Cancer Research Foundation.

Author Contributions: R.B and K.E. developed the overall review content and wrote the manuscript. R. B. performed the in-depth literature review and created the figures.

Conflicts of Interest: The authors declare no conflict of interest.

References

1. Waters, L.S.; Minesinger, B.K.; Wiltrout, M.E.; D'Souza, S.; Woodruff, R.V.; Walker, G.C. Eukaryotic translesion polymerases and their roles and regulation in dna damage tolerance. *Microbiol. Mol. Biol. Rev.* **2009**, *73*, 134–154. [CrossRef] [PubMed]
2. Boyer, A.S.; Grgurevic, S.; Cazaux, C.; Hoffmann, J.S. The human specialized dna polymerases and non-b dna: Vital relationships to preserve genome integrity. *J. Mol. Biol.* **2013**, *425*, 4767–4781. [CrossRef] [PubMed]
3. Sale, J.E.; Lehmann, A.R.; Woodgate, R. Y-family dna polymerases and their role in tolerance of cellular dna damage. *Nat. Rev. Mol. Cell Biol.* **2012**, *13*, 141–152. [CrossRef] [PubMed]
4. Lujan, S.A.; Williams, J.S.; Kunkel, T.A. Dna polymerases divide the labor of genome replication. *Trends Cell Biol.* **2016**, *26*, 640–654. [CrossRef] [PubMed]
5. Zeman, M.K.; Cimprich, K.A. Causes and consequences of replication stress. *Nat. Cell Biol.* **2014**, *16*, 2–9. [CrossRef] [PubMed]
6. Lehmann, A.R.; Fuchs, R.P. Gaps and forks in dna replication: Rediscovering old models. *DNA Repair (Amst.)* **2006**, *5*, 1495–1498. [CrossRef] [PubMed]
7. de Koning, A.P.; Gu, W.; Castoe, T.A.; Batzer, M.A.; Pollock, D.D. Repetitive elements may comprise over two-thirds of the human genome. *PLoS Genet.* **2011**, *7*, e1002384. [CrossRef] [PubMed]
8. Durkin, S.G.; Glover, T.W. Chromosome fragile sites. *Annu. Rev. Genet.* **2007**, *41*, 169–192.
9. Hile, S.E.; Wang, X.; Lee, M.Y.; Eckert, K.A. Beyond translesion synthesis: Polymerase κ fidelity as a potential determinant of microsatellite stability. *Nucleic Acids Res.* **2012**, *40*, 1636–1647. [CrossRef] [PubMed]

10. Shah, S.N.; Opresko, P.L.; Meng, X.; Lee, M.Y.; Eckert, K.A. Dna structure and the werner protein modulate human dna polymerase delta-dependent replication dynamics within the common fragile site fra16d. *Nucleic Acids Res.* **2010**, *38*, 1149–1162. [CrossRef] [PubMed]

11. Walsh, E.; Wang, X.; Lee, M.Y.; Eckert, K.A. Mechanism of replicative dna polymerase delta pausing and a potential role for dna polymerase kappa in common fragile site replication. *J. Mol. Biol.* **2013**, *425*, 232–243. [CrossRef] [PubMed]

12. Bergoglio, V.; Boyer, A.S.; Walsh, E.; Naim, V.; Legube, G.; Lee, M.Y.; Rey, L.; Rosselli, F.; Cazaux, C.; Eckert, K.A.; et al. Dna synthesis by pol η promotes fragile site stability by preventing under-replicated dna in mitosis. *J. Cell Biol.* **2013**, *201*, 395–408. [CrossRef] [PubMed]

13. Bhat, A.; Andersen, P.L.; Qin, Z.; Xiao, W. Rev3, the catalytic subunit of polζ, is required for maintaining fragile site stability in human cells. *Nucleic Acids Res.* **2013**, *41*, 2328–2339. [CrossRef] [PubMed]

14. Rey, L.; Sidorova, J.M.; Puget, N.; Boudsocq, F.; Biard, D.S.; Monnat, R.J.; Cazaux, C.; Hoffmann, J.S. Human dna polymerase eta is required for common fragile site stability during unperturbed dna replication. *Mol. Cell. Biol.* **2009**, *29*, 3344–3354. [CrossRef] [PubMed]

15. Ross, A.L.; Simpson, L.J.; Sale, J.E. Vertebrate dna damage tolerance requires the c-terminus but not brct or transferase domains of rev1. *Nucleic Acids Res.* **2005**, *33*, 1280–1289. [CrossRef] [PubMed]

16. Maxwell, B.A.; Suo, Z. Recent insight into the kinetic mechanisms and conformational dynamics of y-family dna polymerases. *Biochemistry* **2014**, *53*, 2804–2814. [CrossRef] [PubMed]

17. Yang, W. An overview of y-family dna polymerases and a case study of human dna polymerase η. *Biochemistry* **2014**, *53*, 2793–2803. [CrossRef] [PubMed]

18. Biertümpfel, C.; Zhao, Y.; Kondo, Y.; Ramón-Maiques, S.; Gregory, M.; Lee, J.Y.; Masutani, C.; Lehmann, A.R.; Hanaoka, F.; Yang, W. Structure and mechanism of human dna polymerase eta. *Nature* **2010**, *465*, 1044–1048.

19. Zhao, Y.; Biertümpfel, C.; Gregory, M.T.; Hua, Y.J.; Hanaoka, F.; Yang, W. Structural basis of human dna polymerase η-mediated chemoresistance to cisplatin. *Proc. Natl. Acad. Sci. USA* **2012**, *109*, 7269–7274.

20. Lone, S.; Townson, S.A.; Uljon, S.N.; Johnson, R.E.; Brahma, A.; Nair, D.T.; Prakash, S.; Prakash, L.; Aggarwal, A.K. Human dna polymerase kappa encircles dna: Implications for mismatch extension and lesion bypass. *Mol. Cell* **2007**, *25*, 601–614. [CrossRef] [PubMed]

21. Liu, Y.; Yang, Y.; Tang, T.S.; Zhang, H.; Wang, Z.; Friedberg, E.; Yang, W.; Guo, C. Variants of mouse dna polymerase κ reveal a mechanism of efficient and accurate translesion synthesis past a benzo[a]pyrene dg adduct. *Proc. Natl. Acad. Sci. USA* **2014**, *111*, 1789–1794. [CrossRef] [PubMed]

22. Jha, V.; Bian, C.; Xing, G.; Ling, H. Structure and mechanism of error-free replication past the major benzo[a]pyrene adduct by human dna polymerase κ. *Nucleic Acids Res.* **2016**, *44*, 4957–4967.

23. Thompson, H.C.; Sheaff, R.J.; Kuchta, R.D. Interactions of calf thymus dna polymerase alpha with primer/templates. *Nucleic Acids Res.* **1995**, *23*, 4109–4115. [CrossRef] [PubMed]

24. Van, C.; Yan, S.; Michael, W.M.; Waga, S.; Cimprich, K.A. Continued primer synthesis at stalled replication forks contributes to checkpoint activation. *J. Cell Biol.* **2010**, *189*, 233–246. [CrossRef] [PubMed]

25. Yan, S.; Michael, W.M. Topbp1 and dna polymerase-alpha directly recruit the 9-1-1 complex to stalled dna replication forks. *J. Cell Biol.* **2009**, *184*, 793–804. [CrossRef] [PubMed]

26. Vaara, M.; Itkonen, H.; Hillukkala, T.; Liu, Z.; Nasheuer, H.P.; Schaarschmidt, D.; Pospiech, H.; Syväoja, J.E. Segregation of replicative dna polymerases during s phase: Dna polymerase ε, but not dna polymerases α/δ, are associated with lamins throughout s phase in human cells. *J. Biol. Chem.* **2012**, *287*, 33327–33338.

27. Parsons, J.L.; Preston, B.D.; O'Connor, T.R.; Dianov, G.L. Dna polymerase delta-dependent repair of dna single strand breaks containing 3'-end proximal lesions. *Nucleic Acids Res.* **2007**, *35*, 1054–1063.

28. Longley, M.J.; Pierce, A.J.; Modrich, P. Dna polymerase delta is required for human mismatch repair in vitro. *J. Biol. Chem.* **1997**, *272*, 10917–10921. [PubMed]

29. Sneeden, J.L.; Grossi, S.M.; Tappin, I.; Hurwitz, J.; Heyer, W.D. Reconstitution of recombination-associated dna synthesis with human proteins. *Nucleic Acids Res.* **2013**, *41*, 4913–4925. [CrossRef] [PubMed]

30. Ogi, T.; Limsirichaikul, S.; Overmeer, R.M.; Volker, M.; Takenaka, K.; Cloney, R.; Nakazawa, Y.; Niimi, A.; Miki, Y.; Jaspers, N.G.; et al. Three dna polymerases, recruited by different mechanisms, carry out ner repair synthesis in human cells. *Mol. Cell* **2010**, *37*, 714–727. [CrossRef] [PubMed]

31. McIlwraith, M.J.; Mcllwraith, M.J.; Vaisman, A.; Liu, Y.; Fanning, E.; Woodgate, R.; West, S.C. Human DNA polymerase eta promotes DNA synthesis from strand invasion intermediates of homologous recombination. *Mol. Cell* **2005**, *20*, 783–792. [CrossRef] [PubMed]

32. Peña-Diaz, J.; Bregenhorn, S.; Ghodgaonkar, M.; Follonier, C.; Artola-Borán, M.; Castor, D.; Lopes, M.; Sartori, A.A.; Jiricny, J. Noncanonical mismatch repair as a source of genomic instability in human cells. *Mol. Cell* **2012**, *47*, 669–680. [CrossRef] [PubMed]

33. Zeng, X.; Winter, D.B.; Kasmer, C.; Kraemer, K.H.; Lehmann, A.R.; Gearhart, P.J. Dna polymerase eta is an a-t mutator in somatic hypermutation of immunoglobulin variable genes. *Nat. Immunol.* **2001**, *2*, 537–541.

34. Bétous, R.; Rey, L.; Wang, G.; Pillaire, M.J.; Puget, N.; Selves, J.; Biard, D.S.; Shin-ya, K.; Vasquez, K.M.; Cazaux, C.; et al. Role of tls dna polymerases eta and kappa in processing naturally occurring structured dna in human cells. *Mol. Carcinog.* **2009**, *48*, 369–378. [CrossRef] [PubMed]

35. Ogi, T.; Lehmann, A.R. The y-family dna polymerase kappa (pol kappa) functions in mammalian nucleotide-excision repair. *Nat. Cell Biol.* **2006**, *8*, 640–642. [CrossRef] [PubMed]

36. Zhang, X.; Lv, L.; Chen, Q.; Yuan, F.; Zhang, T.; Yang, Y.; Zhang, H.; Wang, Y.; Jia, Y.; Qian, L.; et al. Mouse dna polymerase kappa has a functional role in the repair of dna strand breaks. *DNA Repair (Amst.)* **2013**, *12*, 377–388. [CrossRef] [PubMed]

37. Bétous, R.; Pillaire, M.J.; Pierini, L.; van der Laan, S.; Recolin, B.; Ohl-Séguy, E.; Guo, C.; Niimi, N.; Grúz, P.; Nohmi, T.; et al. Dna polymerase κ-dependent dna synthesis at stalled replication forks is important for chk1 activation. *EMBO J.* **2013**, *32*, 2172–2185. [CrossRef] [PubMed]

38. Faili, A.; Aoufouchi, S.; Flatter, E.; Guéranger, Q.; Reynaud, C.A.; Weill, J.C. Induction of somatic hypermutation in immunoglobulin genes is dependent on dna polymerase iota. *Nature* **2002**, *419*, 944–947.

39. Velasco-Miguel, S.; Richardson, J.A.; Gerlach, V.L.; Lai, W.C.; Gao, T.; Russell, L.D.; Hladik, C.L.; White, C.L.; Friedberg, E.C. Constitutive and regulated expression of the mouse dinb (polkappa) gene encoding dna polymerase kappa. *DNA Repair (Amst.)* **2003**, *2*, 91–106. [CrossRef]

40. Gerlach, V.L.; Aravind, L.; Gotway, G.; Schultz, R.A.; Koonin, E.V.; Friedberg, E.C. Human and mouse homologs of escherichia coli dinb (dna polymerase iv), members of the umuc/dinb superfamily. *Proc. Natl. Acad. Sci. USA* **1999**, *96*, 11922–11927. [CrossRef] [PubMed]

41. Thakur, M.; Wernick, M.; Collins, C.; Limoli, C.L.; Crowley, E.; Cleaver, J.E. Dna polymerase eta undergoes alternative splicing, protects against uv sensitivity and apoptosis, and suppresses mre11-dependent recombination. *Genes Chromosomes Cancer* **2001**, *32*, 222–235. [CrossRef] [PubMed]

42. Yamada, A.; Masutani, C.; Iwai, S.; Hanaoka, F. Complementation of defective translesion synthesis and uv light sensitivity in xeroderma pigmentosum variant cells by human and mouse dna polymerase eta. *Nucleic Acids Res.* **2000**, *28*, 2473–2480. [CrossRef] [PubMed]

43. Lin, W.; Xin, H.; Zhang, Y.; Wu, X.; Yuan, F.; Wang, Z. The human rev1 gene codes for a dna template-dependent dcmp transferase. *Nucleic Acids Res.* **1999**, *27*, 4468–4475. [CrossRef] [PubMed]

44. McDonald, J.P.; Rapić-Otrin, V.; Epstein, J.A.; Broughton, B.C.; Wang, X.; Lehmann, A.R.; Wolgemuth, D.J.; Woodgate, R. Novel human and mouse homologs of saccharomyces cerevisiae dna polymerase eta. *Genomics* **1999**, *60*, 20–30. [CrossRef] [PubMed]

45. Frank, E.G.; Tissier, A.; McDonald, J.P.; Rapić-Otrin, V.; Zeng, X.; Gearhart, P.J.; Woodgate, R. Altered nucleotide misinsertion fidelity associated with poliota-dependent replication at the end of a dna template. *EMBO J.* **2001**, *20*, 2914–2922. [CrossRef] [PubMed]

46. Akagi, J.; Masutani, C.; Kataoka, Y.; Kan, T.; Ohashi, E.; Mori, T.; Ohmori, H.; Hanaoka, F. Interaction with dna polymerase eta is required for nuclear accumulation of rev1 and suppression of spontaneous mutations in human cells. *DNA Repair (Amst.)* **2009**, *8*, 585–599. [CrossRef] [PubMed]

47. Bermudez, V.P.; Farina, A.; Raghavan, V.; Tappin, I.; Hurwitz, J. Studies on human dna polymerase epsilon and gins complex and their role in dna replication. *J. Biol. Chem.* **2011**, *286*, 28963–28977.

48. Diamant, N.; Hendel, A.; Vered, I.; Carell, T.; Reissner, T.; de Wind, N.; Geacinov, N.; Livneh, Z. Dna damage bypass operates in the s and g2 phases of the cell cycle and exhibits differential mutagenicity. *Nucleic Acids Res.* **2012**, *40*, 170–180. [CrossRef] [PubMed]

49. Beishline, K.; Azizkhan-Clifford, J. Sp1 and the 'hallmarks of cancer'. *FEBS J.* **2015**, *282*, 224–258.

50. Izumi, M.; Yokoi, M.; Nishikawa, N.S.; Miyazawa, H.; Sugino, A.; Yamagishi, M.; Yamaguchi, M.; Matsukage, A.; Yatagai, F.; Hanaoka, F. Transcription of the catalytic 180-kda subunit gene of mouse dna polymerase alpha is controlled by e2f, an ets-related transcription factor, and sp1. *Biochim. Biophys. Acta* **2000**, *1492*, 341–352. [CrossRef]

51. Zhao, L.; Chang, L.S. The human pold1 gene. Identification of an upstream activator sequence, activation by sp1 and sp3, and cell cycle regulation. *J. Biol. Chem.* **1997**, *272*, 4869–4882. [CrossRef] [PubMed]

52. Huang, D.; Jokela, M.; Tuusa, J.; Skog, S.; Poikonen, K.; Syväoja, J.E. E2f mediates induction of the sp1-controlled promoter of the human dna polymerase epsilon b-subunit gene pole2. *Nucleic Acids Res.* **2001**, *29*, 2810–2821. [CrossRef] [PubMed]

53. Lemée, F.; Bavoux, C.; Pillaire, M.J.; Bieth, A.; Machado, C.R.; Pena, S.D.; Guimbaud, R.; Selves, J.; Hoffmann, J.S.; Cazaux, C. Characterization of promoter regulatory elements involved in downexpression of the dna polymerase kappa in colorectal cancer. *Oncogene* **2007**, *26*, 3387–3394. [CrossRef] [PubMed]

54. Zhu, H.; Fan, Y.; Shen, J.; Qi, H.; Shao, J. Characterization of human dna polymerase κ promoter in response to benzo[a]pyrene diol epoxide. *Environ. Toxicol. Pharmacol.* **2012**, *33*, 205–211. [CrossRef] [PubMed]

55. Zhou, J.; Zhang, S.; Xie, L.; Liu, P.; Xie, F.; Wu, J.; Cao, J.; Ding, W.Q. Overexpression of dna polymerase iota (polι) in esophageal squamous cell carcinoma. *Cancer Sci.* **2012**, *103*, 1574–1579. [CrossRef] [PubMed]

56. Zhu, H.; Fan, Y.; Jiang, H.; Shen, J.; Qi, H.; Mei, R.; Shao, J. Response of human dna polymerase ι promoter to n-methyl-n'-nitro-n-nitrosoguanidine. *Environ. Toxicol. Pharmacol.* **2010**, *29*, 79–86. [CrossRef] [PubMed]

57. Qi, H.; Zhu, H.; Lou, M.; Fan, Y.; Liu, H.; Shen, J.; Li, Z.; Lv, X.; Shan, J.; Zhu, L.; et al. Interferon regulatory factor 1 transactivates expression of human dna polymerase η in response to carcinogen n-methyl-n'-nitro-n-nitrosoguanidine. *J. Biol. Chem.* **2012**, *287*, 12622–12633. [CrossRef] [PubMed]

58. Kent, W.J.; Sugnet, C.W.; Furey, T.S.; Roskin, K.M.; Pringle, T.H.; Zahler, A.M.; Haussler, D. The human genome browser at ucsc. *Genome Res.* **2002**, *12*, 996–1006. [PubMed]

59. Liu, G.; Chen, X. Dna polymerase eta, the product of the xeroderma pigmentosum variant gene and a target of p53, modulates the dna damage checkpoint and p53 activation. *Mol. Cell. Biol.* **2006**, *26*, 1398–1413.

60. Durando, M.; Tateishi, S.; Vaziri, C. A non-catalytic role of dna polymerase η in recruiting rad18 and promoting pcna monoubiquitination at stalled replication forks. *Nucleic Acids Res.* **2013**, *41*, 3079–3093.

61. Melanson, B.D.; Bose, R.; Hamill, J.D.; Marcellus, K.A.; Pan, E.F.; McKay, B.C. The role of mRNA decay in p53-induced gene expression. *RNA* **2011**, *17*, 2222–2234. [CrossRef] [PubMed]

62. Wang, Y.; Seimiya, M.; Kawamura, K.; Yu, L.; Ogi, T.; Takenaga, K.; Shishikura, T.; Nakagawara, A.; Sakiyama, S.; Tagawa, M.; et al. Elevated expression of dna polymerase kappa in human lung cancer is associated with p53 inactivation: Negative regulation of polk promoter activity by p53. *Int. J. Oncol.* **2004**, *25*, 161–165. [PubMed]

63. Antoniali, G.; Marcuzzi, F.; Casarano, E.; Tell, G. Cadmium treatment suppresses dna polymerase δ catalytic subunit gene expression by acting on the p53 and sp1 regulatory axis. *DNA Repair (Amst.)* **2015**, *35*, 90–105.

64. Li, B.; Lee, M.Y. Transcriptional regulation of the human dna polymerase delta catalytic subunit gene pold1 by p53 tumor suppressor and sp1. *J. Biol. Chem.* **2001**, *276*, 29729–29739. [CrossRef] [PubMed]

65. Nishikawa, N.S.; Izumi, M.; Uchida, H.; Yokoi, M.; Miyazawa, H.; Hanaoka, F. Cloning and characterization of the 5'-upstream sequence governing the cell cycle-dependent transcription of mouse dna polymerase alpha 68 kda subunit gene. *Nucleic Acids Res.* **2000**, *28*, 1525–1534. [CrossRef] [PubMed]

66. Bolognese, F.; Forni, C.; Caretti, G.; Frontini, M.; Minuzzo, M.; Mantovani, R. The pole3 bidirectional unit is regulated by myc and e2fs. *Gene* **2006**, *366*, 109–116. [CrossRef] [PubMed]

67. Karkhanis, V.; Wang, L.; Tae, S.; Hu, Y.J.; Imbalzano, A.N.; Sif, S. Protein arginine methyltransferase 7 regulates cellular response to dna damage by methylating promoter histones h2a and h4 of the polymerase δ catalytic subunit gene, pold1. *J. Biol. Chem.* **2012**, *287*, 29801–29814. [CrossRef] [PubMed]

68. Ren, C.; Cho, S.J.; Jung, Y.S.; Chen, X. Dna polymerase η is regulated by poly(rc)-binding protein 1 via mrna stability. *Biochem. J.* **2014**, *464*, 377–386. [CrossRef] [PubMed]

69. Czochor, J.R.; Sulkowski, P.; Glazer, P.M. Mir-155 overexpression promotes genomic instability by reducing high-fidelity polymerase delta expression and activating error-prone dsb repair. *Mol. Cancer Res.* **2016**, *14*, 363–373. [CrossRef] [PubMed]

70. Wang, Y.; Huang, J.W.; Calses, P.; Kemp, C.J.; Taniguchi, T. Mir-96 downregulates rev1 and rad51 to promote cellular sensitivity to cisplatin and parp inhibition. *Cancer Res.* **2012**, *72*, 4037–4046. [CrossRef] [PubMed]

71. Guo, J.; Jiang, Z.; Li, X.; Wang, X.I.; Xiao, Y. Mir-20b downregulates polymerases κ and θ in xp-v tumor cells. *Oncol. Lett.* **2016**, *11*, 3790–3794. [CrossRef] [PubMed]

72. Srivastava, A.K.; Han, C.; Zhao, R.; Cui, T.; Dai, Y.; Mao, C.; Zhao, W.; Zhang, X.; Yu, J.; Wang, Q.E. Enhanced expression of dna polymerase eta contributes to cisplatin resistance of ovarian cancer stem cells. *Proc. Natl. Acad. Sci. USA* **2015**, *112*, 4411–4416. [CrossRef] [PubMed]

73. Voitenleitner, C.; Rehfuess, C.; Hilmes, M.; O'Rear, L.; Liao, P.C.; Gage, D.A.; Ott, R.; Nasheuer, H.P.; Fanning, E. Cell cycle-dependent regulation of human dna polymerase alpha-primase activity by phosphorylation. *Mol. Cell. Biol.* **1999**, *19*, 646–656. [CrossRef] [PubMed]

74. Nasheuer, H.P.; Moore, A.; Wahl, A.F.; Wang, T.S. Cell cycle-dependent phosphorylation of human dna polymerase alpha. *J. Biol. Chem.* **1991**, *266*, 7893–7903. [PubMed]

75. Zhou, Y.; Meng, X.; Zhang, S.; Lee, E.Y.; Lee, M.Y. Characterization of human dna polymerase delta and its subassemblies reconstituted by expression in the multibac system. *PLoS ONE* **2012**, *7*, e39156. [CrossRef] [PubMed]

76. Lee, M.Y.W.T.; Zhang, S.; Lin, S.H.S.; Chea, J.; Wang, X.; LeRoy, C.; Wong, A.; Zhang, Z.; Lee, E.Y.C. Regulation of human DNA polymerase delta in the cellular responses to DNA damage. *Environ. Mol. Mutagen.* **2012**, *53*, 683–698. [CrossRef] [PubMed]

77. Zeng, X.R.; Hao, H.; Jiang, Y.; Lee, M.Y. Regulation of human dna polymerase delta during the cell cycle. *J. Biol. Chem.* **1994**, *269*, 24027–24033. [PubMed]

78. Li, H.; Xie, B.; Rahmeh, A.; Zhou, Y.; Lee, M.Y. Direct interaction of p21 with p50, the small subunit of human dna polymerase delta. *Cell Cycle* **2006**, *5*, 428–436. [CrossRef] [PubMed]

79. Ducoux, M.; Urbach, S.; Baldacci, G.; Hubscher, U.; Koundrioukoff, S.; Christensen, J.; Hughes, P. Mediation of proliferating cell nuclear antigen (pcna)-dependent dna replication through a conserved p21(cip1)-like pcna-binding motif present in the third subunit of human dna polymerase delta. *J. Biol. Chem.* **2001**, *276*, 49258–49266. [CrossRef] [PubMed]

80. Lemmens, L.; Urbach, S.; Prudent, R.; Cochet, C.; Baldacci, G.; Hughes, P. Phosphorylation of the c subunit (p66) of human dna polymerase delta. *Biochem. Biophys. Res. Commun.* **2008**, *367*, 264–270.

81. Rahmeh, A.A.; Zhou, Y.; Xie, B.; Li, H.; Lee, E.Y.C.; Lee, M.Y.W.T. Phosphorylation of the p68 subunit of poli´ acts as a molecular switch to regulate its interaction with pcna. *Biochemistry* **2011**, *51*, 416–424.

82. Gao, Y.; Zhou, Y.; Xie, B.; Zhang, S.; Rahmeh, A.; Huang, H.-s.; Lee, M.Y.W.T.; Lee, E.Y.C. Protein phosphatase-1 is targeted to dna polymerase δ via an interaction with the p68 subunit†. *Biochemistry* **2008**, *47*, 11367–11376.

83. Zhao, H.; Zhang, S.; Xu, D.; Lee, M.Y.; Zhang, Z.; Lee, E.Y.; Darzynkiewicz, Z. Expression of the p12 subunit of human dna polymerase δ (pol δ), cdk inhibitor p21(waf1), cdt1, cyclin a, pcna and ki-67 in relation to dna replication in individual cells. *Cell Cycle* **2014**, *13*, 3529–3540. [CrossRef] [PubMed]

84. Zhang, S.; Zhou, Y.; Sarkeshik, A.; Yates, J.R.; Thomson, T.M.; Zhang, Z.; Lee, E.Y.; Lee, M.Y. Identification of rnf8 as a ubiquitin ligase involved in targeting the p12 subunit of dna polymerase δ for degradation in response to dna damage. *J. Biol. Chem.* **2013**, *288*, 2941–2950. [CrossRef] [PubMed]

85. Lee, M.Y.; Zhang, S.; Lin, S.H.; Wang, X.; Darzynkiewicz, Z.; Zhang, Z.; Lee, E.Y. The tail that wags the dog: P12, the smallest subunit of DNA polymerase δ, is degraded by ubiquitin ligases in response to DNA damage and during cell cycle progression. *Cell Cycle* **2014**, *13*, 23–31. [CrossRef] [PubMed]

86. Abbas, T.; Dutta, A. Crl4cdt2: Master coordinator of cell cycle progression and genome stability. *Cell Cycle* **2011**, *10*, 241–249. [CrossRef] [PubMed]

87. Lin, S.H.; Wang, X.; Zhang, S.; Zhang, Z.; Lee, E.Y.; Lee, M.Y. Dynamics of enzymatic interactions during short flap human okazaki fragment processing by two forms of human dna polymerase δ. *DNA Repair (Amst.)* **2013**, *12*, 922–935. [CrossRef] [PubMed]

88. Yang, J.; Chen, Z.; Liu, Y.; Hickey, R.J.; Malkas, L.H. Altered dna polymerase iota expression in breast cancer cells leads to a reduction in dna replication fidelity and a higher rate of mutagenesis. *Cancer Res.* **2004**, *64*, 5597–5607. [CrossRef] [PubMed]

89. Yuan, F.; Xu, Z.; Yang, M.; Wei, Q.; Zhang, Y.; Yu, J.; Zhi, Y.; Liu, Y.; Chen, Z.; Yang, J. Overexpressed DNA polymerase iota regulated by jnk/c-jun contributes to hypermutagenesis in bladder cancer. *PLoS ONE* **2013**, *8*, e69317. [CrossRef] [PubMed]

90. Kannouche, P.; Fernández de Henestrosa, A.R.; Coull, B.; Vidal, A.E.; Gray, C.; Zicha, D.; Woodgate, R.; Lehmann, A.R. Localization of DNA polymerases eta and iota to the replication machinery is tightly co-ordinated in human cells. *EMBO J.* **2003**, *22*, 1223–1233. [CrossRef] [PubMed]

91. Guo, J.; Zhou, G.; Zhang, W.; Song, Y.; Bian, Z. A novel polh mutation causes xp-v disease and xp-v tumor proneness may involve imbalance of numerous dna polymerases. *Oncol. Lett.* **2013**, *6*, 1583–1590. [PubMed]

92. Jung, Y.S.; Liu, G.; Chen, X. Pirh2 e3 ubiquitin ligase targets DNA polymerase eta for 20s proteasomal degradation. *Mol. Cell. Biol.* **2010**, *30*, 1041–1048. [CrossRef] [PubMed]

93. Peng, C.; Chen, Z.; Wang, S.; Wang, H.W.; Qiu, W.; Zhao, L.; Xu, R.; Luo, H.; Chen, Y.; Chen, D.; et al. The error-prone DNA polymerase κ promotes temozolomide resistance in glioblastoma through rad17-dependent activation of atr-chk1 signaling. *Cancer Res.* **2016**, *76*, 2340–2353. [CrossRef] [PubMed]

94. Chen, Y.W.; Cleaver, J.E.; Hatahet, Z.; Honkanen, R.E.; Chang, J.Y.; Yen, Y.; Chou, K.M. Human DNA polymerase eta activity and translocation is regulated by phosphorylation. *Proc. Natl. Acad. Sci. USA* **2008**, *105*, 16578–16583. [CrossRef] [PubMed]

95. Göhler, T.; Sabbioneda, S.; Green, C.M.; Lehmann, A.R. ATR-mediated phosphorylation of DNA polymerase η is needed for efficient recovery from UV damage. *J. Cell Biol.* **2011**, *192*, 219–227. [CrossRef] [PubMed]

96. Dai, X.; You, C.; Wang, Y. The functions of serine 687 phosphorylation of human DNA polymerase η in UV damage tolerance. *Mol. Cell. Proteom.* **2016**, *15*, 1913–1920. [CrossRef] [PubMed]

97. Bienko, M.; Green, C.M.; Sabbioneda, S.; Crosetto, N.; Matic, I.; Hibbert, R.G.; Begovic, T.; Niimi, A.; Mann, M.; Lehmann, A.R.; et al. Regulation of translesion synthesis dna polymerase eta by monoubiquitination. *Mol. Cell* **2010**, *37*, 396–407. [CrossRef] [PubMed]

98. Jung, Y.S.; Hakem, A.; Hakem, R.; Chen, X. Pirh2 e3 ubiquitin ligase monoubiquitinates DNA polymerase eta to suppress translesion DNA synthesis. *Mol. Cell. Biol.* **2011**, *31*, 3997–4006. [CrossRef] [PubMed]

99. Bienko, M.; Green, C.M.; Crosetto, N.; Rudolf, F.; Zapart, G.; Coull, B.; Kannouche, P.; Wider, G.; Peter, M.; Lehmann, A.R.; et al. Ubiquitin-binding domains in y-family polymerases regulate translesion synthesis. *Science* **2005**, *310*, 1821–1824. [CrossRef] [PubMed]

100. Plosky, B.S.; Vidal, A.E.; Fernández de Henestrosa, A.R.; McLenigan, M.P.; McDonald, J.P.; Mead, S.; Woodgate, R. Controlling the subcellular localization of DNA polymerases iota and eta via interactions with ubiquitin. *EMBO J.* **2006**, *25*, 2847–2855. [CrossRef] [PubMed]

101. Guo, C.; Tang, T.S.; Bienko, M.; Dikic, I.; Friedberg, E.C. Requirements for the interaction of mouse polkappa with ubiquitin and its biological significance. *J. Biol. Chem.* **2008**, *283*, 4658–4664. [CrossRef] [PubMed]

102. Guo, C.; Tang, T.S.; Bienko, M.; Parker, J.L.; Bielen, A.B.; Sonoda, E.; Takeda, S.; Ulrich, H.D.; Dikic, I.; Friedberg, E.C. Ubiquitin-binding motifs in rev1 protein are required for its role in the tolerance of DNA damage. *Mol. Cell. Biol.* **2006**, *26*, 8892–8900. [CrossRef] [PubMed]

103. Kim, H.; Yang, K.; Dejsuphong, D.; D'Andrea, A.D. Regulation of rev1 by the fanconi anemia core complex. *Nat. Struct. Mol. Biol.* **2012**, *19*, 164–170. [CrossRef] [PubMed]

104. Wallace, H.A.; Merkle, J.A.; Yu, M.C.; Berg, T.G.; Lee, E.; Bosco, G.; Lee, L.A. Trip/nopo e3 ubiquitin ligase promotes ubiquitylation of dna polymerase η. *Development* **2014**, *141*, 1332–1341. [CrossRef] [PubMed]

105. Liu, G.; Warbrick, E. The p66 and p12 subunits of dna polymerase delta are modified by ubiquitin and ubiquitin-like proteins. *Biochem. Biophys. Res. Commun.* **2006**, *349*, 360–366. [CrossRef] [PubMed]

106. Bursomanno, S.; Beli, P.; Khan, A.M.; Minocherhomji, S.; Wagner, S.A.; Bekker-Jensen, S.; Mailand, N.; Choudhary, C.; Hickson, I.D.; Liu, Y. Proteome-wide analysis of sumo2 targets in response to pathological dna replication stress in human cells. *DNA Repair (Amst.)* **2015**, *25*, 84–96. [CrossRef] [PubMed]

107. Jung, Y.S.; Qian, Y.; Chen, X. Dna polymerase eta is targeted by mdm2 for polyubiquitination and proteasomal degradation in response to ultraviolet irradiation. *DNA Repair (Amst.)* **2012**, *11*, 177–184.

108. Qian, J.; Pentz, K.; Zhu, Q.; Wang, Q.; He, J.; Srivastava, A.K.; Wani, A.A. Usp7 modulates uv-induced pcna monoubiquitination by regulating dna polymerase eta stability. *Oncogene* **2015**, *34*, 4791–4796.

109. Kim, S.H.; Michael, W.M. Regulated proteolysis of dna polymerase eta during the DNA-damage response in c. Elegans. *Mol. Cell* **2008**, *32*, 757–766. [CrossRef] [PubMed]

110. Roerink, S.F.; Koole, W.; Stapel, L.C.; Romeijn, R.J.; Tijsterman, M. A broad requirement for tls polymerases η and κ, and interacting sumoylation and nuclear pore proteins, in lesion bypass during c. Elegans embryogenesis. *PLoS Genet.* **2012**, *8*, e1002800. [CrossRef] [PubMed]

111. Despras, E.; Sittewelle, M.; Pouvelle, C.; Delrieu, N.; Cordonnier, A.M.; Kannouche, P.L. Rad18-dependent sumoylation of human specialized DNA polymerase eta is required to prevent under-replicated dna. *Nat. Commun.* **2016**, *7*, 13326. [CrossRef] [PubMed]

112. Hendriks, I.A.; D'Souza, R.C.; Yang, B.; Verlaan-de Vries, M.; Mann, M.; Vertegaal, A.C. Uncovering global sumoylation signaling networks in a site-specific manner. *Nat. Struct. Mol. Biol.* **2014**, *21*, 927–936. [CrossRef] [PubMed]

113. Shim, H.S.; Wei, M.; Brandhorst, S.; Longo, V.D. Starvation promotes rev1 sumoylation and p53-dependent sensitization of melanoma and breast cancer cells. *Cancer Res.* **2015**, *75*, 1056–1067. [CrossRef] [PubMed]

114. Zhang, S.; Zhao, H.; Darzynkiewicz, Z.; Zhou, P.; Zhang, Z.; Lee, E.Y.; Lee, M.Y. A novel function of crl4(cdt2): Regulation of the subunit structure of dna polymerase δ in response to dna damage and during the s phase. *J. Biol. Chem.* **2013**, *288*, 29550–29561. [CrossRef] [PubMed]

115. Zhang, S.; Zhou, Y.; Trusa, S.; Meng, X.; Lee, E.Y.; Lee, M.Y. A novel DNA damage response: Rapid degradation of the p12 subunit of dna polymerase delta. *J. Biol. Chem.* **2007**, *282*, 15330–15340. [CrossRef] [PubMed]

116. Chea, J.; Zhang, S.; Zhao, H.; Zhang, Z.; Lee, E.Y.; Darzynkiewicz, Z.; Lee, M.Y. Spatiotemporal recruitment of human DNA polymerase delta to sites of uv damage. *Cell Cycle* **2012**, *11*, 2885–2895. [CrossRef] [PubMed]

117. Pozo, F.M.; Oda, T.; Sekimoto, T.; Murakumo, Y.; Masutani, C.; Hanaoka, F.; Yamashita, T. Molecular chaperone hsp90 regulates rev1-mediated mutagenesis. *Mol. Cell. Biol.* **2011**, *31*, 3396–3409. [CrossRef] [PubMed]

118. Sekimoto, T.; Oda, T.; Pozo, F.M.; Murakumo, Y.; Masutani, C.; Hanaoka, F.; Yamashita, T. The molecular chaperone hsp90 regulates accumulation of dna polymerase eta at replication stalling sites in uv-irradiated cells. *Mol. Cell* **2010**, *37*, 79–89. [CrossRef] [PubMed]

119. Baldeck, N.; Janel-Bintz, R.; Wagner, J.; Tissier, A.; Fuchs, R.P.; Burkovics, P.; Haracska, L.; Despras, E.; Bichara, M.; Chatton, B.; et al. Ff483-484 motif of human polη mediates its interaction with the pold2 subunit of polδ and contributes to dna damage tolerance. *Nucleic Acids Res.* **2015**, *43*, 2116–2125. [CrossRef] [PubMed]

120. Masuda, Y.; Kanao, R.; Kaji, K.; Ohmori, H.; Hanaoka, F.; Masutani, C. Different types of interaction between pcna and pip boxes contribute to distinct cellular functions of y-family dna polymerases. *Nucleic Acids Res.* **2015**, *43*, 7898–7910. [CrossRef] [PubMed]

121. Ohashi, E.; Murakumo, Y.; Kanjo, N.; Akagi, J.; Masutani, C.; Hanaoka, F.; Ohmori, H. Interaction of hrev1 with three human y-family dna polymerases. *Genes Cells* **2004**, *9*, 523–531. [CrossRef] [PubMed]

122. Yoon, J.H.; Park, J.; Conde, J.; Wakamiya, M.; Prakash, L.; Prakash, S. Rev1 promotes replication through uv lesions in conjunction with dna polymerases η, ι, and κ but not DNA polymerase ζ. *Genes Dev.* **2015**, *29*, 2588–2602. [PubMed]

123. Pustovalova, Y.; Bezsonova, I.; Korzhnev, D.M. The c-terminal domain of human rev1 contains independent binding sites for DNA polymerase η and rev7 subunit of polymerase ζ. *FEBS Lett.* **2012**, *586*, 3051–3056. [CrossRef] [PubMed]

124. Wojtaszek, J.; Lee, C.J.; D'Souza, S.; Minesinger, B.; Kim, H.; D'Andrea, A.D.; Walker, G.C.; Zhou, P. Structural basis of rev1-mediated assembly of a quaternary vertebrate translesion polymerase complex consisting of rev1, heterodimeric polymerase (pol) ζ, and pol κ. *J. Biol. Chem.* **2012**, *287*, 33836–33846.

125. Pustovalova, Y.; Magalhães, M.T.; D'Souza, S.; Rizzo, A.A.; Korza, G.; Walker, G.C.; Korzhnev, D.M. Interaction between the rev1 c-terminal domain and the pold3 subunit of polζ suggests a mechanism of polymerase exchange upon rev1/polζ-dependent translesion synthesis. *Biochemistry* **2016**, *55*, 2043–2053. [CrossRef] [PubMed]

126. Kosarek, J.N.; Woodruff, R.V.; Rivera-Begeman, A.; Guo, C.; D'Souza, S.; Koonin, E.V.; Walker, G.C.; Friedberg, E.C. Comparative analysis of in vivo interactions between rev1 protein and other y-family dna polymerases in animals and yeasts. *DNA Repair (Amst.)* **2008**, *7*, 439–451. [CrossRef] [PubMed]

127. Tissier, A.; Janel-Bintz, R.; Coulon, S.; Klaile, E.; Kannouche, P.; Fuchs, R.P.; Cordonnier, A.M. Crosstalk between replicative and translesional DNA polymerases: Pdip38 interacts directly with poleta. *DNA Repair (Amst.)* **2010**, *9*, 922–928. [CrossRef] [PubMed]

128. Liu, L.; Rodriguez-Belmonte, E.M.; Mazloum, N.; Xie, B.; Lee, M.Y. Identification of a novel protein, pdip38, that interacts with the p50 subunit of DNA polymerase delta and proliferating cell nuclear antigen. *J. Biol. Chem.* **2003**, *278*, 10041–10047. [CrossRef] [PubMed]

129. Kim, M.S.; Machida, Y.; Vashisht, A.A.; Wohlschlegel, J.A.; Pang, Y.P.; Machida, Y.J. Regulation of error-prone translesion synthesis by spartan/c1orf124. *Nucleic Acids Res.* **2013**, *41*, 1661–1668. [CrossRef] [PubMed]

130. Centore, R.C.; Yazinski, S.A.; Tse, A.; Zou, L. Spartan/c1orf124, a reader of pcna ubiquitylation and a regulator of uv-induced dna damage response. *Mol. Cell* **2012**, *46*, 625–635. [CrossRef] [PubMed]

131. Ghosal, G.; Leung, J.W.; Nair, B.C.; Fong, K.W.; Chen, J. Proliferating cell nuclear antigen (pcna)-binding protein c1orf124 is a regulator of translesion synthesis. *J. Biol. Chem.* **2012**, *287*, 34225–34233.

132. Juhasz, S.; Balogh, D.; Hajdu, I.; Burkovics, P.; Villamil, M.A.; Zhuang, Z.; Haracska, L. Characterization of human spartan/c1orf124, an ubiquitin-pcna interacting regulator of DNA damage tolerance. *Nucleic Acids Res.* **2012**, *40*, 10795–10808. [CrossRef] [PubMed]

133. Davis, E.J.; Lachaud, C.; Appleton, P.; Macartney, T.J.; Näthke, I.; Rouse, J. Dvc1 (c1orf124) recruits the p97 protein segregase to sites of dna damage. *Nat. Struct. Mol. Biol.* **2012**, *19*, 1093–1100. [CrossRef] [PubMed]

134. Machida, Y.; Kim, M.S.; Machida, Y.J. Spartan/c1orf124 is important to prevent uv-induced mutagenesis. *Cell Cycle* **2012**, *11*, 3395–3402. [CrossRef] [PubMed]

135. Fu, D.; Dudimah, F.D.; Zhang, J.; Pickering, A.; Paneerselvam, J.; Palrasu, M.; Wang, H.; Fei, P. Recruitment of DNA polymerase eta by fancd2 in the early response to dna damage. *Cell Cycle* **2013**, *12*, 803–809. [CrossRef] [PubMed]

136. Chen, X.; Bosques, L.; Sung, P.; Kupfer, G.M. A novel role for non-ubiquitinated fancd2 in response to hydroxyurea-induced dna damage. *Oncogene* **2016**, *35*, 22–34. [CrossRef] [PubMed]

137. Yang, Y.; Liu, Z.; Wang, F.; Temviriyanukul, P.; Ma, X.; Tu, Y.; Lv, L.; Lin, Y.F.; Huang, M.; Zhang, T.; et al. Fancd2 and rev1 cooperate in the protection of nascent dna strands in response to replication stress. *Nucleic Acids Res.* **2015**, *43*, 8325–8339. [CrossRef] [PubMed]

138. Mirchandani, K.D.; McCaffrey, R.M.; D'Andrea, A.D. The fanconi anemia core complex is required for efficient point mutagenesis and rev1 foci assembly. *DNA Repair (Amst.)* **2008**, *7*, 902–911.

139. Renaud, E.; Rosselli, F. Fanc pathway promotes uv-induced stalled replication forks recovery by acting both upstream and downstream polη and rev1. *PLoS ONE* **2013**, *8*, e53693. [CrossRef] [PubMed]

140. Tian, F.; Sharma, S.; Zou, J.; Lin, S.Y.; Wang, B.; Rezvani, K.; Wang, H.; Parvin, J.D.; Ludwig, T.; Canman, C.E.; et al. Brca1 promotes the ubiquitination of pcna and recruitment of translesion polymerases in response to replication blockade. *Proc. Natl. Acad. Sci. USA* **2013**, *110*, 13558–13563. [CrossRef] [PubMed]

141. Buisson, R.; Niraj, J.; Pauty, J.; Maity, R.; Zhao, W.; Coulombe, Y.; Sung, P.; Masson, J.Y. Breast cancer proteins palb2 and brca2 stimulate polymerase η in recombination-associated dna synthesis at blocked replication forks. *Cell Rep.* **2014**, *6*, 553–564. [CrossRef] [PubMed]

142. Mansilla, S.F.; Soria, G.; Vallerga, M.B.; Habif, M.; Martínez-López, W.; Prives, C.; Gottifredi, V. UV-triggered p21 degradation facilitates damaged-dna replication and preserves genomic stability. *Nucleic Acids Res.* **2013**, *41*, 6942–6951. [CrossRef] [PubMed]

143. Baranovskiy, A.G.; Lada, A.G.; Siebler, H.M.; Zhang, Y.; Pavlov, Y.I.; Tahirov, T.H. Dna polymerase δ and ζ switch by sharing accessory subunits of dna polymerase δ. *J. Biol. Chem.* **2012**, *287*, 17281–17287.

144. Lee, Y.S.; Gregory, M.T.; Yang, W. Human pol ζ purified with accessory subunits is active in translesion DNA synthesis and complements pol η in cisplatin bypass. *Proc. Natl. Acad. Sci. USA* **2014**, *111*, 2954–2959.

145. Janel-Bintz, R.; Wagner, J.; Haracska, L.; Mah-Becherel, M.C.; Bichara, M.; Fuchs, R.P.; Cordonnier, A.M. Evidence for a rad18-independent frameshift mutagenesis pathway in human cell-free extracts. *PLoS ONE* **2012**, *7*, e36004. [CrossRef] [PubMed]

146. Hendel, A.; Krijger, P.H.; Diamant, N.; Goren, Z.; Langerak, P.; Kim, J.; Reissner, T.; Lee, K.Y.; Geacintov, N.E.; Carell, T.; et al. Pcna ubiquitination is important, but not essential for translesion DNA synthesis in mammalian cells. *PLoS Genet.* **2011**, *7*, e1002262. [CrossRef] [PubMed]

147. Sabbioneda, S.; Gourdin, A.M.; Green, C.M.; Zotter, A.; Giglia-Mari, G.; Houtsmuller, A.; Vermeulen, W.; Lehmann, A.R. Effect of proliferating cell nuclear antigen ubiquitination and chromatin structure on the dynamic properties of the y-family dna polymerases. *Mol. Biol. Cell* **2008**, *19*, 5193–5202.

148. Sebesta, M.; Burkovics, P.; Juhasz, S.; Zhang, S.; Szabo, J.E.; Lee, M.Y.; Haracska, L.; Krejci, L. Role of pcna and tls polymerases in d-loop extension during homologous recombination in humans. *DNA Repair (Amst.)* **2013**, *12*, 691–698. [CrossRef] [PubMed]

149. Wilson, T.M.; Vaisman, A.; Martomo, S.A.; Sullivan, P.; Lan, L.; Hanaoka, F.; Yasui, A.; Woodgate, R.; Gearhart, P.J. Msh2-msh6 stimulates DNA polymerase eta, suggesting a role for A:T mutations in antibody genes. *J. Exp. Med.* **2005**, *201*, 637–645. [CrossRef] [PubMed]

150. Lv, L.; Wang, F.; Ma, X.; Yang, Y.; Wang, Z.; Liu, H.; Li, X.; Liu, Z.; Zhang, T.; Huang, M.; et al. Mismatch repair protein msh2 regulates translesion dna synthesis following exposure of cells to uv radiation. *Nucleic Acids Res.* **2013**, *41*, 10312–10322. [CrossRef] [PubMed]

Review

Regulation and Modulation of Human DNA Polymerase δ Activity and Function

Marietta Y. W. T. Lee *, Xiaoxiao Wang, Sufang Zhang, Zhongtao Zhang and Ernest Y. C. Lee

Department Biochemistry and Molecular Biology, New York Medical College, Valhalla, NY 10595, USA;
drwangx2008@gmail.com (X.W.); Sufang_Zhang@nymc.edu (S.Z.); Zhongtao_Zhang@nymc.edu (Z.Z.);
Ernest_lee@NYMC.edu (E.Y.C.L.)
* Correspondence: Marietta_lee@nymc.edu; Tel.: +1-914-594-4070

Academic Editor: Eishi Noguchi
Received: 2 June 2017; Accepted: 11 July 2017; Published: 24 July 2017

Abstract: This review focuses on the regulation and modulation of human DNA polymerase δ (Pol δ). The emphasis is on the mechanisms that regulate the activity and properties of Pol δ in DNA repair and replication. The areas covered are the degradation of the p12 subunit of Pol δ, which converts it from a heterotetramer (Pol δ4) to a heterotrimer (Pol δ3), in response to DNA damage and also during the cell cycle. The biochemical mechanisms that lead to degradation of p12 are reviewed, as well as the properties of Pol δ4 and Pol δ3 that provide insights into their functions in DNA replication and repair. The second focus of the review involves the functions of two Pol δ binding proteins, polymerase delta interaction protein 46 (PDIP46) and polymerase delta interaction protein 38 (PDIP38), both of which are multi-functional proteins. PDIP46 is a novel activator of Pol δ4, and the impact of this function is discussed in relation to its potential roles in DNA replication. Several new models for the roles of Pol δ3 and Pol δ4 in leading and lagging strand DNA synthesis that integrate a role for PDIP46 are presented. PDIP38 has multiple cellular localizations including the mitochondria, the spliceosomes and the nucleus. It has been implicated in a number of cellular functions, including the regulation of specialized DNA polymerases, mitosis, the DNA damage response, mouse double minute 2 homolog (Mdm2) alternative splicing and the regulation of the NADPH oxidase 4 (Nox4).

Keywords: DNA polymerase δ; PDIP46; Poldip3; PDIP38; Poldip2; DNA replication; enzyme regulation; DNA damage response; p12 subunit; E3 ligases; cell cycle

1. Introduction

Pol δ plays a central role, together with Pol ε and Pol α/primase, as the DNA polymerases that synthesize the daughter DNA strands at the eukaryotic replication fork. The unraveling of the biochemistry of the mammalian DNA polymerases has posed significant experimental challenges. Knowledge of the enzymology of the DNA polymerases is essential to an understanding of their cellular functions. The biochemical approach is critical as pointed out by Arthur Kornberg in the context of the discovery and unraveling of the processes of prokaryotic DNA replication [1]. In the following review, we have focused on the regulation of Pol δ by modification of its subunit structure, and the modulation of its functions by accessory proteins. For a broader view of regulation of Pol δ and other polymerases, see [2,3].

1.1. Brief Historical Background

In the early 1970s, three mammalian DNA polymerases were known: Pol α, Pol β and Pol γ [4]. Pol α was considered to be the replicative polymerase, but did not possess an intrinsic or associated 3′ to 5′ exonuclease activity like the *Escherichia coli* or T4 bacteriophage DNA polymerases, where they function to edit or proofread misincorporated nucleotides [5–7]. Thus, the discovery of a novel mammalian DNA polymerase with an intrinsic 3′ to 5′ exonuclease activity represented a major advance. This enzyme, named Pol δ, was studied by a group of investigators at the University of Miami in Florida, in rabbit bone marrow erythroid cells [8–11], calf thymus [12–14] and human placental tissues [15–19]. Their approach was the rigorous isolation of the enzyme activities. This initially resulted in the characterization of a dimeric enzyme, consisting of a catalytic subunit of 125 kDa that harbored both the polymerase and 3′ to 5′ exonuclease catalytic sites and a p50 subunit. Evidence that Pol δ was a distinct enzyme from Pol α came from their separation by purification, by their immunochemical distinction using antibodies against Pol δ [18,20], and by the molecular cloning of the p125 subunit [21–23]. These studies from the Miami laboratories provided a firm basis for the identification of Pol δ as a novel proofreading DNA polymerase, and removed concerns that this new enzyme was merely Pol α contaminated with a cellular exonuclease.

Studies of human placental [17], calf thymus [24] and HeLa Pol δ [25] led to the discovery of a second human DNA polymerase with an intrinsic 3′ 5′ exonuclease activity, which was named Pol ε [26,27]. The early history of the study of Pol δ is also notable for the discovery of a factor which stimulated its activity, and acted to modify synthesis by Pol δ from a distributive to a processive mode [28]. This protein was identified as proliferating cell nuclear antigen (PCNA), which was subsequently shown to be a platform for many DNA transactions [29].

These early studies defined mammalian Pol δ as having two subunits, p125 and p50. The third and fourth accessory subunits were identified as p66/p68 [30,31] and p12 [32] (Table 1). The four subunits of human Pol δ are encoded by the *POLD1*, *POLD2*, *POLD3* and *POLD4* genes. Pol δ has been extensively studied in yeast [33]. *Saccharomyces cerevisiae* (*S. cerevisiae*) Pol δ consists of the homologs of the p125, p50 and p68 subunits [34]. *Schizosaccharomyces pombe* (*S. pombe*) Pol δ has an additional fourth subunit, Cdm1 [35–37], which has limited homology to p12 [32]. Molecular cloning of the p125 catalytic subunit of human Pol δ showed that the catalytic cores of the p125 subunits share greater than 60% similarity with that of *S. cerevisiae* Pol3 [22,23]. Pol δ and Pol ε are members of the B family of DNA polymerases that include T4 and Rb69 DNA polymerases.

Table 1. Subunit compositions of Pol δ.

Human	p125	p50	p68	p12
Schizosaccharomyces pombe	Pol3	Cdc1	Cdc27	Cdm1
Saccharomyces cerevisiae	Pol3	Pol31	Pol32	-

1.2. Properties of Human Pol δ4 and Its Subassemblies and the Roles of the p68 and p12 Subunits

The p68/Pol32/Cdc27 subunits of both human and yeasts possess PCNA interacting protein-boxes (PIP-boxes) at their C-termini [30,34,38]. The p68 subunit has an extended structure, and is highly charged, suggesting that it is flexible and thus an ideal subunit for mediating PCNA interaction [39,40]. In *S cerevisiae*, the Pol32 subunit is not essential, but Cdc27 is required for viability in *S. pombe* [39,40]. However, the human Pol δ p125 [41–44] and p12 [45] subunits also interact with PCNA. The p50 subunit also interacts with PCNA [46], although this interaction is much weaker [47]. Analysis of Pol δ enzymes in which the PIP-boxes of either the p12 [45] or p68 [48] were mutated show that both are required for full expression of activity.

The human Pol δ heterotetramer (Pol δ4), as well as its subassemblies, have been reconstituted by their expression in the baculovirus system [49–51]. Pol δ4 has also been expressed in an *E. coli* system [52]. The use of the baculovirus expression system allowed for the preparation of highly purified Pol δ4 and its subassemblies for biochemical studies. Initial difficulties were encountered in obtaining reproducible behaviors of the subassemblies, including that of the trimer lacking the p12 subunit [50]. This was traced to its instability during the isolation process; additionally, both the p68 and p12 subunits are more susceptible to proteases than the p125 and p50 subunits [51]. Immunoaffinity chromatography was used as a key component of the purification of the Pol δ subassemblies [53]. The preparations of Pol δ4 and its subassemblies were monitored for the appropriate subunit stoichiometry [51] because of the possibility of subunit loss during isolation and the fact these subassemblies do exhibit significant activities.

The activities of Pol δ and its subassemblies were compared by assay using sparsely primed poly(dA)/oligo(dT) as the substrate in the presence of PCNA [51]. PCNA does not have to be loaded onto this linear template with the replication factor C (RFC) clamp loader. This "standard" assay allows reproducible quantitation of Pol δ activities, and specific activities of ca. 20,000 units/mg were consistently obtained. The relative specific activities and the apparent K_d for PCNA binding are summarized in Figure 1 [51]. The figure also shows the subunit arrangement of the p125, p50, p68 and p12 subunits [45]. Notably, the core enzyme and the two trimeric subassemblies all possess significant activities. The presence of either the p12 or the p68 subunit is able to enhance PCNA binding and activity of the core enzyme (Table 1). However, as described below, these subassemblies exhibit defects in assays that require highly processive synthesis.

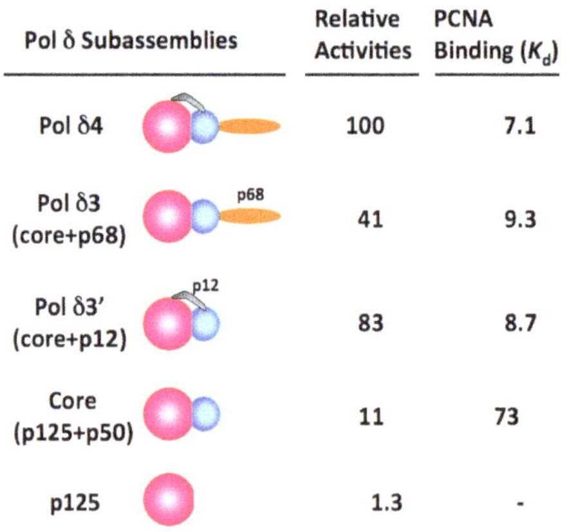

Pol δ Subassemblies		Relative Activities	PCNA Binding (K_d)
Pol δ4		100	7.1
Pol δ3 (core+p68)		41	9.3
Pol δ3' (core+p12)		83	8.7
Core (p125+p50)		11	73
p125		1.3	-

Figure 1. Relative specific activities and proliferating cell nuclear antigen PCNA binding (nM) of Pol δ and its subassemblies. Data from [51].

The second type of assay that has been used is the M13 assay, in which a singly primed M13 single-stranded DNA (ssDNA) is used as the substrate. PCNA is loaded with RFC, together with RPA (single stranded DNA binding protein). This assay monitors synthesis of long strands of DNA up to 7 kb on M13 circular DNA and has been used to demonstrate the processivity of Pol δ [51,54–56]. However, on long circular ssDNA templates, pausing can be observed where Pol δ4 has difficulty synthesizing through regions of secondary structure. In addition, it has been found that Pol δ dissociates frequently during these reactions [52]. Thus, while Pol δ exhibits processivity in the presence of PCNA, the observed processivity is not continuous (i.e., not due to a single binding event) over the entire length of the M13 template. There were marked defects in the abilities of the Pol δ subassemblies to synthesize the full-length products, which could be partially compensated for by increasing the enzyme concentrations, consistent with a more frequent dissociation from the primer template [51].

2. Alteration in Subunit Composition by the Degradation of the p12 Subunit Is the Key Mechanism for the Regulation of Human Pol δ

There is a surprising paucity of literature on the control of eukaryotic DNA polymerase activities by posttranslational modification [2,48,57,58]. The p12 subunit has emerged as a center point for regulation of Pol δ [58,59]. The discovery that the p12 subunit is rapidly degraded by ubiquitination and proteasomal degradation in response to DNA damage opened a new window on the regulation of Pol δ [60]. Later, a similar process was found to take place during the G1/S transition under the control of a key regulator of the entry to S-phase, the E3 ubiquitin ligase CRL4^{Cdt2} [61]. The operational outcome of the degradation of p12 is that the Pol δ4 enzyme is converted in vivo to a trimer, Pol δ3, in synchrony with the S phase (Figure 2). This represents an unusual form of enzyme regulation, whose significance ultimately rests on understanding the comparative properties of the two forms, and how these differences operate to facilitate and/or differentiate their functions in DNA repair or DNA replication. In the following subsections, the mechanisms for the formation of Pol δ3 and the properties of Pol δ4 and Pol δ3 in DNA repair and replication are reviewed.

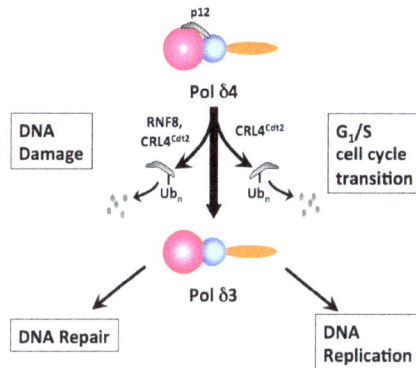

Figure 2. Overview of the regulation of human Pol δ by degradation of the p12 subunit and the formation of Pol δ3.

2.1. The Degradation of the p12 Subunit of Pol δ in Response to DNA Damage

Ultraviolet (UV) damage has been extensively used to study cellular responses to DNA damage. UV treatment of cells triggers global nucleotide excision repair (NER), and activates translesion synthesis (TLS) to deal with the effects of the bulky lesions that are barriers to replicative DNA polymerases. UV exposure also triggers checkpoints that result in the inhibition of cellular DNA synthesis [62] through the activation of Ataxia telangiectasia and Rad3-related (ATR) [63,64]. The intra-S phase checkpoint leads to slowing of progression through the S-phase, and acts by the inhibition of late firing origins of initiation of replication, and also by slowing the rates of replication fork progression [65].

We examined the effects of UV treatment of cells on Pol δ by Western blotting of all four of its subunits to determine if evidence for band-shifts caused by phosphorylation events were detectible. Instead, this led to the discovery that the p12 subunit was rapidly degraded in response to UV damage [60]. This study characterized in a rigorous manner the loss of the p12 subunit of Pol δ in response to genotoxic stress. The significant findings are summarized below:

- p12 is rapidly lost in a variety of cell types, in a UV flux- and time-dependent manner, followed by a slower recovery over 24 h.
- Treatment with alkylating agents such as methyl methanesulfonate (MMS) or agents inducing replication stress (hydroxyurea and aphidicolin) also caused p12 degradation.
- The loss of p12 is due to an accelerated rate of proteasomal degradation initiated by its polyubiquitination.
- Degradation of p12 is dependent on ATR signaling, but not on ATM, as shown by the use of ATR or ATM depleted cells.

- The p12 subunit of Pol δ is selectively targeted, and similar changes are not observed for the other three subunits.
- Loss of the p12 subunit leads to the in vivo conversion of Pol δ4 to the heterotrimer, Pol δ3.

The final observation noted above is of some importance. Prior to these studies, Pol δ4 was considered to be the holoenzyme form, so that the first idea to come to mind was that this might be a way to disable Pol δ4 activity. However, the Pol δ3 isolated from UV-treated cells by immunoaffinity chromatography exhibited significant activity. Direct comparisons of Pol δ3 produced in vivo by UV treatment with recombinant Pol δ3 showed that they had similar properties [51,60] (see Section 1.2 for comparative properties of Pol δ4 and Pol δ3).

Thus, the question is whether Pol δ3 exhibits advantages over Pol δ4 in DNA repair. The route to gaining insights into this possibility came from testing their functionalities utilizing highly purified proteins in specialized assays.

2.2. Pol δ3 Exhibits Altered Behaviors from Pol δ4 in Lesion Bypass and in Extension of Mismatched Primers that Represent a Gain of Function

In order to probe for advantages for the presence of Pol δ3 in cells subjected to genotoxic agents, a comparison of its behavior with that of Pol δ4 was made in two contexts. First, replicative polymerases encounter small lesions that can be bypassed by eukaryotic DNA polymerases in an error prone manner [66,67]. Second, replicative polymerases encounter lesions that act as severe obstacles to chain extension: these include abasic sites and thymine-thymine dimers. Model templates with small lesions were used to study the behavior of Pol δ4 and Pol δ3 [68]; these were O^6-MeG (O^6-Methylguanine), which is produced by alkylating agents [66], and 8-oxoG (7,8-dihydro-8-oxoguanine), which is produced by reactive oxygen species [67]. Templates containing abasic sites and thymine-thymine dimers were used as examples of lesions that are not readily bypassed. Pol δ3 exhibits a decreased tendency for bypass synthesis across these templates. Pol δ3 exhibits a higher exonuclease/polymerase ratio than Pol δ4, suggesting that it was more efficient in proofreading. Further analysis showed that Pol δ3 is less likely to extend mismatched primers or to misincorporate wrong nucleotides in single nucleotide incorporation assays. Overall, this study indicated that Pol δ3 exhibited behavior consistent with it being more discriminatory than Pol δ4, i.e., of having a greater fidelity within the context of these biochemical assays [68]. The inference drawn is that the p12 subunit exerts an influence on the intrinsic properties of Pol δ, which could originate from effects on the polymerase or the exonuclease activities, or both.

The kinetic [69,70] and structural bases [71,72] for the fidelity of replicative polymerases is well understood. The rate constant for polymerization, k_{pol}, plays a major role in the avoidance of misincorporation of wrong nucleotides or in mutagenic bypass [69,70]. This is the so-called kinetic barrier, in which k_{pol} is reduced on encounter with template lesions, the binding of wrong nucleotides, or the presence of a mismatched primer end. This increases the probability for transfer of the primer end to the exonuclease site. The second important kinetic constant is $k_{pol-exo}$, the rate at which the primer end is translocated from the pol active site to the exonuclease site [73,74]. This transfer rate is the rate-limiting step for the exonuclease activity in the kinetic scheme [69]. Thus, for a given polymerase the determination of k_{pol} and $k_{pol-exo}$ provides information on the polymerase and exonuclease, respectively, while the ratio of $k_{pol-exo}$ to k_{pol} may be regarded as a ratio of editing to extension, and an index of its proofreading propensity.

Pre-steady state kinetic analysis was used to determine the kinetic constants for Pol δ4 and Pol δ3 [58,75]. The differences between the two key kinetic constants are summarized in Figure 3, as the ratio of their changes (Pol δ3/Pol δ4). The removal of p12 leads to a nearly five-fold decrease in k_{pol}, and a greater than eight-fold increase in $k_{pol-exo}$, such that the ratio of editing to extension increased by ca. 40 fold (Figure 3). Both polymerase (k_{pol}) and exonuclease ($k_{pol-exo}$) are affected. Thus, by these measures, Pol δ3, compared to Pol δ4, exhibits properties of a polymerase that intrinsically proofreads more frequently and should exhibit greater fidelity. These findings are consistent with the observed behavior of Pol δ3 when tested on lesion containing templates [68].

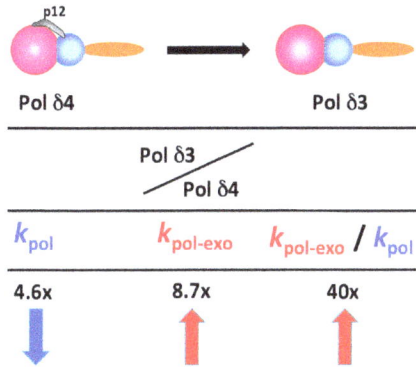

Figure 3. Changes in the kinetic constants of Pol δ3 and Pol δ4. The changes are shown as the ratios of the values for Pol δ3/Pol δ4.

These studies indicated that p12 exerts an influence on the proofreading functions of Pol δ, and that its removal to form Pol δ3 resulted in an apparent gain of function in the form of an increased surveillance against mutagenic synthesis.

The potential significance for the formation of Pol δ3 may be rationalized as a defense against mutagenic DNA synthesis in replicating cells upon genotoxic challenge. The formation of Pol δ3 in response to DNA damage earmarks it as the likely form of Pol δ engaged in DNA repair synthesis, which in the case of UV damage, would primarily involve NER [59,76] and homologous recombination repair of DSBs. (The p12 subunit is also degraded in response to ionizing radiation [59,77]).

2.3. Spatiotemporal Analysis of the Recruitment of Pol δ to Sites of UV Damage Indicates Pol δ3 Is in the Right Place at the Right Time

A hallmark of the cellular response to DNA damage is the recruitment of signaling and repair factors to sites of DNA damage, and the formation of repair foci. The analysis of subcellular localization to these foci has played an important role in dissecting the assembly of proteins involved in the DNA Damage Response [78–82] and DNA damage tolerance pathways [63,83]. A spatiotemporal analysis of the recruitment of all four Pol δ subunits to sites of UV-induced DNA damage provided evidence that Pol δ4 is recruited to sites of DNA damage, and that this is followed by the appearance of Pol δ3 upon loss of the p12 subunit. The loss of p12 from the DNA damage foci was confirmed by chromatin immunoprecipitation (ChIP) analysis with anti-p125 [84].

2.4. Conversion of Pol δ4 to Pol δ3 May Facilitate the Switch between Pol δ and Pol η

In S phase cells, genotoxic agents that introduce bulky lesions lead to the activation of the DNA damage tolerance pathway. The stalled replicative polymerase (usually taken as Pol δ) is switched for a translesion polymerase that bypasses the lesion. The most studied example of translesion synthesis is that performed by Pol η in the bypass of UV-induced CPDs (cyclobutane pyrimidine dimers) [85–89]. The key event that is required for initiation of translesion synthesis is the mono-ubiquitination of PCNA [90], following which Pol δ is switched for Pol η [91,92]. The ubiquitination of PCNA is significant, in that Pol η and other TLS polymerases possess both ubiquitin binding domains and PCNA binding PIP-boxes [91,92]. The switching process in TLS requires the displacement of Pol δ by Pol η in the initial switch, followed by a second switch once TLS is completed. We have proposed a model in which the conversion of Pol δ4 to Pol δ3 facilitates the switch to Pol η, on the basis that Pol δ3 dissociates from the PCNA/DNA primer-template more readily than Pol δ4 [58,93]. This model is consistent with the idea that ubiquitination of PCNA and p12 represent related cooperative events that are involved in TLS at the lesion sites.

The determination of the structure of mono-ubiquitinated PCNA (ub-PCNA) reveals that the ubiquitins are oriented in a radially extended fashion, below the plane of the PCNA trimer, and on the opposite face of PCNA where the PIP-box binding pockets are located [93]. Additionally, the ubiquitins displayed no contacts with PCNA beside the isopeptide linkage, and exhibit the possibility of significant conformational flexibility. Mono-ubiquitinated PCNA was found to lead to inhibition of the combined reactions of Pol δ4 and Fen1 (Flap endonuclease 1) activity in Okazaki fragment processing [93]. This could potentially contribute to the UV-induced inhibition of DNA synthesis.

2.5. Does the Plasticity of Pol δ Subunit Composition Extend to Other Subunits besides p12?

The demonstration that Pol δ is regulated by modification of its quaternary structure raises the question of whether the other subassemblies of Pol δ could also be generated in vivo to serve a functional role. Phenotypic analyses of the deletion of the Pol δ genes should take into account their potential impact on the Pol δ enzyme. In the case of the p68 subunit, its deletion could potentially result in the formation of the Pol δ3 trimer consisting of the core + p12 (Figure 1). This trimer has activity in the standard assay which is comparable to that of Pol δ4. However, deletion of the *POLD3* gene is lethal in the animal system. Conditional knockouts of *POLD3* in mice have shown that it is essential for development, and exhibits haploinsufficiency [94]. Deletion of *POLD3* in B lymphocytes led to severe replication defects and genomic instability. The mechanism was traced to a severe loss of the p125 catalytic subunit, consistent with a loss of stability of the Pol δ complex [94]. It seems unlikely that a regulated conversion of Pol δ4 to yield the Pol δ3' trimer lacking p68 occurs in mammalian cells, as this would coincide during S phase with the degradation of p12, leaving the Pol δ dimer as the major form. However, it is noted that a temporally restricted reversible loss of p68 outside the S phase might occur. In contrast to the effects of gene deletion in mice, DT40 chicken lymphocytes cells in which the *POLD3* gene is deleted are viable, and the cells replicate with a moderate S phase delay, but exhibited increased sensitivity to genotoxic stress [95]. Deletion of the p68 ortholog, Pol32, is not lethal in *S. cerevisiae* [34], but deletion of Cdc27 in *S. pombe* is lethal [38]. Apart from an impact on Pol δ, loss of the *POLD3* gene would also impact Pol ξ, which utilizes the Pol δ p50 and p68 subunits [96,97].

2.6. RNF8 Is Involved in DNA Damage-Induced p12 Degradation

The identification of the E3 ubiquitin ligase(s) that target p12 for degradation is important in understanding how p12 degradation and the ensuing generation of Pol δ3 is integrated into the signaling systems that comprise the DNA Damage Response (DDR) and the DNA Damage Tolerance (DDT) pathways [63,81,85,87]. Two E3 ligases that target p12 for degradation have been identified: RNF8 [98] and CRL4^Cdt2 [61,77]. RNF8 was identified by a classical biochemical approach [98]. An in vitro assay system was devised for the detection of the polyubiquitination of GST-p12. Such in vitro assays require the combined actions of an E1 ubiquitin activating enzyme, an E2 ubiquitin conjugating enzyme and an E3 ubiquitin ligase. UbcH5c, which is active with a number of E3 ligases [99] was used as the E2 enzyme, and GST-p12 as the substrate. An E3 ligase fraction was purified from HeLa cell extracts by conventional chromatographic methods. This preparation was subjected to proteomic analysis by LC/LC/MS/MS; this yielded three peptides that were identified as sequences from RNF8. Western blotting of the column fractions confirmed the presence of RNF8, and in vitro assays of recombinant RNF8 showed that it had a robust activity for the ubiquitination of GST-p12. Depletion of RNF8 confirmed that the rates of p12 degradation by UV or by alkylation with MNNG (*N*-Methyl-*N*'-Nitro-*N*-Nitrosoguanidine) were significantly reduced [98].

RNF8 has a major role in orchestrating the ATM regulated DDR through the noncanonical polyubiquitination of histone H2A [79–81,100]. The discovery that RNF8 mediates the regulation of Pol δ is surprising, as this raises the question as to whether RNF8 also plays significant roles in NER and the DNA damage tolerance pathway that involves activation of translesion synthesis by PCNA ubiquitination. RNF8 is recruited to DNA damage foci induced by UV [98,101]. RNF8, together with UbcH5c, efficiently mono-ubiquitinates PCNA in vitro; mono-ubiquitinated PCNA (ub-PCNA) is further polyubiquitinated via K63 isopeptide linkages by RNF8/UbcH5c and Ubc13/Uev1a [102]. Depletion of RNF8 by shRNA was found to suppress ub-PCNA formation in UV-treated A549 cells [102]. These observations suggest that RNF8 might participate in both modulation of Pol δ and of TLS by PCNA ubiquitination [59]. The possible regulation of Pol δ and Pol η by RNF8 could be a means for cross-talk between the ATR and ATM signaling pathways [59]. However, further work is needed to establish what role RNF8 plays in ub-PCNA formation in vivo.

2.7. Degradation of p12 by CRL4^{Cdt2}

Depletion of RNF8 did not completely block the degradation of p12 in response to UV damage, indicating that more than one E3 ligase is involved. CRL4^{Cdt2} was found to target p12 for degradation in response to UV, and also mediates the degradation of p12 before entry into S phase [59,61]. The CRL4^{Cdt2} ubiquitin ligase plays a critical role in the prevention of re-replication, as the "master coordinator of cell cycle progression and genome stability" [103]. CRL4^{Cdt2} is a member of the Cullin Ring Ligase family of E3 ubiquitin ligases and targets the licensing factors that are involved in the assembly of the pre-replicative complex during G1, so that they are removed during the G1/S transition [104]. The primary targets of CRL4^{Cdt2} are Cdt1, p21 (p21$^{Waf1/CIP1}$) and the histone acetylase Set8. CRL4^{Cdt2} recognition of its substrates depends on their possession of an extended PIP-box, termed a PIP-degron. These PIP-degrons have a higher affinity for PCNA than PIP-boxes, and their degradation also requires that PCNA be loaded onto DNA [105,106]. Studies in *Xenopus* extracts have shown Cdt1 destruction is dependent on the initiation of DNA replication as well as Pol α, indicating that PCNA is loaded onto the primer end [107]. CRL4^{Cdt2} also targets its substrates for destruction in response to DNA damage by UV [108,109]. Both p21 [110] and Cdt1 are degraded in response to UV damage [108,111–114].

The p12 subunit of Pol δ possesses a PIP-degron, and is a substrate for CRL4^{Cdt2} [61,77]. Mutation of the PIP-degron of p12 reduces its UV-induced degradation. Depletion of either of the two isoforms of Cul4 [115] also suppresses UV-induced p12 degradation with similar time courses as for p21 [59,61]. The NEDD8-activating enzyme (NAE) is required for Cullin ligase activity; the NAE inhibitor, MLN4924 [116], blocks UV and IR degradation of p12 [59,77]. CRL4^{Cdt2} has also been shown to be required for the UV-induced inhibition of DNA synthesis; furthermore, replication fork progression is inhibited and is dependent on p12 degradation [77]. The latter findings provide evidence that p12 degradation contributes to the elongation checkpoint that is a component of the intra-S phase checkpoint [65,117].

Analysis of the cell cycle behavior of p12 and its dependence on CRL4^{Cdt2} were examined in synchronized cell populations together with that of p21. These studies showed that p12 levels were reduced during S phase and returned to basal levels during G2/M [59,61]. Depletion of CRL4^{Cdt2} isoforms reduced the S phase degradation of p12 [59,61]. At the same time as p12 undergoes a decrease during the S phase, levels of the other subunits of Pol δ remain fairly constant. Thus, the degradation of p12 by CRL4^{Cdt2} leads to the formation of Pol δ3 in synchrony with the S phase [59,61]. The regulation of Pol δ3 is orchestrated by CRL4^{Cdt2} through common molecular mechanisms by which it controls its other substrates, and speaks to the significance of Pol δ3 as a participant in DNA replication. The cell cycle variations in p12, p21 and Cdt1, broadly follow comparable time courses consistent with their regulation by CRL4^{Cdt2}. This has been demonstrated at the single cell level by laser scanning cytometry, coupled with analysis of DNA replication by 5-Ethynyl-2′-deoxyuridine (EdU) labeling [118,119].

2.8. Mechanism and Characteristics of Okazaki Fragment Processing by Pol δ4 and Pol δ3

Discontinuous DNA synthesis at the lagging strand in eukaryotes involves the synthesis of Okazaki fragments of ca. 200 nucleotides. The process of Okazaki fragment maturation is essentially one where they are joined to the growing lagging strand. The key elements of this process have been characterized by biochemical reconstitution and genetic studies [33,120]. In yeast, Pol δ has been shown to have properties that are conducive to a role in Okazaki fragment processing. One of these is its propensity to idle at a nick, thereby allowing DNA ligase action [121]. On encounter with a 5′ end of the previous Okazaki fragment, Pol δ will advance several nucleotides because of fraying of the primer end and strand displacement, creating short flaps. These short flaps are then cleaved by Flap endonuclease 1 (Fen1) [120,122], so that the primer ends are removed. This process is termed the short flap pathway, and the products are predominantly mononucleotides and short oligonucleotides of 2–10 nucleotides (nt). However, Fen1 does not cleave longer flaps, and the accumulation of longer flaps acts as a barrier to Okazaki fragment maturation and is a potential source of genomic instability. While yeast Pol δ is able to idle at a nick, it does possess a significant ability for strand displacement [123], so that creation of long flaps can take place. A second pathway, the "long flap pathway", cleaves these long flaps via the actions of Pif1 helicase and Dna2 to a length that allows their removal by Fen1 [120,124]. The final step is ligation of the nick by DNA ligase I [125]. This can be contrasted to the situation in prokaryotes, where the removal of the primers is performed by a nick translation mechanism in which Pol I both extends the primer end and excises single nucleotides from the 5′ end of the prior Okazaki fragment by virtue of its 5′ to 3′ exonuclease activity [126].

The behavior of human Pol δ4 and Pol δ3 in the component and complete reactions of Okazaki fragment processing were compared in a reconstituted system [59,127]. The key observations were that Pol δ4 is proficient in strand displacement, and performs Okazaki fragment processing in a manner similar to that of yeast Pol δ in combination with Fen1. The spectrum of flap sizes ranges from 1 to 8 nt, but is dependent on Fen1 concentration. With increasing Fen1 the product spectrum is shifted to 1–3 nt, with the mononucleotide prevailing. Pol δ3 does not perform strand displacement. With Fen1 and Pol δ3 the primary products are single nucleotides and a smaller amount of di- and trinucleotides. The rate and nature of product formation distribution is relatively unaffected by Fen1 concentration, supporting the proposal that mammalian Okazaki processes might involve a PCNA/Pol δ3/Fen1 complex [127], in analogy to that which has been demonstrated in the Archaeal system [128].

The question then arises, why do we need two Pol δ forms for lagging strand synthesis? The answer may lie in the complex nature of genomic DNA. It is possible that there are template regions that Pol δ4 is more capable than Pol δ3 of traversing. In this view, we would assign Pol δ3 as the primary agent for Okazaki fragment synthesis and processing. The preference for the use of Pol δ3 and a nick translation mode of Okazaki fragment processing lies in the avoidance of the generation of long flaps. Pol δ4 could also be used in Okazaki fragment processing, under circumstances discussed below (see Section 3.4).

The large number of Okazaki fragments that needs to be generated during synthesis of the human genome requires that the process be highly efficient, and the tendency of Pol δ to frequently dissociate is compatible with the need for rapid recycling of Pol δ (see Section 1.2). The properties that Pol δ displays in Okazaki fragment processing are similar to those needed for gap filling in DNA repair in terms of the length of DNA synthesis that is required [127]. The role of PCNA may hold more importance in this context as a platform for coordinating the reactions of Pol δ, Fen1 and DNA ligase I than its role as a processivity factor. The kinetic constants for Pol δ4 and Pol δ3 [75] provide for an estimate or a prediction of their processivity, based on the ratio of k_{pol} to k_{off} which can be approximated by k_{cat} [129]. These provide values of 350 and 106 nt for Pol δ4 and Pol δ3 respectively. In contrast, yeast Pol δ is able to sustain DNA synthesis in a strictly processive manner to at least 5 kb [130].

The findings that yeast Pol δ is adapted for Okazaki fragment maturation have led to extensive studies that support a division of labor between Pol δ and Pol ε at the replication fork, where Pol δ is the lagging strand polymerase and Pol ε is the leading strand polymerase. Much of the evidence for a division of labor is based on several genetic studies using mutant polymerases that allow discrimination between leading and lagging strand DNA synthesis (reviewed in [131]). How human Pol δ fits into this concept must now also take into account the presence of Pol δ4 and Pol δ3, although their properties suggest that they are more adapted to lagging than leading strand synthesis (see Sections 3.3 and 3.4 below for further discussion).

A controlled balance between Pol δ4 and Pol δ3 appears to be required in vivo for genomic stability. Reduced expression of the *POLD4* gene has been associated with lung cancer and a poor prognosis for certain lung cancer patients [132,133]. siRNA suppression of p12 in cultured cells was found to lead to cell cycle delay, and an elevated frequency of chromosomal aberrations [132,133].

3. Role of the Pol δ Binding Protein PDIP46/Poldip3 in DNA Replication and Repair

There has been a search for accessory or auxiliary proteins that could modulate Pol δ activity since its initial discovery. Two novel Pol δ interacting proteins of previously unknown functions, PDIP46 (Poldip3) and PDIP38 (Poldip2), were discovered by yeast two hybrid screening with the p50 subunit of Pol δ as the bait [134]. An independent study identified tumor necrosis factor α and interleukin 6 inducible protein (TNFAIP1) as a Pol δ binding protein (PDIP1/Poldip1) [135]. All three Pol δ interacting proteins share in common the abilities to bind to the p50 subunit and PCNA. The functions of Poldip1, PDIP38/Poldip2 and PDIP46/Poldip3 in relation to Pol δ have proven to be enigmatic, and they appear to be multifunctional proteins.

PDIP46 was re-discovered as S6K1 Aly/REF-like target (SKAR) [136]. SKAR possesses a RRM (Figure 4) with strong homology to the Aly/REF RNA binding proteins. The latter are involved in coupling transcription with pre-mRNA splicing and mRNA export [136]. S6K1 (ribosomal protein S6 kinase-1) lies downstream of the mTOR and PI3K signaling pathways that regulate cell growth and proliferation through nutrient, energy and mitogenic signals [137,138]. SKAR is a nuclear protein, and is also present in the nuclear speckles and the EJC (exon junction complex) where it acts to enhance translational efficiency [139–141]. Activation of S6K1 through the mTOR and PI3K signaling pathways leads to phosphorylation of PDIP46 at S383/S385. This phosphorylation is required for the binding of activated (phosphorylated) S6K1 binding to PDIP46 (Figure 4). This leads to their recruitment to the spliceosomes where S6K1 regulates translational efficiency [136,141]. siRNA depletion of S6K1 leads to smaller cell size [138], and this effect is also produced by siRNA depletion of SKAR and 4EBP1/eIF4E [142]. Thus, PDIP46 serves to translocate activated S6K1 to the spliceosome, subsequent to the activation of the mTOR pathway. Whether PDIP46, which possesses an Aly/REF type of RNA binding domain, also independently affects mRNA metabolism is unknown. However, PDIP46 is also a binding partner of enhancer of rudimentary homolog (ERH) [143]. ERH is a transcriptional regulator that affects the expression of a number of genes in the cell cycle as well as genes involved in DNA damage including ATR, and genes involved in DNA replication [144–146].

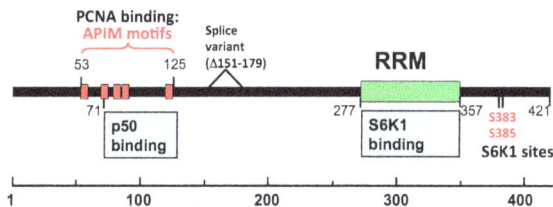

Figure 4. Domain map of PDIP46/SKAR.

Recently, PDIP46 has been shown to have a role in the activation of Pol δ activity [56]. These effects are reviewed below, and point to an important role for modulating Pol δ activity. Its functions in this regard are consistent with its other roles in growth regulation studied as the SKAR protein. Thus, PDIP46 appears to be a multifunctional protein.

3.1. Mapping of the Interaction Sites between PDIP46 and Pol δ /PCNA Reveals that These Are Located in a Region Separate from Those Involved in S6K1 Binding

The interaction sites of PDIP46 with PCNA and the p50 subunit were mapped to residues 71–125 and that for PCNA between residues 53 and 125 (Figure 4). The PCNA binding of PDIP46 is due to its possession of a cluster of five APIM motifs [56]. The APIM (AlkB homologue 2 PCNA-Interacting Motif) is a novel PCNA binding motif that was first identified in the human DNA repair enzyme oxidative demethylase AlkB Homologue 2 (ABH2) [147]. The APIM consists of five residues with the consensus sequence [KR]-[FYW]-[LIVA]-[LIVA]-[KR]. Seven other proteins have been shown to have functional APIM motifs. These include Topo IIα [147], the NER protein XPA [148] and the F-box helicase FBH1 [149] that is involved in homologous recombination. The APIM motif binds to the same regions of PCNA as the PIP-box [150]. The separation of the locations of the PCNA/Pol δ binding regions from the RRM/S6K1 binding domain in the C-terminus (Figure 4) is consistent with the possibility that PDIP46 is a bi-functional protein whose two functions are harbored in two separate structural domains [56].

3.2. Evidence that PDIP46 Is Associated with Pol δ In Vivo

There is supportive evidence that PDIP46 interacts with Pol δ in a cellular context. This has been demonstrated by their co-immunoprecipitation and co-elution during affinity chromatography on immobilized anti-p125 monoclonal antibody. ChIP analysis with antibody against the p125 subunit showed that PDIP46 was present together with two components of the mammalian replisome [56]. These are Mcm2, a component of the Cdc45-MCM-GINS (CMG) helicase [151], and Ctf4, which associates with CMG [152,153]. Thus, the ChIP data supports the idea that PDIP46 is associated with chromatin at or near the replisome.

3.3. PDIP46 Is a Potent Activator of Pol δ

All PCNA binding proteins possess the ability to compete with Pol δ for PCNA, and therefore can inhibit Pol δ in activity assays at sufficiently high concentrations. This was found to be the case for PDIP46 [56] and PDIP38 [154], when assayed using poly(dA)/oligo(dT) as the substrate. More recently, the effects of PDIP46 on Pol δ activity were examined in the M13 assays in which PCNA is pre-loaded onto the primer end with RFC. This assay is more reflective of DNA synthesis in vivo than the standard assay using poly(A)/oligo(dT) as the substrate (see Section 1.2). PDIP46 was revealed to be a remarkably potent activator (ca. 10 fold) of Pol δ4 in the synthesis of the 7 kb M13 DNA, with an apparent K_d of ca. 34 nM [56]. The mechanisms for this activation could be due to several causes. These include an increase in processivity, possibly because PDIP46 may stabilize Pol δ binding to PCNA by a bridging interaction, as well as by a direct activation that involves alteration of the kinetic properties of Pol δ4.

The effects of PDIP46 were examined on model oligonucleotide templates [127] in assays that examined primer extension and strand displacement in order to gain insights into its mechanism(s) of action [56]. In the absence of PCNA, Pol δ4 behaves in a distributive fashion, and PDIP46 clearly stimulates this activity. These results demonstrate that PDIP46 exerts a direct effect on Pol δ4. In the presence of PCNA, the reactions are much faster but it was nevertheless observed that Pol δ4 activity is stimulated. Pol δ3 activity was much less affected than Pol δ4 activity. PDIP46 also stimulated the strand displacement activity of Pol δ4 using model templates with a blocking oligonucleotide, both in the absence and presence of PCNA. Little or no effects were observed on Pol δ3, which does not exhibit strand displacement activity [127].

Next, the effects of PDIP46 on an oligonucleotide substrate with a hairpin/stem-loop (16-nt stem, 8-nt loop) were examined. PDIP46 stimulated Pol δ4 synthesis through the stem-loop by ca. four-fold. While these effects are smaller than those observed with the M13 template, they explain the greatly increased rate of accumulation of full-length products by Pol δ4 on the M13 template in the presence of PDIP46. The M13 template may have many regions of secondary structures. Thus, there would be a cumulative effect on overall rates of Pol δ4 synthesis in the presence of PDIP46 [56]. PDIP46 could act by stabilization of the Pol δ4/PCNA/DNA complex by a bridging interaction (Figure 5), as well as by a direct activation that involves alteration of the kinetic properties of Pol δ4. The effects of PDIP46 on Pol δ4 are highly relevant in the context of chromosomal replication (Section 3.4.2 below).

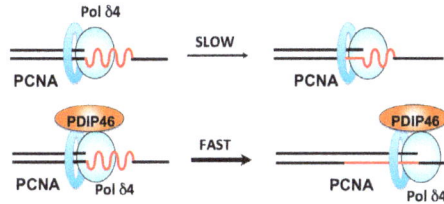

Figure 5. PDIP46 activates the extension of primers across regions of secondary structure in the template (shown in red).

These studies also highlight the connection between strand displacement and the ability of Pol δ to synthesize through a hairpin structure. Once Pol δ encounters the hairpin, further synthesis through the stem portion of the hairpin is analogous to the process of strand displacement [56]. Thus, it is not surprising that Pol δ3 exhibits minimal activity with the hairpin substrate as it does not perform strand displacement activity.

Mutations of PDIP46 in which all of the APIM motifs are mutated abolished the effects of PDIP46 on Pol δ4, validating the assignment of PCNA binding to this region. Deletion of the RRM has no effect on the activation of Pol δ4 by PDIP46, so that PDIP46 appears to have two independent functional domains [56].

These studies are the first to document the effects of PDIP46 on Pol δ4, and obviously raise many more questions regarding its mechanism of action. In particular, kinetic studies are needed to establish whether PDIP46 has any effect on the intrinsic catalytic properties of Pol δ4. Such effects could also involve alterations in fidelity. In addition, characterization of the range of complexity of secondary structures in which PDIP46 can act to facilitate Pol δ4 bypass synthesis is important in understanding the extent to which its functions could facilitate Pol δ4 bypass synthesis.

3.4. Future Horizons: Accommodating Two Forms of Pol δ and PDIP46 at the Replication Fork

Current models for the respective roles of Pol δ and Pol δ at the replication fork are based on both biochemical and genetic approaches in yeast. Several studies [155–157] using error-prone Pol δ and Pol ε support a model where Pol δ and Pol ε function mainly as lagging and leading strand polymerases, respectively (reviewed in [131]). By contrast, it has been argued that genetic approaches also support a model where Pol δ has a major role on both forks [158]. In the case of human DNA replication, the differences in subunit structure and properties between yeast Pol δ and human Pol δ have to be taken into account, in particular the existence of two forms of human Pol δ as well as of PDIP46, which selectively acts on Pol δ4. In the following sections we propose models for their roles in lagging and leading strands.

3.4.1. Roles of Pol δ3, Pol δ4 and PDIP46 in Lagging Strand Synthesis

Biochemical and reconstitution studies have provided strong arguments for an adaptation of Pol δ for Okazaki fragment synthesis and processing [59,127]. In the human system, we have two forms of Pol δ; how do these fit into our current views of the replication fork? While both Pol δ3 and Pol δ4 are capable of Okazaki fragment processing in vitro, Pol δ3 exhibits the more desirable characteristics of acting through a nick translation mode that avoids the generation of long flaps [59,127]. The model we propose is that they are used interchangeably during Okazaki fragment synthesis. This model is based on studies reviewed above in Section 2.8. In this model (Figure 6) Pol δ3 is the default lagging strand polymerase. When regions of secondary structure which act as barriers to Pol δ3 are encountered [56], Pol δ4 is switched with Pol δ3, together with PDIP46 (Figure 6).

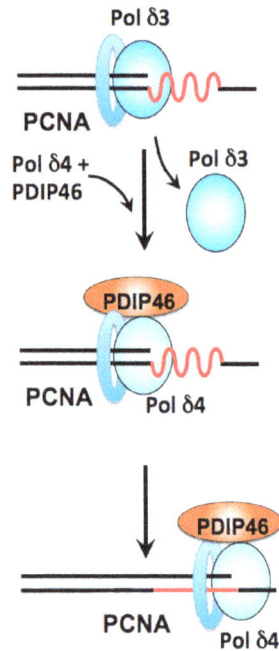

Figure 6. Pol δ3 and Pol δ4/PDIP46 in lagging strand synthesis. Template regions of secondary structure (red) that pose impediments to Pol δ3 leads to dissociation and triggers a polymerase switch to Pol δ4/PDIP46.

There are regions of varying template complexity in chromatin that include simple hairpins, microsatellite regions [159] that contain CFS (common fragile sites), and trinucleotide repeats [160]. These pose potential barriers to the replicative polymerases. There has been a broadening view of polymerase usage during normal DNA replication, e.g., the utilization of translesion polymerases, notably Pol κ [161,162] and Pol η in chromosomal DNA replication to augment the functions of replicative polymerases [88,163–165]. Future characterization of the range of complexity of secondary structures for which PDIP46 may act to facilitate Pol δ4 bypass synthesis is important to understanding its functions.

3.4.2. Roles of Pol δ4 and PDIP46 in Leading Strand Synthesis

It is generally accepted that a leading strand polymerase should have high processivity. The loss of Pol ε function in *S. cerevisiae* is nonlethal, indicating that yeast Pol δ can act at both leading and lagging strands [34]. As previously noted, Pol δ3 appears to be much less processive than Pol δ4, so that it is an unlikely candidate for a role in leading strand synthesis (Sections 1.2 and 2.8). Pol δ4 has been shown to be less processive than Pol ε [166], so that it might be considered also to be a poor candidate for leading strand DNA synthesis. However, PDIP46 could augment Pol δ4 function in leading strand synthesis, in analogy to its effects in the M13 assay that reveal a gain in synthesis rate of about an order of magnitude [56]. Thus, PDIP46 could function as an accessory protein to provide for Pol δ4 with the required speed and processivity in leading strand synthesis.

In addition to a general role as a leading strand polymerase, Pol δ4/PDIP46 could act in an analogous way as proposed above (Figure 6) in lagging strand synthesis. We propose that Pol ε may stall at regions of secondary structure, and then is switched for Pol δ4/PDIP46 (Figure 7). The ability of Pol ε to bypass complex template DNA regions has not been extensively studied. However, Pol ε exhibits only minimal strand displacement activity, like Pol δ3, and has been shown to be unable to perform strand displacement [167]. Thus, it might be predicted that Pol ε, like Pol δ3 [56], could potentially stall at regions of template secondary structures. This model, like that for the lagging strand, views Pol δ4 (with PDIP46) as functioning as a specialized polymerase to deal with regions of secondary structure that stall Pol ε.

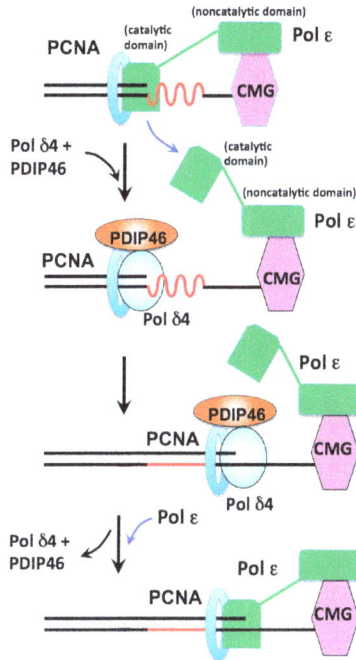

Figure 7. Pol δ4/PDIP46 in leading strand synthesis. Regions of secondary structure that pose impediments to Pol ε (red) lead to disengagement of the catalytic domain of Pol ε. This triggers a polymerase switch to Pol δ4/PDIP46.

The model shown in Figure 7 incorporates recent structural and functional studies of the yeast replisome from the Diffley laboratory [168–171]. These studies show that the catalytic domain of Pol ε is flexibly attached to its non-catalytic domain (which is engaged in complex with the CMG helicase). The catalytic domain adopts two conformations: it is proposed that in one conformation the catalytic domain is actively engaged with the DNA and in the other one it is disengaged [168,169]. In the context of the human replisome, we envisage that encounter with replication blocks stalls Pol ε, leading to the disengagement of the catalytic domain, followed by a switch to Pol δ4/PDIP46 which performs the bypass synthesis (Figure 7). When Pol δ4/PDIP46 encounters the CMG helicase, they dissociate and the Pol ε catalytic subunit re-engages the primer terminus.

The concept that the Pol ε catalytic domain can disengage from the DNA while remaining an integral part of the CMG-helicase leads to a paradigm shift in our thinking of the replisome [168,169]. Thus, where previously disengagement of Pol ε would require a physical uncoupling, this is no longer the case. It was proposed that Pol ε can disengage in response to replication stress, a situation that entertains the possibility of polymerase switching [168,169]. This is the situation where Pol δ4/PDIP46 could come into play in the model shown in Figure 7.

The studies of the yeast replisome have in fact produced evidence for a switch between Pol δ and Pol ε, and a mechanism for dealing with uncoupling events. Reconstitution of the initiation of DNA synthesis supports a model where Pol α forms the primer; this is extended by Pol δ which then "catches up" with the replisome to hand off the 3′ end of the leading strand to Pol ε in the advancing and uncoupled CMG-Pol ε replisome [168,169]. This is essentially a relay of polymerase handoffs where Pol δ has the anchor role of bringing the growing primer terminus to the CMG-Pol ε. This function of Pol δ in the yeast replisome provides a mechanism to deal with uncoupling events in a more general context, as during replication stress [168,169]. In a sense, the view of Pol ε being able to disengage from the primer end without dissociating from the CMG helicase allows it to have its cake and eat it. Other studies have also shown that Pol δ dissociates once it encounters the CMG-helicase by a collision release mechanism; this was taken as a means of selection against Pol δ at the leading strand [172,173]. However, it is noted that this would be dependent on the frequency of disengagement of Pol ε.

Once the possibility of switching of Pol δ and Pol ε is admitted, arguments against the participation of Pol δ in leading strand synthesis based on our previous understanding of the leading strand replisome as a tightly complexed structure are weakened. Polymerase switching involving Pol ε suggests a far more dynamic replisome. Recent kinetic studies have indicated that human Pol δ dissociates much more frequently than was previously thought [174]. The bacterial replisome is the prototype of a fixed structural assembly of both leading and lagging strand polymerases with the clamp loader. However, recent studies indicate that there is a frequent exchange of the bacterial polymerase during replication [175]. In the case of Pol ε, a similar situation could exist in terms of disengagement, such that uncoupling might be more frequent than expected even in the absence of replication stress.

These ideas have significant bearing on the participation of Pol δ4 in leading strand synthesis. There are no comparable studies that bear on the distribution of labor between Pol δ and Pol ε for the replication of the far larger and more complex genome in human cells. However, it is noted that replication of the SV40 genome in reconstituted systems can be achieved with Pol α and Pol δ [176,177]. One study using cross-linking and immunoprecipitation approaches, as well as immuno-electron microscopy, has provided evidence that Pol δ and Pol ε could be functioning independently in early and late S phase in the human system [178]. Taking into account recent views on the interplay between Pol δ and Pol ε during leading strand synthesis in yeast discussed above, it would appear that Pol δ may participate more extensively in leading strand synthesis than previously recognized.

There is a broader significance to the discovery that the catalytic domain of Pol ε is able to "switch" away from the DNA. It was proposed that Pol ε could also disengage from the DNA during replication stress [168,169]. Replication stress, broadly defined as encounter with replication barriers due to template lesions or complex DNA structures might be addressed by similar mechanisms to that which are well established in relation to Pol δ [88,179,180] and are based on the switching of specialized polymerases such as the TLS polymerases. The ability of Pol ε to disengage could also be involved in replication restart mechanisms that involve re-priming by PrimPol [181]. In the case of re-priming, the ability of Pol ε to disengage could also open the possibility of Pol α being able to re-prime, recapitulating the process that occurs during initiation [168,169]. These possibilities point to a convergence and a more unified view of mechanisms that deal with replication stress at both the leading and lagging strands of the replication fork.

There is still much to learn about the functions of PDIP46 in DNA replication. In addition to further biochemical analyses of the mechanisms by which PDIP46 affects Pol δ4, are questions as to whether PDIP46 functions are regulated by the mTOR pathway or by DNA damage signaling pathways. The functions of PDIP46 as SKAR indicates that it is a bifunctional protein, pointing to a need for structural studies, as well as careful dissection of the two functions, to allow design of appropriate mutations that selectively target its effects on Pol δ. The proposed roles of PDIP46 in DNA replication would be expected to yield phenotypes in a cellular context when its functions are disabled that reflect disturbances in DNA replication and genomic stability. Recent studies of genetic alterations in cells derived from high-risk neuroblastoma tissues have identified a group of genes whose alterations in copy number resulted in high tumorigenic capacity. The PDIP46 gene (*POLDIP3*) was one of six genes whose lowered expression was correlated with decreased overall and relapse free survival in a cohort of 88 patients [182]. Along with this, immunohistochemical tissue staining reveals a pattern of lowered expression in over 20 of the most common cancers (Human Protein Atlas, [183]). The catalogue of somatic mutations in cancer (COSMIC) database also showed that significant up- or down-regulation of POLDIP3 as being associated with various cancers as well as a number of mutations [184]. This provides some evidence that PDIP46 function is involved in maintenance of genomic stability.

4. PDIP38/Poldip2: A Multi-Faceted Protein

Recombinant mature PDIP38 at high concentrations (micromolar) inhibit Pol δ activity, an effect likely due to competition for PCNA that is unlikely to be physiologically relevant [154]. Thus despite its binding to Pol δ, the effects of PDIP38 on Pol δ, if any, are as yet not well defined. Studies reviewed below implicate PDIP38 in a number of cellular processes that are diverse, and further complicated by its localization to multiple subcellular organelles and structures, as well as its association with multiple protein partners including a transmembrane enzyme.

4.1. PDIP38 Is a Mitochondrial Protein with Multiple Subcellular Localizations

Analysis of the subcellular localization of PDIP38 revealed that it is primarily a mitochondrial protein. PDIP38 possesses a mitochondrial targeting site located in the N-terminal 30 amino acid residues, and cleavage sites [185] for mitochondrial processing peptidase and mitochondrial intermediate peptidase [154]. The N-terminal 50 residues are efficiently removed to yield a 38 kDa protein rather than the expected 42 kDa precursor. Cell fractionation experiments indicated that the bulk of the PDIP38 in cells was in a mitochondrial pellet, and resistant to proteinase K digestion until the membranes were solubilized with Triton X-100; a smaller amount was present in the nuclear fraction. Immunofluorescence studies of endogenous PDIP38 as well as of ectopically expressed C-terminally-tagged EGFP constructs showed that they are localized to the mitochondria [154]. Similar fractionation and immunofluorescence studies in two other studies confirmed these findings with the further indication that PDIP38 is present in the mitochondria matrix [186,187]. PDIP38 was found to associate with mitochondrial single stranded binding protein (mt SSB) and with the mitochondrial DNA nucleoid/mitochromosome [186,187]. The functions of PDIP38 in mitochondria are still unclear; in addition to potential effects on mitochondrial DNA replication, its depletion affects mitochondrial morphology [187], raising a question of whether the effects of its depletion also impacts mitochondrial energy metabolism.

There are conflicting reports on the subcellular localization of PDIP38. PDIP38 was found to be an interacting protein for CEACAM1, a cell adhesion receptor [188]. Analysis of its subcellular localization using peptide directed antibodies showed that the bulk of the PDIP38 is present in the cytoplasm, but does not co-localize with mitochondrial markers, a result contradictory to the studies described above. The basis of this difference regarding mitochondrial localization from those reported above [154] are unknown, although this could be due to differences in the antibodies used or the fixing of the cells. However, these studies did show significant evidence for PDIP38 in the nuclei. PDIP38 was dynamically localized to the cell surface membranes and the nuclei under influence of CEACAM1 [188]. Further analysis showed that PDIP38 is localized to the mitotic spindle. siRNA depletion of PDIP38 or microinjection of PDIP38 antibodies was associated with the appearance of aberrant spindle formation, chromosome segregation, as well as multinucleate cells [189].

4.2. Interaction of PDIP38/Poldip2 with Pol η and Other TLS Polymerases: Involvement of PDIP38 in the DNA Damage Tolerance Pathway

PDIP38 was found to interact directly with Pol η by a yeast two-hybrid screen with Pol η as the bait. Depletion of Pol η, PDIP38, or both, led to similar degrees of increased sensitivity to UV in cell survival assays. This suggested that PDIP38 plays an integral role in Pol η function [190]. The molecular mechanisms of the connections between PDIP38 and Pol η remain to be elucidated, but it has been suggested that PDIP38 might be a mediator in the switching process between Pol δ and Pol η [190]. In this context, PDIP38 might act to facilitate the recruitment of Pol η to Pol δ stalled at UV lesions. PDIP38 was also found to bind to the specialized polymerases Rev1 and Pol ξ (through interaction with the Rev7 subunit) [190]. These findings indicate that PDIP38 might be involved in the functions of other specialized DNA polymerases. The functional effects of PDIP38 on Rev1 and Pol ξ have not been reported.

Analysis of the effects of PDIP38 on five different DNA polymerases (Pols δ, η, ι, λ and β) showed that only the activities of Pol δ, Pol η and Pol λ were affected, consistent with the finding that Pol λ also physically interacts with PDIP38 [191]. PDIP38 (as the full-length protein) stimulated the processivity and catalytic activities of Pol η and Pol λ at low nanomolar concentrations on oligonucleotide templates containing lesions that included 8-oxoG, abasic sites and thymine-thymine dimers [191]. Additionally, the error-free bypass of 8-oxoG was increased, and a stimulatory effect on Pol δ was also found. Pol λ [192–194] participates in base excision repair of oxidative damage of guanine bases, as well as in a form of nonhomologous end joining repair of DSBs [193,195–197]. It was also demonstrated that depletion of PDIP38 led to an increase in the sensitivity of cultured cells to oxidizing agents [191]. Recently, a sixth polymerase was added to the list of PDIP38 binding proteins, this being PrimPol [198]. PrimPol is a member of the archaeo-eukaryotic primase (AEP) superfamily and exhibits primase, polymerase as well as translesion polymerase activities, and has emerged as having the ability to reprime DNA replication at sites of replication stress [181]. The effects of PDIP38 on PrimPol mirror those found for Pol η and Pol λ, viz., activation, increased processivity and fidelity for bypass of 8-oxoG. Depletion of either PDIP38 or PrimPol (or both) gave rise to replication defects (decrease in replication fork rates) in response to UV damage, suggesting that they are linked in the same pathway in vivo [198].

That PDIP38 is capable of interactions with a diverse group of polymerases raises interesting questions as to how this is achieved. Two similar short amino-acid sequences that are involved in PDIP38 binding were identified in Pol η [190] and in PrimPol [198]. An N-terminal sequence within the mitochondrial targeting sequence of PDIP38 was found to be a binding region for PrimPol. Full length PDIP38, but not the processed form, was able to activate PrimPol and Pol η [198]. These findings raise the question of whether levels of the unprocessed form in the nuclei would be sufficient to achieve functional concentrations in vivo, as most of the cellular PDIP38 is in the processed form [154].

The broad versatility of PDIP38 in the regulation of these polymerases, all of which are involved in the relief of replication stress, makes the elucidation of its structure and the location of its interaction sites an important goal. Furthermore, the apparently wide reach of PDIP38 in modulating activities of polymerases involved in translesion synthesis and relief of replicative stress indicates that it is likely to be under the control of the DNA damage tolerance regulatory pathways, notably those under the apical ATR kinase.

4.3. PDIP38 Responds to Genotoxic and Transcriptional Stress by Translocation to the Spliceosomes/Nuclear Speckles and Is Involved in Regulation of the Alternative Splicing of Mdm2

The potential involvement of PDIP38 in Pol η function suggests that it should be recruited to UV damage foci [190]. Using the technique of UV exposure through UV-opaque polycarbonate filters with 5 or 10 μm pores to create local areas of irradiation [84], it was observed that PDIP38 was not recruited to these DNA damage foci, in contrast to Pol η and PCNA [199]. This finding does not negate the proposal that PDIP38 is involved in Pol η recruitment [190], since the mechanism is unknown and might involve a transient association of PDIP38 to DNA damage sites.

Examination of the nuclear localization of PDIP38 showed that it was nevertheless recruited to nuclear foci in response to UV. These nuclear foci were identified as spliceosomes (nuclear speckles), which are associated with transcription and mRNA splicing processes [200]. Treatment with UV increased the number of cells with visible PDIP38 foci, as well as the number of foci per cell. Thus, the translocation of PDIP38 to the spliceosomes is a novel DNA damage response [199]. In addition to genotoxic stress, transcriptional stress induced by α-amanitin also led to translocation of PDIP38 to the nuclear speckles [199]. Interestingly, the human DNA glycosylase hOGG1 is also translocated to the nuclear speckles under the influence of UVA [201].

The translocation of PDIP38 in response to UV-damage raises the question of its functions in the spliceosomes/nuclear speckles, which are associated with transcription and mRNA splicing. There are a number of genes whose alternative splicing is altered under genotoxic stress [202,203]. One of the more extensively studied of these genes is mouse double minute 2 (*MDM2*) [202,204]. Mdm2 is an E3 ubiquitin ligase that is a negative regulator of p53 [202,205]. Various genotoxic agents, e.g., UV, camptothecin, doxorubicin and cisplatin, lead to skipping of as many as eight exons, resulting in disruption of Mdm2 function and of p53 regulation [205]. Alternative spliced variants of Mdm2 also can exhibit growth regulatory properties independent of p53 and induce tumorigenesis [206–209].

Analysis of the UV induced Mdm2 splice variants in A549 cells showed that this was dependent on PDIP38, as their levels were suppressed in PDIP38 depleted cells [199]. While the extent and mechanisms that underlie the basis for the requirement for PDIP38 in Mdm2 alternative splicing are unknown, it may be another example of the interplay or crosstalk between DNA damage/repair processes and RNA transcription/splicing in the maintenance of genomic stability and cell survival [203,210,211]. This crosstalk has largely focused on RNA binding proteins, but also on the involvement of DNA damage response proteins in regulating splicing factors [210]. The effects of PDIP38 in modulating the splicing of Mdm2, a key regulator of p53, falls into this category, and may represent one of its important functions.

4.4. PDIP38 Binds to p22phox and Regulates the Activity of the Nox4/p22phox NADPH Oxidase

Nox4 (NADPH oxidase 4) is one of seven transmembrane NADPH oxidases that generate reactive oxygen species (ROS): superoxide and H_2O_2 [212–215]. The generation of ROS by the NOX enzymes occurs physiologically in response to various stimuli; these ROS act on signal transduction pathways [212,213,216,217]. Nox4 is widely distributed in tissues, with the highest levels in kidney [213]. Biochemical analysis of partially purified membrane free preparations of Nox4 revealed that Nox4 has a high K_m for O_2, and functions as an oxygen sensor, in that its activity responds to the physiological pO_2 [218]. These studies also demonstrated that the Nox4 reaction generates H_2O_2 as the primary product with a smaller amount of superoxide [218]. This response to pO_2 has relevance to the proposed role for Nox4 as an oxygen sensor that produces H_2O_2 as a signaling molecule [218]. Four of the Nox enzymes including Nox4 are associated with p22phox, which acts as a subunit that interacts with regulatory proteins in response to cellular stimuli [212,219]. However, Nox4 binds p22phox which is required for its activity, and is regarded as being constitutively active [213,215].

The role of PDIP38 in regulating Nox4 functions has been extensively studied in the cardiovascular system [215,219]. PDIP38 was found to bind to p22phox and to activate the Nox4/p22phox enzyme in vascular smooth muscle cells [220]. The latter study supports the view that PDIP38 is involved in Nox4 localization, focal adhesion integrity, stress fiber formation, and plays a role in the maintenance of vascular smooth muscle cell cytoskeletal functions. PDIP38 was found to co-localize with p22phox at focal adhesions and stress fibers [220]. Overexpression of PDIP38 in vascular smooth muscle cells increased NADPH oxidase activity several fold in a Nox4 dependent manner [220]. PDIP38 also regulates vascular smooth muscle cell migration by regulating focal adhesion turnover and traction force generation [221]. PDIP38 knockout in mice has shown that it is essential for development, as this led to perinatal lethality [222]. Analysis of aortas from heterozygous mice showed that these exhibited abnormal structures and decreased contraction and compliance that are consistent with a role in vascular function and integrity. Mouse embryonic fibroblasts derived from the knockout mice exhibit defective growth characteristics, alterations in cell cycle progression and expression of cell cycle proteins [223]. The subcellular localizations of Nox4/p22phox and PDIP38 in vascular smooth muscle cells also raises questions regarding how these integrate into the fact that the Nox enzymes are membrane associated proteins [212].

4.5. Summary

PDIP38 is unusual in that there is evidence for its role in a number of cellular functions, emanating from the discovery of multiple protein partners. In addition to Pol δ and PCNA, PDIP38 interacts with Pol η and other TLS pols. These findings together indicate a role for PDIP38 in regulating translesion synthesis, while its association with Pol δ suggests it may be involved in the mechanisms or regulation of the interchange between the TLS pols and Pol δ. In addition, PDIP38 is likely under regulation from DNA damage signaling pathways and is translocated to the spliceosomes where it affects Mdm2 splicing and thereby p53 regulation.

Nevertheless, the studies of PDIP38 are still in their early stages, and its multifunctional nature poses significant technical challenges to the use of gene depletion or knockouts either in cells or animals, as these approaches may not allow unambiguous cause and effect relationships. Thus, much further investigation is required to establish how these functions are accomplished at the molecular level, as well as the cellular advantages of the investiture of these functions in a single protein. These require biochemical approaches and, in particular, the elucidation of PDIP38 structure and its complexes with its partners. These could lead to strategies for the use of targeted mutations that could provide the means for isolating cause and effect in gene depletion experiments.

Acknowledgments: Studies reported from this laboratory were supported by grants from the National Institutes of Health (GM031793 to Marietta Y. W. T. Lee) and the National Institute of Environmental Health Science (ES014737 to Marietta Y. W. T. Lee and Zhongtao Zhang).

Author Contributions: All authors contributed to the manuscript and approved the final version.

Conflicts of Interest: The authors declare no conflicts of interest.

References

1. Kornberg, A. Ten commandments: Lessons from the enzymology of DNA replication. *J. Bacteriol.* **2000**, *182*, 3613–3618. [CrossRef] [PubMed]
2. Barnes, R.; Eckert, K. Maintenance of Genome Integrity: How Mammalian Cells Orchestrate Genome Duplication by Coordinating Replicative and Specialized DNA Polymerases. *Genes* **2017**, *8*, 19. [CrossRef] [PubMed]
3. Nicolas, E.; Golemis, E.A.; Arora, S. POLD1: Central mediator of DNA replication and repair, and implication in cancer and other pathologies. *Gene* **2016**, *590*, 128–141. [CrossRef] [PubMed]
4. Weissbach, A. Eukaryotic DNA polymerases. *Annu. Rev. Biochem.* **1977**, *46*, 25–47. [CrossRef] [PubMed]
5. Brutlag, D.; Kornberg, A. Enzymatic synthesis of deoxyribonucleic acid. 36. A proofreading function for the 3′ leads to 5′ exonuclease activity in deoxyribonucleic acid polymerases. *J. Biol. Chem.* **1972**, *247*, 241–248. [PubMed]
6. Muzyczka, N.; Poland, R.L.; Bessman, M.J. Studies on the biochemical basis of spontaneous mutation. I. A comparison of the deoxyribonucleic acid polymerases of mutator, antimutator, and wild type strains of bacteriophage T4. *J. Biol. Chem.* **1972**, *247*, 7116–7122. [PubMed]
7. Reha-Krantz, L.J. DNA polymerase proofreading: Multiple roles maintain genome stability. *Biochim. Biophys. Acta* **2010**, *1804*, 1049–1063. [CrossRef] [PubMed]
8. Byrnes, J.J.; Downey, K.M.; Black, V.; Esserman, L.; So, A.G. Selective Inhibition of the 3′ to 5′ Exonuclease Activity Associated with Mammalian DNA Polymerase δ. In *Miami Winter Symposium: Cancer Enzymology*; Schultz, J., Ahmad, F., Eds.; Academic Press: Cambridge, MA, USA, 1976; Volume 12, pp. 245–264.

9. Byrnes, J.J.; Downey, K.M.; Black, V.L.; So, A.G. A new mammalian DNA polymerase with 3′ to 5′ exonuclease activity: DNA polymerase δ. *Biochemistry* **1976**, *15*, 2817–2823. [CrossRef] [PubMed]

10. Byrnes, J.J.; Downey, K.M.; Que, B.G.; Lee, M.Y.; Black, V.L.; So, A.G. Selective inhibition of the 3′ to 5′ exonuclease activity associated with DNA polymerases: A mechanism of mutagenesis. *Biochemistry* **1977**, *16*, 3740–3746. [CrossRef] [PubMed]

11. Lee, M.Y.; Byrnes, J.J.; Downey, K.M.; So, A.G. Mechanism of inhibition of deoxyribonucleic acid synthesis by 1-beta-D-arabinofuranosyladenosine triphosphate and its potentiation by 6-mercaptopurine ribonucleoside 5′-monophosphate. *Biochemistry* **1980**, *19*, 215–219. [CrossRef] [PubMed]

12. Lee, M.Y.; Tan, C.K.; So, A.G.; Downey, K.M. Purification of deoxyribonucleic acid polymerase δ from calf thymus: Partial characterization of physical properties. *Biochemistry* **1980**, *19*, 2096–2101. [CrossRef] [PubMed]

13. Lee, M.Y.; Tan, C.K.; Downey, K.M.; So, A.G. Structural and functional properties of calf thymus DNA polymerase δ. *Prog. Nucleic Acid Res. Mol. Biol.* **1981**, *26*, 83–96. [PubMed]

14. Lee, M.Y.; Tan, C.K.; Downey, K.M.; So, A.G. Further studies on calf thymus DNA polymerase δ purified to homogeneity by a new procedure. *Biochemistry* **1984**, *23*, 1906–1913. [CrossRef] [PubMed]

15. Lee, M.Y.; Toomey, N.L.; Wright, G.E. Differential inhibition of human placental DNA polymerases δ and α by BuPdGTP and BuAdATP. *Nucleic Acids Res.* **1985**, *13*, 8623–8630. [CrossRef] [PubMed]

16. Lee, M.Y.; Toomey, N.L. Human placental DNA polymerase δ: Identification of a 170-kilodalton polypeptide by activity staining and immunoblotting. *Biochemistry* **1987**, *26*, 1076–1085. [CrossRef] [PubMed]

17. Lee, M.Y. Isolation of multiple forms of DNA polymerase δ: Evidence of proteolytic modification during isolation. *Biochemistry* **1988**, *27*, 5188–5193. [CrossRef] [PubMed]

18. Lee, M.Y.; Alejandro, R.; Toomey, N.L. Immunochemical studies of DNA polymerase δ: Relationships with DNA polymerase α. *Arch. Biochem. Biophys.* **1989**, *272*, 1–9. [CrossRef]

19. Lee, M.Y.; Jiang, Y.Q.; Zhang, S.J.; Toomey, N.L. Characterization of human DNA polymerase δ and its immunochemical relationships with DNA polymerase α and epsilon. *J. Biol. Chem.* **1991**, *266*, 2423–2429. [PubMed]

20. Wong, S.W.; Syvaoja, J.; Tan, C.K.; Downey, K.M.; So, A.G.; Linn, S.; Wang, T.S. DNA polymerases α and δ are immunologically and structurally distinct. *J. Biol. Chem.* **1989**, *264*, 5924–5928. [PubMed]

21. Zhang, J.; Chung, D.W.; Tan, C.K.; Downey, K.M.; Davie, E.W.; So, A.G. Primary structure of the catalytic subunit of calf thymus DNA polymerase δ: Sequence similarities with other DNA polymerases. *Biochemistry* **1991**, *30*, 11742–11750. [CrossRef] [PubMed]

22. Hao, H.; Jiang, Y.; Zhang, S.J.; Zhang, P.; Zeng, R.X.; Lee, M.Y. Structural and functional relationships of human DNA polymerases. *Chromosoma* **1992**, *102*, S121–S127. [CrossRef] [PubMed]

23. Yang, C.L.; Chang, L.S.; Zhang, P.; Hao, H.; Zhu, L.; Toomey, N.L.; Lee, M.Y. Molecular cloning of the cDNA for the catalytic subunit of human DNA polymerase δ. *Nucleic Acids Res.* **1992**, *20*, 735–745. [CrossRef] [PubMed]

24. Crute, J.J.; Wahl, A.F.; Bambara, R.A. Purification and characterization of two new high molecular weight forms of DNA polymerase δ. *Biochemistry* **1986**, *25*, 26–36. [CrossRef] [PubMed]

25. Syvaoja, J.; Linn, S. Characterization of a large form of DNA polymerase δ from HeLa cells that is insensitive to proliferating cell nuclear antigen. *J. Biol. Chem.* **1989**, *264*, 2489–2497. [PubMed]

26. Burgers, P.M.; Bambara, R.A.; Campbell, J.L.; Chang, L.M.; Downey, K.M.; Hubscher, U.; Lee, M.Y.; Linn, S.M.; So, A.G.; Spadari, S. Revised nomenclature for eukaryotic DNA polymerases. *Eur. J. Biochem.* **1990**, *191*, 617–618. [CrossRef] [PubMed]

27. Pospiech, H.; Syvaoja, J.E. DNA polymerase epsilon—More than a polymerase. *Sci. World J.* **2003**, *3*, 87–104. [CrossRef] [PubMed]

28. Prelich, G.; Tan, C.K.; Kostura, M.; Mathews, M.B.; So, A.G.; Downey, K.M.; Stillman, B. Functional identity of proliferating cell nuclear antigen and a DNA polymerase-δ auxiliary protein. *Nature* **1987**, *326*, 517–520. [CrossRef] [PubMed]

29. Choe, K.N.; Moldovan, G.L. Forging Ahead through Darkness: PCNA, Still the Principal Conductor at the Replication Fork. *Mol. Cell* **2017**, *65*, 380–392. [CrossRef] [PubMed]

30. Mo, J.; Liu, L.; Leon, A.; Mazloum, N.; Lee, M.Y. Evidence that DNA polymerase δ isolated by immunoaffinity chromatography exhibits high-molecular weight characteristics and is associated with the KIAA0039 protein and RPA. *Biochemistry* **2000**, *39*, 7245–7254. [CrossRef] [PubMed]

31. Hughes, P.; Tratner, I.; Ducoux, M.; Piard, K.; Baldacci, G. Isolation and identification of the third subunit of mammalian DNA polymerase δ by PCNA-affinity chromatography of mouse FM3A cell extracts. *Nucleic Acids Res.* **1999**, *27*, 2108–2114. [CrossRef] [PubMed]

32. Liu, L.; Mo, J.; Rodriguez-Belmonte, E.M.; Lee, M.Y. Identification of a fourth subunit of mammalian DNA polymerase δ. *J. Biol. Chem.* **2000**, *275*, 18739–18744. [CrossRef] [PubMed]

33. Garg, P.; Burgers, P.M. DNA polymerases that propagate the eukaryotic DNA replication fork. *Crit. Rev. Biochem. Mol. Biol.* **2005**, *40*, 115–128. [CrossRef] [PubMed]

34. Gerik, K.J.; Li, X.; Pautz, A.; Burgers, P.M. Characterization of the two small subunits of Saccharomyces cerevisiae DNA polymerase δ. *J. Biol. Chem.* **1998**, *273*, 19747–19755. [CrossRef] [PubMed]

35. Reynolds, N.; Watt, A.; Fantes, P.A.; MacNeill, S.A. Cdm1, the smallest subunit of DNA polymerase d in the fission yeast Schizosaccharomyces pombe, is non-essential for growth and division. *Curr. Genet.* **1998**, *34*, 250–258. [CrossRef] [PubMed]

36. Zuo, S.; Gibbs, E.; Kelman, Z.; Wang, T.S.; O'Donnell, M.; MacNeill, S.A.; Hurwitz, J. DNA polymerase δ isolated from Schizosaccharomyces pombe contains five subunits. *Proc. Natl. Acad. Sci. USA* **1997**, *94*, 11244–11249. [CrossRef]

37. Zuo, S.; Bermudez, V.; Zhang, G.; Kelman, Z.; Hurwitz, J. Structure and activity associated with multiple forms of Schizosaccharomyces pombe DNA polymerase δ. *J. Biol. Chem.* **2000**, *275*, 5153–5162. [CrossRef] [PubMed]

38. MacNeill, S.A.; Moreno, S.; Reynolds, N.; Nurse, P.; Fantes, P.A. The fission yeast Cdc1 protein, a homologue of the small subunit of DNA polymerase δ, binds to Pol3 and Cdc27. *EMBO J.* **1996**, *15*, 4613–4628. [PubMed]

39. Bermudez, V.P.; MacNeill, S.A.; Tappin, I.; Hurwitz, J. The influence of the Cdc27 subunit on the properties of the Schizosaccharomyces pombe DNA polymerase δ. *J. Biol. Chem.* **2002**, *277*, 36853–36862. [CrossRef] [PubMed]

40. Johansson, E.; Majka, J.; Burgers, P.M. Structure of DNA polymerase δ from *Saccharomyces cerevisiae*. *J. Biol. Chem.* **2001**, *276*, 43824–43828. [CrossRef] [PubMed]

41. Zhang, P.; Mo, J.Y.; Perez, A.; Leon, A.; Liu, L.; Mazloum, N.; Xu, H.; Lee, M.Y. Direct interaction of proliferating cell nuclear antigen with the p125 catalytic subunit of mammalian DNA polymerase δ. *J. Biol. Chem.* **1999**, *274*, 26647–26653. [CrossRef] [PubMed]

42. Zhang, P.; Sun, Y.; Hsu, H.; Zhang, L.; Zhang, Y.; Lee, M.Y. The interdomain connector loop of human PCNA is involved in a direct interaction with human polymerase δ. *J. Biol. Chem.* **1998**, *273*, 713–719. [CrossRef] [PubMed]

43. Roos, G.; Jiang, Y.; Landberg, G.; Nielsen, N.H.; Zhang, P.; Lee, M.Y. Determination of the epitope of an inhibitory antibody to proliferating cell nuclear antigen. *Exp. Cell Res.* **1996**, *226*, 208–213. [CrossRef] [PubMed]

44. Zhang, S.J.; Zeng, X.R.; Zhang, P.; Toomey, N.L.; Chuang, R.Y.; Chang, L.S.; Lee, M.Y. A conserved region in the amino terminus of DNA polymerase δ is involved in proliferating cell nuclear antigen binding. *J. Biol. Chem.* **1995**, *270*, 7988–7992. [CrossRef] [PubMed]

45. Li, H.; Xie, B.; Zhou, Y.; Rahmeh, A.; Trusa, S.; Zhang, S.; Gao, Y.; Lee, E.Y.; Lee, M.Y. Functional roles of p12, the fourth subunit of human DNA polymerase δ. *J. Biol. Chem.* **2006**, *281*, 14748–14755. [CrossRef] [PubMed]

46. Lu, X.; Tan, C.K.; Zhou, J.Q.; You, M.; Carastro, L.M.; Downey, K.M.; So, A.G. Direct interaction of proliferating cell nuclear antigen with the small subunit of DNA polymerase δ. *J. Biol. Chem.* **2002**, *277*, 24340–24345. [CrossRef] [PubMed]

47. Wang, Y.; Zhang, Q.; Chen, H.; Li, X.; Mai, W.; Chen, K.; Zhang, S.; Lee, E.Y.; Lee, M.Y.; Zhou, Y. p50, the Small Subunit of DNA Polymerase Δ, Is Required for Mediation of the Interaction of Polymerase Δ Subassemblies with PCNA. *PLoS ONE* **2011**, *6*, e27092. [CrossRef] [PubMed]

48. Rahmeh, A.A.; Zhou, Y.; Xie, B.; Li, H.; Lee, E.Y.; Lee, M.Y. Phosphorylation of the p68 Subunit of Pol Δ Acts as a Molecular Switch to Regulate Its Interaction with PCNA. *Biochemistry* **2012**, *51*, 416–424. [CrossRef] [PubMed]

49. Podust, V.N.; Chang, L.S.; Ott, R.; Dianov, G.L.; Fanning, E. Reconstitution of human DNA polymerase δ using recombinant baculoviruses: The p12 subunit potentiates DNA polymerizing activity of the four-subunit enzyme. *J. Biol. Chem.* **2002**, *277*, 3894–3901. [CrossRef] [PubMed]

50. Xie, B.; Mazloum, N.; Liu, L.; Rahmeh, A.; Li, H.; Lee, M.Y. Reconstitution and characterization of the human DNA polymerase δ four-subunit holoenzyme. *Biochemistry* **2002**, *41*, 13133–13142. [CrossRef] [PubMed]

51. Zhou, Y.; Meng, X.; Zhang, S.; Lee, E.Y.; Lee, M.Y. Characterization of human DNA polymerase δ and its subassemblies reconstituted by expression in the multibac system. *PLoS ONE* **2012**, *7*, e39156. [CrossRef] [PubMed]

52. Masuda, Y.; Suzuki, M.; Piao, J.; Gu, Y.; Tsurimoto, T.; Kamiya, K. Dynamics of human replication factors in the elongation phase of DNA replication. *Nucleic Acids Res.* **2007**, *35*, 6904–6916. [CrossRef] [PubMed]

53. Jiang, Y.; Zhang, S.J.; Wu, S.M.; Lee, M.Y. Immunoaffinity purification of DNA polymerase δ. *Arch. Biochem. Biophys.* **1995**, *320*, 297–304. [CrossRef]

54. Podust, V.N.; Georgaki, A.; Strack, B.; Hubscher, U. Calf thymus RF-C as an essential component for DNA polymerase δ and epsilon holoenzymes function. *Nucleic Acids Res.* **1992**, *20*, 4159–4165. [CrossRef] [PubMed]

55. Burgers, P.M.; Gerik, K.J. Structure and processivity of two forms of *Saccharomyces cerevisiae* DNA polymerase δ. *J. Biol. Chem.* **1998**, *273*, 19756–19762. [CrossRef] [PubMed]

56. Wang, X.; Zhang, S.; Zheng, R.; Yue, F.; Lin, S.H.; Rahmeh, A.A.; Lee, E.Y.; Zhang, Z.; Lee, M.Y. PDIP46 (DNA polymerase δ interacting protein 46) is an activating factor for human DNA polymerase δ. *Oncotarget* **2016**, *7*, 6294–6313. [CrossRef] [PubMed]

57. Gao, Y.; Zhou, Y.; Xie, B.; Zhang, S.; Rahmeh, A.; Huang, H.S.; Lee, M.Y.; Lee, E.Y. Protein phosphatase-1 is targeted to DNA polymerase δ via an interaction with the p68 subunit. *Biochemistry* **2008**, *47*, 11367–11376. [CrossRef] [PubMed]

58. Lee, M.Y.; Zhang, S.; Lin, S.H.; Chea, J.; Wang, X.; LeRoy, C.; Wong, A.; Zhang, Z.; Lee, E.Y. Regulation of human DNA polymerase Δ in the cellular responses to DNA damage. *Environ. Mol. Mutagen.* **2012**, *53*, 683–698. [CrossRef] [PubMed]

59. Lee, M.Y.; Zhang, S.; Lin, S.H.; Wang, X.; Darzynkiewicz, Z.; Zhang, Z.; Lee, E.Y. The tail that wags the dog: p12, the smallest subunit of DNA polymerase δ, is degraded by ubiquitin ligases in response to DNA damage and during cell cycle progression. *Cell Cycle* **2014**, *13*, 23–31. [CrossRef] [PubMed]

60. Zhang, S.; Zhou, Y.; Trusa, S.; Meng, X.; Lee, E.Y.; Lee, M.Y. A novel DNA damage response: Rapid degradation of the p12 subunit of DNA polymerase δ. *J. Biol. Chem.* **2007**, *282*, 15330–15340. [CrossRef] [PubMed]

61. Zhang, S.; Zhao, H.; Darzynkiewicz, Z.; Zhou, P.; Zhang, Z.; Lee, E.Y.; Lee, M.Y. A novel function of CRL4Cdt2: Regulation of the subunit structure of DNA polymerase δ in response to DNA damage and during the S phase. *J. Biol. Chem.* **2013**, *288*, 29950–29961. [CrossRef] [PubMed]

62. Kaufmann, W.K. The human intra-S checkpoint response to UVC-induced DNA damage. *Carcinogenesis* **2010**, *31*, 751–765. [CrossRef] [PubMed]

63. Cimprich, K.A.; Cortez, D. ATR: An essential regulator of genome integrity. *Nat. Rev. Mol. Cell Biol.* **2008**, *9*, 616–627. [CrossRef] [PubMed]

64. Flynn, R.L.; Zou, L. ATR: A master conductor of cellular responses to DNA replication stress. *Trends Biochem. Sci.* **2011**, *36*, 133–140. [CrossRef] [PubMed]

65. Conti, C.; Seiler, J.A.; Pommier, Y. The mammalian DNA replication elongation checkpoint: Implication of Chk1 and relationship with origin firing as determined by single DNA molecule and single cell analyses. *Cell Cycle* **2007**, *6*, 2760–2767. [CrossRef] [PubMed]

66. Warren, J.J.; Forsberg, L.J.; Beese, L.S. The structural basis for the mutagenicity of *O*(6)-methyl-guanine lesions. *Proc. Natl. Acad. Sci. USA* **2006**, *103*, 19701–19706. [CrossRef] [PubMed]

67. Hsu, G.W.; Ober, M.; Carell, T.; Beese, L.S. Error-prone replication of oxidatively damaged DNA by a high-fidelity DNA polymerase. *Nature* **2004**, *431*, 217–221. [CrossRef] [PubMed]

68. Meng, X.; Zhou, Y.; Zhang, S.; Lee, E.Y.; Frick, D.N.; Lee, M.Y. DNA damage alters DNA polymerase δ to a form that exhibits increased discrimination against modified template bases and mismatched primers. *Nucleic Acids Res.* **2009**, *37*, 647–657. [CrossRef] [PubMed]

69. Johnson, K.A. Conformational coupling in DNA polymerase fidelity. *Annu. Rev. Biochem.* **1993**, *62*, 685–713. [CrossRef] [PubMed]

70. Kunkel, T.A.; Bebenek, K. DNA replication fidelity. *Annu. Rev. Biochem.* **2000**, *69*, 497–529. [CrossRef] [PubMed]

71. Steitz, T.A. DNA polymerases: Structural diversity and common mechanisms. *J. Biol. Chem.* **1999**, *274*, 17395–17398. [CrossRef] [PubMed]

72. Shamoo, Y.; Steitz, T.A. Building a replisome from interacting pieces: Sliding clamp complexed to a peptide from DNA polymerase and a polymerase editing complex. *Cell* **1999**, *99*, 155–166. [CrossRef]

73. Donlin, M.J.; Patel, S.S.; Johnson, K.A. Kinetic partitioning between the exonuclease and polymerase sites in DNA error correction. *Biochemistry* **1991**, *30*, 538–546. [CrossRef] [PubMed]
74. Khare, V.; Eckert, K.A. The proofreading 3′-5′ exonuclease activity of DNA polymerases: A kinetic barrier to translesion DNA synthesis. *Mutat. Res.* **2002**, *510*, 45–54. [CrossRef]
75. Meng, X.; Zhou, Y.; Lee, E.Y.; Lee, M.Y.; Frick, D.N. The p12 subunit of human polymerase δ modulates the rate and fidelity of DNA synthesis. *Biochemistry* **2010**, *49*, 3545–3554. [CrossRef] [PubMed]
76. Ogi, T.; Limsirichaikul, S.; Overmeer, R.M.; Volker, M.; Takenaka, K.; Cloney, R.; Nakazawa, Y.; Niimi, A.; Miki, Y.; Jaspers, N.G.; et al. Three DNA polymerases, recruited by different mechanisms, carry out NER repair synthesis in human cells. *Mol. Cell* **2010**, *37*, 714–727. [CrossRef] [PubMed]
77. Terai, K.; Shibata, E.; Abbas, T.; Dutta, A. Degradation of p12 Subunit by CRL4Cdt2 E3 Ligase Inhibits Fork Progression after DNA Damage. *J. Biol. Chem.* **2013**, *288*, 30509–30514. [CrossRef] [PubMed]
78. Bekker-Jensen, S.; Mailand, N. Assembly and function of DNA double-strand break repair foci in mammalian cells. *DNA Repair* **2010**, *9*, 1219–1228. [CrossRef] [PubMed]
79. Mailand, N.; Bekker-Jensen, S.; Faustrup, H.; Melander, F.; Bartek, J.; Lukas, C.; Lukas, J. RNF8 ubiquitylates histones at DNA double-strand breaks and promotes assembly of repair proteins. *Cell* **2007**, *131*, 887–900. [CrossRef] [PubMed]
80. Huen, M.S.; Grant, R.; Manke, I.; Minn, K.; Yu, X.; Yaffe, M.B.; Chen, J. RNF8 transduces the DNA-damage signal via histone ubiquitylation and checkpoint protein assembly. *Cell* **2007**, *131*, 901–914. [CrossRef] [PubMed]
81. Harper, J.W.; Elledge, S.J. The DNA damage response: Ten years after. *Mol. Cell* **2007**, *28*, 739–745. [CrossRef] [PubMed]
82. Yan, J.; Jetten, A.M. RAP80 and RNF8, key players in the recruitment of repair proteins to DNA damage sites. *Cancer Lett.* **2008**, *271*, 179–190. [CrossRef] [PubMed]
83. Sirbu, B.M.; Cortez, D. DNA damage response: Three levels of DNA repair regulation. *Cold Spring Harb. Perspect. Biol.* **2013**, *5*, a012724. [CrossRef] [PubMed]
84. Chea, J.; Zhang, S.; Zhao, H.; Zhang, Z.; Lee, E.Y.; Darzynkiewicz, Z.; Lee, M.Y. Spatiotemporal recruitment of human DNA polymerase δ to sites of UV damage. *Cell Cycle* **2012**, *11*, 2885–2895. [CrossRef] [PubMed]
85. Andersen, P.L.; Xu, F.; Xiao, W. Eukaryotic DNA damage tolerance and translesion synthesis through covalent modifications of PCNA. *Cell Res.* **2008**, *18*, 162–173. [CrossRef] [PubMed]
86. Chen, J.; Bozza, W.; Zhuang, Z. Ubiquitination of PCNA and its essential role in eukaryotic translesion synthesis. *Cell Biochem. Biophys.* **2011**, *60*, 47–60. [CrossRef] [PubMed]
87. Ghosal, G.; Chen, J. DNA damage tolerance: A double-edged sword guarding the genome. *Transl. Cancer Res.* **2013**, *2*, 107–129. [PubMed]
88. Sale, J.E.; Lehmann, A.R.; Woodgate, R. Y-family DNA polymerases and their role in tolerance of cellular DNA damage. *Nat. Rev. Mol. Cell Biol.* **2012**, *13*, 141–152. [CrossRef] [PubMed]
89. Vaisman, A.; Woodgate, R. Translesion DNA polymerases in eukaryotes: What makes them tick? *Crit. Rev. Biochem. Mol. Biol.* **2017**, 1–30. [CrossRef] [PubMed]
90. Hoege, C.; Pfander, B.; Moldovan, G.L.; Pyrowolakis, G.; Jentsch, S. RAD6-dependent DNA repair is linked to modification of PCNA by ubiquitin and SUMO. *Nature* **2002**, *419*, 135–141. [CrossRef] [PubMed]
91. Kannouche, P.L.; Wing, J.; Lehmann, A.R. Interaction of human DNA polymerase eta with monoubiquitinated PCNA: A possible mechanism for the polymerase switch in response to DNA damage. *Mol. Cell* **2004**, *14*, 491–500. [CrossRef]
92. Kannouche, P.L.; Lehmann, A.R. Ubiquitination of PCNA and the polymerase switch in human cells. *Cell Cycle* **2004**, *3*, 1011–1013. [CrossRef] [PubMed]
93. Zhang, Z.; Zhang, S.; Lin, S.H.; Wang, X.; Wu, L.; Lee, E.Y.; Lee, M.Y. Structure of monoubiquitinated PCNA: Implications for DNA polymerase switching and Okazaki fragment maturation. *Cell Cycle* **2012**, *11*, 2128–2136. [CrossRef] [PubMed]
94. Murga, M.; Lecona, E.; Kamileri, I.; Diaz, M.; Lugli, N.; Sotiriou, S.K.; Anton, M.E.; Mendez, J.; Halazonetis, T.D.; Fernandez-Capetillo, O. POLD3 Is Haploinsufficient for DNA Replication in Mice. *Mol. Cell* **2016**, *63*, 877–883. [CrossRef] [PubMed]
95. Hirota, K.; Yoshikiyo, K.; Guilbaud, G.; Tsurimoto, T.; Murai, J.; Tsuda, M.; Phillips, L.G.; Narita, T.; Nishihara, K.; Kobayashi, K.; et al. The POLD3 subunit of DNA polymerase δ can promote translesion synthesis independently of DNA polymerase ζ. *Nucleic Acids Res.* **2015**, *43*, 1671–1683. [CrossRef] [PubMed]

96. Johnson, R.E.; Prakash, L.; Prakash, S. Pol31 and Pol32 subunits of yeast DNA polymerase δ are also essential subunits of DNA polymerase ζ. *Proc. Natl. Acad. Sci. USA* **2012**, *109*, 12455–12460. [CrossRef] [PubMed]

97. Baranovskiy, A.G.; Lada, A.G.; Siebler, H.; Zhang, Y.; Pavlov, Y.I.; Tahirov, T.H. DNA polymerases δ and ζ switching by sharing the accessory subunits of DNA polymerase δ. *J. Biol. Chem.* **2012**, *287*, 17281–17287. [CrossRef] [PubMed]

98. Zhang, S.; Zhou, Y.; Sarkeshik, A.; Yates, J.R.; Thomson, T.; Zhang, Z.; Lee, E.Y.; Lee, M.Y. Identification of RNF8 as a Ubiquitin Ligase Involved in Targeting the p12 Subunit of DNA Polymerase δ for Degradation in Response to DNA Damage. *J. Biol. Chem.* **2013**, *288*, 2941–2950. [CrossRef] [PubMed]

99. Brzovic, P.S.; Klevit, R.E. Ubiquitin transfer from the E2 perspective: Why is UbcH5 so promiscuous? *Cell Cycle* **2006**, *5*, 2867–2873. [CrossRef] [PubMed]

100. Wang, B.; Elledge, S.J. Ubc13/Rnf8 ubiquitin ligases control foci formation of the Rap80/Abraxas/Brca1/Brcc36 complex in response to DNA damage. *Proc. Natl. Acad. Sci. USA* **2007**, *104*, 20759–20763. [CrossRef] [PubMed]

101. Marteijn, J.A.; Bekker-Jensen, S.; Mailand, N.; Lans, H.; Schwertman, P.; Gourdin, A.M.; Dantuma, N.P.; Lukas, J.; Vermeulen, W. Nucleotide excision repair-induced H2A ubiquitination is dependent on MDC1 and RNF8 and reveals a universal DNA damage response. *J. Cell Biol.* **2009**, *186*, 835–847. [CrossRef] [PubMed]

102. Zhang, S.; Chea, J.; Meng, X.; Zhou, Y.; Lee, E.Y.; Lee, M.Y. PCNA is ubiquitinated by RNF8. *Cell Cycle* **2008**, *7*, 3399–3404. [CrossRef] [PubMed]

103. Abbas, T.; Dutta, A. CRL4Cdt2: Master coordinator of cell cycle progression and genome stability. *Cell Cycle* **2011**, *10*, 241–249. [CrossRef] [PubMed]

104. Hannah, J.; Zhou, P. Distinct and overlapping functions of the cullin E3 ligase scaffolding proteins CUL4A and CUL4B. *Gene* **2015**, *573*, 33–45. [CrossRef] [PubMed]

105. Havens, C.G.; Shobnam, N.; Guarino, E.; Centore, R.C.; Zou, L.; Kearsey, S.E.; Walter, J.C. Direct Role for proliferating cell nuclear antigen (PCNA) in substrate recognition by the E3 Ubiquitin ligase CRL4-Cdt2. *J. Biol. Chem.* **2012**, *287*, 11410–11421. [CrossRef] [PubMed]

106. Havens, C.G.; Walter, J.C. Mechanism of CRL4(Cdt2), a PCNA-dependent E3 ubiquitin ligase. *Genes Dev.* **2011**, *25*, 1568–1582. [CrossRef] [PubMed]

107. Arias, E.E.; Walter, J.C. Replication-dependent destruction of Cdt1 limits DNA replication to a single round per cell cycle in Xenopus egg extracts. *Genes Dev.* **2005**, *19*, 114–126. [CrossRef] [PubMed]

108. Nishitani, H.; Sugimoto, N.; Roukos, V.; Nakanishi, Y.; Saijo, M.; Obuse, C.; Tsurimoto, T.; Nakayama, K.I.; Nakayama, K.; Fujita, M.; et al. Two E3 ubiquitin ligases, SCF-Skp2 and DDB1-Cul4, target human Cdt1 for proteolysis. *EMBO J.* **2006**, *25*, 1126–1136. [CrossRef] [PubMed]

109. Higa, L.A.; Banks, D.; Wu, M.; Kobayashi, R.; Sun, H.; Zhang, H. L2DTL/CDT2 interacts with the CUL4/DDB1 complex and PCNA and regulates CDT1 proteolysis in response to DNA damage. *Cell Cycle* **2006**, *5*, 1675–1680. [CrossRef] [PubMed]

110. Soria, G.; Gottifredi, V. PCNA-coupled p21 degradation after DNA damage: The exception that confirms the rule? *DNA Repair* **2010**, *9*, 358–364. [CrossRef] [PubMed]

111. Bendjennat, M.; Boulaire, J.; Jascur, T.; Brickner, H.; Barbier, V.; Sarasin, A.; Fotedar, A.; Fotedar, R. UV irradiation triggers ubiquitin-dependent degradation of p21(WAF1) to promote DNA repair. *Cell* **2003**, *114*, 599–610. [CrossRef] [PubMed]

112. Higa, L.A.; Mihaylov, I.S.; Banks, D.P.; Zheng, J.; Zhang, H. Radiation-mediated proteolysis of CDT1 by CUL4-ROC1 and CSN complexes constitutes a new checkpoint. *Nat. Cell Biol.* **2003**, *5*, 1008–1015. [CrossRef] [PubMed]

113. Abbas, T.; Sivaprasad, U.; Terai, K.; Amador, V.; Pagano, M.; Dutta, A. PCNA-dependent regulation of p21 ubiquitylation and degradation via the CRL4Cdt2 ubiquitin ligase complex. *Genes Dev.* **2008**, *22*, 2496–2506. [CrossRef] [PubMed]

114. Hu, J.; McCall, C.M.; Ohta, T.; Xiong, Y. Targeted ubiquitination of CDT1 by the DDB1-CUL4A-ROC1 ligase in response to DNA damage. *Nat. Cell Biol.* **2004**, *6*, 1003–1009. [CrossRef] [PubMed]

115. Hannah, J.; Zhou, P. Regulation of DNA damage response pathways by the cullin-RING ubiquitin ligases. *DNA Repair* **2009**, *8*, 536–543. [CrossRef] [PubMed]

116. Lin, J.J.; Milhollen, M.A.; Smith, P.G.; Narayanan, U.; Dutta, A. NEDD8-targeting drug MLN4924 elicits DNA rereplication by stabilizing Cdt1 in S phase, triggering checkpoint activation, apoptosis, and senescence in cancer cells. *Cancer Res.* **2010**, *70*, 10310–10320. [CrossRef] [PubMed]

117. Seiler, J.A.; Conti, C.; Syed, A.; Aladjem, M.I.; Pommier, Y. The intra-S-phase checkpoint affects both DNA replication initiation and elongation: Single-cell and -DNA fiber analyses. *Mol. Cell Biol.* **2007**, *27*, 5806–5818. [CrossRef] [PubMed]

118. Zhao, H.; Zhang, S.; Xu, D.; Lee, M.Y.; Zhang, Z.; Lee, E.Y.; Darzynkiewicz, Z. Expression of the p12 subunit of human DNA polymerase δ (Pol δ), CDK inhibitor p21(WAF1), Cdt1, cyclin A, PCNA and Ki-67 in relation to DNA replication in individual cells. *Cell Cycle* **2014**, *13*, 3529–3540. [CrossRef] [PubMed]

119. Darzynkiewicz, Z.; Zhao, H.; Zhang, S.; Lee, M.Y.; Lee, E.Y.; Zhang, Z. Initiation and termination of DNA replication during S phase in relation to cyclins D1, E and A, p21WAF1, Cdt1 and the p12 subunit of DNA polymerase δ revealed in individual cells by cytometry. *Oncotarget* **2015**, *6*, 11735–11750. [CrossRef] [PubMed]

120. Balakrishnan, L.; Bambara, R.A. Okazaki fragment metabolism. *Cold Spring Harb. Perspect. Biol.* **2013**, *5*, a10173. [CrossRef] [PubMed]

121. Garg, P.; Stith, C.M.; Sabouri, N.; Johansson, E.; Burgers, P.M. Idling by DNA polymerase δ maintains a ligatable nick during lagging-strand DNA replication. *Genes Dev.* **2004**, *18*, 2764–2773. [CrossRef] [PubMed]

122. Balakrishnan, L.; Bambara, R.A. FLAP Endonuclease 1. *Ann. Rev. Biochem.* **2013**, *82*, 119–138. [CrossRef] [PubMed]

123. Jin, Y.H.; Ayyagari, R.; Resnick, M.A.; Gordenin, D.A.; Burgers, P.M. Okazaki fragment maturation in yeast. II. Cooperation between the polymerase and 3′-5′-exonuclease activities of Pol δ in the creation of a ligatable nick. *J. Biol. Chem.* **2003**, *278*, 1626–1633. [CrossRef] [PubMed]

124. Burgers, P.M. It's all about flaps: DNA2 and checkpoint activation. *Cell Cycle* **2011**, *10*, 2417–2418. [CrossRef] [PubMed]

125. Howes, T.R.; Tomkinson, A.E. DNA ligase I, the replicative DNA ligase. *Subcell. Biochem.* **2012**, *62*, 327–341. [PubMed]

126. Kelly, R.B.; Cozzarelli, N.R.; Deutscher, M.P.; Lehman, I.R.; Kornberg, A. Enzymatic synthesis of deoxyribonucleic acid. XXXII. Replication of duplex deoxyribonucleic acid by polymerase at a single strand break. *J. Biol. Chem.* **1970**, *245*, 39–45. [PubMed]

127. Lin, S.H.; Wang, X.; Zhang, S.; Zhang, Z.; Lee, E.Y.; Lee, M.Y. Dynamics of Enzymatic Interactions during Short Flap Human Okazaki Fragment Processing by Two Forms of Human DNA Polymerase δ. *DNA Repair* **2013**, *12*, 922–935. [CrossRef] [PubMed]

128. Beattie, T.R.; Bell, S.D. Coordination of multiple enzyme activities by a single PCNA in archaeal Okazaki fragment maturation. *EMBO J.* **2012**, *31*, 1556–1567. [CrossRef] [PubMed]

129. Patel, S.S.; Wong, I.; Johnson, K.A. Pre-steady-state kinetic analysis of processive DNA replication including complete characterization of an exonuclease-deficient mutant. *Biochemistry* **1991**, *30*, 511–525. [CrossRef] [PubMed]

130. Langston, L.D.; O'Donnell, M. DNA polymerase δ is highly processive with proliferating cell nuclear antigen and undergoes collision release upon completing DNA. *J. Biol. Chem.* **2008**, *283*, 29522–29531. [CrossRef] [PubMed]

131. Burgers, P.M.; Kunkel, T.A. Eukaryotic DNA Replication Fork. *Annu. Rev. Biochem.* **2017**, *86*, 417–438. [CrossRef] [PubMed]

132. Huang, Q.M.; Akashi, T.; Masuda, Y.; Kamiya, K.; Takahashi, T.; Suzuki, M. Roles of POLD4, smallest subunit of DNA polymerase δ, in nuclear structures and genomic stability of human cells. *Biochem. Biophys. Res. Commun.* **2010**, *391*, 542–546. [CrossRef] [PubMed]

133. Huang, Q.; Suzuki, M.; Zeng, Y.; Zhang, H.; Yang, D.; Lin, H. Downregulation of POLD4 in Calu6 cells results in G1-S blockage through suppression of the Akt-Skp2-p27 pathway. *Bioorg. Med. Chem. Lett.* **2014**, *24*, 1780–1783. [CrossRef] [PubMed]

134. Liu, L.; Rodriguez-Belmonte, E.M.; Mazloum, N.; Xie, B.; Lee, M.Y. Identification of a novel protein, PDIP38, that interacts with the p50 subunit of DNA polymerase δ and proliferating cell nuclear antigen. *J. Biol. Chem.* **2003**, *278*, 10041–10047. [CrossRef] [PubMed]

135. He, H.; Tan, C.K.; Downey, K.M.; So, A.G. A tumor necrosis factor α- and interleukin 6-inducible protein that interacts with the small subunit of DNA polymerase δ and proliferating cell nuclear antigen. *Proc. Natl. Acad. Sci. USA* **2001**, *98*, 11979–11984. [CrossRef] [PubMed]

136. Richardson, C.J.; Broenstrup, M.; Fingar, D.C.; Julich, K.; Ballif, B.A.; Gygi, S.; Blenis, J. SKAR is a specific target of S6 kinase 1 in cell growth control. *Curr. Biol.* **2004**, *14*, 1540–1549. [CrossRef] [PubMed]

137. Magnuson, B.; Ekim, B.; Fingar, D.C. Regulation and function of ribosomal protein S6 kinase (S6K) within mTOR signalling networks. *Biochem. J.* **2012**, *441*, 1–21. [CrossRef] [PubMed]

138. Richardson, C.J.; Schalm, S.S.; Blenis, J. PI3-kinase and TOR: PIKTORing cell growth. *Semin. Cell Dev. Biol.* **2004**, *15*, 147–159. [CrossRef] [PubMed]

139. Banko, M.I.; Krzyzanowski, M.K.; Turcza, P.; Maniecka, Z.; Kulis, M.; Kozlowski, P. Identification of amino acid residues of ERH required for its recruitment to nuclear speckles and replication foci in HeLa cells. *PLoS ONE* **2013**, *8*, e74885. [CrossRef] [PubMed]

140. Ma, X.M.; Yoon, S.O.; Richardson, C.J.; Julich, K.; Blenis, J. SKAR links pre-mRNA splicing to mTOR/S6K1-mediated enhanced translation efficiency of spliced mRNAs. *Cell* **2008**, *133*, 303–313. [CrossRef] [PubMed]

141. Ma, X.M.; Blenis, J. Molecular mechanisms of mTOR-mediated translational control. *Nat. Rev. Mol. Cell Biol.* **2009**, *10*, 307–318. [CrossRef] [PubMed]

142. Fingar, D.C.; Salama, S.; Tsou, C.; Harlow, E.; Blenis, J. Mammalian cell size is controlled by mTOR and its downstream targets S6K1 and 4EBP1/eIF4E. *Genes Dev.* **2002**, *16*, 1472–1487. [CrossRef] [PubMed]

143. Smyk, A.; Szuminska, M.; Uniewicz, K.A.; Graves, L.M.; Kozlowski, P. Human enhancer of rudimentary is a molecular partner of PDIP46/SKAR, a protein interacting with DNA polymerase δ and S6K1 and regulating cell growth. *FEBS J.* **2006**, *273*, 4728–4741. [CrossRef] [PubMed]

144. Weng, M.T.; Luo, J. The enigmatic ERH protein: Its role in cell cycle, RNA splicing and cancer. *Protein Cell* **2013**, *4*, 807–812. [CrossRef] [PubMed]

145. Weng, M.T.; Tung, T.H.; Lee, J.H.; Wei, S.C.; Lin, H.L.; Huang, Y.J.; Wong, J.M.; Luo, J.; Sheu, J.C. Enhancer of rudimentary homolog regulates DNA damage response in hepatocellular carcinoma. *Sci. Rep.* **2015**, *5*, 9357. [CrossRef] [PubMed]

146. Kavanaugh, G.; Zhao, R.; Guo, Y.; Mohni, K.N.; Glick, G.; Lacy, M.E.; Hutson, M.S.; Ascano, M.; Cortez, D. Enhancer of Rudimentary Homolog affects the replication stress response through regulation of RNA processing. *Mol. Cell Biol.* **2015**, *35*, 2979–2990. [CrossRef] [PubMed]

147. Gilljam, K.M.; Feyzi, E.; Aas, P.A.; Sousa, M.M.; Muller, R.; Vagbo, C.B.; Catterall, T.C.; Liabakk, N.B.; Slupphaug, G.; Drablos, F.; et al. Identification of a novel, widespread, and functionally important PCNA-binding motif. *J. Cell Biol.* **2009**, *186*, 645–654. [CrossRef] [PubMed]

148. Gilljam, K.M.; Muller, R.; Liabakk, N.B.; Otterlei, M. Nucleotide excision repair is associated with the replisome and its efficiency depends on a direct interaction between XPA and PCNA. *PLoS ONE* **2012**, *7*, e49199. [CrossRef] [PubMed]

149. Bacquin, A.; Pouvelle, C.; Siaud, N.; Perderiset, M.; Salome-Desnoulez, S.; Tellier-Lebegue, C.; Lopez, B.; Charbonnier, J.B.; Kannouche, P.L. The helicase FBH1 is tightly regulated by PCNA via CRL4(Cdt2)-mediated proteolysis in human cells. *Nucleic Acids Res.* **2013**, *41*, 6501–6513. [CrossRef] [PubMed]

150. Fu, D.; Samson, L.D.; Hubscher, U.; van Loon, B. The interaction between ALKBH2 DNA repair enzyme and PCNA is direct, mediated by the hydrophobic pocket of PCNA and perturbed in naturally-occurring ALKBH2 variants. *DNA Repair* **2015**, *35*, 13–18. [CrossRef] [PubMed]

151. O'Donnell, M.; Li, H. The Eukaryotic Replisome Goes Under the Microscope. *Curr. Biol.* **2016**, *26*, R247–R256. [CrossRef] [PubMed]

152. Kang, Y.H.; Farina, A.; Bermudez, V.P.; Tappin, I.; Du, F.; Galal, W.C.; Hurwitz, J. Interaction between human Ctf4 and the Cdc45/Mcm2-7/GINS (CMG) replicative helicase. *Proc. Natl. Acad. Sci. USA* **2013**, *110*, 19760–19765. [CrossRef] [PubMed]

153. Villa, F.; Simon, A.C.; Ortiz Bazan, M.A.; Kilkenny, M.L.; Wirthensohn, D.; Wightman, M.; Matak-Vinkovic, D.; Pellegrini, L.; Labib, K. Ctf4 Is a Hub in the Eukaryotic Replisome that Links Multiple CIP-Box Proteins to the CMG Helicase. *Mol. Cell* **2016**, *63*, 385–396. [CrossRef] [PubMed]

154. Xie, B.; Li, H.; Wang, Q.; Xie, S.; Rahmeh, A.; Dai, W.; Lee, M.Y. Further characterization of human DNA polymerase δ interacting protein 38. *J. Biol. Chem.* **2005**, *280*, 22375–22384. [CrossRef] [PubMed]

155. Pursell, Z.F.; Isoz, I.; Lundstrom, E.B.; Johansson, E.; Kunkel, T.A. Yeast DNA polymerase epsilon participates in leading-strand DNA replication. *Science* **2007**, *317*, 127–130. [CrossRef] [PubMed]

156. Nick McElhinny, S.A.; Gordenin, D.A.; Stith, C.M.; Burgers, P.M.; Kunkel, T.A. Division of labor at the eukaryotic replication fork. *Mol. Cell* **2008**, *30*, 137–144. [CrossRef] [PubMed]

157. Larrea, A.A.; Lujan, S.A.; Nick McElhinny, S.A.; Mieczkowski, P.A.; Resnick, M.A.; Gordenin, D.A.; Kunkel, T.A. Genome-wide model for the normal eukaryotic DNA replication fork. *Proc. Natl. Acad. Sci. USA* **2010**, *107*, 17674–17679. [CrossRef] [PubMed]

158. Johnson, R.E.; Klassen, R.; Prakash, L.; Prakash, S. A Major Role of DNA Polymerase δ in Replication of Both the Leading and Lagging DNA Strands. *Mol. Cell* **2015**, *59*, 163–175. [CrossRef] [PubMed]

159. Eckert, K.A.; Hile, S.E. Every microsatellite is different: Intrinsic DNA features dictate mutagenesis of common microsatellites present in the human genome. *Mol. Carcinog.* **2009**, *48*, 379–388. [CrossRef] [PubMed]

160. Le, H.P.; Masuda, Y.; Tsurimoto, T.; Maki, S.; Katayama, T.; Furukohri, A.; Maki, H. Short CCG repeat in huntingtin gene is an obstacle for replicative DNA polymerases, potentially hampering progression of replication fork. *Genes Cells* **2015**, *20*, 817–833. [CrossRef] [PubMed]

161. Hile, S.E.; Wang, X.; Lee, M.Y.; Eckert, K.A. Beyond translesion synthesis: Polymerase k fidelity as a potential determinant of microsatellite stability. *Nucleic Acids Res.* **2012**, *40*, 1636–1647. [CrossRef] [PubMed]

162. Baptiste, B.A.; Eckert, K.A. DNA polymerase k microsatellite synthesis: Two distinct mechanisms of slippage-mediated errors. *Environ. Mol. Mutagen.* **2012**, *53*, 787–796. [CrossRef] [PubMed]

163. Rey, L.; Sidorova, J.M.; Puget, N.; Boudsocq, F.; Biard, D.S.; Monnat, R.J.J.; Cazaux, C.; Hoffmann, J.S. Human DNA polymerase eta is required for common fragile site stability during unperturbed DNA replication. *Mol. Cell Biol.* **2009**, *29*, 3344–3354. [CrossRef] [PubMed]

164. Bergoglio, V.; Boyer, A.S.; Walsh, E.; Naim, V.; Legube, G.; Lee, M.Y.; Rey, L.; Rosselli, F.; Cazaux, C.; Eckert, K.A.; et al. DNA synthesis by Pol eta promotes fragile site stability by preventing under-replicated DNA in mitosis. *J. Cell Biol.* **2013**, *201*, 395–408. [CrossRef] [PubMed]

165. Garcia-Exposito, L.; Bournique, E.; Bergoglio, V.; Bose, A.; Barroso-Gonzalez, J.; Zhang, S.; Roncaioli, J.L.; Lee, M.; Wallace, C.T.; Watkins, S.C.; et al. Proteomic Profiling Reveals a Specific Role for Translesion DNA Polymerase eta in the Alternative Lengthening of Telomeres. *Cell Rep.* **2016**, *17*, 1858–1871. [CrossRef] [PubMed]

166. Bermudez, V.P.; Farina, A.; Raghavan, V.; Tappin, I.; Hurwitz, J. Studies on Human DNA Polymerase epsilon and GINS Complex and Their Role in DNA Replication. *J. Biol. Chem.* **2011**, *286*, 28963–28977. [CrossRef] [PubMed]

167. Ganai, R.A.; Zhang, X.P.; Heyer, W.D.; Johansson, E. Strand displacement synthesis by yeast DNA polymerase epsilon. *Nucleic Acids Res.* **2016**, *44*, 8229–8240. [CrossRef] [PubMed]

168. Yeeles, J.T.; Janska, A.; Early, A.; Diffley, J.F. How the Eukaryotic Replisome Achieves Rapid and Efficient DNA Replication. *Mol. Cell* **2017**, *65*, 105–116. [CrossRef] [PubMed]

169. Zhou, J.C.; Janska, A.; Goswami, P.; Renault, L.; Abid Ali, F.; Kotecha, A.; Diffley, J.F.X.; Costa, A. CMG-Pol epsilon dynamics suggests a mechanism for the establishment of leading-strand synthesis in the eukaryotic replisome. *Proc. Natl. Acad. Sci. USA* **2017**, *114*, 4141–4146. [CrossRef] [PubMed]

170. Kurat, C.F.; Yeeles, J.T.; Patel, H.; Early, A.; Diffley, J.F. Chromatin Controls DNA Replication Origin Selection, Lagging-Strand Synthesis, and Replication Fork Rates. *Mol. Cell* **2017**, *65*, 117–130. [CrossRef] [PubMed]

171. Yeeles, J.T.; Deegan, T.D.; Janska, A.; Early, A.; Diffley, J.F. Regulated eukaryotic DNA replication origin firing with purified proteins. *Nature* **2015**, *519*, 431–435. [CrossRef] [PubMed]

172. Schauer, G.D.; O'Donnell, M.E. Quality control mechanisms exclude incorrect polymerases from the eukaryotic replication fork. *Proc. Natl. Acad. Sci. USA* **2017**, *114*, 675–680. [CrossRef] [PubMed]

173. Yurieva, O.; O'Donnell, M. Reconstitution of a eukaryotic replisome reveals the mechanism of asymmetric distribution of DNA polymerases. *Nucleus* **2016**, *7*, 360–368. [CrossRef] [PubMed]

174. Hedglin, M.; Pandey, B.; Benkovic, S.J. Stability of the human polymerase δ holoenzyme and its implications in lagging strand DNA synthesis. *Proc. Natl. Acad. Sci. USA* **2016**, *113*, E1777–E1786. [CrossRef] [PubMed]

175. Beattie, T.R.; Kapadia, N.; Nicolas, E.; Uphoff, S.; Wollman, A.J.; Leake, M.C.; Reyes-Lamothe, R. Frequent exchange of the DNA polymerase during bacterial chromosome replication. *Elife* **2017**, *6*, e21763. [CrossRef] [PubMed]

176. Li, J.J.; Kelly, T.J. Simian virus 40 DNA replication in vitro. *Proc. Natl. Acad. Sci. USA* **1984**, *81*, 6973–6977. [CrossRef] [PubMed]

177. Waga, S.; Stillman, B. Anatomy of a DNA replication fork revealed by reconstitution of SV40 DNA replication in vitro. *Nature* **1994**, *369*, 207–212. [CrossRef] [PubMed]

178. Rytkonen, A.K.; Vaara, M.; Nethanel, T.; Kaufmann, G.; Sormunen, R.; Laara, E.; Nasheuer, H.P.; Rahmeh, A.; Lee, M.Y.; Syvaoja, J.E.; et al. Distinctive activities of DNA polymerases during human DNA replication. *FEBS J.* **2006**, *273*, 2984–3001. [CrossRef] [PubMed]

179. Friedberg, E.C.; Lehmann, A.R.; Fuchs, R.P. Trading places: How do DNA polymerases switch during translesion DNA synthesis? *Mol. Cell* **2005**, *18*, 499–505. [CrossRef] [PubMed]

180. Lehmann, A.R. Translesion synthesis in mammalian cells. *Exp. Cell Res.* **2006**, *312*, 2673–2676. [CrossRef] [PubMed]

181. Guilliam, T.A.; Doherty, A.J. PrimPol-Prime Time to Reprime. *Genes* **2017**, *8*, 20. [CrossRef] [PubMed]

182. Khan, F.H.; Pandian, V.; Ramraj, S.; Natarajan, M.; Aravindan, S.; Herman, T.S.; Aravindan, N. Acquired genetic alterations in tumor cells dictate the development of high-risk neuroblastoma and clinical outcomes. *BMC Cancer* **2015**, *15*, 514. [CrossRef] [PubMed]

183. Human Protein Atlas. Available online: http://www.proteinatlas.org/ENSG00000100227-POLDIP3/cancer (accessed on 9 April 2017).

184. COSMIC (Catalogue of Somatic Mutations in Cancer). Available online: http://cancer.sanger.ac.uk/cosmic (accessed on 9 April 2017).

185. Gakh, O.; Cavadini, P.; Isaya, G. Mitochondrial processing peptidases. *Biochim. Biophys. Acta* **2002**, *1592*, 63–77. [CrossRef]

186. Cheng, X.; Kanki, T.; Fukuoh, A.; Ohgaki, K.; Takeya, R.; Aoki, Y.; Hamasaki, N.; Kang, D. PDIP38 associates with proteins constituting the mitochondrial DNA nucleoid. *J. Biochem.* **2005**, *138*, 673–678. [CrossRef] [PubMed]

187. Arakaki, N.; Nishihama, T.; Kohda, A.; Owaki, H.; Kuramoto, Y.; Abe, R.; Kita, T.; Suenaga, M.; Himeda, T.; Kuwajima, M.; et al. Regulation of mitochondrial morphology and cell survival by Mitogenin I and mitochondrial single-stranded DNA binding protein. *Biochim. Biophys. Acta* **2006**, *1760*, 1364–1372. [CrossRef] [PubMed]

188. Klaile, E.; Muller, M.M.; Kannicht, C.; Otto, W.; Singer, B.B.; Reutter, W.; Obrink, B.; Lucka, L. The cell adhesion receptor carcinoembryonic antigen-related cell adhesion molecule 1 regulates nucleocytoplasmic trafficking of DNA polymerase δ-interacting protein 38. *J. Biol. Chem.* **2007**, *282*, 26629–26640. [CrossRef] [PubMed]

189. Klaile, E.; Kukalev, A.; Obrink, B.; Muller, M.M. PDIP38 is a novel mitotic spindle-associated protein that affects spindle organization and chromosome segregation. *Cell Cycle* **2008**, *7*, 3180–3186. [CrossRef] [PubMed]

190. Tissier, A.; Janel-Bintz, R.; Coulon, S.; Klaile, E.; Kannouche, P.; Fuchs, R.P.; Cordonnier, A.M. Crosstalk between replicative and translesional DNA polymerases: PDIP38 interacts directly with Poleta. *DNA Repair* **2010**, *9*, 922–928. [CrossRef] [PubMed]

191. Maga, G.; Crespan, E.; Markkanen, E.; Imhof, R.; Furrer, A.; Villani, G.; Hubscher, U.; van Loon, B. DNA polymerase δ-interacting protein 2 is a processivity factor for DNA polymerase lambda during 8-oxo-7,8-dihydroguanine bypass. *Proc. Natl. Acad. Sci. USA* **2013**, *110*, 18850–18855. [CrossRef] [PubMed]

192. Bebenek, K.; Pedersen, L.C.; Kunkel, T.A. Structure-Function Studies of DNA Polymerase lambda. *Biochemistry* **2014**, *53*, 2781–2792. [CrossRef] [PubMed]

193. Mentegari, E.; Kissova, M.; Bavagnoli, L.; Maga, G.; Crespan, E. DNA Polymerases lambda and beta: The Double-Edged Swords of DNA Repair. *Genes* **2016**, *7*, 57. [CrossRef] [PubMed]

194. Waters, L.S.; Minesinger, B.K.; Wiltrout, M.E.; D'Souza, S.; Woodruff, R.V.; Walker, G.C. Eukaryotic translesion polymerases and their roles and regulation in DNA damage tolerance. *Microbiol. Mol. Biol. Rev.* **2009**, *73*, 134–154. [CrossRef] [PubMed]

195. Braithwaite, E.K.; Kedar, P.S.; Stumpo, D.J.; Bertocci, B.; Freedman, J.H.; Samson, L.D.; Wilson, S.H. DNA polymerases beta and lambda mediate overlapping and independent roles in base excision repair in mouse embryonic fibroblasts. *PLoS ONE* **2010**, *5*, e12229. [CrossRef] [PubMed]

196. Braithwaite, E.K.; Kedar, P.S.; Lan, L.; Polosina, Y.Y.; Asagoshi, K.; Poltoratsky, V.P.; Horton, J.K.; Miller, H.; Teebor, G.W.; Yasui, A.; et al. DNA polymerase lambda protects mouse fibroblasts against oxidative DNA damage and is recruited to sites of DNA damage/repair. *J. Biol. Chem.* **2005**, *280*, 31641–31647. [CrossRef] [PubMed]

197. Hubscher, U.; Maga, G. DNA replication and repair bypass machines. *Curr. Opin. Chem. Biol.* **2011**, *15*, 627–635. [CrossRef] [PubMed]

198. Guilliam, T.A.; Bailey, L.J.; Brissett, N.C.; Doherty, A.J. PolDIP2 interacts with human PrimPol and enhances its DNA polymerase activities. *Nucleic Acids Res.* **2016**, *44*, 3317–3329. [CrossRef] [PubMed]

199. Wong, A.; Zhang, S.; Mordue, D.; Wu, J.M.; Zhang, Z.; Darzynkiewicz, Z.; Lee, E.Y.; Lee, M.Y. PDIP38 is translocated to the spliceosomes/nuclear speckles in response to UV-induced DNA damage and is required for UV-induced alternative splicing of MDM2. *Cell Cycle* **2013**, *12*, 3184–3193. [CrossRef] [PubMed]

200. Spector, D.L.; Lamond, A.I. Nuclear speckles. *Cold Spring Harb. Perspect. Biol.* **2011**, *3*, a000646. [CrossRef] [PubMed]

201. Campalans, A.; Amouroux, R.; Bravard, A.; Epe, B.; Radicella, J.P. UVA irradiation induces relocalisation of the DNA repair protein hOGG1 to nuclear speckles. *J. Cell Sci.* **2007**, *120*, 23–32. [CrossRef] [PubMed]

202. Dutertre, M.; Sanchez, G.; Barbier, J.; Corcos, L.; Auboeuf, D. The emerging role of pre-messenger RNA splicing in stress responses: Sending alternative messages and silent messengers. *RNA Biol.* **2011**, *8*, 740–747. [CrossRef] [PubMed]

203. Giono, L.E.; Nieto Moreno, N.; Cambindo Botto, A.E.; Dujardin, G.; Munoz, M.J.; Kornblihtt, A.R. The RNA Response to DNA Damage. *J. Mol. Biol.* **2016**, *428*, 2636–2651. [CrossRef] [PubMed]

204. Bartel, F.; Taubert, H.; Harris, L.C. Alternative and aberrant splicing of MDM2 mRNA in human cancer. *Cancer Cell* **2002**, *2*, 9–15. [CrossRef]

205. Chandler, D.S.; Singh, R.K.; Caldwell, L.C.; Bitler, J.L.; Lozano, G. Genotoxic stress induces coordinately regulated alternative splicing of the p53 modulators MDM2 and MDM4. *Cancer Res.* **2006**, *66*, 9502–9508. [CrossRef] [PubMed]

206. Jeyaraj, S.; O'Brien, D.M.; Chandler, D.S. MDM2 and MDM4 splicing: An integral part of the cancer spliceome. *Front. Biosci.* **2009**, *14*, 2647–2656. [CrossRef]

207. Okoro, D.R.; Rosso, M.; Bargonetti, J. Splicing up mdm2 for cancer proteome diversity. *Genes Cancer* **2012**, *3*, 311–319. [CrossRef] [PubMed]

208. Jones, S.N.; Hancock, A.R.; Vogel, H.; Donehower, L.A.; Bradley, A. Overexpression of Mdm2 in mice reveals a p53-independent role for Mdm2 in tumorigenesis. *Proc. Natl. Acad. Sci. USA* **1998**, *95*, 15608–15612. [CrossRef] [PubMed]

209. Matsumoto, R.; Tada, M.; Nozaki, M.; Zhang, C.L.; Sawamura, Y.; Abe, H. Short alternative splice transcripts of the mdm2 oncogene correlate to malignancy in human astrocytic neoplasms. *Cancer Res.* **1998**, *58*, 609–613. [PubMed]

210. Naro, C.; Bielli, P.; Pagliarini, V.; Sette, C. The interplay between DNA damage response and RNA processing: the unexpected role of splicing factors as gatekeepers of genome stability. *Front. Genet.* **2015**, *6*, 142. [CrossRef] [PubMed]

211. Montecucco, A.; Biamonti, G. Pre-mRNA processing factors meet the DNA damage response. *Front. Genet.* **2013**, *4*, 102. [CrossRef] [PubMed]

212. Lambeth, J.D.; Neish, A.S. Nox enzymes and new thinking on reactive oxygen: A double-edged sword revisited. *Annu. Rev. Pathol.* **2014**, *9*, 119–145. [CrossRef] [PubMed]

213. Bedard, K.; Krause, K.H. The NOX family of ROS-generating NADPH oxidases: Physiology and pathophysiology. *Physiol. Rev.* **2007**, *87*, 245–313. [CrossRef] [PubMed]

214. Lambeth, J.D.; Kawahara, T.; Diebold, B. Regulation of Nox and Duox enzymatic activity and expression. *Free Radic. Biol. Med.* **2007**, *43*, 319–331. [CrossRef]

215. Brown, D.I.; Griendling, K.K. Regulation of signal transduction by reactive oxygen species in the cardiovascular system. *Circ. Res.* **2015**, *116*, 531–549. [CrossRef]

216. Valko, M.; Leibfritz, D.; Moncol, J.; Cronin, M.T.; Mazur, M.; Telser, J. Free radicals and antioxidants in normal physiological functions and human disease. *Int J. Biochem. Cell Biol.* **2007**, *39*, 44–84. [CrossRef]

217. Lambeth, J.D. Nox enzymes, ROS, and chronic disease: An example of antagonistic pleiotropy. *Free Radic. Biol. Med.* **2007**, *43*, 332–347. [CrossRef]

218. Nisimoto, Y.; Diebold, B.A.; Cosentino-Gomes, D.; Lambeth, J.D. Nox4: A hydrogen peroxide-generating oxygen sensor. *Biochemistry* **2014**, *53*, 5111–5120. [CrossRef] [PubMed]

219. Lassegue, B.; Griendling, K.K. NADPH Oxidases: Functions and Pathologies in the Vasculature. *Arterioscler. Thromb. Vasc. Biol.* **2010**, *30*, 653–661. [CrossRef] [PubMed]

220. Lyle, A.N.; Deshpande, N.N.; Taniyama, Y.; Seidel-Rogol, B.; Pounkova, L.; Du, P.; Papaharalambus, C.; Lassegue, B.; Griendling, K.K. Poldip2, a novel regulator of Nox4 and cytoskeletal integrity in vascular smooth muscle cells. *Circ. Res.* **2009**, *105*, 249–259. [CrossRef] [PubMed]

221. Datla, S.R.; McGrail, D.J.; Vukelic, S.; Huff, L.P.; Lyle, A.N.; Pounkova, L.; Lee, M.; Seidel-Rogol, B.; Khalil, M.K.; Hilenski, L.L.; et al. Poldip2 controls vascular smooth muscle cell migration by regulating focal adhesion turnover and force polarization. *Am. J. Physiol. Heart Circ. Physiol.* **2014**, *307*, H945–H957. [CrossRef] [PubMed]

222. Sutliff, R.L.; Hilenski, L.L.; Amanso, A.M.; Parastatidis, I.; Dikalova, A.E.; Hansen, L.; Datla, S.R.; Long, J.S.; El-Ali, A.M.; Joseph, G.; et al. Polymerase Delta Interacting Protein 2 Sustains Vascular Structure and Function. *Arterioscler. Thromb. Vasc. Biol.* **2013**, *33*, 2154–2161. [CrossRef] [PubMed]

223. Brown, D.I.; Lassegue, B.; Lee, M.; Zafari, R.; Long, J.S.; Saavedra, H.I.; Griendling, K.K. Poldip2 knockout results in perinatal lethality, reduced cellular growth and increased autophagy of mouse embryonic fibroblasts. *PLoS ONE* **2014**, *9*, e96657. [CrossRef] [PubMed]

GCAT
TACG
GCAT

genes

MDPI

Review

Mechanisms of Post-Replication DNA Repair

Yanzhe Gao [1,*], Elizabeth Mutter-Rottmayer [1,2], Anastasia Zlatanou [1], Cyrus Vaziri [1] and Yang Yang [1]

[1] Department of Pathology and Laboratory Medicine, University of North Carolina at Chapel Hill, Chapel Hill, NC 27599, USA; evmutter@email.unc.edu (E.M.-R.); anastasia_zlatanou@med.unc.edu (A.Z.); cyrus_vaziri@med.unc.edu (C.V.); yang_yang1@med.unc.edu (Y.Y.)

[2] Curriculum in Toxicology, University of North Carolina at Chapel Hill, Chapel Hill, NC 27599, USA

* Correspondence: yanzhe_gao@med.unc.edu

Academic Editor: Eishi Noguchi
Received: 5 December 2016; Accepted: 3 February 2017; Published: 8 February 2017

Abstract: Accurate DNA replication is crucial for cell survival and the maintenance of genome stability. Cells have developed mechanisms to cope with the frequent genotoxic injuries that arise from both endogenous and environmental sources. Lesions encountered during DNA replication are often tolerated by post-replication repair mechanisms that prevent replication fork collapse and avert the formation of DNA double strand breaks. There are two predominant post-replication repair pathways, trans-lesion synthesis (TLS) and template switching (TS). TLS is a DNA damage-tolerant and low-fidelity mode of DNA synthesis that utilizes specialized 'Y-family' DNA polymerases to replicate damaged templates. TS, however, is an error-free 'DNA damage avoidance' mode of DNA synthesis that uses a newly synthesized sister chromatid as a template in lieu of the damaged parent strand. Both TLS and TS pathways are tightly controlled signaling cascades that integrate DNA synthesis with the overall DNA damage response and are thus crucial for genome stability. This review will cover the current knowledge of the primary mediators of post-replication repair and how they are regulated in the cell.

Keywords: DNA damage tolerance; post replication repair; DNA damage response; trans-lesion synthesis; template switching

1. Introduction

Accurate and efficient DNA replication is crucial for the health and survival of all living organisms. Under optimal conditions, the replicative DNA polymerases ε, δ, and α can work in concert to ensure that the genome is replicated efficiently with high accuracy in every cell cycle [1]. However, DNA is constantly challenged by exogenous and endogenous genotoxic threats, including solar ultraviolet (UV) radiation and reactive oxygen species (ROS) generated as a byproduct of cellular metabolism. Damaged DNA can act as a steric block to replicative polymerases, thereby leading to incomplete DNA replication or the formation of secondary DNA strand breaks at the sites of replication stalling. Incomplete DNA synthesis and DNA strand breaks are both potential sources of genomic instability [2]. As discussed elsewhere in this special issue, an arsenal of DNA repair mechanisms exists to repair various forms of damaged DNA and minimize genomic instability. Most DNA repair mechanisms require an intact DNA strand as template to fix the damaged strand. In this review, we will discuss the mechanisms behind Post-Replication Repair (PRR) that specifically help cells tolerate damage on the single stranded DNA template.

2. DNA Damage Repair and Complications at the Replication Fork

DNA damage can be categorized by structural changes in the DNA such as base alteration, single stranded break (SSB), and double stranded break (DSB), each repaired via a distinct mechanism [3]. As summarized in Figure 1, a broad spectrum of DNA repair mechanisms has evolved to remove lesions that occur on double stranded DNA. Most DNA repair mechanisms rely on information from an undamaged DNA strand, either the complementary strand of the double helix (nucleotide excision repair (NER), base excision repair (BER) and SSB repair) or the sister chromatid and homologous allele (homologous recombination). Utilizing an undamaged template prevents aberrant alteration of the genetic coding on the damaged DNA strand. A major limitation to template-based repair mechanisms is that sometimes an undamaged DNA template strand is unavailable. This problem is frequently encountered during DNA replication, in the synthesis (S) phase of the cell cycle.

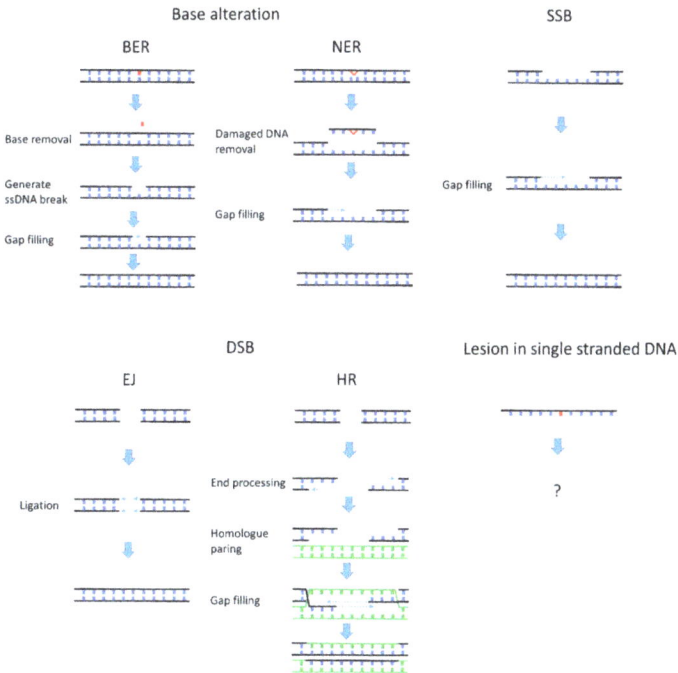

Figure 1. Many mechanisms efficiently repair DNA damage on the DNA double helix. Lesions in the double stranded DNA can be efficiently repaired by mechanisms corresponding to the specific type of DNA damage. Base-specific damage can be directly reversed by particular enzymes such as photolyases and O6-methylguanine DNA methyltransferase (MGMT) (reviewed in [4]). The majority of base-specific damage is repaired by base excision repair (BER) and nucleotide excision repair (NER). In BER and NER, the damaged base or surrounding DNA is excised from the double stranded DNA. The gap left behind is then filled by a DNA polymerase. Single stranded breaks (SSBs) are recognized by poly(ADP-ribose) polymerase 1 (PARP1), which activates downstream signaling that leads to gap-filling by DNA polymerases. Double stranded breaks (DSBs) are repaired by end joining (EJ) or by homologous recombination (HR). EJ directly ligates the exposed DSB with DNA ligase, while, during HR, break sites are replicated using undamaged homologous sequences of sister chromatid templates. In contrast, DNA lesions in single stranded DNA (ssDNA) cannot be repaired by BER, NER, HR, or EJ and must be remediated using alternative mechanisms (as suggested by the question mark in the figure). Post replication repair is a mechanism specialized in tolerating lesions in single stranded template.

DNA replication is a multistep process with two key events; (1) unwinding of the annealed double helix to expose ssDNA and (2) using this ssDNA as template to synthesize daughter strands. During an unperturbed S phase, DNA unwinding, carried out by the replicative helicase (the CDC45-MCM2-7-GINS or "CMG" complex), is strictly coupled with polymerase activity at replication forks (reviewed in [5]). Three replicative polymerases, pol ε on the leading strand and pol δ and pol α on the lagging strand, copy the template DNA with an error rate less than 10^{-4} [1]. The compact catalytic sites of replicative DNA polymerases confer high fidelity but preclude DNA damage-tolerant synthesis when using templates harboring bulky DNA lesions. As a result, replicative polymerases stall when a lesion is encountered (Figure 2). Fork-stalling DNA lesions are very prevalent in cells. In the human body there are approximately 30,000 lesions in every cell at any given time due to aerobic metabolism and endogenous depurination and deamination events [6]. It is inevitable that replication forks will be challenged by fork stalling lesions during DNA synthesis. Lesions encountered at replicating DNA are unique because the DNA in the vicinity of a replication fork is not double-helical. Excising the lesion from the ssDNA, as seen in BER and NER, will generate DNA strand breaks and result in fork collapse.

To survive fork-stalling DNA lesions, cells have developed post replication DNA repair mechanisms (PRR), which allow replication forks to progress through the lesions on damaged templates. The main role of PRR is to "patch" ssDNA gaps in the daughter strand and restore DNA to its double-stranded state for subsequent DNA repair via other mechanisms (covered in Figures 2 and 3 and later sections of the paper). There are two mechanisms of PRR; trans-lesion synthesis (TLS), which employs TLS polymerases to directly replicate across the DNA lesion [7], and template switching (TS), which "borrows" the genetic information from the newly synthesized sister chromatid as a replication template [8] and thus avoids the lesion (Figure 3). As described in detail below, the TLS and TS pathways are coordinated to facilitate ongoing DNA synthesis on damaged genomes.

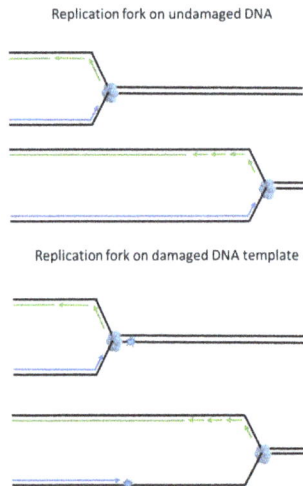

Figure 2. DNA lesions in single stranded DNA (ssDNA) are detrimental to the replication fork. DNA lesions on an ssDNA template act as road blocks for replicative polymerases but not for the replicative DNA helicase. An uncoupling of replicative DNA polymerase and DNA helicase activities generates single stranded DNA tracts. Persistent ssDNA is fragile and prone to breakage, generating lethal DSB.

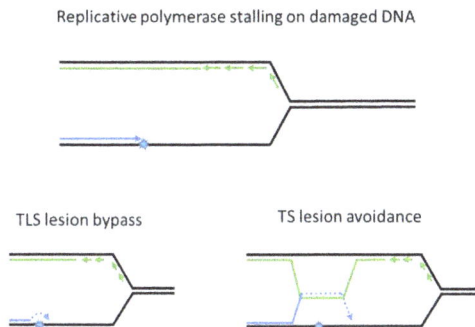

Replicative polymerase stalling on damaged DNA

TLS lesion bypass   TS lesion avoidance

Figure 3. Post replication repair efficiently returns lesions in ssDNA to double helix. PRR utilizes trans-lesion synthesis (TLS) or template switching (TS) to bypass or avoid DNA lesions and prevent accumulation of ssDNA gaps. After being restored to its double-stranded state, damaged DNA may be repaired via the mechanisms described in Figure 1.

3. Activation of Post Replication Repair

Proliferating cell nuclear antigen (PCNA) is a ring shaped homo-trimeric protein complex that surrounds the DNA and is a central player in PRR. During the initiation of DNA replication, PCNA is loaded onto the chromatin by the Replication Factor C (RFC) clamp loader [9]. Upon completion of DNA replication, ATAD5 (Elg1 in yeast) unloads PCNA from chromatin [10,11]. Chromatin-bound PCNA slides along the DNA strand and serves as a processivity factor for DNA polymerases. In addition to tethering polymerases to template DNA, PCNA is a platform for a wide variety of proteins that participate in DNA replication and damage repair [12]. The interactions between PCNA and its binding partners are typically mediated by the PCNA inter-domain connecting loop (IDCL) and the PCNA interacting peptide (PIP) motif on its binding partner [13]. PCNA can also be modified by ubiquitin and SUMO (Small Ubiquitin Modifier) to create additional interfaces for binding partners during the S-phase or when the replication fork is under stress [12,14,15]. These post-translational modifications of PCNA are crucial events in PRR.

When the replication fork encounters a bulky DNA lesion, replicative polymerases stall but the MCM helicases continue unwinding the double helix ahead of the polymerase. The uncoupling of DNA polymerase and helicase at the stalled replication fork generates long stretches of ssDNA covered by replication protein A (RPA) that activates the DNA replication checkpoint [16]. RPA-coated ssDNA generated by fork-stalling recruits Rad18 (a PCNA-directed E3 ubiquitin ligase) to the vicinity of the DNA lesion [17–19]. Chromatin-bound Rad18 and its associated E2 ubiquitin-conjugating enzyme (Rad6) mono-ubiquitinate PCNA at the conserved residue, K164 [14,15]. Mono-ubiquitinated PCNA initiates PRR by recruiting TLS polymerases to replace the activity of replicative polymerase at the stalled replication fork [20–22]. Although RPA-coated ssDNA is necessary for Rad18 chromatin-binding, multiple regulators have been shown to modulate the recruitment of Rad18 to PCNA. For example, TLS Pol η can facilitate the Rad18-PCNA interaction by binding to both proteins with its C-terminus domain and enhancing PCNA ubiquitination [23]. NBS1 (mutated in Nijmegen Breakage Syndrome) interacts with Rad18 at the Rad6-interacting domain to help recruit Rad18 to damaged DNA [24]. Additionally, BRCA1 facilitates efficient recruitment of RPA and Rad18 to damaged DNA and promotes PRR [25]. SIVA1 physically bridges chromatin-bound Rad18 and its substrate PCNA and promotes PCNA ubiquitination. However, SIVA1 is not required for Rad18 recruitment to DNA damage sites [26]. Spartan/DVC1 interacts with Rad18 and PCNA and is necessary for UV-tolerance, although the molecular mechanism by which Spartan/DVC1 regulates PRR is unclear [27–31]. Of note, Rad18 might not be the only enzyme that mono-ubiquitinates PCNA.

PCNA mono-ubiquitination has been observed in both Rad18$^{-/-}$ DT40 cells and in Rad18 KO mice, suggesting other E3 ubiquitin ligases may use PCNA as a substrate [32–34].

PCNA K164 mono-ubiquitination can be further extended to K63-linked poly-ubiquitin chains by another E3 ubiquitin-ligase, which in yeast is Rad5 [14]. The interaction between Rad18 and Rad5 brings Ubc13/Mms2-Rad5 to the vicinity of stalled replication forks [35]. Ubc13/Mms2 and Rad5-mediated poly-ubiquitination of PCNA directs lesion avoidance using the TS pathway [36]. In *Xenopus laevis*, PCNA poly-ubiquitination is induced by DNA damage, although it is unclear why PCNA is also modified when replicating undamaged DNA in this system [37]. In humans, the two human Rad5 orthologues, SNF2 histone-linker PHD-finger RING-finger helicase (SHPRH) [38,39] and helicase-like transcription factor (HLTF) [40,41], mediate PCNA poly-ubiquitination. Although PCNA poly-ubiquitination is a less abundant modification than PCNA mono-ubiquitination [42], it is clear that the human RAD5 homologues do contribute to DNA damage tolerance [43]. There are also some studies suggesting that Rad5 has an Mms2-Ubc13-independent role in the TLS pathway [44–46].

SUMOylation of PCNA, catalyzed by UBC9 and Siz1/2, has also been observed on lysines K127 and/or K164 during normal DNA replication or following sub-lethal DNA-damaging treatments [14,15]. SUMOylated PCNA interacts with Srs2, a helicase that displaces the Rad51 recombinase from ssDNA. Since Rad51 is essential for DNA repair via homologous recombination, Rad51 displacement prevents recombinational repair [47,48]. Consequently, the inhibition of homologous recombination by Srs2 at the replication fork further limits the pathway choices to PRR when a DNA lesion is encountered on the single stranded template [49]. Interestingly, PCNA SUMOylation also has been shown to facilitate Rad18 E3 ligase activity towards PCNA by physically linking Rad18 and PCNA in yeast [50]. Although this is not an evolutionarily conserved mechanism for Rad18 activation, it still exemplifies the cross talk between different PCNA modifications.

4. Trans-Lesion Synthesis

The Trans-Lesion Synthesis branch of PRR employs specialized DNA polymerases to perform replicative bypass of DNA lesions. In a process termed "polymerase switching" the TLS polymerases are recruited to stalled replication forks where they transiently replace the replicative polymerases. There are three TLS polymerases RAD30 (η), Rev1, and ζ, in budding yeast and two additional TLS polymerases, κ and ι, in vertebrates. Of these TLS polymerases, η, κ, ι, and Rev1 belong to the Y-family, while Pol ζ belongs to the B family [51]. The unique structure of TLS polymerases allows them to synthesize across lesions that block the conventional replicative polymerases. Compared to replicative polymerases, TLS polymerases have larger catalytic sites that are able to make loose contact with the template DNA and incoming nucleotide. This structure makes TLS polymerases more promiscuous in their selection of template DNA and allows them to accommodate templates with bulky adducts and abasic sites [52]. Furthermore, TLS polymerases lack the proofreading exonuclease domain that is present in the replicative ones and which is critical for accurate DNA synthesis. Therefore, utilizing TLS polymerases to replicate damaged templates can confer damage-tolerant DNA synthesis at the cost of reduced replication accuracy [7].

Although TLS polymerases are inherently error prone, TLS can be relatively error-free in instances when the "correct" Y-family polymerase(s) are recruited to bypass a cognate lesion. For example, in the presence of a UV radiation-induced thymine-thymine cis-syn cyclobutane dimer (CPD), DNA Pol η preferentially incorporates two adenines (A) opposite the thymine-thymine dimer to accurately bypass the lesion [53]. Similarly, Pol κ accurately bypasses bulky (BP)-7,8-diol-9, 10-epoxide-N(2)-deoxyguanosine (BPDE-dG) adducts induced by the environmental carcinogen Benzo[a]pyrene [54,55]. Moreover, Pol ι frequently incorporates a correct base (cytosine, C) following the oxidative lesion 8-oxoguanine as well as 2-Acetylaminofluorene (AAF) adducted guanine [56]. Thus TLS polymerases are capable of contributing to DNA damage tolerance and S-phase progression without compromising genome stability.

Due to the intrinsic ability of TLS polymerases to accommodate a wide variety of lesions, sometimes lesion bypass can be carried out by a "non-ideal" error-prone polymerase, especially when the "correct" polymerase is not available. This phenomenon of compensatory error-prone lesion bypass by inappropriate DNA polymerases is exemplified by xeroderma pigmentosum variant (XP-V) patients, in which Pol η is mutated [57]. XP-V patients experience extreme sunlight sensitivity and have an increased incidence of skin cancer. In XP-V patients, UV-induced DNA damage is bypassed by other Y-family DNA polymerases such as Pol ι [58] and Pol κ [59], resulting in high mutation rates. These studies suggest that, despite the presence of intact nucleotide excision repair, selecting the correct TLS polymerase to accurately bypass the DNA lesion is crucial for the prevention of elevated mutagenesis.

As mentioned previously, the recruitment of TLS polymerases to stalled replication forks is facilitated by Rad18-mediated PCNA mono-ubiquitination. TLS polymerases possess a higher affinity towards PCNA in its mono-ubiquinated state [20,21] and may displace the processive DNA polymerases to replicate through damaged DNA [60]. Interestingly, lysine 164 of PCNA is not located at the IDCL, the protein-protein interacting domain on PCNA that mediates the interaction with the PIP motif of target proteins [13]. Instead, the K164-linked ubiquitin is attached to the back face of PCNA, creating a distinct interacting motif for TLS polymerases [22] (Figure 4). In addition to a PIP-motif, all Y family polymerases contain at least one Ubiquitin-Binding Zinc finger (UBZ) or Ubiquitin-Binding Motif (UBM) at the C-terminus of the protein [61]. The ubiquitin-binding domain, together with the PIP motif on TLS polymerases, mediates the preferential interaction with mono-ubiquitinated PCNA. In this structure, the binding of Pol η to PCNA does not interfere with the binding of Pol δ. Instead, Pol η is resting at the back face of PCNA, while Pol δ is contacting the front surface of PCNA [22] (Figure 4). It is important to note that PCNA is a trimeric ring, which in theory could interact with three DNA polymerases at the same time. In fact, the structural study by Freudenthal et al. favors the notion that the PCNA ring acts as a molecular 'tool belt', carrying both TLS and replicative polymerases to cope with damage on ssDNA, similar to the β sliding clamp in E.coli [62]. Unlike the Y-family TLS polymerases (η, κ, ι, REV1), Pol ζ (a B-family DNA polymerase) does not contain a UBZ domain. However, Pol ζ recruitment to stalled replication forks is mediated by the Y family polymerases, such as REV1 [63], and therefore might have some dependency on PCNA mono-ubiquitination.

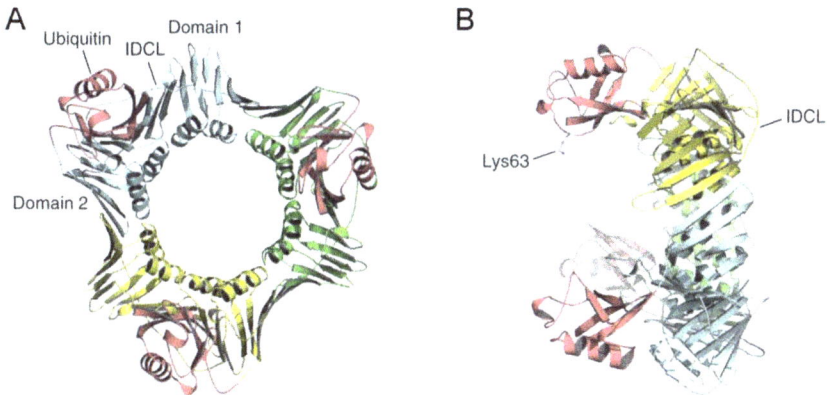

Figure 4. Structure of a monoubiquitinated PCNA ring (picture from reference [22]). (**A**) Back view of monoubiquitinated PCNA ring showing the two domains of a single PCNA subunit and the inter-domain connecting loop (IDCL). Ubiquitin is shown in red. Three individual PCNA molecules (shown in blue, green, and yellow) constitute the ring shape; (**B**) Side view of the monoubiquitinated PCNA where the back surface is to the left and the front surface is to the right of the figure. Notice that the ubiquitin is located on the back surface of the PCNA ring while IDCL is to the front side of the PCNA ring.

Despite the extensive studies suggesting that PCNA and its ubiquitination facilitate TLS activation, the absolute requirement of PCNA mono-ubiquitination is still being debated. Several lines of evidence suggest that TLS can, in some instances, proceed without the need for PCNA mono-ubiquitination. Gueranger and collegues showed that the Pol η PIP-box mutant could completely restore the UV resistance in a pol η-deficient cell line [64]. Acharya and colleagues additionally found that the ubiquitin-binding domain of pol η is dispensable for its TLS function [65,66]. Embryonic fibroblasts from a genetically-engineered PCNA K164R "knock-in" mouse show some attenuation of TLS activity, yet retain lesion bypass activity [67]. In a recent study, pol η was shown to interact with unmodified and mono-ubiquitinated PCNA with equivalent affinities in vitro. Furthermore, mono-ubiquitinated PCNA did not enhance the lesion bypass activity of pol η [68]. Together, these studies suggest that PCNA ubiquitination is important for high-capacity TLS and efficient recruitment of TLS polymerases in normal cells. However, TLS may also occur in the absence of PCNA mono-ubiquitination, most likely due to residual UBZ-independent interactions between PCNA and the PIP motifs.

Even with our mechanistic understanding of trans-lesion synthesis, it is still not known how cells recruit appropriate and specific TLS polymerases to their cognate DNA lesions. It is formally possible that unknown factors "read" the structure of distorted DNA and then signal for the recruitment of the specific polymerases. However, as clearly documented in XP-V cells, CPD lesions can be bypassed by non-cognate TLS polymerases when Pol η is absent. Therefore, it is possible that mono-ubiquitinated PCNA does not discriminate between different TLS polymerases and serves as a recruitment platform that interacts with all TLS polymerases equally. Perhaps all the TLS polymerases are recruited to the damage site randomly and attempt to replicate through the lesion. In this "trial-and-error" mechanism, the "correct" polymerase bypasses a cognate lesion with the lowest energy expenditure. Non-ideal TLS polymerases would only perform a bypass when the "correct" polymerase is unavailable. Therefore, it might be the nature of the lesion itself that determines which polymerase engages and replicates across a specific type of DNA damage.

5. Template Switching

The Template Switching branch of PRR enables the stalled replication fork to use the newly synthesized daughter strand as template to avoid damaged DNA (Figure 3). Similar to TLS, TS is also mediated by PCNA post-translational modifications, specifically poly-ubiquitination and SUMOylation. The extension of Rad18-induced K164 mono-ubiquitination to poly-ubiquitination by Ubc13-Mms2 and Rad5 redirects the PRR mode to TS.

Most of our understanding of TS was generated from a series of elegant studies in yeast, which provided the basis for the existence of an error-free form of PRR [69,70]. This error-free mechanism requires Rad5 [71,72], Ubc13/Mms2 [73,74], DNA Pol δ [75], a subset of the RAD52 epistasis group [76] and involves recombination between partially replicated sister strands [77]. A groundbreaking study by Branzei and colleagues combining 2D gel electrophoresis of DNA replication intermediates and yeast genetics identified TS intermediates and defined their relationship with the previously mentioned TS factors [78].

Template switching, involves the formation of an X-shaped recombination intermediate like structure consisting of sister chromatic junctions (SCJs) close to the stalled replication fork. SCJ formation requires Rad51 and the resolution of SCJs structures depends on the Sgs1 helicase [78]. The SCJs generated during TS resemble the properties of a DNA crossover intermediate in homologous recombination [79]. A recent study visualized the recombination intermediates using electron microscopy and proved that the undamaged sister chromatid is used as template using a recombination-based mechanism [80]. During TS, the ssDNA template containing the DNA lesion anneals with the newly synthesized double stranded sister chromatid to form a three-strand duplex. This intermediate then releases the newly synthesized, undamaged daughter strand from the parental strand so that it can be used as template for damage avoidance. The structures formed in this process are later resolved by the Sgs1-Top3-Rim1 complex [80]. In addition to the core TS participants such as Rad5, Rad51, and Sgs1, there is a growing body of evidence that other DNA replication and damage repair factors are also involved in TS. For example, the 9-1-1 complex and Exo1 nuclease are essential for the initiation of TS [81], and Ctf4 was found to establish the connection between Pol α/primase and the MCM helicase to protect the replication fork structure that favors TS [82].

Although Ubc13-Mms2 and Rad5-mediated poly-ubiquitination of PCNA at K164 is a crucial event in TS, the structure and function of the PCNA poly-ubiquitin chain formed during TS remains elusive. In contrast, the significance of PCNA SUMOylation has been studied in more detail and is better understood. PCNA SUMOylation occurs both during normal, unperturbed DNA replication and in response to DNA damage [14,15]. SUMOylated PCNA provides an interaction platform for the recruitment of the Srs2 helicase [47,48]. Similar to other PCNA-interacting proteins, Srs2 contains a non-canonical PIP box motif that mediates PCNA binding. However, the interaction between the Srs2 PIP box and PCNA is fairly weak until a second interaction is established between SUMOylated PCNA and a SUMO-interacting motif at the C terminus of Srs2 [83]. PCNA-bound Srs2 helicase functions as a safeguard that limits unscheduled recombination at the replication fork by disrupting Rad51 filament formation on ssDNA during normal replication [84,85]. A small controversy still exists regarding why PCNA SUMOylation is required for the TS pathway; Srs2 actively removes Rad51 from the replication fork while template switching requires Rad51 activity. For this reason, it is generally believed that SUMOylation antagonizes the effect of PCNA ubiquitination and inhibits the TS pathway [14,15].

A recent study may help resolve this paradox; a SUMO-like domain protein, Esc2, was found to be recruited to stalled replication forks and displace Srs2, thereby creating a microenvironment that is permissive for Rad51 chromatin-binding [86]. Therefore, PCNA SUMOylation facilitates the usage of PRR on a challenged replication fork by suppressing homologous recombination. When TS is initiated by PCNA poly-ubiquitination, replication fork binding factors such as Esc2 alleviate the inhibition of recombination by PCNA SUMOylation and allow DNA damage avoidance [47].

Error-free DNA damage avoidance is a conserved PRR mechanism in metazoans [43,87]. Although the identity of the human Srs2 orthologue is still being debated, human PCNA SUMOylation has also been shown to suppress unscheduled DNA recombination via PARI (PCNA-associated recombination inhibitor), suggesting a conserved mechanism of regulating HR at the replication fork [88–90]. PCNA is also poly-ubiquitinated in human cells in response to DNA damage. Blocking K63 linked poly-ubiquitination chain formation sensitizes cells to DNA damage, increases UV-induced mutagenesis, and increases the reliance of cells on TLS for DNA damage tolerance [43]. Rad5 has evolved into two orthologues, SHPRH [38,39] and HLTF, in higher organisms [40,41]. Both SHPRH and HLTF can poly-ubiquitinate PCNA in vitro but via distinct mechanisms. SHPRH extends Rad18 mediated PCNA mono-ubiquitination, while HLTF transfers the pre-assembled poly-ubiquitin chain to Rad6-Rad18 and eventually onto unmodified PCNA [38,39,91]. Depletion of SHPRH and HLTF sensitizes the cell to DNA damaging agents and reduces PCNA poly-ubiquitination; however, $SHPRH^{-/-}HLTF^{-/-}$ double knockout mouse embryonic fibroblasts are still able to poly-ubiquitinate PCNA, suggesting that other Rad5 orthologues might exist in higher organisms [92]. In addition to poly-ubiquitination of PCNA, HLTF has acquired additional functions in DNA damage tolerance. In response to UV damage, HLTF is able to mono-ubiquitinate PCNA and promote Pol η recruitment [93]. Furthermore, HLTF can also facilitate DNA strand invasion and D-loop formation in a Rad51-independent manner [94].

In addition to TS, there are other recombination-based mechanisms, such as complementary strand transfer repair (CSTR) [95] and replication fork reversal [96–99], that have also been shown to contribute to DNA damage avoidance.

6. Timing of Post Replication Repair

Both major modes of PRR are used to cope with collisions between DNA polymerases and lesions on the single-stranded DNA template. For this reason, PRR is critically important during S-phase of the cell cycle, when the DNA duplex is unwound and vulnerable to injury. In fact, cells have developed sophisticated mechanisms to control the timing of DNA post replication repair by limiting the availability of crucial PRR factors [100–102]. Interestingly, two studies using temporally controlled expression of Rad18 or Pol η found that it is possible to delay the onset of PRR without significantly affecting cell viability. Moreover, limiting PRR in the G2/M phase of the cell cycle does not significantly delay the progression of the S-phase [103,104]. These studies suggest that it is possible to detach the PRR with bulk DNA synthesis in the S-phase without compromising its function.

Nevertheless, the delayed onset of PRR during S-phase could potentially lead to the accumulation of dangerously long and fragile ssDNA stretches, especially on the leading strand. Exposed ssDNA in cells is frequently observed when the replicative polymerase is blocked. However, these ssDNA gaps are usually small in size and are located inside a single replicon, regardless of whether they are on the leading or the lagging strand. However extremely long ssDNA gaps (>3 kb) are rarely observed. This suggests that the leading strand is also synthesized discontinuously when replicating a damaged DNA template, similar to the discontinuous synthesis of the lagging strand [105].

Restart of replication requires a de-novo re-priming mechanism downstream (3′) of the stalled leading strand DNA polymerase. This repriming activity is carried out by DnaG in *E. coli* [106], and by a specialized polymerase PrimPol in higher organisms [107–110]. This repriming mechanism of PRR explains why UV-induced lesions only cause a slight reduction in fork speed even when Pol η is mutated in human cells [111]. The ability of PRR to function distal (5′) to a newly-primed leading strand may provide ample time to select the optimal DNA damage tolerance mechanism. It is also well established that TLS is functional outside the S phase of the cell cycle and can patch ssDNA arising in the G0 and G1 phases [112–114].

7. Conclusions and Outlook

Although neither TLS nor TS directly repair DNA damage, both PRR mechanisms enable an immediate response to polymerase stalling DNA lesions. PRR during S phase prevents gross chromosomal rearrangements and ensures that replication is completed in a timely manner.

A deficiency in PRR could lead to replication fork collapse and the accumulation of DNA DSBs. In the absence of PRR, DSB repair mechanisms could allow for tolerance of replication-associated DNA damage. However, DSB repair pathways have limitations; DNA end-joining frequently results in mutations, while HR serves as the salvage pathway and creates complex and unstable repair intermediates through the use of a homologous strand from another DNA molecule. (For more insight into salvage and other homologous recombination-mediated DNA damage tolerance, we invite readers to read a recent review on this topic [115]) Therefore, PRR is perhaps the least genome-destabilizing option for the tolerance of DNA lesions arising in S-phase.

Both branches of PRR are important for cells to tolerate and survive DNA damage. In terms of maintaining genome stability, TS has a great advantage over TLS because it does not induce base mutations. Because of its intrinsic error-propensity, TLS has been linked to both increased mutation rates and might, therefore, fuel carcinogenesis [116–118]. In established cancers, TLS is also suggested to be responsible for a high mutation frequency and elevated treatment resistance [116,119,120]. However, it is not known why untransformed cells would utilize the error-prone TLS pathway when error-free TS is available.

Many DNA damaging chemotherapy agents cause lesions that can be tolerated by PRR. Such lesions include bulky adducts generated by alkylating agents and the DNA crosslinks produced by platinating agents. For this reason, cancer cells could upregulate PRR to survive therapy-induced DNA damage. Interestingly, a recent report identified a cancer cell-specific mechanism of TLS activation that might provide a general paradigm for how tumors acquire both mutability and DNA damage tolerance via pathological PRR [121]. Therefore, targeting PRR pathways individually, or in combination with compensatory genome maintenance mechanisms, could sensitize cancer cells to intrinsic and therapy-induced replicative stress.

Acknowledgments: This manuscript was supported by grant R01 ES009558 from the National Institutes of Health to Cyrus Vaziri, a Tier 1 Pilot Award from the University of North Carolina Lineberger Comprehensive Cancer

Center to Yanzhe Gao, the Leon and Bertha Golberg Postdoctoral Fellowship to Yanzhe Gao, a PhRMA Foundation Pre-Doctoral Award in Pharmacology/Toxicology to Elizabeth Mutter-Rottmayer, and a NC TraCS Institute CTSA grant UL1TR001111 to Yang Yang.

Conflicts of Interest: The authors declare no conflict of interest.

References

1. Burgers, P.M.J. Polymerase Dynamics at the Eukaryotic DNA Replication Fork. *J. Biol. Chem.* **2009**, *284*, 4041–4045. [CrossRef] [PubMed]

2. Lehmann, A.R. Replication of UV-damaged DNA: New insights into links between DNA polymerases, mutagenesis and human disease. *Gene* **2000**, *253*, 1–12. [CrossRef]

3. Ciccia, A.; Elledge, S.J. The DNA Damage Response: Making It Safe to Play with Knives. *Mol. Cell* **2010**, *40*, 179–204. [CrossRef] [PubMed]

4. Yi, C.; He, C. DNA Repair by Reversal of DNA Damage. *Cold Spring Harb. Perspect. Biol.* **2013**.

5. Masai, H.; Matsumoto, S.; You, Z.; Yoshizawa-Sugata, N.; Oda, M. Eukaryotic Chromosome DNA Replication: Where, When, and How? *Ann. Rev. Biochem.* **2010**, *79*, 89–130. [CrossRef] [PubMed]

6. Nakamura, J.; Mutlu, E.; Sharma, V.; Collins, L.; Bodnar, W.; Yu, R.; Lai, Y.; Moeller, B.; Lu, K.; Swenberg, J. The Endogenous Exposome. *DNA Repair* **2014**, *19*, 3–13. [CrossRef] [PubMed]

7. Sale, J.E.; Lehmann, A.R.; Woodgate, R. Y-family DNA polymerases and their role in tolerance of cellular DNA damage. *Nat. Rev. Mol. Cell Biol.* **2012**, *13*, 141–152. [CrossRef] [PubMed]

8. Branzei, D. Ubiquitin family modifications and template switching. *FEBS Lett.* **2011**, *585*, 2810–2817.

9. Majka, J.; Burgers, P.M.J. The PCNA–RFC Families of DNA Clamps and Clamp Loaders. *Prog. Nucleic Acid Res. Mol. Biol.* **2004**, *78*, 227–260. [PubMed]

10. Lee, K.Y.; Fu, H.; Aladjem, M.I.; Myung, K. ATAD5 regulates the lifespan of DNA replication factories by modulating PCNA level on the chromatin. *J. Cell Biol.* **2013**, *200*, 31–44. [CrossRef] [PubMed]

11. Kubota, T.; Nishimura, K.; Kanemaki, M.T.; Donaldson, A.D. The Elg1 replication factor C-like complex functions in PCNA unloading during DNA replication. *Mol. Cell* **2013**, *50*, 273–280. [CrossRef] [PubMed]

12. Moldovan, G.-L.; Pfander, B.; Jentsch, S. PCNA, the Maestro of the Replication Fork. *Cell* **2007**, *129*, 665–679.

13. Gulbis, J.M.; Kelman, Z.; Hurwitz, J.; O'Donnell, M.; Kuriyan, J. Structure of the C-terminal region of p21(WAF1/CIP1) complexed with human PCNA. *Cell* **1996**, *87*, 297–306. [CrossRef]

14. Hoege, C.; Pfander, B.; Moldovan, G.L.; Pyrowolakis, G.; Jentsch, S. RAD6-dependent DNA repair is linked to modification of PCNA by ubiquitin and SUMO. *Nature* **2002**, *419*, 135–141. [CrossRef] [PubMed]

15. Stelter, P.; Ulrich, H.D. Control of spontaneous and damage-induced mutagenesis by SUMO and ubiquitin conjugation. *Nature* **2003**, *425*, 188–191. [CrossRef] [PubMed]

16. Byun, T.S.; Pacek, M.; Yee, M.-c.; Walter, J.C.; Cimprich, K.A. Functional uncoupling of MCM helicase and DNA polymerase activities activates the ATR-dependent checkpoint. *Genes Dev.* **2005**, *19*, 1040–1052.

17. Davies, A.A.; Huttner, D.; Daigaku, Y.; Chen, S.; Ulrich, H.D. Activation of ubiquitin-dependent DNA damage bypass is mediated by replication protein a. *Mol. Cell* **2008**, *29*, 625–636. [CrossRef] [PubMed]

18. Huttner, D.; Ulrich, H.D. Cooperation of replication protein A with the ubiquitin ligase Rad18 in DNA damage bypass. *Cell Cycle* **2008**, *7*, 3629–3633. [CrossRef]

19. Tsuji, Y.; Watanabe, K.; Araki, K.; Shinohara, M.; Yamagata, Y.; Tsurimoto, T.; Hanaoka, F.; Yamamura, K.; Yamaizumi, M.; Tateishi, S. Recognition of forked and single-stranded DNA structures by human RAD18 complexed with RAD6B protein triggers its recruitment to stalled replication forks. *Genes Cells Devot. Mol. Cell. Mech.* **2008**, *13*, 343–354. [CrossRef] [PubMed]

20. Watanabe, K.; Tateishi, S.; Kawasuji, M.; Tsurimoto, T.; Inoue, H.; Yamaizumi, M. Rad18 guides polη to replication stalling sites through physical interaction and PCNA monoubiquitination. *EMBO J.* **2004**, *23*, 3886–3896. [CrossRef] [PubMed]

21. Kannouche, P.L.; Wing, J.; Lehmann, A.R. Interaction of Human DNA Polymerase η with Monoubiquitinated PCNA: A Possible Mechanism for the Polymerase Switch in Response to DNA Damage. *Mol. Cell* **2004**, *14*, 491–500. [CrossRef]

22. Freudenthal, B.D.; Gakhar, L.; Ramaswamy, S.; Washington, M.T. Structure of monoubiquitinated PCNA and implications for translesion synthesis and DNA polymerase exchange. *Nat. Struct. Mol. Biol.* **2010**, *17*, 479–484.

23. Durando, M.; Tateishi, S.; Vaziri, C. A non-catalytic role of DNA polymerase eta in recruiting Rad18 and promoting PCNA monoubiquitination at stalled replication forks. *Nucleic Acids Res.* **2013**, *41*, 3079–3093. [CrossRef] [PubMed]

24. Yanagihara, H.; Kobayashi, J.; Tateishi, S.; Kato, A.; Matsuura, S.; Tauchi, H.; Yamada, K.; Takezawa, J.; Sugasawa, K.; Masutani, C.; et al. NBS1 recruits RAD18 via a RAD6-like domain and regulates Pol eta-dependent translesion DNA synthesis. *Mol. Cell* **2011**, *43*, 788–797. [CrossRef] [PubMed]

25. Tian, F.; Sharma, S.; Zou, J.; Lin, S.Y.; Wang, B.; Rezvani, K.; Wang, H.; Parvin, J.D.; Ludwig, T.; Canman, C.E.; et al. BRCA1 promotes the ubiquitination of PCNA and recruitment of translesion polymerases in response to replication blockade. *Proc. Natl. Acad. Sci. USA* **2013**, *110*, 13558–13563. [CrossRef] [PubMed]

26. Han, J.; Liu, T.; Huen, M.S.; Hu, L.; Chen, Z.; Huang, J. SIVA1 directs the E3 ubiquitin ligase RAD18 for PCNA monoubiquitination. *J. Cell Biol.* **2014**, *205*, 811–827. [CrossRef] [PubMed]

27. Centore, R.C.; Yazinski, S.A.; Tse, A.; Zou, L. Spartan/C1orf124, a Reader of PCNA Ubiquitylation and a Regulator of UV-Induced DNA Damage Response. *Mol. Cell* **2012**, *46*, 625–635. [CrossRef] [PubMed]

28. Mosbech, A.; Gibbs-Seymour, I.; Kagias, K.; Thorslund, T.; Beli, P.; Povlsen, L.; Nielsen, S.V.; Smedegaard, S.; Sedgwick, G.; Lukas, C.; et al. DVC1 (C1orf124) is a DNA damage-targeting p97 adaptor that promotes ubiquitin-dependent responses to replication blocks. *Nat. Struct. Mol. Biol.* **2012**, *19*, 1084–1092.

29. Juhasz, S.; Balogh, D.; Hajdu, I.; Burkovics, P.; Villamil, M.A.; Zhuang, Z.; Haracska, L. Characterization of human Spartan/C1orf124, an ubiquitin-PCNA interacting regulator of DNA damage tolerance. *Nucleic Acids Res.* **2012**, *40*, 10795–10808. [CrossRef] [PubMed]

30. Davis, E.J.; Lachaud, C.; Appleton, P.; Macartney, T.J.; Nathke, I.; Rouse, J. DVC1 (C1orf124) recruits the p97 protein segregase to sites of DNA damage. *Nat. Struct. Mol. Biol.* **2012**, *19*, 1093–1100. [CrossRef] [PubMed]

31. Machida, Y.; Kim, M.S.; Machida, Y.J. Spartan/C1orf124 is important to prevent UV-induced mutagenesis. *Cell Cycle* **2012**, *11*, 3395–3402. [CrossRef] [PubMed]

32. Arakawa, H.; Moldovan, G.-L.; Saribasak, H.; Saribasak, N.N.; Jentsch, S.; Buerstedde, J.-M. A Role for PCNA Ubiquitination in Immunoglobulin Hypermutation. *PLoS Biol.* **2006**, *4*, e366. [CrossRef] [PubMed]

33. Simpson, L.J.; Ross, A.L.; Szuts, D.; Alviani, C.A.; Oestergaard, V.H.; Patel, K.J.; Sale, J.E. RAD18-independent ubiquitination of proliferating-cell nuclear antigen in the avian cell line DT40. *EMBO Rep.* **2006**, *7*, 927–932.

34. Shimizu, T.; Tateishi, S.; Tanoue, Y.; Azuma, T.; Ohmori, H. Somatic hypermutation of immunoglobulin genes in Rad18 knockout mice. *DNA Repair (Amst)* **2017**, *50*, 54–60. [CrossRef] [PubMed]

35. Ulrich, H.D.; Jentsch, S. Two RING finger proteins mediate cooperation between ubiquitin-conjugating enzymes in DNA repair. *Embo. J.* **2000**, *19*, 3388–3397. [CrossRef]

36. Haracska, L.; Torres-Ramos, C.A.; Johnson, R.E.; Prakash, S.; Prakash, L. Opposing effects of ubiquitin conjugation and SUMO modification of PCNA on replicational bypass of DNA lesions in Saccharomyces cerevisiae. *Mol. Cell Biol.* **2004**, *24*, 4267–4274. [CrossRef] [PubMed]

37. Leach, C.A.; Michael, W.M. Ubiquitin/SUMO modification of PCNA promotes replication fork progression in Xenopus laevis egg extracts. *J. Cell Biol.* **2005**, *171*, 947–954. [CrossRef] [PubMed]

38. Motegi, A.; Sood, R.; Moinova, H.; Markowitz, S.D.; Liu, P.P.; Myung, K. Human SHPRH suppresses genomic instability through proliferating cell nuclear antigen polyubiquitination. *J. Cell Biol.* **2006**, *175*, 703–708.

39. Unk, I.; Hajdu, I.; Fatyol, K.; Szakal, B.; Blastyak, A.; Bermudez, V.; Hurwitz, J.; Prakash, L.; Prakash, S.; Haracska, L. Human SHPRH is a ubiquitin ligase for Mms2-Ubc13-dependent polyubiquitylation of proliferating cell nuclear antigen. *Proc. Natl. Acad. Sci. USA* **2006**, *103*, 18107–18112. [CrossRef] [PubMed]

40. Unk, I.; Hajdu, I.; Fatyol, K.; Hurwitz, J.; Yoon, J.H.; Prakash, L.; Prakash, S.; Haracska, L. Human HLTF functions as a ubiquitin ligase for proliferating cell nuclear antigen polyubiquitination. *Proc. Natl. Acad. Sci. USA* **2008**, *105*, 3768–3773. [CrossRef] [PubMed]

41. Motegi, A.; Liaw, H.J.; Lee, K.Y.; Roest, H.P.; Maas, A.; Wu, X.; Moinova, H.; Markowitz, S.D.; Ding, H.; Hoeijmakers, J.H.; et al. Polyubiquitination of proliferating cell nuclear antigen by HLTF and SHPRH prevents genomic instability from stalled replication forks. *Proc. Natl. Acad. Sci. USA* **2008**, *105*, 12411–12416. [CrossRef] [PubMed]

42. Kannouche, P.L.; Lehmann, A.R. Ubiquitination of PCNA and the polymerase switch in human cells. *Cell Cycle* **2004**, *3*, 1011–1013. [CrossRef] [PubMed]

43. Chiu, R.K.; Brun, J.; Ramaekers, C.; Theys, J.; Weng, L.; Lambin, P.; Gray, D.A.; Wouters, B.G. Lysine 63-polyubiquitination guards against translesion synthesis-induced mutations. *PLoS Genet.* **2006**, *2*, e116. [CrossRef]

44. Gangavarapu, V.; Haracska, L.; Unk, I.; Johnson, R.E.; Prakash, S.; Prakash, L. Mms2-Ubc13-dependent and -independent roles of Rad5 ubiquitin ligase in postreplication repair and translesion DNA synthesis in Saccharomyces cerevisiae. *Mol. Cell. Biol.* **2006**, *26*, 7783–7790. [CrossRef]

45. Pages, V.; Bresson, A.; Acharya, N.; Prakash, S.; Fuchs, R.P.; Prakash, L. Requirement of Rad5 for DNA polymerase zeta-dependent translesion synthesis in Saccharomyces cerevisiae. *Genetics* **2008**, *180*, 73–82. [CrossRef] [PubMed]

46. Xu, X.; Lin, A.; Zhou, C.; Blackwell, S.R.; Zhang, Y.; Wang, Z.; Feng, Q.; Guan, R.; Hanna, M.D.; Chen, Z.; et al. Involvement of budding yeast Rad5 in translesion DNA synthesis through physical interaction with Rev1. *Nucleic Acids Res.* **2016**, *44*, 5231–5245. [CrossRef] [PubMed]

47. Pfander, B.; Moldovan, G.L.; Sacher, M.; Hoege, C.; Jentsch, S. SUMO-modified PCNA recruits Srs2 to prevent recombination during S phase. *Nature* **2005**, *436*, 428–433. [CrossRef] [PubMed]

48. Papouli, E.; Chen, S.; Davies, A.A.; Huttner, D.; Krejci, L.; Sung, P.; Ulrich, H.D. Crosstalk between SUMO and ubiquitin on PCNA is mediated by recruitment of the helicase Srs2p. *Mol. Cell* **2005**, *19*, 123–133. [CrossRef] [PubMed]

49. Schiestl, R.H.; Prakash, S.; Prakash, L. The SRS2 suppressor of rad6 mutations of Saccharomyces cerevisiae acts by channeling DNA lesions into the RAD52 DNA repair pathway. *Genetics* **1990**, *124*, 817–831. [PubMed]

50. Parker, J.L.; Ulrich, H.D. A SUMO-interacting motif activates budding yeast ubiquitin ligase Rad18 towards SUMO-modified PCNA. *Nucleic Acids Res.* **2012**, *40*, 11380–11388. [CrossRef] [PubMed]

51. Sale, J.E. Translesion DNA synthesis and mutagenesis in eukaryotes. *Cold Spring Harb. Perspect. Biol.* **2013**, *5*, a012708. [CrossRef] [PubMed]

52. Prakash, S.; Johnson, R.E.; Prakash, L. Eukaryotic translesion synthesis DNA polymerases: Specificity of structure and function. *Annu. Rev. Biochem.* **2005**, *74*, 317–353. [CrossRef] [PubMed]

53. Johnson, R.E.; Prakash, S.; Prakash, L. Efficient Bypass of a Thymine-Thymine Dimer by Yeast DNA Polymerase, Polη. *Science* **1999**, *283*, 1001–1004. [CrossRef] [PubMed]

54. Zhang, Y.; Yuan, F.; Wu, X.; Wang, M.; Rechkoblit, O.; Taylor, J.-S.; Geacintov, N.E.; Wang, Z. Error-free and error-prone lesion bypass by human DNA polymerase κ in vitro. *Nucleic Acids Res.* **2000**, *28*, 4138–4146. [CrossRef] [PubMed]

55. Bi, X.; Slater, D.M.; Ohmori, H.; Vaziri, C. DNA polymerase kappa is specifically required for recovery from the benzo[a]pyrene-dihydrodiol epoxide (BPDE)-induced S-phase checkpoint. *J. Biol. Chem.* **2005**, *280*, 22343–22355. [CrossRef] [PubMed]

56. Zhang, Y.; Yuan, F.; Wu, X.; Taylor, J.-S.; Wang, Z. Response of human DNA polymerase ι to DNA lesions. *Nucleic Acids Res.* **2001**, *29*, 928–935. [CrossRef] [PubMed]

57. Masutani, C.; Kusumoto, R.; Yamada, A.; Dohmae, N.; Yokoi, M.; Yuasa, M.; Araki, M.; Iwai, S.; Takio, K.; Hanaoka, F. The XPV (xeroderma pigmentosum variant) gene encodes human DNA polymerase [eta]. *Nature* **1999**, *399*, 700–704. [PubMed]

58. Wang, Y.; Woodgate, R.; McManus, T.P.; Mead, S.; McCormick, J.J.; Maher, V.M. Evidence that in Xeroderma Pigmentosum Variant Cells, which Lack DNA Polymerase η, DNA Polymerase ι Causes the Very High Frequency and Unique Spectrum of UV-Induced Mutations. *Cancer Res.* **2007**, *67*, 3018–3026. [CrossRef] [PubMed]

59. Ziv, O.; Geacintov, N.; Nakajima, S.; Yasui, A.; Livneh, Z. DNA polymerase zeta cooperates with polymerases kappa and iota in translesion DNA synthesis across pyrimidine photodimers in cells from XPV patients. *Proc. Natl. Acad. Sci. USA* **2009**, *106*, 11552–11557. [CrossRef] [PubMed]

60. Zhuang, Z.; Johnson, R.E.; Haracska, L.; Prakash, L.; Prakash, S.; Benkovic, S.J. Regulation of polymerase exchange between Polη and Polδ by monoubiquitination of PCNA and the movement of DNA polymerase holoenzyme. *Proc. Natl. Acad. Sci. USA* **2008**, *105*, 5361–5366. [CrossRef] [PubMed]

61. Bienko, M.; Green, C.M.; Crosetto, N.; Rudolf, F.; Zapart, G.; Coull, B.; Kannouche, P.; Wider, G.; Peter, M.; Lehmann, A.R.; et al. Ubiquitin-binding domains in Y-family polymerases regulate translesion synthesis. *Science* **2005**, *310*, 1821–1824. [CrossRef] [PubMed]

62. Indiani, C.; McInerney, P.; Georgescu, R.; Goodman, M.F.; O'Donnell, M. A sliding-clamp toolbelt binds high- and low-fidelity DNA polymerases simultaneously. *Mol. Cell* **2005**, *19*, 805–815. [CrossRef] [PubMed]

63. Kikuchi, S.; Hara, K.; Shimizu, T.; Sato, M.; Hashimoto, H. Structural basis of recruitment of DNA polymerase zeta by interaction between REV1 and REV7 proteins. *J. Biol. Chem.* **2012**, *287*, 33847–33852. [CrossRef] [PubMed]

64. Gueranger, Q.; Stary, A.; Aoufouchi, S.; Faili, A.; Sarasin, A.; Reynaud, C.-A.; Weill, J.-C. Role of DNA polymerases η, ι and ζ in UV resistance and UV-induced mutagenesis in a human cell line. *DNA Repair* **2008**, *7*, 1551–1562. [CrossRef] [PubMed]

65. Acharya, N.; Yoon, J.H.; Gali, H.; Unk, I.; Haracska, L.; Johnson, R.E.; Hurwitz, J.; Prakash, L.; Prakash, S. Roles of PCNA-binding and ubiquitin-binding domains in human DNA polymerase eta in translesion DNA synthesis. *Proc. Natl. Acad. Sci. USA* **2008**, *105*, 17724–17729. [CrossRef] [PubMed]

66. Acharya, N.; Yoon, J.H.; Hurwitz, J.; Prakash, L.; Prakash, S. DNA polymerase eta lacking the ubiquitin-binding domain promotes replicative lesion bypass in humans cells. *Proc. Natl. Acad. Sci. USA* **2010**, *107*, 10401–10405. [CrossRef] [PubMed]

67. Hendel, A.; Krijger, P.H.L.; Diamant, N.; Goren, Z.; Langerak, P.; Kim, J.; Reißner, T.; Lee, K.-Y.; Geacintov, N.E.; Carell, T.; et al. PCNA Ubiquitination Is Important, But Not Essential for Translesion DNA Synthesis in Mammalian Cells. *PLoS Genet.* **2011**, *7*, e1002262. [CrossRef] [PubMed]

68. Hedglin, M.; Pandey, B.; Benkovic, S.J. Characterization of human translesion DNA synthesis across a UV-induced DNA lesion. *Elife* **2016**. [CrossRef] [PubMed]

69. Prakash, L. Characterization of postreplication repair in Saccharomyces cerevisiae and effects of rad6, rad18, rev3 and rad52 mutations. *Mol. Gen. Genet. MGG* **1981**, *184*, 471–478. [CrossRef] [PubMed]

70. Prakash, S.; Sung, P.; Prakash, L. DNA repair genes and proteins of Saccharomyces cerevisiae. *Annu. Rev. Genet.* **1993**, *27*, 33–70. [CrossRef] [PubMed]

71. Xiao, W.; Chow, B.L.; Broomfield, S.; Hanna, M. The Saccharomyces cerevisiae RAD6 group is composed of an error-prone and two error-free postreplication repair pathways. *Genetics* **2000**, *155*, 1633–1641. [PubMed]

72. Minca, E.C.; Kowalski, D. Multiple Rad5 Activities Mediate Sister Chromatid Recombination to Bypass DNA Damage at Stalled Replication Forks. *Mol. Cell* **2010**, *38*, 649–661. [CrossRef] [PubMed]

73. Hofmann, R.M.; Pickart, C.M. Noncanonical *MMS2*-Encoded Ubiquitin-Conjugating Enzyme Functions in Assembly of Novel Polyubiquitin Chains for DNA Repair. *Cell* **1999**, *96*, 645–653. [CrossRef]

74. Broomfield, S.; Chow, B.L.; Xiao, W. MMS2, encoding a ubiquitin-conjugating-enzyme-like protein, is a member of the yeast error-free postreplication repair pathway. *Proc. Natl. Acad. Sci. USA* **1998**, *95*, 5678–5683. [CrossRef] [PubMed]

75. Torres-Ramos, C.A.; Prakash, S.; Prakash, L. Requirement of Yeast DNA Polymerase δ in Post-replicational Repair of UV-damaged DNA. *J. Biol. Chem.* **1997**, *272*, 25445–25448. [CrossRef] [PubMed]

76. Gangavarapu, V.; Prakash, S.; Prakash, L. Requirement of RAD52 group genes for postreplication repair of UV-damaged DNA in Saccharomyces cerevisiae. *Mol. Cell Biol.* **2007**, *27*, 7758–7764. [CrossRef] [PubMed]

77. Zhang, H.; Lawrence, C.W. The error-free component of the RAD6/RAD18 DNA damage tolerance pathway of budding yeast employs sister-strand recombination. *Proc. Natl. Acad. Sci. USA* **2005**, *102*, 15954–15959. [CrossRef] [PubMed]

78. Branzei, D.; Vanoli, F.; Foiani, M. SUMOylation regulates Rad18-mediated template switch. *Nature* **2008**, *456*, 915–920. [CrossRef] [PubMed]

79. Wu, L.; Hickson, I.D. The Bloom's syndrome helicase suppresses crossing over during homologous recombination. *Nature* **2003**, *426*, 870–874. [CrossRef] [PubMed]

80. Giannattasio, M.; Zwicky, K.; Follonier, C.; Foiani, M.; Lopes, M.; Branzei, D. Visualization of recombination-mediated damage bypass by template switching. *Nat. Struct. Mol. Biol.* **2014**, *21*, 884–892. [CrossRef] [PubMed]

81. Karras, G.I.; Fumasoni, M.; Sienski, G.; Vanoli, F.; Branzei, D.; Jentsch, S. Noncanonical Role of the 9-1-1 Clamp in the Error-Free DNA Damage Tolerance Pathway. *Mol. Cell* **2013**, *49*, 536–546. [CrossRef] [PubMed]

82. Fumasoni, M.; Zwicky, K.; Vanoli, F.; Lopes, M.; Branzei, D. Error-Free DNA Damage Tolerance and Sister Chromatid Proximity during DNA Replication Rely on the Polα/Primase/Ctf4 Complex. *Mol. Cell* **2015**, *57*, 812–823. [CrossRef] [PubMed]

83. Kim, S.O.; Yoon, H.; Park, S.O.; Lee, M.; Shin, J.-S.; Ryu, K.-S.; Lee, J.-O.; Seo, Y.-S.; Jung, H.S.; Choi, B.-S. Srs2 possesses a non-canonical PIP box in front of its SBM for precise recognition of SUMOylated PCNA. *J. Mol. Cell Biol.* **2012**, *4*, 258–261. [CrossRef] [PubMed]

84. Krejci, L.; Van Komen, S.; Li, Y.; Villemain, J.; Reddy, M.S.; Klein, H.; Ellenberger, T.; Sung, P. DNA helicase Srs2 disrupts the Rad51 presynaptic filament. *Nature* **2003**, *423*, 305–309. [CrossRef] [PubMed]

85. Veaute, X.; Jeusset, J.; Soustelle, C.; Kowalczykowski, S.C.; Le Cam, E.; Fabre, F. The Srs2 helicase prevents recombination by disrupting Rad51 nucleoprotein filaments. *Nature* **2003**, *423*, 309–312. [CrossRef] [PubMed]

86. Urulangodi, M.; Sebesta, M.; Menolfi, D.; Szakal, B.; Sollier, J.; Sisakova, A.; Krejci, L.; Branzei, D. Local regulation of the Srs2 helicase by the SUMO-like domain protein Esc2 promotes recombination at sites of stalled replication. *Genes Dev.* **2015**, *29*, 2067–2080. [CrossRef] [PubMed]

87. Izhar, L.; Ziv, O.; Cohen, I.S.; Geacintov, N.E.; Livneh, Z. Genomic assay reveals tolerance of DNA damage by both translesion DNA synthesis and homology-dependent repair in mammalian cells. *Proc. Natl. Acad. Sci. USA* **2013**, *110*, E1462–E1469. [CrossRef] [PubMed]

88. Moldovan, G.-L.; Dejsuphong, D.; Petalcorin, M.I.R.; Hofmann, K.; Takeda, S.; Boulton, S.J.; D'Andrea, A.D. Inhibition of Homologous Recombination by the PCNA-Interacting Protein PARI. *Mol. Cell* **2012**, *45*, 75–86. [CrossRef] [PubMed]

89. Gali, H.; Juhasz, S.; Morocz, M.; Hajdu, I.; Fatyol, K.; Szukacsov, V.; Burkovics, P.; Haracska, L. Role of SUMO modification of human PCNA at stalled replication fork. *Nucleic Acids Res.* **2012**, *40*, 6049–6059. [CrossRef] [PubMed]

90. Burkovics, P.; Dome, L.; Juhasz, S.; Altmannova, V.; Sebesta, M.; Pacesa, M.; Fugger, K.; Sorensen, C.S.; Lee, M.Y.W.T.; Haracska, L.; et al. The PCNA-associated protein PARI negatively regulates homologous recombination via the inhibition of DNA repair synthesis. *Nucleic Acids Res.* **2016**, *44*, 3176–3189. [CrossRef] [PubMed]

91. Masuda, Y.; Suzuki, M.; Kawai, H.; Hishiki, A.; Hashimoto, H.; Masutani, C.; Hishida, T.; Suzuki, F.; Kamiya, K. En bloc transfer of polyubiquitin chains to PCNA in vitro is mediated by two different human E2-E3 pairs. *Nucleic Acids Res.* **2012**, *40*, 10394–10407. [CrossRef] [PubMed]

92. Krijger, P.H.; Lee, K.Y.; Wit, N.; van den Berk, P.C.; Wu, X.; Roest, H.P.; Maas, A.; Ding, H.; Hoeijmakers, J.H.; Myung, K.; et al. HLTF and SHPRH are not essential for PCNA polyubiquitination, survival and somatic hypermutation: Existence of an alternative E3 ligase. *DNA Repair (Amst)* **2011**, *10*, 438–444. [CrossRef] [PubMed]

93. Lin, J.-R.; Zeman, M.K.; Chen, J.-Y.; Yee, M.-C.; Cimprich, K.A. SHPRH and HLTF Act in a Damage-Specific Manner to Coordinate Different Forms of Postreplication Repair and Prevent Mutagenesis. *Mol. Cell* **2011**, *42*, 237–249. [CrossRef] [PubMed]

94. Burkovics, P.; Sebesta, M.; Balogh, D.; Haracska, L.; Krejci, L. Strand invasion by HLTF as a mechanism for template switch in fork rescue. *Nucleic Acids Res.* **2014**, *42*, 1711–1720. [CrossRef] [PubMed]

95. Adar, S.; Izhar, L.; Hendel, A.; Geacintov, N.; Livneh, Z. Repair of gaps opposite lesions by homologous recombination in mammalian cells. *Nucleic Acids Res.* **2009**, *37*, 5737–5748. [CrossRef] [PubMed]

96. Ciccia, A.; Nimonkar, A.V.; Hu, Y.; Hajdu, I.; Achar, Y.J.; Izhar, L.; Petit, S.A.; Adamson, B.; Yoon, J.C.; Kowalczykowski, S.C.; et al. Polyubiquitinated PCNA Recruits the ZRANB3 Translocase to Maintain Genomic Integrity after Replication Stress. *Mol. Cell* **2012**, *47*, 396–409. [CrossRef] [PubMed]

97. Weston, R.; Peeters, H.; Ahel, D. ZRANB3 is a structure-specific ATP-dependent endonuclease involved in replication stress response. *Genes Dev.* **2012**, *26*, 1558–1572. [CrossRef] [PubMed]

98. Kile, A.C.; Chavez, D.A.; Bacal, J.; Eldirany, S.; Korzhnev, D.M.; Bezsonova, I.; Eichman, B.F.; Cimprich, K.A. HLTF's Ancient HIRAN Domain Binds 3' DNA Ends to Drive Replication Fork Reversal. *Mol. Cell* **2015**, *58*, 1090–1100. [CrossRef] [PubMed]

99. Achar, Y.J.; Balogh, D.; Neculai, D.; Juhasz, S.; Morocz, M.; Gali, H.; Dhe-Paganon, S.; Venclovas, Č.; Haracska, L. Human HLTF mediates postreplication repair by its HIRAN domain-dependent replication fork remodelling. *Nucleic Acids Res.* **2015**, *43*, 10277–10291. [CrossRef] [PubMed]

100. Waters, L.S.; Walker, G.C. The critical mutagenic translesion DNA polymerase Rev1 is highly expressed during G(2)/M phase rather than S phase. *Proc. Natl. Acad. Sci. USA* **2006**, *103*, 8971–8976. [CrossRef] [PubMed]

101. Edmunds, C.E.; Simpson, L.J.; Sale, J.E. PCNA Ubiquitination and REV1 Define Temporally Distinct Mechanisms for Controlling Translesion Synthesis in the Avian Cell Line DT40. *Mol. Cell* **2008**, *30*, 519–529. [CrossRef] [PubMed]

102. Ortiz-Bazán, M.Á.; Gallo-Fernández, M.; Saugar, I.; Jiménez-Martín, A.; Vázquez, M.V.; Tercero, J.A. Rad5 Plays a Major Role in the Cellular Response to DNA Damage during Chromosome Replication. *Cell Rep.* **2014**, *9*, 460–468. [CrossRef] [PubMed]

103. Daigaku, Y.; Davies, A.A.; Ulrich, H.D. Ubiquitin-dependent DNA damage bypass is separable from genome replication. *Nature* **2010**, *465*, 951–955. [CrossRef] [PubMed]

104. Karras, G.I.; Jentsch, S. The RAD6 DNA damage tolerance pathway operates uncoupled from the replication fork and is functional beyond S phase. *Cell* **2010**, *141*, 255–267. [CrossRef] [PubMed]
105. Lopes, M.; Foiani, M.; Sogo, J.M. Multiple Mechanisms Control Chromosome Integrity after Replication Fork Uncoupling and Restart at Irreparable UV Lesions. *Mol. Cell* **2006**, *21*, 15–27. [CrossRef] [PubMed]
106. Yeeles, J.T.P.; Marians, K.J. The Escherichia coli replisome is inherently DNA damage tolerant. *Science* **2011**, *334*, 235–238. [CrossRef] [PubMed]
107. Wan, L.; Lou, J.; Xia, Y.; Su, B.; Liu, T.; Cui, J.; Sun, Y.; Lou, H.; Huang, J. hPrimpol1/CCDC111 is a human DNA primase-polymerase required for the maintenance of genome integrity. *EMBO Rep.* **2013**, *14*, 1104–1112.
108. García-Gómez, S.; Reyes, A.; Martínez-Jiménez, M.I.; Chocrón, E.S.; Mourón, S.; Terrados, G.; Powell, C.; Salido, E.; Méndez, J.; Holt, I.J.; et al. PrimPol, an Archaic Primase/Polymerase Operating in Human Cells. *Mol. Cell* **2013**, *52*, 541–553.
109. Bianchi, J.; Rudd, S.G.; Jozwiakowski, S.K.; Bailey, L.J.; Soura, V.; Taylor, E.; Stevanovic, I.; Green, A.J.; Stracker, T.H.; Lindsay, H.D.; et al. PrimPol bypasses UV photoproducts during eukaryotic chromosomal DNA replication. *Mol. Cell* **2013**, *52*, 566–573. [CrossRef] [PubMed]
110. Mourón, S.; Rodriguez-Acebes, S.; Martínez-Jiménez, M.I.; García-Gómez, S.; Chocrón, S.; Blanco, L.; Méndez, J. Repriming of DNA synthesis at stalled replication forks by human PrimPol. *Nat. Struct. Mol. Biol.* **2013**, *20*, 1383–1389. [CrossRef] [PubMed]
111. Elvers, I.; Johansson, F.; Groth, P.; Erixon, K.; Helleday, T. UV stalled replication forks restart by re-priming in human fibroblasts. *Nucleic Acids Res.* **2011**, *39*, 7049–7057. [CrossRef] [PubMed]
112. Zlatanou, A.; Despras, E.; Braz-Petta, T.; Boubakour-Azzouz, I.; Pouvelle, C.; Stewart, G.S.; Nakajima, S.; Yasui, A.; Ishchenko, A.A.; Kannouche, P.L. The hMsh2-hMsh6 Complex Acts in Concert with Monoubiquitinated PCNA and Pol η in Response to Oxidative DNA Damage in Human Cells. *Mol. Cell* **2011**, *43*, 649–662. [CrossRef] [PubMed]
113. Yang, Y.; Durando, M.; Smith-Roe, S.L.; Sproul, C.; Greenwalt, A.M.; Kaufmann, W.; Oh, S.; Hendrickson, E.A.; Vaziri, C. Cell cycle stage-specific roles of Rad18 in tolerance and repair of oxidative DNA damage. *Nucleic Acids Res.* **2013**, *41*, 2296–2312. [CrossRef] [PubMed]
114. Nakajima, S.; Lan, L.; Kanno, S.-i.; Usami, N.; Kobayashi, K.; Mori, M.; Shiomi, T.; Yasui, A. Replication-dependent and -independent Responses of RAD18 to DNA Damage in Human Cells. *J. Biol. Chem.* **2006**, *281*, 34687–34695. [CrossRef] [PubMed]
115. Lambert, S.; Carr, A.M. Replication stress and genome rearrangements: Lessons from yeast models. *Curr. Opin. Genet. Dev.* **2013**, *23*, 132–139. [CrossRef] [PubMed]
116. Xie, K.; Doles, J.; Hemann, M.T.; Walker, G.C. Error-prone translesion synthesis mediates acquired chemoresistance. *Proc. Natl. Acad. Sci. USA* **2010**, *107*, 20792–20797. [CrossRef] [PubMed]
117. Bavoux, C.; Leopoldino, A.M.; Bergoglio, V.; O-Wang, J.; Ogi, T.; Bieth, A.; Judde, J.-G.; Pena, S.D.J.; Poupon, M.-F.; Helleday, T.; et al. Up-Regulation of the Error-Prone DNA Polymerase κ Promotes Pleiotropic Genetic Alterations and Tumorigenesis. *Cancer Res.* **2005**, *65*, 325–330. [PubMed]
118. Yang, Y.; Poe, J.C.; Yang, L.; Fedoriw, A.; Desai, S.; Magnuson, T.; Li, Z.; Fedoriw, Y.; Araki, K.; Gao, Y.; et al. Rad18 confers hematopoietic progenitor cell DNA damage tolerance independently of the Fanconi Anemia pathway in vivo. *Nucleic Acids Res.* **2016**, *44*, 4174–4188. [CrossRef] [PubMed]
119. Ceppi, P.; Novello, S.; Cambieri, A.; Longo, M.; Monica, V.; Lo Iacono, M.; Giaj-Levra, M.; Saviozzi, S.; Volante, M.; Papotti, M.; et al. Polymerase eta mRNA expression predicts survival of non-small cell lung cancer patients treated with platinum-based chemotherapy. *Clin. Cancer Res.* **2009**, *15*, 1039–1045.
120. Doles, J.; Oliver, T.G.; Cameron, E.R.; Hsu, G.; Jacks, T.; Walker, G.C.; Hemann, M.T. Suppression of Rev3, the catalytic subunit of Polζ, sensitizes drug-resistant lung tumors to chemotherapy. *Proc. Natl. Acad. Sci. USA* **2010**, *107*, 20786–20791. [CrossRef] [PubMed]
121. Gao, Y.; Mutter-Rottmayer, E.; Greenwalt, A.M.; Goldfarb, D.; Yan, F.; Yang, Y.; Martinez-Chacin, R.C.; Pearce, K.H.; Tateishi, S.; Major, M.B.; et al. A neomorphic cancer cell-specific role of MAGE-A4 in trans-lesion synthesis. *Nat. Commun.* **2016**. [CrossRef] [PubMed]

![genes logo] **genes**

MDPI

Review

Translesion Synthesis: Insights into the Selection and Switching of DNA Polymerases

Linlin Zhao [1,2,*] and M. Todd Washington [3]

[1]   Department of Chemistry and Biochemistry, Central Michigan University, Mount Pleasant, MI 48859, USA
[2]   Science of Advanced Materials Program, Central Michigan University, Mount Pleasant, MI 48859, USA
[3]   Department of Biochemistry, Carver College of Medicine, University of Iowa, Iowa City, IA 52242, USA;
       todd-washington@uiowa.edu
*   Correspondence: linlin.zhao@cmich.edu; Tel.: +1-989-774-3252

Academic Editor: Eishi Noguchi
Received: 30 November 2016; Accepted: 4 January 2017; Published: 10 January 2017

Abstract: DNA replication is constantly challenged by DNA lesions, noncanonical DNA structures and difficult-to-replicate DNA sequences. Two major strategies to rescue a stalled replication fork and to ensure continuous DNA synthesis are: (1) template switching and recombination-dependent DNA synthesis; and (2) translesion synthesis (TLS) using specialized DNA polymerases to perform nucleotide incorporation opposite DNA lesions. The former pathway is mainly error-free, and the latter is error-prone and a major source of mutagenesis. An accepted model of translesion synthesis involves DNA polymerase switching steps between a replicative DNA polymerase and one or more TLS DNA polymerases. The mechanisms that govern the selection and exchange of specialized DNA polymerases for a given DNA lesion are not well understood. In this review, recent studies concerning the mechanisms of selection and switching of DNA polymerases in eukaryotic systems are summarized.

Keywords: DNA damage; DNA lesion bypass; DNA polymerase; genomic instability; mutagenesis; translesion synthesis

1. Introduction

DNA is susceptible to numerous endogenous and exogenous chemicals, producing a wide variety of DNA lesions. Unrepaired DNA lesions are potential sources of replication and transcription errors, replication fork arrest, and cell death, which together contribute to genomic instability and pathogenesis. Two strategies exist to counteract replication fork stalling. One involves template switching, in which the undamaged template from the sister chromatid is used for recombination-dependent DNA synthesis; this process is usually error-free. A second strategy is to use one or more of the translesion synthesis (TLS) DNA polymerases (pols) to accomplish nucleotide incorporation opposite and past the DNA lesion before a replicative DNA polymerase (pol ε or pol δ in eukaryotes) resumes its function. This process—which is intrinsically error-prone—is a major source of DNA damage-induced mutagenesis [1].

Genetic studies in the 1970s showed that mutations in the UV nonmutable (*umu*) locus in *Escherichia coli* (*E. coli*) [2,3] and the reversionless (*REV*) locus in *Saccharomyces cerevisiae* (*S. cerevisiae*) [4,5] were associated with deficiencies in mutagenesis in these organisms upon treatment with DNA-damaging agents. Around the same time, cells from patients with a variant form of a cancer predisposition syndrome *xeroderma pigmentosum* (XP-V) were found to be deficient in synthesizing daughter DNA strands after UV irradiation [6]. It was not until the 1990s that the products of these and related genes were purified and biochemically characterized. The product of the yeast *REV1* gene was found to be a dCMP transferase [7], and the product of the yeast *REV3* gene was shown to be the catalytic subunit of pol ζ, which is able to bypass a common UV-induced cyclobutane pyrimidine dimer (CPD) DNA lesion with low efficiency [8]. In 1999, the yeast Rad30 protein was shown to be able to replicate past a thymine–thymine CPD as efficiently and accurately as with undamaged thymines [9]. Shortly after, defects in the human gene encoding Rad30 was shown to cause the XP-V syndrome [10,11]. By 2000, the arsenal of TLS polymerases had expanded rapidly with the discovery of *E. coli* pol IV (DinB) [12] and pol V (UmuC) [13,14], pol ι (a second human ortholog of Rad30) [15–18], and pol κ (a human ortholog of *E. coli* DinB) [19–22]. These findings led to the realization that TLS is a conserved process from bacteria to humans [23], which involves a large family of proteins, known as TLS DNA polymerases.

Today, 17 human DNA polymerases have been purified and biochemically characterized, and these proteins are classified into A, B, X, Y, and AEP (archaeo-eukaryotic primase superfamily) families according to their sequence homology and structural similarities [24–26]. The best-characterized Y-family DNA polymerases include pol η, pol ι, pol κ, and Rev1, which, together with B-family enzyme pol ζ, are the principle TLS pols in humans. Pols of A and X families also have TLS activities and contribute to mutagenesis in DNA repair pathways such as base excision repair and non-homologous end joining (NHEJ) [27]. The most recently discovered DNA polymerase/primase PrimPol (AEP superfamily) has the capability of bypassing a number of DNA lesions [26,28–31]. More importantly, PrimPol has primase activity that can perform de novo DNA synthesis using deoxyribonucleotide triphosphates (dNTPs), which is important for replication re-start downstream of a stalled fork [32–35]. Nowadays, the understanding of TLS polymerases has evolved from their conventional lesion bypass activities to myriad roles in organismal fitness and disease, such as to increase the diversity of the immunoglobulin gene during hypermutation, to overcome secondary DNA structures during DNA copying, to participate in DNA repair, and to contribute to mutagenesis in tumors [25,27,36,37].

Translesion synthesis is thought to occur via two non-mutually exclusive processes. One is for TLS pols to participate at a replication fork, and the other is to fill post-replicative gaps [38]. The first process involves several polymerase-switching processes, including dissociation of a stalled replicative polymerase from the replication fork, binding of one or two TLS polymerases to the replication terminus for nucleotide insertion and extension, and eventually displacement of TLS pols with a replicative polymerase downstream of the DNA lesion [38,39]. The latter pathway requires fewer switching events. A major unanswered question is how polymerase switching occurs at the replication factories (reviewed in [40–42]). Deciphering the mechanisms of the polymerase exchange is not only fundamental for the understanding of translesion synthesis, but also important for the development of chemotherapy to control TLS activities [25,38,43]. This is because many cancer chemotherapies work by damaging DNA, and inhibiting TLS pols that affect DNA repair capability holds promise for improving responses to treatments [25,43]. This review aims to summarize recent studies on the mechanistic aspects of TLS in eukaryotic systems. For detailed discussions on the biochemical properties, regulation, and functions of TLS DNA polymerases, please see these excellent reviews [24,27,38,44–46]. Readers interested in TLS in bacteria are referred to the following reviews [42,47].

2. Selection and Switching of Specialized DNA Polymerases

DNA is susceptible to a variety of chemicals from endogenous and exogenous sources, which generates up to 100,000 DNA lesions per cell each day [48]. Selection of the most appropriate specialized DNA polymerase to bypass a given lesion is dictated by a number of possible factors. One obvious factor is the identity of DNA lesions. A second potential factor is the interactions of specialized polymerases with hub proteins such as PCNA and Rev1. Other potential factors include the availability of TLS polymerases in the vicinity of stalled replication forks owing to cell cycle and transcription regulation or protein degradation.

2.1. Selection of the "Right" TLS Pol for Benzo[a]pyrene-Derived DNA Lesions: A Case Study

In eukaryotes, various TLS polymerases have evolved to accommodate different types of DNA damage. When a polymerase is recruited to a stalled fork, it can only be used if it is able to accommodate the damaged primer-template in its active site and is able to catalyze the nucleotide-incorporation reaction [38]. Certain DNA modifications can be bypassed by replicative DNA polymerases [49,50], whereas bulky DNA lesions, such as carcinogen benzo[*a*]pyrene (BaP)-derived DNA damage, often require one or more TLS DNA polymerases to facilitate the fork progression [51]. Knowledge concerning cognate DNA lesions of each TLS pol has been reviewed [27,44,52]. TLS pols often act redundantly in the bypass of a given DNA lesion, and it is challenging to firmly identify the most biologically relevant DNA lesion for some pols. A few structurally distinct DNA lesions, such as cyclobutane pyrimidine dimer (CPD) and BaP-derived lesions, require specific polymerase activities [53,54]. Multiple factors including the chemistry of DNA lesion and DNA polymerase structure affect the selection of TLS pols. In the following section, DNA lesions derived from BaP, a prototypical carcinogen, will be used as an example to discuss how the chemistry of DNA lesions affects the enzymatic activities of DNA polymerases.

2.1.1. BaP-Induced DNA Damage

BaP is a ubiquitous environmental pollutant that exists in overcooked meat, vehicular exhaust, coal tar, and tobacco smoke. BaP is a Group 1 carcinogen classified by the International Agency for Research on Cancer (IARC), and has been associated with skin, lung, and colon cancers in humans [55,56]. The carcinogenicity of BaP is attributed in part to its ability to form the ultimate tumorigenic metabolites (+)-7β,8α-dihydroxy-9α,10α-epoxy-7,8,9,10-tetrahydrobenzo[*a*]pyrene [7R,8S,9S,10R steric configuration; the most distant hydroxyl group is *anti* relative to the orientation of the epoxide group, and is hereinafter referred to as (+)-*anti*-BPDE] and (−)-7α,8β-dihydroxy-9β,10β-epoxy-7,8,9,10-tetrahydrobenzo[*a*]pyrene [7S,8R,9R,10S steric configuration; (−)-*anti*-BPDE] (Figure 1). (+)-*anti*-BPDE is more tumorigenic than its enantiomer (−)-*anti*-BPDE [57–59]. Both metabolites react with the N^2 exocyclic amino group of guanine (Figure 1) and to a lesser extent with the N^6 exocyclic amino groups of adenine and the N^4 exocyclic amino groups of cytosine to form DNA adducts [60,61]. Due to the carcinogenic potency, BPDE-derived DNA lesions are among the best-studied DNA lesions in terms of their toxicological mechanisms. Alternative bioactivation routes can convert BaP to radical cations that are reactive towards the C8 or N7 atoms of guanine and the N3 or N7 positions of adenine, some of which can form mutagenic apuridinic/apyrimidinic (AP) sites due to the unstable glycosidic linkage [62]. Other pathways involve biotransformation via aldo-keto reductase to yield reactive quinone-derived DNA adducts that are chemically labile or stable [63,64]. BaP-derived DNA lesions block DNA synthesis by replicative pols and induce mutagenic replication products via TLS. A prevalent mutation resulting from BaP exposure is a G to T transversion, a common mutation found in BaP-treated mammalian cells and the *p53* gene of lung cancers of smokers [55,56]. The local sequence context of BaP-induced DNA damage also plays a role in the resulting mutation pattern [65–68].

2.1.2. Accurate Bypass of BaP-Derived DNA Lesions

Major BPDE-derived DNA lesions include the stereoisomeric 2′-deoxyguanine (dG) adducts (+)-*trans-anti*-BPDE-N^2-dG, (+)-*cis-anti*-BPDE-N^2-dG, (−)-*trans-anti*-BPDE-N^2-dG and (−)-*cis-anti*-BPDE-N^2-dG (Figure 1), as well as the 2′-deoxyadenosine (dA) adducts (+)-*trans-anti*-BPDE-N^6-dA, (+)-*cis-anti*-BPDE-N^6-dA, (−)-*trans-anti*-BPDE-N^6-dA and (−)-*cis-anti*-BPDE-N^6-dA. These lesions are able to assume a variety of conformations depending on the local sequence context, as evidenced by solution nuclear magnetic resonance (NMR) structures (reviewed in [69] and references therein). Consequently, there is no universal TLS pol to bypass all lesions due to their structural diversity and the varying bypass capabilities of TLS pols. In addition, effects of the host cell and the local sequence context contribute to the varying degrees of bypass efficiencies and the resulting mutations [70]. Pol κ is well known for its role in the accurate bypass of BPDE-N^2-dG DNA lesions. Pol κ is capable of replicating past all four BPDE-derived N^2-dG lesions in a primarily error-free fashion in vitro and in vivo [22,71–74] and is protective against the mutagenic effects of BaP in cells [54,75]. However, pol κ is unable to bypass (+)-*trans-anti*-BPDE-N^6-dA or (−)-*trans-anti*-BPDE-N^6-dA lesions [76], and these lesions are thought to contribute to the mutagenicity of low-dose BaP exposure [77–79]. The extent of the involvement of pol κ in the accurate replication across the (+)-*trans-anti*-BPDE-N^2-dG lesion in cells remains controversial, mostly likely due to the different sequences and cell lines used in respective laboratories. Using a quantitative bypass assay, Avkin et al. demonstrated that approximately 60% of the (+)-*trans-anti*-BPDE-N^2-dG adducts require pol κ for accurate bypass [75]. On the other hand, Hashimoto et al. showed that the error-free products account for less than 10% of total TLS products with the same DNA lesion in mouse embryonic fibroblasts [80]. Pol ι, which is the least accurate TLS pol, is known for preferentially misincorporating T opposite unmodified G [81]. Interestingly, in vitro pol ι incorporates a correct nucleotide opposite stereoisomeric BPDE-N^2-dA adducts, although it is unable to insert nucleotides opposite BPDE-N^2-dG adducts or to extend the primer beyond the lesion [76,82]. Further experiments are needed to confirm the biological significance of this particular bypass activity of pol ι.

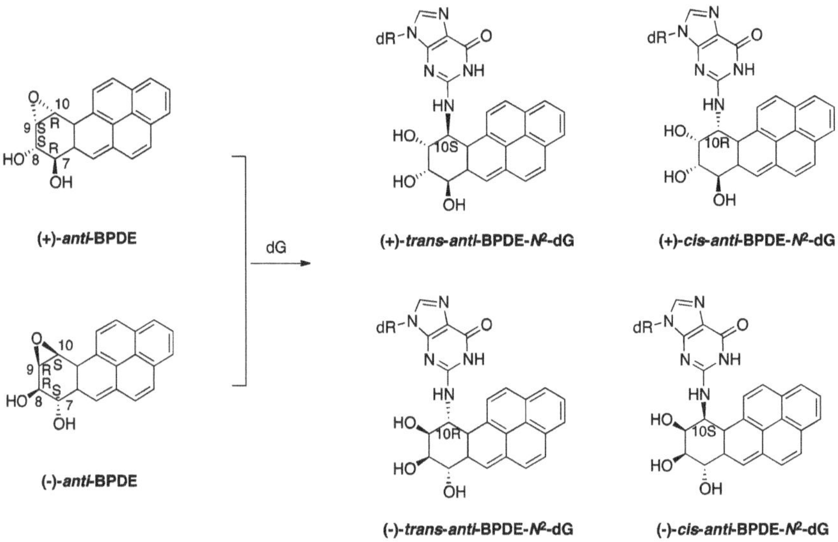

Figure 1. Structures of stereoisomers of BPDE-derived N^2-dG DNA adducts.

A recent X-ray crystal structure of pol κ:(+)-*trans-anti*-BPDE-N^2-dG-DNA:dCTP (pol κ-BPDE) complex has provided insights into why pol κ is adept at bypassing bulky BPDE-induced DNA lesions [83]. The overall structure of pol κ-BPDE closely resembles the structure of a pol κ complex with an unmodified DNA substrate, indicating that pol κ accommodates the (+)-*trans-anti*-BPDE-N^2-dG lesion at the active site (Figure 2A,B). The BPDE-adduced substrate adopts a standard B-form of DNA, and the BPDE-N^2-dG adduct retains the *anti* conformation. The BPDE ring is positioned in the minor groove and forms an additional H-bond with the incoming dCTP (Figure 2C). The BPDE ring points towards the 5′-end of the template strand, consistent with the solution NMR structures of DNA containing BPDE-derived dG lesions [84,85]. This conformation of the adduct is accommodated by an open DNA binding cleft in pol κ (Figure 2D), which is not found in pol η or pol ι. Modeling this conformation of the BPDE-adduced DNA into the structures of pol η (Figure 2E) and pol ι (Figure 2F) results in steric clash with both pols. In addition, the unique N-clasp domain of pol κ (not found in other Y-family TLS pols) supports an open conformation of the protein and stabilizes the single-stranded template for the efficient and error-free bypass of BPDE-dG DNA lesions [83].

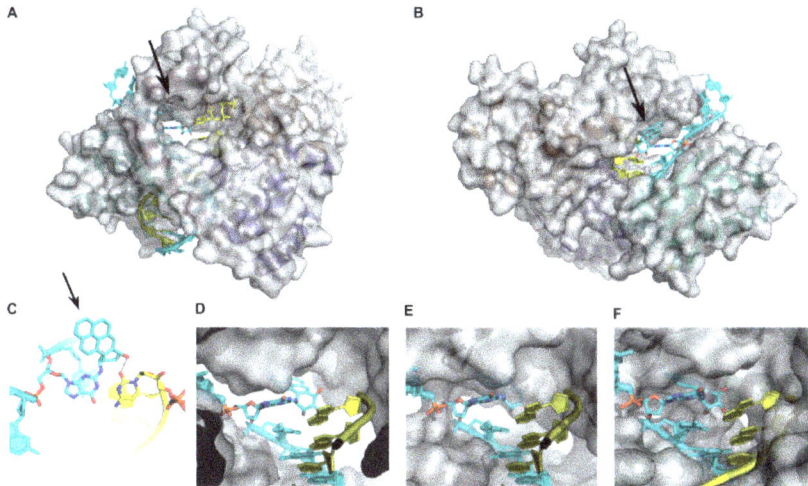

Figure 2. The structures of DNA polymerase complexes with a BPDE-dG lesion-containing duplex. The adducted template is shown in cyan, and the primer and incoming dCTP are shown in yellow. The black arrows are pointing at the BPDE ring. (**A,B**) Different views of the X-ray crystal structure of pol κ:(+)-*trans*-dG-N^2-BPDE-DNA:dCTP (pol κ-BPDE) complex (PDB: 4U7C). The major groove of DNA is facing the viewer in (**A**); and the minor groove of DNA is facing the viewer in (**B**). (**C**) Base pairing of (+)-*trans*-dG-N^2-BPDE lesion and the incoming dCTP at the active site of pol κ. An additional hydrogen bond formed between a hydroxyl group of BPDE and the O_2 atom of cytidine is shown with a dashed line. (**D**) Zoomed-in view of pol κ accommodating the BPDE ring in an open DNA binding cleft. (**E**) Structural model of pol ι (PDB: 4FS2) with an adducted substrate. The conformation of the DNA is adopted from the pol κ-BPDE structure. (**F**) Structural model of pol η (PDB: 3MR2) with an adducted substrate. The conformation of the DNA is from the pol κ-BPDE structure. For simplicity, the incoming dCTP is omitted in (**D–F**).

The fact that pol ι is able to incorporate the correct dTTP opposite BPDE-N^6-dA DNA lesions in vitro suggests that pol ι can accommodate certain conformers of BPDE-N^6-dA DNA lesions at the active site. Although a ternary structure of pol ι with the BPDE-N^6-dA lesion and an incoming nucleotide is unavailable, molecular dynamics simulations have demonstrated that a BPDE-N^6-dA lesion assumes an *anti* or *syn* conformation at the active site of pol ι depending on the adjacent nucleotides forming a Watson–Crick or Hoogsteen base pair with the incoming dTTP, respectively [86]. The BPDE ring is positioned in the major groove due to the relatively narrow active site of pol ι, and forms additional H-bonds with nearby nucleotides [86].

2.1.3. Error-Prone Bypass of BaP-Derived DNA Lesions

BaP-induced mutations are fueled at least in part by error-prone DNA replication across BPDE-derived DNA adducts [87]. More than 90% of the bypass events across the (+)-*trans*-*anti*-BPDE-N^2-dG DNA lesion are error-prone in mouse embryonic fibroblasts [80]. For pol η, DNA synthesis is almost completely blocked by (−)-*trans*-*anti*-BPDE-N^6-dA adduct, whereas weak and error-prone bypass activities exist for both stereoisomeric BPDE-N^2-dG adducts and (+)-*trans*-*anti*-BPDE-N^6-dA adduct [76]. Using human XP-V fibroblasts that express a truncated and non-functional pol η [9], Avkin et al. found that the bypass of the (+)-*trans*-*anti*-BPDE-N^2-dG-DNA lesion is largely accurate and concluded that pol η is not essential for TLS across this particular lesion with the template sequence they used [75]. On the other hand, pol ζ plays an important role in the mutagenic bypass of the (+)-*trans*-*anti*-BPDE-N^2-dG-DNA lesion, which is likely due to its function as an extender DNA polymerase [51,80]. The importance of pol ζ in error-free bypass of BPDE-derived lesions remains controversial [51,80]. Rev1 is known for its deoxycytidyl transferase activity and its role as a scaffold protein to interact with other Y-family DNA polymerases [7,88–93]. Although Rev1 is capable of inserting dCTP opposite (+)-*trans-anti*-BPDE-N^2-dG and (−)-*trans-anti*-BPDE-N^2-dG DNA lesions in vitro [88], its role in error-free bypass seems to be nonessential in mouse cells [80]. Instead, the non-catalytic function of Rev1 is important for pol κ-mediated BPDE resistance of mouse embryonic fibroblast cells [94], and for the erroneous bypass of the (+)-*trans-anti*-BPDE-N^2-dG lesion by pol ζ [80].

Together, it is apparent that the identity of BaP-derived DNA lesions drives the selection of TLS pols. Multiple factors, including the steric effects, tautomerization, the ability to form base pairs with the incoming nucleotide and local sequence context, seem to affect the selection of TLS pols. Apart from BaP-derived DNA lesions, a variety of DNA lesions have been assayed in vitro and in cellular experiments to identify the most biologically relevant TLS pol(s); however, in many cases, different TLS pols act redundantly during TLS [52], and it remains a challenge to generate a list of cognate lesions for each TLS pol. It seems logical for backup enzymes to exist for DNA replication and repair. The fact that a respective TLS pol has evolved to protect against the mutagenic effects of BPDE and CPD-derived DNA damage underscores the importance of these carcinogens.

2.2. PCNA: An Interaction Hub for Many Partners

PCNA is known for orchestrating a variety of components in DNA metabolism. PCNA was first discovered as an auxiliary protein that stimulates the activity of DNA polymerase δ [95,96], and was subsequently recognized for its remarkable abilities in coordinating multiple cellar processes such as unperturbed DNA replication, translesion synthesis, Okazaki fragment maturation, DNA repair, chromatin remodeling, and cell cycle regulation [97–100]. PCNA promotes the access of specialized pols to the replication factories through physical and functional interactions with these proteins. PCNA interacts with purified Y-family TLS pols and stimulates the catalytic efficiencies of these polymerases in vitro [101–104]. The understanding of the importance of these interactions in vivo was obtained primarily from nuclear focus-formation assays with DNA damaging reagent-treated cells ([105] and references therein). However, care should be taken in interpreting these results because the composition of these foci and whether they represent direct interactions are not known [105]. In this section, the biochemical basis of interactions between PCNA and different DNA polymerases is discussed.

2.2.1. Interactions between PCNA and DNA Polymerases

Eukaryotic PCNA comprises three identical subunits, and each subunit has two similarly folded domains joined by an interdomain connector loop (Figure 3A) [106,107]. The homotrimeric eukaryotic PCNA is assembled into a circular ring with a central hole that is wide enough to encircle the DNA and to allow diffusion of PCNA along the DNA [108]. The PCNA ring has one side facing the direction of DNA synthesis and the other side pointing away (hereinafter referred to as the front side and the back side of PCNA, respectively). The front side contains the C-terminus of each monomer and the interdomain-connecting loop. A hydrophobic pocket (Figure 3A) near the interdomain-connecting loop on the front side of each monomer serves as a platform to interact with DNA polymerases. In vitro, interactions between PCNA and purified TLS pols (e.g., human pol η, pol ι and pol κ) stimulate the catalytic efficiencies of these polymerases with unmodified and damaged DNA substrates via lowering the K_m of the incoming nucleotide [101–104]. Pol η and pol ι, but not pol κ, have elevated processivity in the presence of PCNA, replication factor C (RFC) and replication protein A (RPA) [101–104]. Pol ζ is stimulated by PCNA with lesion-bearing DNA, but not with unmodified substrates [109,110]. PCNA stimulates the catalytic efficiency of Rev1 and does so to a greater extent when the PCNA is monoubiquitinated [111].

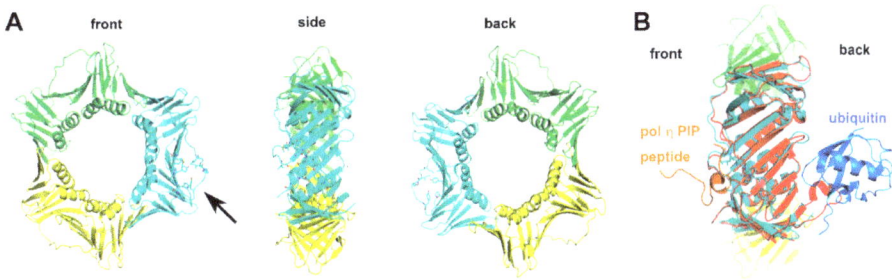

Figure 3. Structures of human PCNA and yeast ubiquitinated PCNA. (**A**) Front, side, and back views of human PCNA (PDB:2ZVK). Three subunits are shown in green, yellow, and cyan. In one subunit (cyan), amino acid residues surrounding the hydrophobic pocket near the interdomain-connecting loop are shown in stick. The black arrow is pointing at the hydrophobic pocket. For simplicity, the pol η PIP peptide is omitted from the original crystal structure. (**B**) A subunit of yeast ubiquitinated PCNA (red; PDB:3L10) is superimposed with a subunit (cyan) of human PCNA (PDB:2ZVK). The pol η PIP peptide (orange) interacts with the hydrophobic pocket on the front side of PCNA, and ubiquitin (blue) interacts with the back side of PCNA.

2.2.2. Biochemical Basis of PCNA-Pol Interactions

The interacting partners of PCNA in eukaryotes generally contain one or more PCNA-interacting protein (PIP) motifs. Based on the amino acid sequence of these motifs, PIPs are classified into canonical and non-canonical PIPs, which differ in their sequence and binding affinity for PCNA. Canonical PIPs, found in p21[WAF1/CIP1] [107], the p66 subunit of pol δ [112] and FEN1 [113], have a consensus sequence Qxx[L/I/M]xx[F/Y][F/Y/W] featuring high-affinity interactions with PCNA. Non-canonical PIPs, on the other hand, have alternative residues at the first and last positions, lowering the binding affinity for PCNA relative to the consensus sequence. The difference in the binding affinities for PCNA potentially contributes to affinity-driven polymerase switching [98]. For example, the PIP peptide (QVSITGFF, canonical) of the p66 subunit of human pol δ has a higher affinity for PCNA relative to pol η (MQTLESFF, non-canonical) [114]. Changing the first amino acid residue of the PIP peptide of pol η to a glutamine (QQTLESFF) results in a four-fold increase in its affinity for PCNA [114]. The apparent dissociation constant (K_d) of human pol δ, pol η, pol κ, and pol ι PIP peptides with PCNA are summarized in Table 1. Although affinity-driven competition has been proposed as a mechanism for polymerase switching, the molecular mechanism of this model remains to be studied in much detail. Rev1, on the other hand, has no PIP motifs, but interacts with PCNA through its N-terminal BRCA1 C-terminus (BRCT) domain [115,116] and/or polymerase-associated domain (PAD) [117]. This interaction between the PAD domain of Rev1 and PCNA observed in yeast remains to be confirmed in vertebrates. Importantly, several recent studies have discovered non-conventional interacting partners of the PIP motif as well as the related Rev1-interacting region (RIR, see below). For example, yeast pol η uses its PIP motif to interact with both PCNA and Rev1 [118], and human pol η uses one of its RIR motifs to interact with Rev1 and pol δ [119]. In fact, the very notion of a PIP motif as a distinct entity has recently been questioned, and it has been proposed that these and other related motifs be renamed PIP-like motifs to better reflect their broader roles in the network of interacting proteins responsible for DNA replication and repair [120].

Table 1. Apparent dissociation constants (K_d) of DNA polymerase holoenzymes or PIP peptides with PCNA. Conserved amino acid residues relative to a consensus sequence are in bold. Italic cysteines indicate that these amino acid residues were included in addition to the native RIR peptide to facilitate the measurement.

DNA Polymerase	Sequence	K_d (μM)
pol δ PIP	[451]GKANRQVSITGFFQRK	16 [1]
pol δ holoenzyme		<0.010 [2]
pol η PIP2	C[694]KRPRPEGMQTLESFFKPLTH	0.40 [3]
pol η holoenzyme		0.12 [4]
pol κ PIP + PLTH	C[856]IKPNNPKHTLDIFFK*PLTH*	4.9 [3]
pol ι PIP	C[419]AKKGLIDYYLMPSLST	0.39 [3]

[1] Measured by isothermal titration calorimetry [112]; [2] Estimated using a binding assay containing forked DNA-PCNA complex as substrate and pol δ as ligand. Values in [2] and [4] are from ref. [121]; [3] Obtained from surface plasmon resonance (SPR) assays [114].

2.2.3. Ubiquitination of PCNA

Post-translational modifications of PCNA play an important role in DNA damage tolerance pathways [97,98,122]. Ubiquitination of PCNA, in particular, is known to participate in a variety of pathways during DNA replication and repair [122]. Ubiquitination of PCNA, mediated by the Rad6–Rad18 ubiquitination system, occurs in response to fork stalling near a lesion or an unusual DNA structure. Generally, the monoubiquitinated PCNA serves as an interacting platform for TLS DNA polymerases, whereas the polyubiquitinated PCNA is involved in error-free bypass via recombination-dependent pathways [122]. Ubiquitination of PCNA occurs primarily at K164 and to a lesser extent at other lysine residues [123,124]. One or two ubiquitin-binding motifs (UBMs; pol ι and Rev1) or ubiquitin-binding zinc-fingers (UBZs; pol η and pol κ) are present in Y-family DNA polymerases [125], which increase the affinity of DNA polymerases for monoubiquitinated PCNA and potentially facilitate the recruitment of TLS pols. In *S. cerevisiae*, it is established that the monoubiquitination of PCNA is essential for optimal TLS and TLS polymerase switching. For example, in vitro studies using recombinant yeast enzymes show that both unmodified and monoubiquitinated PCNA stimulates the efficiencies of nucleotide incorporation by pol η and REV1; however, a stronger stimulatory effect is observed when the PCNA is monoubiquitinated [111,126,127]. In addition, upon replication stalling, the exchange of yeast pol η and pol δ occurs in the presence of monoubiquitinated PCNA but not with the unmodified PCNA [128]. In yeast cells, Rad6-mediated monoubiquitination of PCNA is required to activate TLS by pol η [129,130].

On the contrary, in mammalian systems, whether a direct interaction between pol η and ubiquitinated PCNA is required (or even occurs) during TLS remains controversial. In human cells, UBMs are needed for foci formation of Y-family polymerases and for physical interactions between polymerases and ubiquitinated PCNA [125,131]. However, as mentioned earlier, the foci formation should not be used to conclude that a direct interaction between pol η and ubiquitinated PCNA is required (or even occurs) during TLS in mammalian systems. On the other hand, physical and specific interactions of pol η with ubiquitinated PCNA have been demonstrated with co-immunoprecipitation using cell extracts [132,133]. While Acharya et al. reported that a direct binding of the UBZ domain of pol η with ubiquitinated PCNA is not required during TLS [134], this conclusion has been questioned because the dispensability of the pol η UBZ domain is thought to be due to an artificially increased PCNA expression [135]. Other in vivo evidence suggests ubiquitination of PCNA is in fact dispensable. For example, pol η localizes into replication foci during unperturbed DNA replication [125] as well as upon treatment with UV irradiation [133,136,137] independently of PCNA monoubiquitination. Hendel et al. have shown that the ubiquitination of PCNA is important, but not essential for TLS in mouse cells [138]. Using photobleaching techniques, Sabbioneda et al. have demonstrated that PCNA ubiquitination is not required for the pol η foci formation, but increases the residence time of pol η in foci in human cells [136]. In addition, studies from several laboratories have demonstrated that

PCNA ubiquitination is dispensable during lesion bypass [136,138,139], in which TLS pols may be recruited via interactions with Rev1 (discussed in Section 2.3) [140,141]. The interactions between TLS polymerases and PCNA are considered to be highly dynamic judging by the times of immobilization of pol η and pol ι (100–200 ms) upon DNA damage [136]. Therefore, it is proposed that pol η transiently and continually probes the exposed DNA for suitable substrates [136]. Recently, using quantitative kinetic assays and a reconstituted lagging-strand replication system, Hedglin et al. have shown that the binding of pol η to PCNA and pol η-catalyzed DNA synthesis occur without PCNA monoubiquitination, and that efficient exchange of pol η with pol δ happens owing to the intrinsic DNA binding properties of these pols [121]. Additional studies are warranted to unequivocally determine the biological functions of PCNA ubiquitination in vivo.

2.2.4. Structure of Monoubiquitinated PCNA

The X-ray crystal structure of monoubiquitinated *S. cerevisiae* PCNA has provided additional insights into PCNA–polymerase interactions [126]. The expression of yeast monoubiquitinated PCNA is achieved by splitting the protein into two self-assembling polypeptides [126]. As shown in Figure 3B, the ubiquitin moiety uses its canonical hydrophobic surface to interact specifically but weakly with PCNA via electrostatic and hydrogen-bonding interactions. The attachment of ubiquitin does not alter the conformation of PCNA, suggesting that there is no or minimal conformational change of PCNA upon ubiquitin binding [126]. The ubiquitin molecule is located on the back side of PCNA, presumably leaving the hydrophobic pocket on the front side to interact with the PIPs of other DNA polymerases, which is consistent with a tool belt model of translesion synthesis. A PCNA tool belt is a structure with multiple TLS polymerases directly interacting with PCNA without directly interacting with one another. Based on the structure of ubiquitinated PCNA, it is proposed that when pol δ stalls at a DNA lesion, the ubiquitination of PCNA facilitates the recruitment of pol η to the back side of PCNA [126]. The catalytic core of pol η then displaces pol δ since it is connected to the C-terminus of pol η by a long, flexible linker. A recent structural model derived from low-resolution single-particle electron microscopy suggests that pol η can associate with the front face of the PCNA in the editing mode [142]. Additional structures of eukaryotic multi-protein complexes with DNA, PCNA and TLS pols are needed to fully understand how multiple TLS pols are coordinated.

2.2.5. Additional Structural Motifs for Stabilizing PCNA-Pol Complexes

In eukaryotes, B-family DNA polymerases include pol α, pol δ, pol ε, and pol ζ. The former three polymerases are the major players responsible for the bulk of DNA synthesis, and pol ζ is a major error-prone DNA polymerase. In *S. cerevisiae*, pol δ is a three-subunit complex comprised of the catalytic subunit pol3 and accessory subunits pol31 and pol32. The pol δ holoenzyme is formed via interactions between pol31 and the C-terminal segment of pol3, and between pol32 and pol31 [143]. Subunits pol 31 and pol32 are also components of a four-subunit pol ζ4 (discussed in Section 2.4). In addition to the aforementioned structural motifs (PIPs and RIRs) that are important for protein–protein interactions, two conserved cysteine-rich metal-binding motifs (CysA and CysB) within the C-terminal segment of the catalytic subunits of all four B-family DNA polymerases are important for DNA replication and stabilizing multi-protein complexes in *S. cerevisiae* [144]. The Zn-binding motif, CysA of pol3 (the catalytic subunit of yeast pol δ) plays a critical role in PCNA-pol δ complex formation, whereas [4Fe-4S]-binding motif CysB is imperative for the formation of a highly processive yeast pol δ holoenzyme [144]. Mutation of the conserved cysteine residues in the CysA motif significantly decreases the processivity of yeast pol δ; processive DNA replication can be partially restored by adding wild-type pol δ into the system but cannot be restored by adding a mutant form of pol δ without the PIP motif on pol32 (pol32−ΔPIP). By contrast, fully proficient DNA replication was observed for mutant pol δ with pol32−ΔPIP. These results suggest that PIPs may be more relevant for recruiting pols to replication foci in the nucleus, whereas the conserved cysteine-rich metal-binding motifs are important for the formation and/or stability of the PCNA–pol δ complex in processive DNA replication [144]. This is consistent with the previously proposed two-stage recruitment model for TLS polymerases—first, to increase the local concentration of TLS pol(s) at the replication factories, and second, to load TLS pol(s) to the replication termini [38].

2.3. Rev1: A Scaffold Protein

REV1, along with *REV3* and *REV7*, is among the first translesion synthesis DNA polymerase genes discovered in yeast mutagenesis experiments [4]. Rev1 is the most intriguing Y-family polymerase because of its deoxycytidyl transferase activity [7,88] and its protein template-directed nucleotide incorporation [145]. Yeast genetic studies led to the suggestion that Rev1 has a "second function" separate from its catalytic activity [89]. Subsequent biochemical and cellular studies augmented this proposal by demonstrating that human and mouse Rev1 physically interacts with pol η, pol ι, pol κ, and Rev7 (an accessory subunit of pol ζ) [90–93], and that the catalytic-null mutant of Rev1 does not affect the levels of mutagenesis induced by DNA-damaging agents [146,147].

2.3.1. Interactions between Rev1 and Other Pols

The interactions of Rev1 with its protein partners are critically dependent on its C-terminal domain (CTD) [90–92,147]. Rev1-interacting proteins contain RIRs that are centered around conserved phenylalanine residues (FF). These interacting proteins include B-family pol δ [110,148,149] and pol ζ [93, 150]; Y-family pol ι, pol κ, and pol η [90–92,94,119]; base excision repair protein XRCC1 [151]; and yeast Rad5 (a multi-functional protein involved in template switching) [152]. Recent NMR and X-ray crystallographic data have provided a structural basis of the interactions between Rev1 and its partners. According to the solution NMR structures of the mouse Rev1 CTD–pol κ RIR peptide complex and the human Rev1 CTD–pol η RIR peptide complex [153,154], the overall core helix-bundle structure of the RIR-bound human Rev1 CTD is similar to that of the free Rev1 CTD (Figure 4A). Rev1 CTD folds into a four-helix bundle (α1—α4), mediated by a network of interacting residues from individual helices. A majority of these residues are conserved from yeast to human, which contribute to the stability of the CTD of Rev1 across species [154]. Six residues at the N-terminus of α1 helix fold into a structurally defined β-hairpin, and together with the shallow hydrophobic surface between α1 and α2, create a deep hydrophobic cavity for high-affinity binding with RIR peptides [153,154]. The disordered RIR peptides of pol η and κ arrange into a three-turn α-helix upon binding with Rev1 CTD. Two phenylalanine residues of the RIR peptides (of pol η, pol κ, and p66) interact with the hydrophobic cavity of Rev1 CTD (Figure 4B). These two conserved phenylalanine residues are essential for the formation of the protein complex as evidenced by mutational studies in yeast two-hybrid assays [94,153].

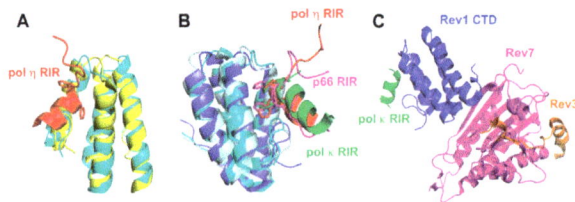

Figure 4. Interactions between Rev1 CTD and the RIR peptides or the interacting fragments of pol ζ. (**A**) Superimposed structures of free human Rev1 CTD (yellow, PDB: 2LSY) and human pol η RIR-bound Rev1 CTD (cyan, PDB: 2LSK). The pol η RIR is in red with the side chains of the conserved phenylalanine residues shown in stick. (**B**) Superimposed structural complexes of mouse Rev1 CTD (blue; PDB:2LSJ) with the pol κ RIR peptide (green), human Rev1 CTD (cyan; PDB:2LSK) with the pol η RIR peptide (red) and human Rev1 CTD (pale cyan; PDB:2N1G) with the p66 (a subunit of pol ζ4) RIR peptide (magenta). Three RIR peptides interact with the same region of Rev1 CTD. The side chains of conserved phenylalanines are shown in stick. (**C**) Mouse Rev1 CTD in complex with Rev7, a fragment of Rev3 and the pol κ RIR peptide (PDB: 4FJO).

2.3.2. Interactions between Rev1 and Pol ζ

Pol ζ is considered as an "extender polymerase" in the generally accepted two-step bypass mechanism in mammals [39,155]. In the first step, an "inserter" polymerase (e.g., pol η, pol ι or pol κ) incorporates a nucleotide opposite the lesion, and in the second step an "extender" polymerase (e.g., pol ζ) extends beyond the base pair that involved the lesion before a replicative polymerase takes over the DNA synthesis. It is well documented that the Rev7 subunit of human pol ζ interacts with Rev1 [93], and that the interaction is functionally important for translesion synthesis across a (6–4) thymine-thymine photoproduct [156]. Since the discovery of a four-subunit complex of pol ζ4 (Rev3-Rev7-p50-p66; p50 and p66 are also subunits of human pol δ) [110,148,149], an additional RIR has been mapped on the p66 subunit of pol ζ, which could also facilitate the formation of Rev1-pol ζ complex [150]. Together, interactions of Rev1 with both Rev7 and p66 potentially contribute to the recruitment of pol ζ via Rev1 and the functional linkage between pol ζ and Rev1.

2.3.3. Coordination of Multiple Binding Partners by Rev1

Recent X-ray crystallographic data have illuminated the molecular mechanisms of the interactions of Rev1 with a number of proteins. Wojtaszek et al. reported a crystal structure of mouse Rev1 CTD in complex with Rev7, an interacting fragment of Rev3 and the pol κ RIR peptide (Figure 4C) [157]. Shortly after, Xie et al. reported the structure of a similar protein complex from humans [158]. In addition, Kikuchi et al. solved the crystal structure of a ternary complex containing the C-terminal domain of human Rev1 CTD, Rev7, and a Rev3 fragment [159]. Collectively, these studies have demonstrated that mammalian Rev1 CTD uses different binding regions to interact with Y-family pols and the Rev7 subunit of pol ζ. As noted earlier, RIRs of pol η and pol κ target the same binding region of Rev1 CTD (Figure 4B), which involves the N-terminal β-hairpin, α1 and α2 helices, and α1-α2 loop [153,154]. On the other hand, Rev7 interacts with a distinct and non-overlapping region of CTD diagonal to the binding site of other Y-family pols (Figure 4C) [157,158], presumably to minimize the chance of steric clash between an "inserter" polymerase and pol ζ at the "insertion" step during Rev1/polζ-dependent TLS [150]. Incidentally, the recently mapped RIR on the p66 subunit of pol ζ interacts with the same site on Rev1 CTD as RIRs of pol η and pol κ do (Figure 4B). Although the dissociation constants (K_d) of Rev1 with RIR peptides vary slightly based on the different techniques used (summarized in Table 2) [94,150,151], RIRs of pol κ and p66 bind to the Rev1 CTD approximately an order of magnitude stronger relative to RIRs of pol ι and pol η. The high affinity between p66 RIR and Rev1 CTD may be a contributing factor to the "inserter" to "extender" polymerase switching in a two-step Rev1/Polζ-dependent TLS [150]. In summary, this body of work has provided structural mechanisms for the interactions between Rev1 and other TLS pols, and such information is important for designing inhibitors to disrupt these interactions [43].

Table 2. Apparent dissociation constants (K_d) of human p66 (a subunit of pol δ), pol η, pol κ, and pol ι RIR peptides with human Rev1. Conserved phenylalanine residues are in bold.

DNA Polymerase	Sequence	K_d (μM) SPR	Fluorescence [3]
p66	231KGNMMSN**FF**GKAAMNK	2.3 [1]	
pol η	524QSTGTEP**FF**KQKSLLL	13 [2]	4.4
pol κ	560EMSHKKS**FF**DKKRSER	7.6 [2]	
pol κ	560EMSHKKS**FF**DKKRSER	1.7 [1]	0.28
pol ι PIP	539ASRGVLS**FF**SKKQMQD	69 [2]	5.5

[1,2] Values are from references [94,150], respectively, and are obtained with surface plasma resonance (SPR) assays; [3] Calculated from fluorescence titration assays [151].

2.3.4. PCNA Tool Belts and Rev1 Bridges

Based on the ways in which TLS polymerases interact with one another and with PCNA, it seems likely that multiple TLS polymerases and PCNA can form higher ordered complexes with different molecular architectures. For example, multiple TLS polymerases can directly interact with PCNA without directly interacting with one another, and form a PCNA tool belt. By contrast, Rev1 can serve as a bridging molecule to link PCNA (via BRCT and/or PAD domains) and another TLS polymerase (via CTD) without PCNA and this other TLS polymerase directly interacting. Such an arrangement is called a Rev1 bridge. Recently, single-molecule studies using yeast PCNA, pol η, and Rev1 have shown that both PCNA tool belts and Rev1 bridges form in approximately equal proportions [160]. Surprisingly, it was observed that these higher ordered complexes were dynamic, meaning that PCNA tool belts can switch to Rev1 bridges and vice versa without dissociation. The dynamic nature of these complexes likely permits rapid sampling of multiple TLS polymerases to find the one that is most appropriate for bypassing a given DNA lesion [160].

2.3.5. Physiological Functions of Rev1-Mediated Protein Interactions

The functional importance of Rev1-mediated protein–protein interactions appears to be polymerase- and lesion-specific. In the case of pol η-mediated CPD bypass, the formation of pol η foci is dependent on the interactions between PCNA and pol η (via PIPs and UBZ of pol η) [125,131], but not on the interactions between Rev1 and pol η (via RIRs) [161]. In keeping with these data, complementation with a variant form of pol η with a F to A mutation in the RIRs resulted in a similar extent of suppression of UV-induced mutagenesis in XP-V fibroblasts relative to cells complemented with wild-type pol η [162]. On the other hand, transient expression of wild-type pol κ in pol κ-knockout mouse embryonic fibroblast cells restored the resistance to BPDE, whereas complementation with pol κ bearing substitutions of phenylalanine residues in RIR fails to correct BPDE-sensitivity [94]. Together, Rev1 plays an important role in interacting with multiple TLS pols, but the biological significance of these interactions remains to be firmly established.

2.4. Subunits Sharing between Pol δ and Pol ζ

2.4.1. The Subunit Organization of Pol ζ

In 2012, two groups discovered that yeast pol31 and pol32 proteins (previously recognized subunits of pol δ) together with the Rev3-Rev7 complex of pol ζ form a four-subunit pol ζ4 [110,149]. Similarly, p50 (POLD2) and p66 (POLD3) (human counterparts of yeast pol31 and pol32, respectively) are also components of pol ζ4 in humans [148]. Pol ζ4 has a higher catalytic activity than the minimally functional Rev3-Rev7 complex [110,149,163], and its activity is further enhanced in the presence of PCNA [110]. The pol ζ4 complex is organized via interactions between Rev3 and Rev7, Rev3 and pol31, pol31 and pol32, and pol32 and Rev7 [110,148,149,163]. In addition, pol32 is known to interact with PCNA, which is important for processive DNA replication by pol δ [164]. Analogous to the interaction between pol31 and the C-terminal segment of pol3, CysB of Rev3 (one of the two conserved cysteine-rich metal binding motifs) is essential for Rev3-pol31 interactions [110,148]. A structural model of yeast pol ζ4 based on electron microscopy reconstruction has been reported [165]. In this model, pol ζ4 adopts an elongated bilobal architecture, whereby Rev3 occupies a large lobe of the electron microscopy density map, and accessory subunits (Rev7, pol31, and pol32) locate in a small lobe connected to Rev3 via a longer amino acid linker.

2.4.2. Switching between Pol δ and Pol ζ

Baranovskiy et al. proposed that the subunits sharing between pol δ and pol ζ may be a mechanism to facilitate polymerase switching [148]. Specifically, when pol δ stalls at a DNA lesion, p125 (the catalytic subunit of pol δ) dissociates from p50-p66 for pol ζ to gain access to the replication fork. A caveat is that this proposal does not explain how pol ζ may operate on the leading strand (replicated by pol ε) [148]. Although this proposal remains to be explicitly tested, it provides a basis for further hypothesis generation and testing. Two possible pathways have been postulated for this polymerase switching model [166]. First, the p50-p66 complex remains attached to PCNA to interact with Rev3-Rev7 for pol ζ4 to gain access to the fork. Subunits p125 and p12 (an accessory subunit of pol δ) can be degraded by proteolysis [167]. Second, p50-p66 dissociates from the fork together with p125, and a pre-assembled pol ζ4 complex is recruited for translesion synthesis. The latter pathway is augmented by the observation that p50-p66 complex binds to Rev3 fairly strongly, which withstands stringent washing with 1.0 M NaCl solution [163]. Interestingly, a recent proteomic analysis discovered significant changes of the levels of multiple components of pol δ when comparing wild-type cells to POLD3-deficient mouse cells, and that the levels of pol ζ constituents remain unchanged [168], which implies that p50-p66 may be preferentially associated with pol δ under normal conditions without DNA damage. The concentrations of p50 and p66 at the fork and their preferential association with pol δ or pol ζ under different cellular conditions remain to be determined. Remarkably, Stepchenkova et al. observed that a defect in the catalytic subunit of pol δ that affects the [4Fe-4S] cluster binding leads to suppressed UV-induced mutagenesis and enhanced pol ζ-dependent spontaneous mutagenesis in a yeast strain. On the basis of this finding, the authors proposed that the conserved [4Fe-4S] cluster in pol3 and Rev3 plays a role in pol δ-pol ζ switching [169]. It is imperative to decipher the functional importance of Fe-S clusters in various aspects of DNA metabolism, including polymerase switching, and this question is being actively pursued in the field.

2.5. Proteasomal Degradation of DNA Polymerases

2.5.1. Regulation of the Steady-State Levels of TLS Pols

The error-prone nature of TLS polymerases means their access to the replication fork must be carefully regulated. Controlling the steady-state levels of DNA polymerases is a simple way to restrict enzymatic activities of low fidelity DNA polymerases. In *E. coli*, TLS pols are regulated via the global SOS response [42,170,171]. The levels of *E. coli* pol II, pol IV and pol V increase dramatically following LexA inactivation, which contributes to the polymerase switching ([42] and references therein). On the contrary, eukaryotes do not seem to use the overall expression level of TLS pols to respond to genotoxic stress [27], likely due to a larger number of TLS pols in eukaryotes compared to *E. coli*. Nonetheless, the steady-state levels of TLS pols are under strict regulation throughout the cell cycle in eukaryotes. In *S. cerevisiae*, the steady-state levels of both pol η and Rev1 peak at G2/M phase relative to G1 phase and early S phase, whereby a 3-fold increase is observed for pol η and a 50-fold increase is observed for Rev1 [172–175]. In *Schizosaccharomyces pombe*, Rev1 exists at the highest level in G1 phase and is down-regulated at the entry of S phase of the cell cycle [176]. The exact reason of such regulations remains unknown.

Contradictory results exist regarding whether the overexpression of TLS pols is associated with increased mutagenicity. While King et al. showed no mutagenic effects upon overproducing pol η in diploid XP-V fibroblasts [177], other studies using yeast and mammalian systems demonstrated that overproduction and deletion of *RAD30/POLH* result in mutator phenotypes. For instance, overexpression of *POLH* in a multicopy episomal vector has been shown to be toxic to human cells [178]. Abnormal up-regulation of human pol η through IRF1 transactivation leads to an elevated mutation frequency and carcinogenesis in human cells upon exposure to the alkylating agent *N*-methyl-*N'*-nitro-*N*-nitrosoguanidine [179]. When *RAD30* gene is compromised [10,15,180] or overexpressed [181,182] in *S. cerevisiae*, replication infidelity and genomic instability are observed. Similarly, overexpression of Rev1 confers sensitivity to cisplatin in fission yeast [176]. In addition, TLS pols are over-expressed in a number of cancers, which is considered to be a contributing factor to mutagenicity and resistance to chemotherapies [25,43].

2.5.2. Proteasomal Degradation of TLS Pols

TLS regulation can be achieved in part by proteasomal degradation orchestrated by posttranslational modifications. Posttranslational modifications with ubiquitin or ubiquitin-like modifiers play a critical role in the regulation of normal DNA replication and DNA damage tolerance pathways [183,184]. The attachment of ubiquitin to substrates is achieved via an enzymatic cascade by first attaching ubiquitin to an E1 ubiquitin-activating enzyme, then by transfer of ubiquitin to an E2 ubiquitin-conjugating enzyme, and by finally binding of E2 and substrate together with an E3 ubiquitin ligase, which completes the ubiquitin transfer from the E2 enzyme to the substrate [183]. There have been several reports of different E3 ligases being involved in the ubiquitination of pol η, which include Pirh2 (RING-H2 type E3 ligase) [185,186], mdm2 (murine double minute) [187], TRIP (human TNF receptor associated factor (TRAF)-interacting protein) in humans [188], and NOPO (homolog of human TRIP) in *Drosophila* [188]. For example, Pirh2 physically interacts with and monoubiquitinates human pol η and is involved in the 20S proteasomal degradation of pol η [185,186]. Mdm2 physically interacts with pol η in vivo and in vitro and facilitates pol η degradation via ubiquitin-dependent proteolysis [187]. On the other hand, TRIP and NOPO E3 ligases promote the ubiquitination of pol η, and enhance the localization of pol η in replication foci [188]. Apparently, unlike *E. coli*, eukaryotes prefer to regulate the local concentrations of pols at the fork by modulating the interactions of TLS pols with multiple binding partners. It should be kept in mind that the proteosomal degradation of TLS pols does not necessarily indicate their activities at the replication factories, and whether the degradation targets the soluble pool or chromatin-bound TLS pols remains to be elucidated. Nonetheless, a decrease in the concentration of a given TLS pol is likely to limit its access to the replication fork or to facilitate its removal after TLS.

2.5.3. Protein Degradation Creates Binding Sites for TLS Pols

CRL4^{Cdt2} (*Cullin 4-RING Ligase* (CRL4)-Ddb1-Cdt2) is an E3 ubiquitin ligase that targets PCNA binding partners for proteasomal degradation and is known as a master regulator for genomic stability [189]. CRL4^{Cdt2} mediates the degradation of replication licensing factor Cdt1, which prevents DNA re-replication and genome instability [189]. In addition, CRL4^{Cdt2} facilitates the rapid degradation of Cdt1 after DNA damage [190,191]. In *Caenorhabditis elegans*, CRL4^{Cdt2} participates in the degradation of pol η [192]; however, whether CRL4^{Cdt2} is involved in the degradation of pol η in humans is yet to be tested. A number of CRL4^{Cdt2} substrates including Cdt1 contain specialized PIP modules (PIP degrons), which are important for protein degradation [189,193,194]. Compared to a canonical PIP sequence, a PIP degron contains both a TD motif and a basic amino acid four residues downstream ([Q/N]xxɸTD[F/Y][F/Y]xxx[R/K]); the conserved TD motif confers stronger PCNA binding relative to canonical PIPs [193,194]. The conserved threonine residue within the Cdt1 PIP degron is important for interfering with pol η foci formation after UV damage [195]. Importantly, CRL4^{Cdt2}-mediated proteolysis facilitates pol η and pol κ focus formation after UV-induced DNA damage [195]. Thus, it is proposed that CRL4^{Cdt2}-mediated Cdt1 degradation unmasks the site on PCNA for the binding of TLS pols [195], although the molecular basis of this model remains to be established.

2.5.4. Proteasomal Degradation of Pol δ

Protein degradation is an important means to regulate multi-subunit replicative DNA polymerase δ, which potentially contributes to the displacement of pol δ at a stalled fork. Human pol δ is a four subunit complex (p125-p50-p66-p12, herein after referred to as pol δ4) [196]. Collective studies by Lee and colleagues have shown that the p12 subunit of pol δ holoenzyme is subject to rapid proteolysis in human cells triggered by DNA damage or replication stress [167]. The loss of p12 leads to the formation of a trimeric form of pol δ3 (p125-p50-p66), which has impaired catalytic activities relative to pol δ4 [167,196]. Detailed kinetic characterizations revealed that such a compromise in catalytic activity is mainly attributed to a decreased burst rate (a function of the rates of phosphodiester bond formation and conformational change) and a greater proofreading activity of pol δ3 [197]. As a result, pol δ3 has an increased tendency to stall at DNA lesions, which may facilitate the exchange of TLS pols [198]. Interestingly, subsequent studies indicate that pol δ3 also functions during unperturbed DNA replication [199,200], and the level of p12 subunit remains at a baseline level during unperturbed growth in unsynchronized cells [201]. As the authors pointed out, these studies measure the nuclear pool of p12 and pol δ3, and do not provide direct information on the assembly of pol δ at the replication fork [199]. Therefore, future studies are needed to fully understand the biological functions of pol δ3 and pol δ4, as well as the partition between the two. A recent study by Hedglin et al. demonstrates that human pol δ4 maintains a loose association with PCNA when replicating DNA, and that pol δ4 holoenzyme is relatively unstable and rapidly dissociates upon stalling [202]. These authors suggest that on a lagging strand it may not be necessary for polymerases to engage in active polymerase switching in humans [128]. It is likely that p12 maintains a dynamic equilibrium between association and dissociation during lagging strand DNA synthesis, especially considering that pol δ has to continually replace pol α at primed sites [203,204].

3. Concluding Remarks

In summary, the understanding of the selection and switching of DNA polymerases has substantially advanced over the past decade. Nonetheless, questions remain regarding the molecular mechanisms of these processes. First, structures of multi-protein complexes with one or more specialized DNA polymerases, DNA, PCNA and a replicative DNA polymerase need to be solved. Such structures will be useful to further understand the coordination of multiple factors at the fork. Although protein complexes are often recalcitrant for crystallization, recent advances in cryo-electron microscopy holds promise for solving the problem. Second, the dynamics of multi-protein assembly remain poorly understood. Single molecule techniques together with rapid kinetics can potentially tackle this problem. Third, novel approaches are needed to systematically understand the coordination of multiple components during the selection and switching of DNA polymerases. Modern omics-based approaches in combination with bioinformatics may offer new solutions to this challenging task.

Acknowledgments: This work is supported in part by Central Michigan University start-up funds (to Linlin Zhao), US NIH R15 GM117522 (to Linlin Zhao), and US NIH R01 GM081433 (to M. Todd Washington). Authors are grateful to the anonymous reviewers who made constructive suggestions during revision of this manuscript, and to Benjamin Swarts for critically reading the manuscript.

Author Contributions: Linlin Zhao created the first draft; and Linlin Zhao and M. Todd Washington revised the paper.

Conflicts of Interest: The authors declare no conflict of interest. The funding sponsors had no role in the writing of the manuscript.

References

1. Hoeijmakers, J.H.J. Genome maintenance mechanisms for preventing cancer. *Nature* **2001**, *411*, 366–374. [CrossRef] [PubMed]
2. Kato, T.; Shinoura, Y. Isolation and characterization of mutants of *Escherichia coli* deficient in induction of mutations by ultraviolet light. *Mol. Gen. Genet.* **1977**, *156*, 121–131. [CrossRef] [PubMed]
3. Steinborn, G. *Uvm* mutants of *Escherichia coli* K 12 deficient in UV mutagenesis. *Mol. Gen. Genet.* **1978**, *165*, 87–93. [CrossRef] [PubMed]
4. Lemontt, J.F. Mutants of yeast defective in mutation induced by ultraviolet light. *Genetics* **1971**, *68*, 21–33. [PubMed]
5. Lawrence, C.W.; Christensen, R. UV mutagenesis in radiation-sensitive strains of yeast. *Genetics* **1976**, *82*, 207–232. [PubMed]
6. Lehmann, A.R.; Kirk-Bell, S.; Arlett, C.F.; Paterson, M.C.; Lohman, P.H.; de Weerd-Kastelein, E.A.; Bootsma, D. Xeroderma pigmentosum cells with normal levels of excision repair have a defect in DNA synthesis after UV-irradiation. *Proc. Natl. Acad. Sci. USA* **1975**, *72*, 219–223. [CrossRef] [PubMed]
7. Nelson, J.R.; Lawrence, C.W.; Hinkle, D.C. Deoxycytidyl transferase activity of yeast REV1 protein. *Nature* **1996**, *382*, 729–731. [CrossRef] [PubMed]
8. Nelson, J.R.; Lawrence, C.W.; Hinkle, D.C. Thymine-thymine dimer bypass by yeast DNA polymerase zeta. *Science* **1996**, *272*, 1646–1649. [CrossRef] [PubMed]
9. Johnson, R.E.; Prakash, S.; Prakash, L. Efficient bypass of a thymine-thymine dimer by yeast DNA polymerase, Polη. *Science* **1999**, *283*, 1001–1004. [CrossRef] [PubMed]

10. Johnson, R.E.; Kondratick, C.M.; Prakash, S.; Prakash, L. hRAD30 mutations in the variant form of xeroderma pigmentosum. *Science* **1999**, *285*, 263–265. [CrossRef] [PubMed]

11. Masutani, C.; Kusumoto, R.; Yamada, A.; Dohmae, N.; Yokoi, M.; Yuasa, M.; Araki, M.; Iwai, S.; Takio, K.; Hanaoka, F. The XPV (xeroderma pigmentosum variant) gene encodes human DNA polymerase η. *Nature* **1999**, *399*, 700–704. [PubMed]

12. Wagner, J.; Gruz, P.; Kim, S.; Yamada, M.; Matsui, K.; Fuchs, R.P.P.; Nohmi, T. The dinB gene encodes a novel *E. coli* DNA polymerase, DNA pol IV, involved in mutagenesis. *Mol. Cell* **1999**, *4*, 281–286. [CrossRef]

13. Reuven, N.B.; Arad, G.; Maor-Shoshani, A.; Livneh, Z. The mutagenesis protein UmuC is a DNA polymerase activated by UmuD′, RecA, and SSB and is specialized for translesion replication. *J. Biol. Chem.* **1999**, *274*, 31763–31766. [CrossRef] [PubMed]

14. Tang, M.; Shen, X.; Frank, E.G.; O'Donnell, M.; Woodgate, R.; Goodman, M.F. UmuD′ 2C is an error-prone DNA polymerase, *Escherichia coli* pol V. *Proc. Natl. Acad. Sci. USA* **1999**, *96*, 8919–8924. [CrossRef] [PubMed]

15. McDonald, J.P.; Levine, A.S.; Woodgate, R. The *Saccharomyces cerevisiae* RAD30 gene, a homologue of *Escherichia coli dinB* and *umuC*, is DNA damage inducible and functions in a novel error-free postreplication repair mechanism. *Genetics* **1997**, *147*, 1557–1568. [PubMed]

16. Tissier, A.; McDonald, J.P.; Frank, E.G.; Woodgate, R. Pol ι, a remarkably error-prone human DNA polymerase. *Genes Dev.* **2000**, *14*, 1642–1650. [PubMed]

17. Johnson, R.E.; Washington, M.T.; Haracska, L.; Prakash, S.; Prakash, L. Eukaryotic polymerases ι and ζ act sequentially to bypass DNA lesions. *Nature* **2000**, *406*, 1015–1019. [PubMed]

18. Zhang, Y.; Yuan, F.; Wu, X.; Wang, Z. Preferential incorporation of G opposite template T by the low-fidelity human DNA polymerase ι. *Mol. Cell. Biol.* **2000**, *20*, 7099–7108. [CrossRef] [PubMed]

19. Ogi, T.; Kato, T.; Kato, T.; Ohmori, H. Mutation enhancement by DINB1, a mammalian homologue of the *Escherichia coli* mutagenesis protein DinB. *Genes Cells* **1999**, *4*, 607–618. [CrossRef] [PubMed]

20. Gerlach, V.L.; Aravind, L.; Gotway, G.; Schultz, R.A.; Koonin, E.V.; Friedberg, E.C. Human and mouse homologs of *Escherichia coli* DinB (DNA polymerase IV), members of the UmuC/DinB superfamily. *Proc. Natl. Acad. Sci. USA* **1999**, *96*, 11922–11927. [CrossRef] [PubMed]

21. Ohashi, E.; Ogi, T.; Kusumoto, R.; Iwai, S.; Masutani, C.; Hanaoka, F.; Ohmori, H. Error-prone bypass of certain DNA lesions by the human DNA polymerase κ. *Genes Dev.* **2000**, *14*, 1589–1594. [PubMed]

22. Zhang, Y.; Yuan, F.; Wu, X.; Wang, M.; Rechkoblit, O.; Taylor, J.-S.; Geacintov, N.E.; Wang, Z. Error-free and error-prone lesion bypass by human DNA polymerase κ in vitro. *Nucleic Acids Res.* **2000**, *28*, 4138–4146. [CrossRef] [PubMed]

23. Woodgate, R. A plethora of lesion-replicating DNA polymerases. *Genes Dev.* **1999**, *13*, 2191–2195. [CrossRef] [PubMed]

24. Hübscher, U.; Maga, G.; Spadari, S. Eukaryotic DNA polymerases. *Annu. Rev. Biochem.* **2002**, *71*, 133–163. [CrossRef] [PubMed]

25. Lange, S.S.; Takata, K.; Wood, R.D. DNA polymerases and cancer. *Nat. Rev. Cancer* **2011**, *11*, 96–110. [CrossRef] [PubMed]

26. Rudd, S.G.; Bianchi, J.; Doherty, A.J. PrimPol—A new polymerase on the block. *Mol. Cell Oncol.* **2014**, *1*, e960754. [CrossRef] [PubMed]

27. Sale, J.E. Translesion DNA synthesis and mutagenesis in eukaryotes. *Cold Spring Harb. Perspect. Biol.* **2013**, *5*, a012708. [CrossRef] [PubMed]

28. Bianchi, J.; Rudd, S.G.; Jozwiakowski, S.K.; Bailey, L.J.; Soura, V.; Taylor, E.; Stevanovic, I.; Green, A.J.; Stracker, T.H.; Lindsay, H.D. PrimPol bypasses UV photoproducts during eukaryotic chromosomal DNA replication. *Mol. Cell* **2013**, *52*, 566–573. [CrossRef] [PubMed]

29. García-Gómez, S.; Reyes, A.; Martínez-Jiménez, M.I.; Chocrón, E.S.; Mourón, S.; Terrados, G.; Powell, C.; Salido, E.; Méndez, J.; Holt, I.J. PrimPol, an archaic primase/polymerase operating in human cells. *Mol. Cell* **2013**, *52*, 541–553. [CrossRef] [PubMed]

30. Wan, L.; Lou, J.; Xia, Y.; Su, B.; Liu, T.; Cui, J.; Sun, Y.; Lou, H.; Huang, J. hPrimpol1/CCDC111 is a human DNA primase-polymerase required for the maintenance of genome integrity. *EMBO Rep.* **2013**, *14*, 1104–1112. [CrossRef] [PubMed]

31. Zafar, M.K.; Ketkar, A.; Lodeiro, M.F.; Cameron, C.E.; Eoff, R.L. Kinetic analysis of human PrimPol DNA polymerase activity reveals a generally error-prone enzyme capable of accurately bypassing 7,8-dihydro-8-oxo-2′-deoxyguanosine. *Biochemistry* **2014**, *53*, 6584–6594. [CrossRef] [PubMed]

32. Mourón, S.; Rodriguez-Acebes, S.; Martínez-Jiménez, M.I.; García-Gómez, S.; Chocrón, S.; Blanco, L.; Méndez, J. Repriming of DNA synthesis at stalled replication forks by human PrimPol. *Nat. Struct. Mol. Biol.* **2013**, *20*, 1383–1389. [CrossRef] [PubMed]

33. Martínez-Jiménez, M.I.; García-Gómez, S.; Bebenek, K.; Sastre-Moreno, G.; Calvo, P.A.; Díaz-Talavera, A.; Kunkel, T.A.; Blanco, L. Alternative solutions and new scenarios for translesion DNA synthesis by human PrimPol. *DNA Repair* **2015**, *29*, 127–138. [CrossRef] [PubMed]

34. Schiavone, D.; Jozwiakowski, S.K.; Romanello, M.; Guilbaud, G.; Guilliam, T.A.; Bailey, L.J.; Sale, J.E.; Doherty, A.J. PrimPol is required for replicative tolerance of G quadruplexes in vertebrate cells. *Mol. Cell* **2016**, *61*, 161–169. [CrossRef] [PubMed]

35. Kobayashi, K.; Guilliam, T.A.; Tsuda, M.; Yamamoto, J.; Bailey, L.J.; Iwai, S.; Takeda, S.; Doherty, A.J.; Hirota, K. Repriming by PrimPol is critical for DNA replication restart downstream of lesions and chain-terminating nucleosides. *Cell Cycle* **2016**, *15*, 1997–2008. [CrossRef] [PubMed]

36. Jansen, J.G.; Tsaalbi-Shtylik, A.; de Wind, N. Roles of mutagenic translesion synthesis in mammalian genome stability, health and disease. *DNA Repair* **2015**, *29*, 56–64. [CrossRef] [PubMed]

37. Wickramasinghe, C.M.; Arzouk, H.; Frey, A.; Maiter, A.; Sale, J.E. Contributions of the specialised DNA polymerases to replication of structured DNA. *DNA Repair* **2015**, *29*, 83–90. [CrossRef] [PubMed]

38. Sale, J.E.; Lehmann, A.R.; Woodgate, R. Y-family DNA polymerases and their role in tolerance of cellular DNA damage. *Nat. Rev. Mol. Cell Biol.* **2012**, *13*, 141–152. [CrossRef] [PubMed]

39. Prakash, S.; Prakash, L. Translesion DNA synthesis in eukaryotes: A one-or two-polymerase affair. *Genes Dev.* **2002**, *16*, 1872–1883. [CrossRef] [PubMed]

40. Lehmann, A.R.; Niimi, A.; Ogi, T.; Brown, S.; Sabbioneda, S.; Wing, J.F.; Kannouche, P.L.; Green, C.M. Translesion synthesis: Y-family polymerases and the polymerase switch. *DNA Repair* **2007**, *6*, 891–899. [CrossRef] [PubMed]

41. Friedberg, E.C.; Lehmann, A.R.; Fuchs, R.P.P. Trading places: How do DNA polymerases switch during translesion DNA synthesis? *Mol. Cell* **2005**, *18*, 499–505. [CrossRef] [PubMed]

42. Sutton, M.D. Coordinating DNA polymerase traffic during high and low fidelity synthesis. *BAA Protein Proteom.* **2010**, *1804*, 1167–1179. [CrossRef] [PubMed]

43. Korzhnev, D.M.; Hadden, M.K. Targeting the translesion synthesis pathway for the development of anti-cancer chemotherapeutics. *J. Med. Chem.* **2016**, *59*, 9321–9336. [CrossRef] [PubMed]

44. Prakash, S.; Johnson, R.E.; Prakash, L. Eukaryotic translesion synthesis DNA polymerases: Specificity of structure and function. *Annu. Rev. Biochem.* **2005**, *74*, 317–353. [CrossRef] [PubMed]

45. Yang, W.; Woodgate, R. What a difference a decade makes: Insights into translesion DNA synthesis. *Proc. Natl. Acad. Sci. USA* **2007**, *104*, 15591–15598. [CrossRef] [PubMed]

46. Washington, M.T.; Carlson, K.D.; Freudenthal, B.D.; Pryor, J.M. Variations on a theme: Eukaryotic Y-family DNA polymerases. *BAA Protein Proteom.* **2010**, *1804*, 1113–1123. [CrossRef] [PubMed]

47. Goodman, M.F.; Woodgate, R. Translesion DNA polymerases. *Cold Spring Harb. Perspect. Biol.* **2013**, *5*, a010363. [CrossRef] [PubMed]

48. Hübscher, U.; Maga, G. DNA replication and repair bypass machines. *Curr. Opin. Chem. Biol.* **2011**, *15*, 627–635. [CrossRef] [PubMed]

49. Einolf, H.J.; Guengerich, F.P. Fidelity of nucleotide insertion at 8-oxo-7,8-dihydroguanine by mammalian DNA polymerase δ steady-state and pre-steady-state kinetic analysis. *J. Biol. Chem.* **2001**, *276*, 3764–3771. [CrossRef] [PubMed]

50. Hirota, K.; Tsuda, M.; Mohiuddin; Tsurimoto, T.; Cohen, I.S.; Livneh, Z.; Kobayashi, K.; Narita, T.; Nishihara, K.; Murai, J.; et al. In vivo evidence for translesion synthesis by the replicative DNA polymerase δ. *Nucleic Acids Res.* **2016**, *44*, 7242–7250. [PubMed]

51. Shachar, S.; Ziv, O.; Avkin, S.; Adar, S.; Wittschieben, J.; Reißner, T.; Chaney, S.; Friedberg, E.C.; Wang, Z.; Carell, T.; et al. Two-polymerase mechanisms dictate error-free and error-prone translesion DNA synthesis in mammals. *EMBO J.* **2009**, *28*, 383–393. [CrossRef] [PubMed]

52. Waters, L.S.; Minesinger, B.K.; Wiltrout, M.E.; D'Souza, S.; Woodruff, R.V.; Walker, G.C. Eukaryotic translesion polymerases and their roles and regulation in DNA damage tolerance. *Microbiol. Mol. Biol. Rev.* **2009**, *73*, 134–154. [CrossRef] [PubMed]

53. Matsuda, T.; Bebenek, K.; Masutani, C.; Hanaoka, F.; Kunkel, T.A. Low fidelity DNA synthesis by human DNA polymerase η. *Nature* **2000**, *404*, 1011–1013. [PubMed]

54. Ogi, T.; Shinkai, Y.; Tanaka, K.; Ohmori, H. Polκ protects mammalian cells against the lethal and mutagenic effects of benzo[*a*]pyrene. *Proc. Natl. Acad. Sci. USA* **2002**, *99*, 15548–15553. [CrossRef]

55. Denissenko, M.F.; Pao, A.; Tang, M.-S.; Pfeifer, G.P. Preferential formation of benzo[*a*]pyrene adducts at lung cancer mutational hotspots in p53. *Science* **1996**, *274*, 430–432. [CrossRef] [PubMed]

56. Pfeifer, G.P.; Denissenko, M.F.; Olivier, M.; Tretyakova, N.; Hecht, S.S.; Hainaut, P. Tobacco smoke carcinogens, DNA damage and p53 mutations in smoking-associated cancers. *Oncogene* **2002**, *21*, 7435–7451. [CrossRef] [PubMed]

57. Slaga, T.J.; Bracken, W.J.; Gleason, G.; Levin, W.; Yagi, H.; Jerina, D.M.; Conney, A.H. Marked differences in the skin tumor-initiating activities of the optical enantiomers of the diastereomeric benzo[*a*]pyrene 7,8-diol-9,10-epoxides. *Cancer Res.* **1979**, *39*, 67–71. [PubMed]

58. Thakker, D.R.; Yagi, H.; Lu, A.Y.; Levin, W.; Conney, A.H. Metabolism of benzo[*a*]pyrene: Conversion of (+/−)-*trans*-7,8-dihydroxy-7,8-dihydrobenzo[*a*]pyrene to highly mutagenic 7,8-diol-9,10-epoxides. *Proc. Natl. Acad. Sci. USA* **1976**, *73*, 3381–3385. [CrossRef] [PubMed]

59. Kapitulnik, J.; Wislocki, P.G.; Levin, W.; Yagi, H.; Thakker, D.R.; Akagi, H.; Koreeda, M.; Jerina, D.M.; Conney, A.H. Marked differences in the carcinogenic activity of optically pure (+)-and (−)-*trans*-7,8-dihydroxy-7,8-dihydrobenzo[*a*]pyrene in newborn mice. *Cancer Res.* **1978**, *38*, 2661–2665. [PubMed]

60. Straub, K.M.; Meehan, T.; Burlingame, A.L.; Calvin, M. Identification of the major adducts formed by reaction of benzo[*a*]pyrene diol epoxide with DNA in vitro. *Proc. Natl. Acad. Sci. USA* **1977**, *74*, 5285–5289. [CrossRef] [PubMed]

61. Cheng, S.C.; Hilton, B.D.; Roman, J.M.; Dipple, A. DNA adducts from carcinogenic and noncarcinogenic enantiomers of benzo[*a*]pyrenedihydrodiol epoxide. *Chem. Res. Toxicol.* **1989**, *2*, 334–340.

62. Devanesan, P.D.; RamaKrishna, N.V.S.; Todorovic, R.; Rogan, E.G.; Cavalieri, E.L.; Jeong, H.; Jankowiak, R.; Small, G.J. Identification and quantitation of benzo[*a*]pyrene-DNA adducts formed by rat liver microsomes in vitro. *Chem. Res. Toxicol.* **1992**, *5*, 302–309. [CrossRef] [PubMed]

63. McCoull, K.D.; Rindgen, D.; Blair, I.A.; Penning, T.M. Synthesis and characterization of polycyclic aromatic hydrocarbon *o*-quinone depurinating N7-guanine adducts. *Chem. Res. Toxicol.* **1999**, *12*, 237–246.

64. Balu, N.; Padgett, W.T.; Lambert, G.R.; Swank, A.E.; Richard, A.M.; Nesnow, S. Identification and characterization of novel stable deoxyguanosine and deoxyadenosine adducts of benzo[*a*]pyrene-7, 8-quinone from reactions at physiological pH. *Chem. Res. Toxicol.* **2004**, *17*, 827–838. [CrossRef] [PubMed]

65. Shukla, R.; Geacintov, N.E.; Loechler, E.L. The major, N^2-dG adduct of (+)-*anti*-B[a]PDE induces G→A mutations in a 5′-AGA-3′ sequence context. *Carcinogenesis* **1999**, *20*, 261–268. [CrossRef] [PubMed]

66. Seo, K.-Y.; Jelinsky, S.A.; Loechler, E.L. Factors that influence the mutagenic patterns of DNA adducts from chemical carcinogens. *Mutat. Res.* **2000**, *463*, 215–246. [CrossRef]

67. Chary, P.; Stone, M.P.; Lloyd, R.S. Sequence context modulation of polycyclic aromatic hydrocarbon-induced mutagenesis. *Environ. Mol. Mutagen.* **2013**, *54*, 652–658. [CrossRef] [PubMed]

68. Menzies, G.E.; Reed, S.H.; Brancale, A.; Lewis, P.D. Base damage, local sequence context and TP53 mutation hotspots: A molecular dynamics study of benzo[*a*]pyrene induced DNA distortion and mutability. *Nucleic Acids Res.* **2015**, *43*, 9133–9146. [CrossRef] [PubMed]

69. Geacintov, N.E.; Cosman, M.; Hingerty, B.E.; Amin, S.; Broyde, S.; Patel, D.J. NMR solution structures of stereoisomeric covalent polycyclic aromatic carcinogen-DNA adducts: Principles, patterns, and diversity. *Chem. Res. Toxicol.* **1997**, *10*, 111–146. [CrossRef] [PubMed]

70. Moriya, M.; Spiegel, S.; Fernandes, A.; Amin, S.; Liu, T.; Geacintov, N.; Grollman, A.P. Fidelity of translesional synthesis past benzo[*a*]pyrene diol epoxide-2′-deoxyguanosine DNA adducts: Marked effects of host cell, sequence context, and chirality. *Biochemistry* **1996**, *35*, 16646–16651. [CrossRef] [PubMed]

71. Zhang, Y.; Wu, X.; Guo, D.; Rechkoblit, O.; Wang, Z. Activities of human DNA polymerase κ in response to the major benzo[*a*]pyrene DNA adduct: Error-free lesion bypass and extension synthesis from opposite the lesion. *DNA Repair* **2002**, *1*, 559–569. [CrossRef]

72. Choi, J.-Y.; Angel, K.C.; Guengerich, F.P. Translesion synthesis across bulky N2-alkyl guanine DNA adducts by human DNA polymerase κ. *J. Biol. Chem.* **2006**, *281*, 21062–21072. [CrossRef] [PubMed]

73. Suzuki, N.; Ohashi, E.; Kolbanovskiy, A.; Geacintov, N.E.; Grollman, A.P.; Ohmori, H.; Shibutani, S. Translesion synthesis by human DNA polymerase κ on a DNA template containing a single stereoisomer of dG-(+)-or dG-(−)-*anti*-N^2-BPDE (7,8-dihydroxy-*anti*-9,10-epoxy-7,8,9,10-tetrahydrobenzo[*a*]pyrene). *Biochemistry* **2002**, *41*, 6100–6106. [CrossRef] [PubMed]

74. Huang, X.; Kolbanovskiy, A.; Wu, X.; Zhang, Y.; Wang, Z.; Zhuang, P.; Amin, S.; Geacintov, N.E. Effects of base sequence context on translesion synthesis past a bulky (+)-*trans-anti*-B[*a*]P-N^2-dG lesion catalyzed by the Y-family polymerase pol κ. *Biochemistry* **2003**, *42*, 2456–2466. [CrossRef] [PubMed]

75. Avkin, S.; Goldsmith, M.; Velasco-Miguel, S.; Geacintov, N.; Friedberg, E.C.; Livneh, Z. Quantitative analysis of translesion DNA synthesis across a benzo[*a*]pyrene-guanine adduct in mammalian cells: The role of DNA polymerase κ. *J. Biol. Chem.* **2004**, *279*, 53298–53305. [CrossRef] [PubMed]

76. Rechkoblit, O.; Zhang, Y.; Guo, D.; Wang, Z.; Amin, S.; Krzeminsky, J.; Louneva, N.; Geacintov, N.E. *trans*-Lesion synthesis past bulky benzo[*a*]pyrene diol epoxide N^2-dG and N^6-dA lesions catalyzed by DNA bypass polymerases. *J. Biol. Chem.* **2002**, *277*, 30488–30494. [CrossRef] [PubMed]

77. Wei, S.J.; Chang, R.L.; Wong, C.-Q.; Bhachech, N.; Cui, X.X.; Hennig, E.; Yagi, H.; Sayer, J.M.; Jerina, D.M.; Preston, B.D. Dose-dependent differences in the profile of mutations induced by an ultimate carcinogen from benzo[*a*]pyrene. *Proc. Natl. Acad. Sci. USA* **1991**, *88*, 11227–11230. [CrossRef] [PubMed]

78. Wei, S.C.; Chang, R.L.; Bhachech, N.; Cui, X.X.; Merkler, K.A.; Wong, C.Q.; Hennig, E.; Yagi, H.; Jerina, D.M.; Conney, A.H. Dose-dependent differences in the profile of mutations induced by (+)-7*R*,8*S*-dihydroxy-9*S*,10*R*-epoxy-7,8,9,10-tetrahydrobenzo[*a*]pyrene in the coding region of the hypoxanthine (guanine) phosphoribosyltransferase gene in Chinese hamster V-79 cells. *Cancer Res.* **1993**, *53*, 3294–3301. [PubMed]

79. Conney, A.H.; Chang, R.L.; Jerina, D.M.; Caroline Wei, S.J. Studies on the metabolism of benzo[*a*]pyrene and dose-dependent differences in the mutagenic profile of its ultimate carcinogenic metabolite. *Drug Metab. Rev.* **1994**, *26*, 125–163. [CrossRef] [PubMed]

80. Hashimoto, K.; Cho, Y.; Yang, I.-Y.; Akagi, J.-I.; Ohashi, E.; Tateishi, S.; De Wind, N.; Hanaoka, F.; Ohmori, H.; Moriya, M. The vital role of polymerase ζ and REV1 in mutagenic, but not correct, DNA synthesis across benzo[*a*]pyrene-dG and recruitment of polymerase ζ by REV1 to replication-stalled site. *J. Biol. Chem.* **2012**, *287*, 9613–9622. [CrossRef]

81. Frank, E.G.; Woodgate, R. Increased catalytic activity and altered fidelity of human DNA polymerase ι in the presence of manganese. *J. Biol. Chem.* **2007**, *282*, 24689–24696. [CrossRef] [PubMed]

82. Frank, E.G.; Sayer, J.M.; Kroth, H.; Ohashi, E.; Ohmori, H.; Jerina, D.M.; Woodgate, R. Translesion replication of benzo[*a*]pyrene and benzo[*c*]phenanthrene diol epoxide adducts of deoxyadenosine and deoxyguanosine by human DNA polymerase ι. *Nucleic Acids Res.* **2002**, *30*, 5284–5292. [CrossRef] [PubMed]

83. Jha, V.; Bian, C.; Xing, G.; Ling, H. Structure and mechanism of error-free replication past the major benzo[*a*]pyrene adduct by human DNA polymerase κ. *Nucleic Acids Res.* **2016**, *44*, 4957–4967. [CrossRef] [PubMed]

84. Cosman, M.; De Los Santos, C.; Fiala, R.; Hingerty, B.E.; Singh, S.B.; Ibanez, V.; Margulis, L.A.; Live, D.; Geacintov, N.E.; Broyde, S. Solution conformation of the major adduct between the carcinogen (+)-*anti*-benzo[*a*]pyrene diol epoxide and DNA. *Proc. Natl. Acad. Sci. USA* **1992**, *89*, 1914–1918. [CrossRef] [PubMed]

85. Feng, B.; Gorin, A.; Hingerty, B.E.; Geacintov, N.E.; Broyde, S.; Patel, D.J. Structural alignment of the (+)-*trans-anti*-benzo[*a*]pyrene-dG adduct positioned opposite dC at a DNA template-primer junction. *Biochemistry* **1997**, *36*, 13769–13779. [CrossRef] [PubMed]

86. Donny-Clark, K.; Broyde, S. Influence of local sequence context on damaged base conformation in human DNA polymerase ι: Molecular dynamics studies of nucleotide incorporation opposite a benzo[*a*]pyrene-derived adenine lesion. *Nucleic Acids Res.* **2009**, *37*, 7095–7109. [CrossRef] [PubMed]

87. Beland, F.A.; Churchwell, M.I.; Von Tungeln, L.S.; Chen, S.; Fu, P.P.; Culp, S.J.; Schoket, B.; Gyorffy, E.; Minárovits, J.; Poirier, M.C. High-performance liquid chromatography electrospray ionization tandem mass spectrometry for the detection and quantitation of benzo[*a*]pyrene-DNA adducts. *Chem. Res. Toxicol.* **2005**, *18*, 1306–1315. [CrossRef] [PubMed]

88. Zhang, Y.; Wu, X.; Rechkoblit, O.; Geacintov, N.E.; Taylor, J.-S.; Wang, Z. Response of human REV1 to different DNA damage: Preferential dCMP insertion opposite the lesion. *Nucleic Acids Res.* **2002**, *30*, 1630–1638. [CrossRef] [PubMed]

89. Nelson, J.R.; Gibbs, P.E.M.; Nowicka, A.M.; Hinkle, D.C.; Lawrence, C.W. Evidence for a second function for *Saccharomyces cerevisiae* Rev1p. *Mol. Microbiol.* **2000**, *37*, 549–554. [CrossRef] [PubMed]

90. Guo, C.; Fischhaber, P.L.; Luk-Paszyc, M.J.; Masuda, Y.; Zhou, J.; Kamiya, K.; Kisker, C.; Friedberg, E.C. Mouse Rev1 protein interacts with multiple DNA polymerases involved in translesion DNA synthesis. *EMBO J.* **2003**, *22*, 6621–6630. [CrossRef] [PubMed]

91. Ohashi, E.; Murakumo, Y.; Kanjo, N.; Akagi, J.i.; Masutani, C.; Hanaoka, F.; Ohmori, H. Interaction of hREV1 with three human Y-family DNA polymerases. *Genes Cells* **2004**, *9*, 523–531. [CrossRef] [PubMed]

92. Tissier, A.; Kannouche, P.; Reck, M.-P.; Lehmann, A.R.; Fuchs, R.P.P.; Cordonnier, A. Co-localization in replication foci and interaction of human Y-family members, DNA polymerase polη and REV1 protein. *DNA Repair* **2004**, *3*, 1503–1514. [CrossRef] [PubMed]

93. Murakumo, Y.; Ogura, Y.; Ishii, H.; Numata, S.-I.; Ichihara, M.; Croce, C.M.; Fishel, R.; Takahashi, M. Interactions in the error-prone postreplication repair proteins hREV1, hREV3, and hREV7. *J. Biol. Chem.* **2001**, *276*, 35644–35651. [CrossRef] [PubMed]

94. Ohashi, E.; Hanafusa, T.; Kamei, K.; Song, I.; Tomida, J.; Hashimoto, H.; Vaziri, C.; Ohmori, H. Identification of a novel REV1-interacting motif necessary for DNA polymerase κ function. *Genes Cells* **2009**, *14*, 101–111. [CrossRef] [PubMed]

95. Prelich, G.; Tan, C.-K.; Kostura, M.; Mathews, M.B.; So, A.G.; Downey, K.M.; Stillman, B. Functional identity of proliferating cell nuclear antigen and a DNA polymerase-delta auxiliary protein. *Nature* **1986**, *326*, 517–520. [CrossRef] [PubMed]

96. Tan, C.-K.; Castillo, C.; So, A.G.; Downey, K.M. An auxiliary protein for DNA polymerase-delta from fetal calf thymus. *J. Biol. Chem.* **1986**, *261*, 12310–12316. [PubMed]

97. Maga, G.; Hübscher, U. Proliferating cell nuclear antigen (PCNA): A dancer with many partners. *J. Cell Sci.* **2003**, *116*, 3051–3060. [CrossRef] [PubMed]

98. Moldovan, G.-L.; Pfander, B.; Jentsch, S. PCNA, the maestro of the replication fork. *Cell* **2007**, *129*, 665–679. [CrossRef] [PubMed]

99. Boehm, E.M.; Gildenberg, M.S.; Washington, M.T. The many roles of PCNA in eukaryotic DNA replication. In *The Enzymes*; Laurie, S.K., Marcos Túlio, O., Eds.; Academic Press: London, UK, 2016; Volume 39, pp. 231–254.

100. Mailand, N.; Gibbs-Seymour, I.; Bekker-Jensen, S. Regulation of PCNA–protein interactions for genome stability. *Nat. Rev. Mol. Cell Biol.* **2013**, *14*, 269–282. [CrossRef] [PubMed]

101. Haracska, L.; Johnson, R.E.; Unk, I.; Phillips, B.; Hurwitz, J.; Prakash, L.; Prakash, S. Physical and functional interactions of human DNA polymerase η with PCNA. *Mol. Cell. Biol.* **2001**, *21*, 7199–7206. [CrossRef] [PubMed]

102. Haracska, L.; Johnson, R.E.; Unk, I.; Phillips, B.B.; Hurwitz, J.; Prakash, L.; Prakash, S. Targeting of human DNA polymerase ι to the replication machinery via interaction with PCNA. *Proc. Natl. Acad. Sci. USA* **2001**, *98*, 14256–14261. [CrossRef] [PubMed]

103. Haracska, L.; Kondratick, C.M.; Unk, I.; Prakash, S.; Prakash, L. Interaction with PCNA is essential for yeast DNA polymerase η function. *Mol. Cell* **2001**, *8*, 407–415. [CrossRef]

104. Haracska, L.; Unk, I.; Johnson, R.E.; Phillips, B.B.; Hurwitz, J.; Prakash, L.; Prakash, S. Stimulation of DNA synthesis activity of human DNA polymerase κ by PCNA. *Mol. Cell. Biol.* **2002**, *22*, 784–791. [CrossRef] [PubMed]

105. Ohmori, H.; Hanafusa, T.; Ohashi, E.; Vaziri, C. Separate roles of structured and unstructured regions of Y-family DNA polymerases. *Adv. Protein Chem. Struct. Biol.* **2009**, *78*, 99–146. [PubMed]

106. Krishna, T.S.R.; Kong, X.-P.; Gary, S.; Burgers, P.M.; Kuriyan, J. Crystal structure of the eukaryotic DNA polymerase processivity factor PCNA. *Cell* **1994**, *79*, 1233–1243. [CrossRef]

107. Gulbis, J.M.; Kelman, Z.; Hurwitz, J.; O'Donnell, M.; Kuriyan, J. Structure of the C-terminal region of p21 WAF1/CIP1 complexed with human PCNA. *Cell* **1996**, *87*, 297–306. [CrossRef]

108. Kochaniak, A.B.; Habuchi, S.; Loparo, J.J.; Chang, D.J.; Cimprich, K.A.; Walter, J.C.; van Oijen, A.M. Proliferating cell nuclear antigen uses two distinct modes to move along DNA. *J. Biol. Chem.* **2009**, *284*, 17700–17710. [CrossRef] [PubMed]

109. Garg, P.; Stith, C.M.; Majka, J.; Burgers, P.M.J. Proliferating cell nuclear antigen promotes translesion synthesis by DNA polymerase ζ. *J. Biol. Chem.* **2005**, *280*, 23446–23450. [CrossRef] [PubMed]

110. Makarova, A.V.; Stodola, J.L.; Burgers, P.M. A four-subunit DNA polymerase ζ complex containing Pol δ accessory subunits is essential for PCNA-mediated mutagenesis. *Nucleic Acids Res.* **2012**, *40*, 11618–11626. [CrossRef] [PubMed]

111. Garg, P.; Burgers, P.M. Ubiquitinated proliferating cell nuclear antigen activates translesion DNA polymerases η and REV1. *Proc. Natl. Acad. Sci. USA* **2005**, *102*, 18361–18366. [CrossRef] [PubMed]

112. Bruning, J.B.; Shamoo, Y. Structural and thermodynamic analysis of human PCNA with peptides derived from DNA polymerase-δ p66 subunit and flap endonuclease-1. *Structure* **2004**, *12*, 2209–2219. [CrossRef] [PubMed]

113. Sakurai, S.; Kitano, K.; Yamaguchi, H.; Hamada, K.; Okada, K.; Fukuda, K.; Uchida, M.; Ohtsuka, E.; Morioka, H.; Hakoshima, T. Structural basis for recruitment of human flap endonuclease 1 to PCNA. *EMBO J.* **2005**, *24*, 683–693. [CrossRef] [PubMed]

114. Hishiki, A.; Hashimoto, H.; Hanafusa, T.; Kamei, K.; Ohashi, E.; Shimizu, T.; Ohmori, H.; Sato, M. Structural basis for novel interactions between human translesion synthesis polymerases and proliferating cell nuclear antigen. *J. Biol. Chem.* **2009**, *284*, 10552–10560. [CrossRef] [PubMed]

115. Guo, C.; Sonoda, E.; Tang, T.-S.; Parker, J.L.; Bielen, A.B.; Takeda, S.; Ulrich, H.D.; Friedberg, E.C. REV1 protein interacts with PCNA: Significance of the REV1 BRCT domain in vitro and in vivo. *Mol. Cell* **2006**, *23*, 265–271. [CrossRef] [PubMed]

116. Pustovalova, Y.; Maciejewski, M.W.; Korzhnev, D.M. NMR mapping of PCNA interaction with translesion synthesis DNA polymerase Rev1 mediated by Rev1-BRCT domain. *J. Mol. Biol.* **2013**, *425*, 3091–3105. [CrossRef] [PubMed]

117. Sharma, N.M.; Kochenova, O.V.; Shcherbakova, P.V. The non-canonical protein binding site at the monomer-monomer interface of yeast proliferating cell nuclear antigen (PCNA) regulates the Rev1-PCNA interaction and Polζ/Rev1-dependent translesion DNA synthesis. *J. Biol. Chem.* **2011**, *286*, 33557–33566. [CrossRef] [PubMed]

118. Boehm, E.M.; Powers, K.T.; Kondratick, C.M.; Spies, M.; Houtman, J.C.D.; Washington, M.T. The proliferating cell nuclear antigen (PCNA)-interacting protein (PIP) motif of DNA polymerase η mediates its interaction with the C-terminal domain of Rev1. *J. Biol. Chem.* **2016**, *291*, 8735–8744. [CrossRef] [PubMed]

119. Baldeck, N.; Janel-Bintz, R.; Wagner, J.; Tissier, A.; Fuchs, R.P.; Burkovics, P.; Haracska, L.; Despras, E.; Bichara, M.; Chatton, B. FF483-484 motif of human Polη mediates its interaction with the POLD2 subunit of Polδ and contributes to DNA damage tolerance. *Nucleic Acids Res.* **2015**, 2116–2125. [CrossRef] [PubMed]

120. Boehm, E.M.; Washington, M.T. R.I.P. to the PIP: PCNA-binding motif no longer considered specific. *Bioessays* **2016**, *38*, 1117–1122. [CrossRef] [PubMed]

121. Hedglin, M.; Pandey, B.; Benkovic, S.J. Characterization of human translesion DNA synthesis across a UV-induced DNA lesion. *eLife* **2016**, *5*, e19788. [CrossRef] [PubMed]

122. Yang, K.; Weinacht, C.P.; Zhuang, Z. Regulatory role of ubiquitin in eukaryotic DNA translesion synthesis. *Biochemistry* **2013**, *52*, 3217–3228. [CrossRef] [PubMed]

123. Kim, W.; Bennett, E.J.; Huttlin, E.L.; Guo, A.; Li, J.; Possemato, A.; Sowa, M.E.; Rad, R.; Rush, J.; Comb, M.J. Systematic and quantitative assessment of the ubiquitin-modified proteome. *Mol. Cell* **2011**, *44*, 325–340. [CrossRef] [PubMed]

124. Wagner, S.A.; Beli, P.; Weinert, B.T.; Nielsen, M.L.; Cox, J.; Mann, M.; Choudhary, C. A proteome-wide, quantitative survey of in vivo ubiquitylation sites reveals widespread regulatory roles. *Mol. Cell. Proteom.* **2011**. [CrossRef] [PubMed]

125. Bienko, M.; Green, C.M.; Crosetto, N.; Rudolf, F.; Zapart, G.; Coull, B.; Kannouche, P.; Wider, G.; Peter, M.; Lehmann, A.R. Ubiquitin-binding domains in Y-family polymerases regulate translesion synthesis. *Science* **2005**, *310*, 1821–1824. [CrossRef] [PubMed]

126. Freudenthal, B.D.; Gakhar, L.; Ramaswamy, S.; Washington, M.T. Structure of monoubiquitinated PCNA and implications for translesion synthesis and DNA polymerase exchange. *Nat. Struct. Mol. Biol.* **2010**, *17*, 479–484. [CrossRef] [PubMed]

127. Dieckman, L.M.; Washington, M.T. PCNA trimer instability inhibits translesion synthesis by DNA polymerase η and by DNA polymerase δ. *DNA Repair* **2013**, *12*, 367–376. [CrossRef] [PubMed]

128. Zhuang, Z.; Johnson, R.E.; Haracska, L.; Prakash, L.; Prakash, S.; Benkovic, S.J. Regulation of polymerase exchange between Pol η and Pol δ by monoubiquitination of PCNA and the movement of DNA polymerase holoenzyme. *Proc. Natl. Acad. Sci. USA* **2008**, *105*, 5361–5366. [CrossRef] [PubMed]

129. Hoege, C.; Pfander, B.; Moldovan, G.-L.; Pyrowolakis, G.; Jentsch, S. RAD6-dependent DNA repair is linked to modification of PCNA by ubiquitin and SUMO. *Nature* **2002**, *419*, 135–141. [CrossRef] [PubMed]

130. Stelter, P.; Ulrich, H.D. Control of spontaneous and damage-induced mutagenesis by SUMO and ubiquitin conjugation. *Nature* **2003**, *425*, 188–191. [CrossRef] [PubMed]

131. Kannouche, P.L.; Wing, J.; Lehmann, A.R. Interaction of human DNA polymerase η with monoubiquitinated PCNA: A possible mechanism for the polymerase switch in response to DNA damage. *Mol. Cell* **2004**, *14*, 491–500. [CrossRef]

132. Watanabe, K.; Tateishi, S.; Kawasuji, M.; Tsurimoto, T.; Inoue, H.; Yamaizumi, M. Rad18 guides polη to replication stalling sites through physical interaction and PCNA monoubiquitination. *EMBO J.* **2004**, *23*, 3886–3896. [CrossRef] [PubMed]

133. Despras, E.; Delrieu, N.; Garandeau, C.; Ahmed-Seghir, S.; Kannouche, P.L. Regulation of the specialized DNA polymerase eta: Revisiting the biological relevance of its PCNA-and ubiquitin-binding motifs. *Environ. Mol. Mutagen.* **2012**, *53*, 752–765. [CrossRef] [PubMed]

134. Acharya, N.; Yoon, J.-H.; Gali, H.; Unk, I.; Haracska, L.; Johnson, R.E.; Hurwitz, J.; Prakash, L.; Prakash, S. Roles of PCNA-binding and ubiquitin-binding domains in human DNA polymerase η in translesion DNA synthesis. *Proc. Natl. Acad. Sci. USA* **2008**, *105*, 17724–17729. [CrossRef] [PubMed]

135. Sabbioneda, S.; Green, C.M.; Bienko, M.; Kannouche, P.; Dikic, I.; Lehmann, A.R. Ubiquitin-binding motif of human DNA polymerase η is required for correct localization. *Proc. Natl. Acad. Sci. USA* **2009**, *106*. [CrossRef] [PubMed]

136. Sabbioneda, S.; Gourdin, A.M.; Green, C.M.; Zotter, A.; Giglia-Mari, G.; Houtsmuller, A.; Vermeulen, W.; Lehmann, A.R. Effect of proliferating cell nuclear antigen ubiquitination and chromatin structure on the dynamic properties of the Y-family DNA polymerases. *Mol. Biol. Cell* **2008**, *19*, 5193–5202. [CrossRef] [PubMed]

137. Göhler, T.; Sabbioneda, S.; Green, C.M.; Lehmann, A.R. ATR-mediated phosphorylation of DNA polymerase η is needed for efficient recovery from UV damage. *J. Cell Biol.* **2011**, *192*, 219–227. [CrossRef] [PubMed]

138. Hendel, A.; Krijger, P.H.L.; Diamant, N.; Goren, Z.; Langerak, P.; Kim, J.; Reißner, T.; Lee, K.-y.; Geacintov, N.E.; Carell, T. PCNA ubiquitination is important, but not essential for translesion DNA synthesis in mammalian cells. *PLoS Genet.* **2011**, *7*, e1002262. [CrossRef] [PubMed]

139. Langerak, P.; Nygren, A.O.H.; Krijger, P.H.L.; van den Berk, P.C.M.; Jacobs, H. A/T mutagenesis in hypermutated immunoglobulin genes strongly depends on PCNA^{K164} modification. *J. Exp. Med.* **2007**, *204*, 1989–1998. [CrossRef] [PubMed]

140. Szüts, D.; Marcus, A.P.; Himoto, M.; Iwai, S.; Sale, J.E. REV1 restrains DNA polymerase ζ to ensure frame fidelity during translesion synthesis of UV photoproducts in vivo. *Nucleic Acids Res.* **2008**, *36*, 6767–6780. [CrossRef] [PubMed]

141. Edmunds, C.E.; Simpson, L.J.; Sale, J.E. PCNA ubiquitination and REV1 define temporally distinct mechanisms for controlling translesion synthesis in the avian cell line DT40. *Mol. Cell* **2008**, *30*, 519–529. [CrossRef] [PubMed]

142. Lau, W.C.; Li, Y.; Zhang, Q.; Huen, M.S. Molecular architecture of the Ub-PCNA/Pol η complex bound to DNA. *Sci. Rep.* **2015**. [CrossRef] [PubMed]

143. Prindle, M.J.; Loeb, L.A. DNA polymerase delta in DNA replication and genome maintenance. *Environ. Mol. Mutagen.* **2012**, *53*, 666–682. [CrossRef] [PubMed]

144. Netz, D.J.A.; Stith, C.M.; Stümpfig, M.; Köpf, G.; Vogel, D.; Genau, H.M.; Stodola, J.L.; Lill, R.; Burgers, P.M.J.; Pierik, A.J. Eukaryotic DNA polymerases require an iron-sulfur cluster for the formation of active complexes. *Nat. Chem. Biol.* **2012**, *8*, 125–132. [CrossRef] [PubMed]

145. Nair, D.T.; Johnson, R.E.; Prakash, L.; Prakash, S.; Aggarwal, A.K. Rev1 employs a novel mechanism of DNA synthesis using a protein template. *Science* **2005**, *309*, 2219–2222. [CrossRef] [PubMed]

146. Haracska, L.; Unk, I.; Johnson, R.E.; Johansson, E.; Burgers, P.M.J.; Prakash, S.; Prakash, L. Roles of yeast DNA polymerases δ and ζ and of Rev1 in the bypass of abasic sites. *Genes Dev.* **2001**, *15*, 945–954. [CrossRef] [PubMed]

147. Ross, A.-L.; Simpson, L.J.; Sale, J.E. Vertebrate DNA damage tolerance requires the C-terminus but not BRCT or transferase domains of REV1. *Nucleic Acids Res.* **2005**, *33*, 1280–1289. [CrossRef] [PubMed]

148. Baranovskiy, A.G.; Lada, A.G.; Siebler, H.M.; Zhang, Y.; Pavlov, Y.I.; Tahirov, T.H. DNA polymerase δ and ζ switch by sharing accessory subunits of DNA polymerase δ. *J. Biol. Chem.* **2012**, *287*, 17281–17287. [CrossRef] [PubMed]

149. Johnson, R.E.; Prakash, L.; Prakash, S. Pol31 and Pol32 subunits of yeast DNA polymerase δ are also essential subunits of DNA polymerase ζ. *Proc. Natl. Acad. Sci. USA* **2012**, *109*, 12455–12460. [CrossRef] [PubMed]

150. Pustovalova, Y.; Magalhães, M.T.Q.; D'Souza, S.; Rizzo, A.A.; Korza, G.; Walker, G.C.; Korzhnev, D.M. Interaction between the Rev1 C-terminal domain and the PolD3 subunit of Polζ suggests a mechanism of polymerase exchange upon Rev1/Polζ-dependent translesion synthesis. *Biochemistry* **2016**, *55*, 2043–2053. [CrossRef] [PubMed]

151. Gabel, S.A.; DeRose, E.F.; London, R.E. XRCC1 interaction with the REV1 C-terminal domain suggests a role in post replication repair. *DNA Repair* **2013**, *12*, 1105–1113. [CrossRef] [PubMed]

152. Xu, X.; Lin, A.; Zhou, C.; Blackwell, S.R.; Zhang, Y.; Wang, Z.; Feng, Q.; Guan, R.; Hanna, M.D.; Chen, Z.; et al. Involvement of budding yeast Rad5 in translesion DNA synthesis through physical interaction with Rev1. *Nucleic Acids Res.* **2016**, *44*, 5231–5245. [CrossRef] [PubMed]

153. Pozhidaeva, A.; Pustovalova, Y.; D'Souza, S.; Bezsonova, I.; Walker, G.C.; Korzhnev, D.M. NMR structure and dynamics of the C-terminal domain from human Rev1 and its complex with Rev1 interacting region of DNA polymerase η. *Biochemistry* **2012**, *51*, 5506–5520. [CrossRef] [PubMed]

154. Wojtaszek, J.; Liu, J.; D'Souza, S.; Wang, S.; Xue, Y.; Walker, G.C.; Zhou, P. Multifaceted recognition of vertebrate Rev1 by translesion polymerases ζ and κ. *J. Biol. Chem.* **2012**, *287*, 26400–26408. [CrossRef] [PubMed]

155. Livneh, Z.; Shachar, S. Multiple two-polymerase mechanisms in mammalian translesion DNA synthesis. *Cell Cycle* **2010**, *9*, 729–735. [CrossRef] [PubMed]

156. Yoon, J.-H.; Prakash, L.; Prakash, S. Error-free replicative bypass of (6–4) photoproducts by DNA polymerase ζ in mouse and human cells. *Genes Dev.* **2010**, *24*, 123–128. [CrossRef] [PubMed]

157. Wojtaszek, J.; Lee, C.-J.; D'Souza, S.; Minesinger, B.; Kim, H.; D'Andrea, A.D.; Walker, G.C.; Zhou, P. Structural basis of Rev1-mediated assembly of a quaternary vertebrate translesion polymerase complex consisting of Rev1, heterodimeric polymerase (Pol) ζ, and Pol κ. *J. Biol. Chem.* **2012**, *287*, 33836–33846. [CrossRef] [PubMed]

158. Xie, W.; Yang, X.; Xu, M.; Jiang, T. Structural insights into the assembly of human translesion polymerase complexes. *Protein Cell* **2012**, *3*, 864–874. [CrossRef] [PubMed]

159. Kikuchi, S.; Hara, K.; Shimizu, T.; Sato, M.; Hashimoto, H. Structural basis of recruitment of DNA polymerase ζ by interaction between REV1 and REV7 proteins. *J. Biol. Chem.* **2012**, *287*, 33847–33852. [CrossRef] [PubMed]

160. Boehm, E.M.; Spies, M.; Washington, M.T. PCNA tool belts and polymerase bridges form during translesion synthesis. *Nucleic Acids Res.* **2016**, *44*, 8250–8260. [CrossRef] [PubMed]

161. Andersen, P.L.; Xu, F.; Ziola, B.; McGregor, W.G.; Xiao, W. Sequential assembly of translesion DNA polymerases at UV-induced DNA damage sites. *Mol. Biol. Cell* **2011**, *22*, 2373–2383. [CrossRef] [PubMed]

162. Akagi, J.-I.; Masutani, C.; Kataoka, Y.; Kan, T.; Ohashi, E.; Mori, T.; Ohmori, H.; Hanaoka, F. Interaction with DNA polymerase η is required for nuclear accumulation of REV1 and suppression of spontaneous mutations in human cells. *DNA Repair* **2009**, *8*, 585–599. [CrossRef] [PubMed]

163. Lee, Y.-S.; Gregory, M.T.; Yang, W. Human Pol ζ purified with accessory subunits is active in translesion DNA synthesis and complements Pol η in cisplatin bypass. *Proc. Natl. Acad. Sci. USA* **2014**, *111*, 2954–2959. [CrossRef] [PubMed]

164. Johansson, E.; Garg, P.; Burgers, P.M.J. The Pol32 subunit of DNA polymerase δ contains separable domains for processive replication and proliferating cell nuclear antigen (PCNA) binding. *J. Biol. Chem.* **2004**, *279*, 1907–1915. [CrossRef] [PubMed]

165. Gómez-Llorente, Y.; Malik, R.; Jain, R.; Choudhury, J.R.; Johnson, R.E.; Prakash, L.; Prakash, S.; Ubarretxena-Belandia, I.; Aggarwal, A.K. The architecture of yeast DNA polymerase ζ. *Cell reports* **2013**, *5*, 79–86. [CrossRef] [PubMed]

166. Makarova, A.V.; Burgers, P.M. Eukaryotic DNA polymerase ζ. *DNA Repair* **2015**, *29*, 47–55. [CrossRef] [PubMed]

167. Zhang, S.; Zhou, Y.; Trusa, S.; Meng, X.; Lee, E.Y.C.; Lee, M.Y.W.T. A novel DNA damage response: Rapid degradation of the p12 subunit of DNA polymerase δ. *J. Biol. Chem.* **2007**, *282*, 15330–15340. [CrossRef] [PubMed]

168. Murga, M.; Lecona, E.; Kamileri, I.; Díaz, M.; Lugli, N.; Sotiriou, S.K.; Anton, M.E.; Méndez, J.; Halazonetis, T.D.; Fernandez-Capetillo, O. POLD3 is haploinsufficient for DNA replication in mice. *Mol. Cell* **2016**, *63*, 877–883. [CrossRef] [PubMed]

169. Stepchenkova, E.I.; Tarakhovskaya, E.R.; Siebler, H.M.; Pavlov, Y.I. Defect of Fe-S cluster binding by DNA polymerase δ in yeast suppresses UV-induced mutagenesis, but enhances DNA polymerase ζ—Dependent spontaneous mutagenesis. *DNA Repair* **2016**. [CrossRef]

170. Little, J.W.; Mount, D.W. The SOS regulatory system of *Escherichia coli*. *Cell* **1982**, *29*, 11–22. [CrossRef]

171. Goodman, M.F.; McDonald, J.P.; Jaszczur, M.M.; Woodgate, R. Insights into the complex levels of regulation imposed on *Escherichia coli* DNA polymerase V. *DNA Repair* **2016**, *44*, 42–50. [CrossRef] [PubMed]

172. Wiltrout, M.E.; Walker, G.C. Proteasomal regulation of the mutagenic translesion DNA polymerase, Saccharomyces cerevisiae Rev1. *DNA Repair* **2011**, *10*, 169–175. [CrossRef] [PubMed]

173. Pabla, R.; Rozario, D.; Siede, W. Regulation of *Saccharomyces cerevisiae* DNA polymerase η transcript and protein. *Radiat. Environ. Biophys.* **2008**, *47*, 157–168. [CrossRef] [PubMed]

174. Waters, L.S.; Walker, G.C. The critical mutagenic translesion DNA polymerase Rev1 is highly expressed during G_2/M phase rather than S phase. *Proc. Natl. Acad. Sci. USA* **2006**, *103*, 8971–8976. [CrossRef] [PubMed]

175. Plachta, M.; Halas, A.; McIntyre, J.; Sledziewska-Gojska, E. The steady-state level and stability of TLS polymerase eta are cell cycle dependent in the yeast *S. cerevisiae*. *DNA Repair* **2015**, *29*, 147–153. [CrossRef] [PubMed]

176. Uchiyama, M.; Terunuma, J.; Hanaoka, F. The protein level of Rev1, a TLS polymerase in fission yeast, is strictly regulated during the cell cycle and after DNA damage. *PLoS ONE* **2015**, *10*, e0130000. [CrossRef] [PubMed]

177. King, N.M.; Nikolaishvili-Feinberg, N.; Bryant, M.F.; Luche, D.D.; Heffernan, T.P.; Simpson, D.A.; Hanaoka, F.; Kaufmann, W.K.; Cordeiro-Stone, M. Overproduction of DNA polymerase eta does not raise the spontaneous mutation rate in diploid human fibroblasts. *DNA Repair* **2005**, *4*, 714–724. [CrossRef] [PubMed]

178. Thakur, M.; Wernick, M.; Collins, C.; Limoli, C.L.; Crowley, E.; Cleaver, J.E. DNA polymerase η undergoes alternative splicing, protects against UV sensitivity and apoptosis, and suppresses Mre11-dependent recombination. *Genes Chromosomes Cancer* **2001**, *32*, 222–235. [CrossRef] [PubMed]

179. Qi, H.; Zhu, H.; Lou, M.; Fan, Y.; Liu, H.; Shen, J.; Li, Z.; Lv, X.; Shan, J.; Zhu, L. Interferon regulatory factor 1 transactivates expression of human DNA polymerase η in response to carcinogen N-methyl-N'-nitro-N-nitrosoguanidine. *J. Biol. Chem.* **2012**, *287*, 12622–12633. [CrossRef] [PubMed]

180. Yu, S.-L.; Johnson, R.E.; Prakash, S.; Prakash, L. Requirement of DNA polymerase η for error-free bypass of UV-induced CC and TC photoproducts. *Mol. Cell. Biol.* **2001**, *21*, 185–188. [CrossRef] [PubMed]

181. Haruta, N.; Kubota, Y.; Hishida, T. Chronic low-dose ultraviolet-induced mutagenesis in nucleotide excision repair-deficient cells. *Nucleic Acids Res.* **2012**, *40*, 8406–8415. [CrossRef] [PubMed]

182. Pavlov, Y.I.; Nguyen, D.; Kunkel, T.A. Mutator effects of overproducing DNA polymerase η (Rad30) and its catalytically inactive variant in yeast. *Mutat. Res. Fundam. Mol. Mech. Mutag.* **2001**, *478*, 129–139. [CrossRef]

183. McIntyre, J.; Woodgate, R. Regulation of translesion DNA synthesis: Posttranslational modification of lysine residues in key proteins. *DNA Repair* **2015**, *29*, 166–179. [CrossRef] [PubMed]

184. García-Rodríguez, N.; Wong, R.P.; Ulrich, H.D. Functions of ubiquitin and SUMO in DNA replication and replication stress. *Front. Genet.* **2016**. [CrossRef] [PubMed]

185. Jung, Y.-S.; Liu, G.; Chen, X. Pirh2 E3 ubiquitin ligase targets DNA polymerase eta for 20S proteasomal degradation. *Mol. Cell. Biol.* **2010**, *30*, 1041–1048. [CrossRef] [PubMed]

186. Jung, Y.-S.; Hakem, A.; Hakem, R.; Chen, X. Pirh2 E3 ubiquitin ligase monoubiquitinates DNA polymerase eta to suppress translesion DNA synthesis. *Mol. Cell. Biol.* **2011**, *31*, 3997–4006. [CrossRef] [PubMed]

187. Jung, Y.-S.; Qian, Y.; Chen, X. DNA polymerase eta is targeted by Mdm2 for polyubiquitination and proteasomal degradation in response to ultraviolet irradiation. *DNA Repair* **2012**, *11*, 177–184. [CrossRef] [PubMed]

188. Wallace, H.A.; Merkle, J.A.; Yu, M.C.; Berg, T.G.; Lee, E.; Bosco, G.; Lee, L.A. TRIP/NOPO E3 ubiquitin ligase promotes ubiquitylation of DNA polymerase η. *Development* **2014**, *141*, 1332–1341. [CrossRef] [PubMed]

189. Havens, C.G.; Walter, J.C. Mechanism of $CRL4^{Cdt2}$, a PCNA-dependent E3 ubiquitin ligase. *Genes Dev.* **2011**, *25*, 1568–1582. [CrossRef] [PubMed]

190. Higa, L.A.A.; Mihaylov, I.S.; Banks, D.P.; Zheng, J.; Zhang, H. Radiation-mediated proteolysis of CDT1 by CUL4–ROC1 and CSN complexes constitutes a new checkpoint. *Nat. Cell Biol.* **2003**, *5*, 1008–1015. [CrossRef] [PubMed]

191. Shiomi, Y.; Hayashi, A.; Ishii, T.; Shinmyozu, K.; Nakayama, J.-I.; Sugasawa, K.; Nishitani, H. Two different replication factor C proteins, Ctf18 and RFC1, separately control PCNA-CRL4^{Cdt2}-mediated Cdt1 proteolysis during S phase and following UV irradiation. *Mol. Cell. Biol.* **2012**, *32*, 2279–2288. [CrossRef] [PubMed]

192. Kim, S.-H.; Michael, W.M. Regulated proteolysis of DNA polymerase η during the DNA-damage response in *C. elegans*. *Mol. Cell* **2008**, *32*, 757–766. [CrossRef] [PubMed]

193. Havens, C.G.; Walter, J.C. Docking of a specialized PIP box onto chromatin-bound PCNA creates a degron for the ubiquitin ligase CRL4^{Cdt2}. *Mol. Cell* **2009**, *35*, 93–104. [CrossRef] [PubMed]

194. Michishita, M.; Morimoto, A.; Ishii, T.; Komori, H.; Shiomi, Y.; Higuchi, Y.; Nishitani, H. Positively charged residues located downstream of PIP box, together with TD amino acids within PIP box, are important for CRL4^{Cdt2}-mediated proteolysis. *Genes Cells* **2011**, *16*, 12–22. [CrossRef] [PubMed]

195. Tsanov, N.; Kermi, C.; Coulombe, P.; van der Laan, S.; Hodroj, D.; Maiorano, D. PIP degron proteins, substrates of CRL4^{Cdt2}, and not PIP boxes, interfere with DNA polymerase η and κ focus formation on UV damage. *Nucleic Acids Res.* **2014**, *42*, 3692–3706. [CrossRef] [PubMed]

196. Li, H.; Xie, B.; Zhou, Y.; Rahmeh, A.; Trusa, S.; Zhang, S.; Gao, Y.; Lee, E.Y.; Lee, M.Y. Functional roles of p12, the fourth subunit of human DNA polymerase δ. *J. Biol. Chem.* **2006**, *281*, 14748–14755. [CrossRef] [PubMed]

197. Meng, X.; Zhou, Y.; Lee, E.Y.C.; Lee, M.Y.W.T.; Frick, D.N. The p12 subunit of human polymerase δ modulates the rate and fidelity of DNA synthesis. *Biochemistry* **2010**, *49*, 3545–3554. [CrossRef] [PubMed]

198. Meng, X.; Zhou, Y.; Zhang, S.; Lee, E.Y.C.; Frick, D.N.; Lee, M.Y.W.T. DNA damage alters DNA polymerase δ to a form that exhibits increased discrimination against modified template bases and mismatched primers. *Nucleic Acids Res.* **2009**, *37*, 647–657. [CrossRef] [PubMed]

199. Zhao, H.; Zhang, S.; Xu, D.; Lee, M.Y.W.T.; Zhang, Z.; Lee, E.Y.C.; Darzynkiewicz, Z. Expression of the p12 subunit of human DNA polymerase δ (Pol δ), CDK inhibitor p21WAF1, Cdt1, cyclin A, PCNA and Ki-67 in relation to DNA replication in individual cells. *Cell Cycle* **2014**, *13*, 3529–3540. [CrossRef] [PubMed]

200. Lee, M.Y.W.T.; Zhang, S.; Hua Lin, S.; Wang, X.; Darzynkiewicz, Z.; Zhang, Z.; Lee, E. The tail that wags the dog: P12, the smallest subunit of DNA polymerase δ, is degraded by ubiquitin ligases in response to DNA damage and during cell cycle progression. *Cell Cycle* **2014**, *13*, 23–31. [CrossRef] [PubMed]

201. Darzynkiewicz, Z.; Zhao, H.; Zhang, S.; Marietta, Y.W.T.L.; Ernest, Y.C.L.; Zhang, Z. Initiation and termination of DNA replication during S phase in relation to cyclins D1, E and A, p21WAF1, Cdt1 and the p12 subunit of DNA polymerase δ revealed in individual cells by cytometry. *Oncotarget* **2015**, *6*, 11735–11750. [CrossRef] [PubMed]

202. Hedglin, M.; Pandey, B.; Benkovic, S.J. Stability of the human polymerase δ holoenzyme and its implications in lagging strand DNA synthesis. *Proc. Natl. Acad. Sci. USA* **2016**, *113*, E1777–E1786. [CrossRef] [PubMed]

203. Tsurimoto, T.; Stillman, B. Replication factors required for SV40 DNA replication in vitro. II. Switching of DNA polymerase alpha and delta during initiation of leading and lagging strand synthesis. *J. Biol. Chem.* **1991**, *266*, 1961–1968. [PubMed]

204. Yurieva, O.; O'Donnell, M. Reconstitution of a eukaryotic replisome reveals the mechanism of asymmetric distribution of DNA polymerases. *Nucleus* **2016**, *7*, 360–368. [CrossRef] [PubMed]

![genes logo] *genes*

MDPI

Review
PrimPol—Prime Time to Reprime

Thomas A. Guilliam and Aidan J. Doherty *

Genome Damage and Stability Centre, School of Life Sciences, University of Sussex, Brighton BN1 9RQ, UK;
T.Guilliam@sussex.ac.uk
* Correspondence: ajd21@sussex.ac.uk; Tel.: +44-1273-877-500

Academic Editor: Eishi Noguchi
Received: 4 November 2016; Accepted: 16 December 2016; Published: 6 January 2017

Abstract: The complex molecular machines responsible for genome replication encounter many obstacles during their progression along DNA. Tolerance of these obstructions is critical for efficient and timely genome duplication. In recent years, primase-polymerase (PrimPol) has emerged as a new player involved in maintaining eukaryotic replication fork progression. This versatile replicative enzyme, a member of the archaeo-eukaryotic primase (AEP) superfamily, has the capacity to perform a range of template-dependent and independent synthesis activities. Here, we discuss the emerging roles of PrimPol as a leading strand repriming enzyme and describe the mechanisms responsible for recruiting and regulating the enzyme during this process. This review provides an overview and update of the current PrimPol literature, as well as highlighting unanswered questions and potential future avenues of investigation.

Keywords: primase; polymerase; AEP; PrimPol; DNA replication; priming; translesion synthesis; damage tolerance

1. Introduction

The eukaryotic replisome is a highly co-ordinated complex of molecular machines, tasked with efficiently duplicating the genome whilst maintaining a near-perfect level of accuracy. At the heart of the replisome lie the classical DNA polymerases (Pols) α, δ, and ε, which together perform the bulk of the "reading" and copying during replication [1]. These enzymes are exceptionally specialised to faithfully duplicate intact DNA but, consequently, are also highly sensitive to perturbations in the DNA template, resulting in the slowing and stalling of replication forks in the presence of replication stress [2].

Many endogenous and exogenous sources contribute to replication stress. Unrepaired DNA lesions generated by inherent metabolic processes within the cell, in addition to external chemical and physical mutagens, serve as potent blocks to the progression of the canonical replicative DNA Pols [3]. Furthermore, non-B DNA secondary structures and collisions between the replisome and transcription machinery also lead to fork stalling and potentially collapse [4,5]. Aside from direct blockages, repetitive DNA sequences, common fragile sites, ribonucleotides incorporated in the template strand, and limiting pools of nucleotides, can all also act as sources of replication stress [6–8].

In order to maintain replication in the presence of stalled and damaged replication forks, eukaryotes possess a number of distinct damage tolerance and fork restart mechanisms [9]. The deployment of these mechanisms differs depending on which template strand is affected. Obstacles encountered on the lagging strand are easily overcome due to the discontinuous nature of lagging strand replication, where primers are repeatedly generated for Okazaki fragment synthesis. This allows resumption of lagging strand synthesis downstream of a lesion, or obstacle, through use of a newly generated primer to reinitiate replication, leaving behind a single-stranded DNA (ssDNA) gap [10,11].

The situation on the leading strand is more complex with numerous restart pathways available [9]. However, in the last decade evidence has emerged indicating that repriming downstream of lesions and secondary structures also occurs on the leading strand, suggesting that eukaryotic leading strand replication is not exclusively continuous as first thought [11,12]. Indeed, repriming of leading strand replication in prokaryotes is now well documented [13,14]. In addition to repriming, stalled replication forks can also utilise translesion synthesis (TLS) to directly replicate over damaged nucleobases in the template strand. Here, specialised, but error-prone, damage tolerant Pols, predominantly of the Y-family, replace the replicative Pol and synthesise a short section of DNA over the lesion, before handing back over to the replicase [15,16]. It is widely believed that TLS can occur both at the replication fork and post-replicatively to fill in ssDNA gaps left opposite damaged bases as a result of repriming [17–20]. Such gaps may also appear due to the firing of dormant replication origins downstream of a stalled fork, a process which can itself rescue replication [21]. Aside from TLS, ssDNA gaps left opposite lesions can also be filled in an error-free manner through recombination-mediated template switching. Here, the newly synthesised undamaged sister chromatid is used as a template for extension [22]. Template switching may additionally occur at the replication fork via fork reversal; remodelling of the stalled fork generates a four-way junction through annealing of the two nascent DNA strands, thus providing an undamaged template for continued extension [23].

DNA damage tolerance, in particular TLS, has long been associated with the specialised damage tolerant Pols of the Y-family. However, more recently it is becoming clear that members of the archaeo-eukaryotic primase (AEP) superfamily also display novel roles in DNA damage tolerance and repair pathways [24]. In archaea, where many species lack Y-family TLS Pols, the replicative primase is inherently TLS proficient [25]. This review will focus on a new player in eukaryotic nuclear and mitochondrial DNA damage tolerance, and only the second human AEP to be identified, primase-polymerase (PrimPol) (alternative names CCDC111, FLJ33167, EukPrim2 or PrimPol1), encoded by the *PRIMPOL* gene on chromosome 4q35.1 [26–28]. After an overview and evolutionary history, we will describe the domain architecture and biochemical features of the enzyme before moving on to discuss recent advances in our understanding of its roles, recruitment and regulation in vertebrate cells.

2. Discovery and Evolutionary History of PrimPol

The AEP superfamily is evolutionarily and structurally distinct from the bacterial DnaG-type primases which, like AEPs in archaea and eukarya, are absolutely required for DNA replication initiation in bacteria [24]. Nevertheless, DnaG-like primases are also present in archaea, and likewise, AEPs have been identified in bacteria [29]. In each case, these enzymes have diverged to fulfil alternative roles, for example in bacteria a member of the AEP superfamily is employed, together with Ku and DNA ligase homologues, in a non homologous end-joining (NHEJ) DNA break repair pathway [30,31]. It is likely that the presence of AEPs in bacteria is a result of horizontal gene transfer (HGT), with the enzymes originally being recruited for replication initiation by the archaeo-eukaryotic lineage following their divergence from bacteria [24,32]. The catalytic core of AEPs is defined by two structural modules; an N-terminal module with an $(\alpha\beta)_2$ unit, and a C-terminal RNA recognition module-like (RRM-like) fold. These two modules pack together, with the active site residues located in between them [32].

In 2005, detailed *in silico* analyses divided the AEP superfamily into 13 major families, which were further organised into three higher order clades; the AEP proper clade, the nucleo-cytoplasmic large DNA virus (NCLDV)-herpesvirus primase clade, and the primpol clade [32]. These analyses also identified PrimPol and assigned it to the NCLDV-herpesvirus clade, whose members are only present in eukaryotes and their viruses. This clade encompasses the iridovirus primase and herpes-pox primase families, PrimPol belonging to the latter. Members of the herpes-pox primase family possess a conserved C-terminal β-strand-rich region, which replaces the Primase C Terminal (PriCT) domain of the iridovirus primase family [32]. The NCLDV-herpesvirus primase clade is suggested to have originated from bacteriophage or bacterial proteins possessing a fused AEP and PriCT-2 domain. Herpes viruses likely acquired their primase from the NCLDV class, before replacing the C-terminal PriCT domain with the characteristic β-strand-rich region [32].

PrimPol orthologues are conserved across vertebrates, plants, and primitive eukaryotes including species of fungi, algae, and protists, such as apicomplexans and the slime mold *Dictyostelium*. However, PrimPol is notably absent from prokaryotes and a number of fungi and animal species, including *Caenorhabditis elegans* and *Drosophila* [26,27,32]. This interrupted distribution of PrimPol, coupled with the diversity of AEPs observed in mobile elements such as viruses and plasmids [24], suggests that PrimPol was originally obtained through HGT by an early eukaryote and then lost on multiple separate occasions. Importantly, PrimPol is not closely related to the eukaryotic replicative DNA primase small subunit (Prim1), a member of the AEP-proper clade, and is dispensable for DNA replication in higher eukaryotes [26,32]. It has been speculated that PrimPol may have originated as a DNA repair enzyme in NCLDVs, potentially required due to their large genome size and lack of access to cellular DNA repair enzymes during replication [32]. Likewise, PrimPol may play a role in DNA replication initiation in these viruses.

3. What Can PrimPol Do? The Domain Architecture and Catalytic Activities of PrimPol

Since the initial identification of PrimPol in 2005 [32], a number of groups have purified and characterised the recombinant protein, permitting insight into the architectural and biochemical properties of the enzyme [26–28]. These studies revealed PrimPol's impressive range of nucleotidyl transferase activities, suggesting a number of potential roles in vivo. In this section, we will describe these activities and the domain architecture of the protein, which underpins its catalytic flexibility (Figure 1).

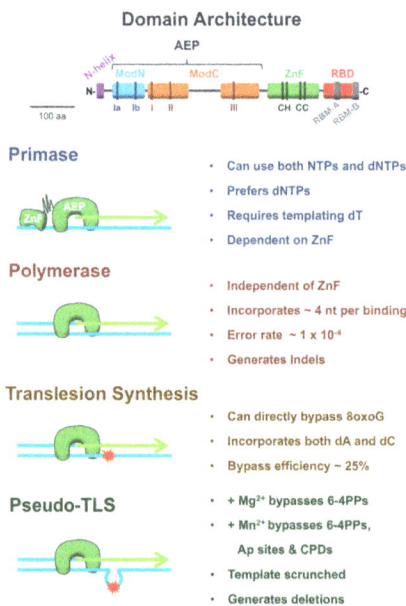

Figure 1. Domain architecture and catalytic activities of primase-polymerase (PrimPol). The domain architecture of PrimPol is depicted in the top panel. A helix (**purple**) located at the N-terminus is connected to ModN by a flexible linker and contacts the DNA major groove. ModN (**blue**) and ModC (**orange**) comprise the archaeo-eukaryotic primase (AEP) domain and contain motifs Ia, Ib, I, II, and III, required for template binding and catalytic activity. The zinc finger (ZnF) (**green**) contains three conserved cysteines and a histidine which coordinate a zinc ion and are required for primase, but not polymerase, activity. The replication protein A (RPA) binding domain (RBD) (**red**) containing RPA binding motif-A (RBM-A) and RBM-B (**grey**) is located at the C-terminus. A 100 amino acid (aa) scale bar is shown to the right. The catalytic activities of PrimPol are displayed below.

3.1. Domain Architecture and Structure

Previously, an alignment of PrimPol homologues identified 14 conserved regions within the protein, including three characteristic AEP catalytic motifs (motifs I, II, and III) towards the N-terminus, forming the AEP domain [32]. Interestingly, motif I displays the variant DxE, rather than the typical DxD motif possessed by most AEPs. Motif I and motif III (xD) together form the divalent metal ion binding site and are essential for the catalytic activity of the enzyme. Motif II (SxH) was predicted to form part of the nucleotide binding site, and is again required for all catalytic activity [26,27,33,34]. Recently, the crystal structure of a ternary complex of the AEP domain of PrimPol (residues 1–354) bound to a DNA template-primer and incoming nucleotide was elucidated, confirming the existence and role of these motifs and two additional motifs, Ia (RQ) and Ib (QRhY/F), which interact with the template DNA strand [35]. The structure reveals that PrimPol's catalytic core encloses the 3′-end of the primer with two α/β modules, ModN and ModC, lining the cavity. ModN primarily interacts with the template strand, whilst ModC contains the catalytic residues and interacts with the incoming nucleotide, as well as the template strand. Intriguingly, the structure of PrimPol's AEP domain does not resemble a typical polymerase fold in any way. There is no thumb domain to hold the primer-template, in fact the primer DNA strand almost completely lacks protein contacts, and ModC was shown to function as both the finger and palm domains [35].

PrimPol also possesses a second conserved domain, a C-terminal UL52-like zinc finger (ZnF) containing three conserved cysteines and a histidine, as is typical for herpes-pox primase family members [32]. The first conserved cysteine and histidine residues of this domain coordinate a zinc ion and are critical for the primase, but not polymerase, activity of the enzyme [34,36].

3.2. Primase Activity

As predicted by the initial *in silico* identification, PrimPol is an active primase that is able to utilise both NTPs and dNTPs for primer synthesis, a unique ability amongst eukaryotic enzymes [26–28]. Surprisingly, PrimPol actually displays a preference for dNTPs over NTPs during primer synthesis, a feature more typically associated with archaeal primases [26]. Similar to the requirement of templated pyrimidines for dinucleotide synthesis by Prim1, PrimPol only generates primers on dT containing templates [26,37]. This primase activity is dependent upon an intact ZnF domain, which is consistent with previous studies on the herpes simplex virus type I (HSV1) helicase/primase complex. Here, primase activity was lost when key residues in the UL52 zinc-binding domain were mutated [38,39]. Interestingly, the ZnF domain of PrimPol has been shown to bind single-stranded (ss) but not double-stranded (ds) DNA, suggesting that this module may be important for stabilising PrimPol on ssDNA templates to allow synthesis of the initial dinucleotide [34]. The ability of PrimPol to synthesise DNA primers *de novo* gives it the potential to reprime and restart replication downstream of DNA damage lesions and fork-stalling obstacles in vivo.

3.3. Polymerase and Lesion Bypass Activities

In addition to its DNA and RNA primase activity, PrimPol is also a template-dependent DNA polymerase, with an ability to bypass a number of DNA damage lesions. Notably, PrimPol can bypass both oxidative and ultraviolet (UV)-induced lesions, including 8-oxo-guanine (8oxoG), and pyrimidine (6-4) pyrimidone photoproducts (6-4PPs) [26,27]. A recent study analysing the kinetics of 8oxoG bypass by PrimPol found that the enzyme incorporates dC (error free) opposite the lesion with 6-fold higher efficiency than dA (error prone). Incorporation of dC opposite 8oxoG occurred at ≈25% efficiency compared to an unmodified templating dG, suggesting that PrimPol has the potential to function as an efficient TLS Pol in vivo [40]. However, the accuracy of bypass differs in other reports, in some instances being only 50% error-free [26,34,41,42]. In the case of 6-4PPs, PrimPol bypasses the lesion in an error-prone manner [26]. Although unable to directly traverse a cyclobutane pyrimidine dimer (CPD), PrimPol can extend from mismatched bases opposite a CPD [26]. Additionally, a truncated form of PrimPol, lacking the ZnF domain, can facilitate TLS past a CPD [34]. In contrast, in the presence of manganese, PrimPol's TLS activity is altered allowing the full-length enzyme to extend past cyclobutane pyrimidine dimers (CPDs) and abasic sites (Ap sites), in addition to 6-4PPs and 8oxoG lesions [36]. However, the usage of either magnesium or manganese as the primary cofactor for PrimPol in vivo remains unclear.

3.4. Lesion Skipping and Template Independent Extension

Despite displaying the ability to directly read through some damaged nucleobases, such as 8oxoG, it appears that PrimPol's bypass of more bulky or distorting lesions is facilitated through a pseudo-TLS mechanism. Here, PrimPol is able to re-anneal the primer to a new position downstream of the lesion prior to extension, thus looping out the templating lesion and generating a shorter extension product than would be produced from strict template-dependent extension [27,36,43]. This activity is enhanced in the presence of manganese, permitting bypass of 6-4PPs, CPDs, and Ap sites by pseudo-TLS [27,36]. Intriguingly, this characteristic is reminiscent of the Ap site bypass strategy employed by the primase/polymerisation domain (PolDom) of *Mycobacterium tuberculosis* DNA ligase (LigD) [44,45]. The ability of manganese to stimulate primer-realignment and template scrunching by PrimPol offers a clear explanation for the altered TLS ability of the enzyme in the presence of this metal ion. It has also been reported that manganese increases both PrimPol's polymerase activity and affinity for DNA, compared to magnesium [40].

Notably however, manganese also promotes promiscuous template-independent extension by PrimPol, resulting in the generation of non-complementary homopolymeric strands [34]. The mutagenic effect of manganese on polymerase activity, through increased reactivity and promotion of non-template-directed nucleotidyl transfer, has been clear for several decades [46–50]. Moreover, the bypass of lesions via template scrunching is potentially more detrimental than beneficial to genomic integrity, due to the high risk of generating frame-shift mutations. Therefore, it seems likely that more low-risk mechanisms would be employed in vivo where available.

The lower affinity of PrimPol for DNA and incoming nucleotides in the presence of magnesium is often taken as support for manganese as the enzyme's primary metal ion cofactor in vivo [40]. However, PrimPol's inherent low affinity for DNA and dNTPs, when using magnesium as a cofactor, may actually act as an important mechanism to regulate its activity. In support of this, it has previously been shown that dNTP levels in yeast are increased 6–8-fold in the presence of DNA damage [51]. Importantly, TLS Pols often require $\approx$10 times greater dNTP concentrations for nucleotide binding opposite a lesion, compared to a replicative Pol at an undamaged site [52,53]. Increased intracellular dNTP concentrations have been found to correlate with an increase in damage tolerance, but also increased mutation rates, potentially due to the unregulated participation of TLS Pols in 'normal' replication [51]. Thus, in yeast it appears that the in vivo activity of TLS Pols is partly regulated by dNTP levels, which increase after DNA damage, consequently restricting the contribution of these Pols to 'normal' DNA replication. Intriguingly, ribonucleotide reductase has been found to be up-regulated in response to DNA damage in all studied organisms, suggesting that increased dNTP synthesis in response to damage may be a conserved mechanism across all domains of life [54]. Similarly, PrimPol's relatively poor affinity for DNA may be overcome in vivo by association with other factors, such as replication protein A (RPA) and polymerase-delta interacting protein 2 (PolDIP2 or PDIP38), again acting to regulate the enzyme by only recruiting it to loci where it is actually required [28,41,55].

Additionally, it is not clear whether the relatively low intracellular concentrations of manganese (0.1 to 40 μM) [56–58], compared to magnesium (0.21 to 0.24 mM) [59,60], are sufficient to support the manganese-dependent TLS activities of PrimPol in vivo. Indeed, PrimPol required manganese concentrations of 200–1000 μM to facilitate pseudo-TLS bypass of an abasic site in vitro, whilst 100 μM did not permit any observable bypass [43]. Thus, the cellular relevance of these activities is not immediately clear. One intriguing possibility is that PrimPol utilises manganese in the mitochondria only [40]. Here, dNTP concentrations are lower than those in the cytosol, there is a dearth of TLS Pols, manganese uptake is increased in response to oxidative stress, and the high copy number nature of mitochondrial DNA (mtDNA) may allow more promiscuous lesion bypass mechanisms to be employed [40,61]. That said, more recent in vitro reconstitution experiments argue against a TLS-like role for PrimPol in oxidative damage bypass during mitochondrial DNA replication [42].

3.5. Fidelity, Mutagenic Signature, and Processivity

Typically, the price paid by Pols for DNA damage tolerance is a significant decrease in both fidelity and processivity. Whilst the structural features of replicative Pols confer extremely efficient and high fidelity DNA synthesis, TLS Pols possess more spacious active sites, altered finger and thumb domains, and lack proofreading exonuclease capabilities. These characteristics permit bypass of bulky lesions, but result in greatly decreased fidelity and processivity on undamaged DNA templates [16]. Likewise, the eukaryotic replicative primase exhibits poor fidelity compared to replicative Pols [62–64]. Rather unsurprisingly, PrimPol, which combines both TLS and primase capabilities, exhibits high error rates of $\approx 1 \times 10^{-4}$, comparable with Y and X-family Pols [55]. Unlike these Pols however, PrimPol generates insertion-deletion (indel) errors at a much higher frequency than substitution mutations, which may be a result of its template scrunching ability [55]. Manganese acts to further decrease PrimPol's fidelity on undamaged DNA and even more so on 8oxoG containing templates [40]. In addition to poor fidelity, PrimPol shares the characteristic of low processivity with canonical TLS Pols, incorporating only 1–4 nucleotides per binding event [34]. Intriguingly, the enzyme's processivity was found to be negatively regulated by its ZnF domain, which may act to stabilise DNA binding and allow primer synthesis, whilst additionally limiting primer extension. Removal of the ZnF domain has also been found to lower PrimPol's fidelity, suggesting the domain acts to regulate processivity and fidelity, as well as enabling primase activity [34].

4. What Does PrimPol Do? The Role of PrimPol in DNA Replication

The biochemical classification of PrimPol as both a RNA/DNA primase and a TLS Pol clearly suggests a role in DNA replication and damage tolerance. Moreover, these two characteristics give PrimPol the potential to assist the replisome in two different ways; through TLS or repriming. In this section, we will describe the in vivo characterisation of the enzyme, as well as the consequences of its deletion on the cell. Using this information, we will discuss recent advances in our understanding of the cellular roles of PrimPol (Figure 2).

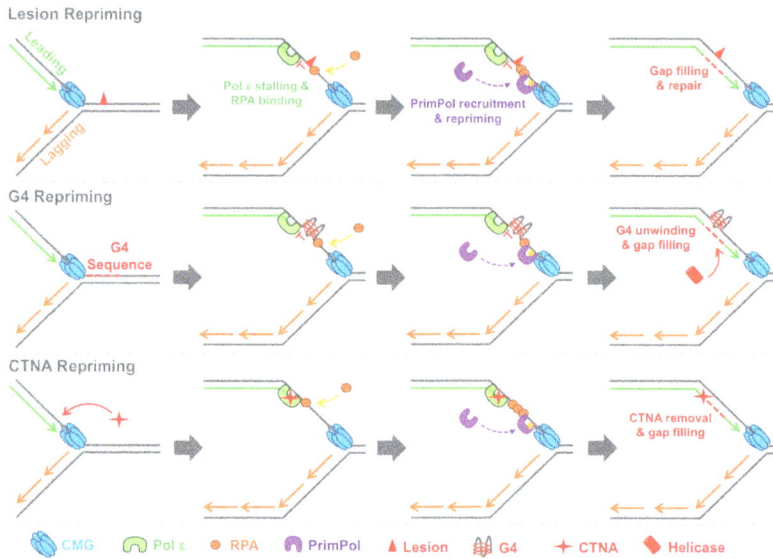

Figure 2. Repriming roles of PrimPol in nuclear DNA replication. PrimPol is able to reprime and reinitiate leading strand replication downstream of a range of replicase stalling obstacles. Here, the ability of PrimPol to reprime downstream of DNA lesions, G4 secondary structures, and chain-terminating nucleotide analogues, is highlighted. Following repriming, replication can proceed and the resulting single-stranded DNA (ssDNA) gap is filled through translesion synthesis (TLS) or template switching mechanisms, permitting subsequent repair or removal of the obstacle. Only the CDC45, MCM, GINS (CMG) complex, Pol ε, PrimPol, and RPA, are shown for simplicity. A key for identifying each factor is shown below. CTNA: chain-terminating nucleoside analogues.

4.1. PrimPol—A DNA Damage Tolerance Enzyme

DNA damage tolerance is critical to support continued replisome progression in the presence of unrepaired DNA damage. An inability to tolerate this damage can lead to prolonged fork stalling, collapse and, ultimately, genome instability and/or cell death. The importance of DNA damage tolerance in preserving genomic integrity is highlighted by the consequences on human health of dysfunction in these mechanisms. An obvious example is the variant form of *xeroderma pigmentosum* (XPV). Here, mutation of Pol η, one of many Y-family TLS Pols, causes increased sensitivity to sunlight and a predisposition to skin cancer [16]. This is thought to occur due to mutagenic bypass of UV-induced CPDs by alternative TLS Pols.

Interestingly, loss of PrimPol in human XPV cells leads to a synergistic increase in UV sensitivity, with the enzyme performing a distinct role from Pol η during this process [26]. In line with this, PrimPol forms sub-nuclear foci, and is recruited to chromatin, in response to UV irradiation [26,36]. Both human MRC5 and avian DT40 cells lacking PrimPol (PrimPol$^{-/-}$) also accumulate an increased number of stalled forks, or a reduced ability to restart stalled forks, following UV damage [26,28,36,41]. Unlike human cells, DT40 cells are hypersensitive to UV irradiation in the absence of PrimPol only, potentially due to the faster doubling times and increased S-phase population of these cells [26]. Interestingly, it has recently been shown that PrimPol$^{-/-}$ avian cells are even more sensitive to UV damage than previously appreciated [65]. In fact, these cells were found to be more sensitive than those lacking Pol η when analysed by colony formation assays. This effect was determined to be due to an extended G2 arrest, which prevented cell cycle progression, rather than an increase in apoptosis [65].

These reports clearly implicate PrimPol in the maintenance of replisome progression, or restart of stalled replication forks, in the presence of UV damage lesions.

However, PrimPol is also involved in the tolerance of other types of DNA damage. PrimPol$^{-/-}$ DT40 cells are hypersensitivity to methylmethane sulfonate (MMS), cisplatin, and hydroxyurea (HU) [66]. Further deletion of Pols ζ and η in these cells leads to an additional increase in damage sensitivity to a similar extent as in wild-type cells, again indicating an independent role for PrimPol in DNA damage tolerance [66]. PrimPol is also required for recovery of stalled replication forks following HU treatment in HeLa cells [28,36]. Notably, each of these DNA damaging agents acts to stall the progression of replication forks. MMS causes the generation of abasic sites in the template strand, cisplatin crosslinks DNA, and HU acts to inhibit ribonucleotide reductase and, consequently, dNTP production. In contrast, loss of PrimPol does not sensitise cells to ICRF193, camptothecin, or γ-rays, agents that produce DNA strand breaks. This is suggestive of a broad role for the enzyme in damage tolerance, but not in break repair [66]. PrimPol associates with chromatin during G1 and S-phase and PrimPol$^{-/-}$ mouse embryonic fibroblasts (MEFs) present chromosome aberrations indicative of S-phase defects, which are enhanced after aphidicolin treatment [26,36]. Collectively, these findings place PrimPol at the replication fork during S-phase and indicate a role in the tolerance of replicase-stalling DNA damage.

4.2. PrimPol Reprimes and Restarts Stalled Replication Forks

The DNA damage tolerance defects observed in the absence of PrimPol potentially indicates that it acts as both a TLS polymerase and a repriming enzyme. However, more recent reports clearly support the latter function [34,36,66–68]. Although PrimPol is described as a TLS Pol, the spectrum of DNA damage types it can traverse by 'true' TLS is actually rather limited. Discounting pseudo-TLS bypass, which may or may not be relevant in vivo, PrimPol is essentially only able to directly bypass 8oxoG lesions [26,27,34,36]. Moreover, a number of other Pols are also able to efficiently and accurately bypass these lesions [69–71]. If PrimPol's primary role were as a TLS Pol, this observation would be at odds with the range of replicase-stalling DNA damaging agents it is involved in tolerating [66]. This implies that PrimPol most likely acts as a repriming enzyme for the tolerance of DNA damage and this is supported by the study of separation of function mutants [34,36,66,67].

Mutation of PrimPol's ZnF domain abolishes primase, but actually enhances polymerase activity [34,36]. This important observation has permitted investigation into the requirement of primase activity for the enzyme's role during DNA replication in vivo. In each case, when PrimPol$^{-/-}$ cells were complemented with the primase-deficient/polymerase-proficient ZnF mutant, it was unable to rescue any of the observed damage tolerance defects [34,36,66]. In contrast, complementation with a primase-proficient/reduced-polymerase mutant of PrimPol restored DNA damage tolerance to wild-type levels [66,72]. In agreement with this, PrimPol was able to facilitate close-coupled repriming downstream of lesions in vitro, which it cannot bypass by TLS [66]. Aside from increased sensitivity to DNA damaging agents and decreased replication fork rates, PrimPol$^{-/-}$ and knockdown cells exhibit persistent RPA foci and increased phosphorylation of Chk1 [28,36]. Both of these stress response markers are indicative of the generation of stretches of ssDNA [11,73]. This would be an expected consequence of a lack of repriming by PrimPol, resulting in the uncoupling of leading/lagging strand replication and excessive strand-specific unwinding by MCM [11]. In agreement, cells compensate for the loss of PrimPol by increasing both homologous recombination (HR) mediated fork rescue and dormant origin firing [26,33,36]. These compensatory back-up mechanisms, in addition to redundancy between PrimPol and TLS polymerases, may explain why PrimPol is dispensable for viability in human cells and mouse models [27]. These observations give further credibility to a requirement for PrimPol at the progressing replication fork during S-phase, which might not necessarily be the case if a TLS-like role was being performed.

As previously mentioned, TLS can potentially occur both at the replication fork, as well as post-replicatively, to fill in gaps left opposite lesions following repriming or dormant origin firing [16]. Each of these possibilities are not mutually exclusive, but a number of studies point to post-replicative gap-filling as the predominant role for TLS. In yeast, DNA damage tolerance mechanisms, including TLS, have been found to operate effectively in a post-replicative manner, and ssDNA gaps, indicative of repriming, accumulate following UV-damage [11,17,18]. Likewise, in human cells DNA replication fork progression in the presence of UV damage was found to be independent of TLS and ssDNA gaps opposite UV lesions were identified. It was concluded that these gaps were likely a result of repriming downstream of lesions rather than dormant origin firing [12]. Importantly, mutation of Pol η or other TLS factors does not appear to significantly alter replication fork rates in the presence of damage [12,20]. This is in stark contrast to the effect of loss of PrimPol on replication fork progression following damage, further supporting a repriming, rather than TLS, role for this enzyme in vivo.

4.3. PrimPol Bypasses Non-Canonical Replication Impediments

Whilst DNA damage lesions are some of the best characterised replication impediments, they are not the only obstacles replication forks must overcome during their progression. In addition to the B-form of dsDNA we have become familiar with since Watson and Crick's famous model [74], genomic DNA can also adopt a number of other secondary structures as a result of specific sequence motifs and protein interactions [75]. One alternative DNA secondary structure, which has received increasing attention as evidence for its formation in vivo grows, is the G-quadruplex (G4) [76]. G4s are produced by the stacking of G-quartets, which form through alternative Hoogsten base-pairing between guanine bases. These structures may potentially play an important role in transcription and DNA replication in the cell, but they can also pose as major impediments to replisome progression [77–80]. Consequently, cells possess a number of specialised helicases and Pols to replicate past G4s [81].

Previously, cells lacking fanconi anemia complementation group J (FANCJ) or REV1 DNA-directed polymerase were found to stochastically lose Bu-1a protein expression [82,83]. Importantly, the *BU-1A* locus contains a G4, which was determined to stall replication in these cells. This stalling causes uncoupling of replication from histone recycling at the *BU-1A* locus and consequently leads to the deletion of epigenetic marks, manifesting in loss of Bu-1a expression. It was recently identified using Bu-1a read-out assays that PrimPol also plays a critical role in the bypass of these structures during DNA replication [67]. Consistent with PrimPol's behaviour at most DNA damage lesions, in vitro analysis revealed that the enzyme is unable to directly read through G4s, but can bind to and facilitate close-coupled repriming downstream of these structures. Vitally, close-coupled repriming ≈6 nt ahead of the G4 would permit the appropriate recycling of histones, and thus maintain epigenetic marks and Bu-1a expression. Bypass of G4 structures through repriming by PrimPol was confirmed in vivo using the ZnF primase-deficient mutant discussed previously. Here, complementation of PrimPol$^{-/-}$ cells with the ZnF mutant failed to prevent instability of Bu-1a expression, in contrast to the wild-type protein, confirming that PrimPol's primase activity is critical for G4 bypass [67]. Intriguingly, PrimPol was found to only be required for G4 bypass during leading strand replication. Presumably, this is because primers are constantly generated on the lagging strand due to the discontinuous nature of DNA synthesis on this strand.

Further evidence supporting a general role for PrimPol in repriming replication downstream of fork-stalling obstacles is provided by studies of chain-terminating nucleoside analogues (CTNAs) [66]. CTNAs cause replication to stall, when incorporated into the 3′-termini of growing DNA polymers, by preventing further extension as they lack the 3′ hydroxyl required for phosphodiester bond formation [84,85]. Loss of PrimPol has been shown to cause hypersensitivity to a wide range of CTNAs [66]. Critically, the inability of Pols to extend from CTNAs rules out bypass by direct extension. PrimPol was found to be important for the tolerance of CTNAs by repriming downstream. This role was confirmed by both in vivo characterisation of the ZnF mutant and in vitro analysis of repriming synthesis after CTNAs [66].

These critical findings not only establish that PrimPol deploys a repriming mechanism to bypass G4s and CTNAs, in a similar manner to DNA damage lesions, they also point to the possibility that PrimPol is able to bypass a wide range of leading strand obstacles during normal and perturbed replication. This is in contrast to canonical TLS Pols, which are typically highly specialised in the lesions they can bypass. Consequently, it is likely PrimPol is broadly employed as a general mechanism to reprime and restart replication ahead of many different leading strand replication impediments.

4.4. A Role for PrimPol in Mitochondrial DNA Replication?

The majority of genetic information in mammalian cells is stored in the nucleus. However, a small proportion of DNA is also located in the mitochondria. Despite being only ≈16.6 kb long and encoding just 13 polypeptides, mutation of the mitochondrial genome is responsible for a number of mitochondriopathies and is implicated in various other pathologies including cancer, cardiovascular diseases, and neurodegenerative disorders [86]. Unlike nuclear DNA, cells possess many copies of mtDNA making it highly redundant. In line with this, the rate of mutagenesis is ≈10-fold greater in the mitochondria than the nucleus [86]. A major function of mitochondria is the generation of ATP through oxidative phosphorylation (OXPHOS). This process produces reactive oxygen species (ROS) which can induce damage lesions, including 8oxoG and Ap sites, in mtDNA [86].

A significant proportion of PrimPol has been found to localise to the mitochondria where it interacts with mitochondrial single-strand binding proteins (mtSSB), suggesting a potential role in the tolerance of mtDNA damage [26,27,55,87]. This is supported by defects in mtDNA replication and copy number observed in cells lacking PrimPol [27,87]. However, the ability to generate viable PrimPol$^{-/-}$ mice demonstrates that this role is redundant. Indeed, mitochondrial RNA Pol (POLRMT) is likely responsible for generating the initial primers essential for mtDNA replication [88]. These primers are then extended by Pol γ, which until recently was thought to be the only mitochondrial DNA Pol [89]. In addition to PrimPol, more recent reports indicate that Pol θ and Pol ζ are also involved in human mtDNA replication [90,91].

Given that few TLS Pols appear to localise to the mitochondria, in addition to the high levels of ROS there, it was speculated that PrimPol may be involved in TLS bypass of mitochondrial 8oxoG lesions and Ap sites [27]. In order to investigate this, a recent study analysed the ability of PrimPol to assist the mitochondrial replisome in oxidative damage bypass by TLS [42]. Here, it was found that the mitochondrial replisome is completely stalled by Ap sites and pauses significantly at 8oxoG lesions. PrimPol did not enhance the bypass of either of these lesions, disagreeing with a TLS role in oxidative damage bypass in the mitochondria [42]. Thus, it seems more likely that PrimPol functions to reprime mtDNA replication downstream of blocking lesions, similar to its role in the nucleus. In addition to oxidative damage, mtDNA is also subject to deletions. Intriguingly, these deletions map in close proximity to G4-forming sequences [92]. In light of the role of PrimPol in repriming after G4s in nuclear DNA replication, it would not be surprising if the enzyme fulfilled the same role in the mitochondria. However, further work is required to confirm a repriming role for PrimPol here. The potential role for PrimPol in the mitochondria has recently been reviewed in more detail [93].

4.5. Is PrimPol Involved in Somatic Hypermutation?

Generally, mutagenesis during DNA replication is avoided at all costs in order to preserve genomic stability. However, an exception to this is during the development of the immune system. Here, mutagenesis occurs in immunoglobulin (Ig) genes to enable variation in the generated antibodies. This programmed mutagenesis is driven by activation-induced deaminase (AID), which deaminates dC to dU [94]. Replication of dU facilitates C>T transitions. Additionally, dU may be further processed by uracil DNA glycosylase (UNG) to generate Ap sites. TLS bypass of these Ap sites can alternatively create C>A/G/T mutations due to the non-instructive nature of the lesion [95].

The involvement of TLS Pols in somatic hypermutation (SHM) at Ap sites led to speculation that, if PrimPol functions as a TLS Pol in vivo, it might also modulate this mutagenesis. Analysis of DT40 cells found that hypermutation and gene-conversion events are similar in wild-type and PrimPol$^{-/-}$ cells [66]. Moreover, loss of PrimPol in wild-type and Pol η$^{-/-}$/Pol ζ$^{-/-}$ avian cells did not significantly alter the mutation spectrum of the studied Ig gene. Intriguingly, another report, which analysed large mutational data sets in mice, identified that PrimPol does have a subtle effect on SHM outcome [68]. In this study, loss of PrimPol was found to selectively increase C>G transversions, but did not affect other G/C or A/T mutations. Interestingly, PrimPol was found to specifically prevent the generation of C>G transversions in the leading strand, potentially explaining the G>C over C>G strand bias of somatically mutated IgH loci [68]. However, this anti-mutagenic activity of PrimPol was attributed to the enzyme's primase, rather than TLS polymerase, activity. It was concluded that PrimPol preferentially reprimes downstream of Ap sites on the leading strand, therefore maintaining fork progression and preventing error-prone TLS. The resulting ssDNA gap opposite the Ap site could then be filled in by error-free homology directed repair. Fascinatingly, in the same report, studies of invasive breast cancers suggested that this leading strand anti-mutagenic activity of PrimPol may be genome wide.

Together, these reports establish that PrimPol does not act as a canonical TLS polymerase during SHM. Rather, PrimPol affects the mutational outcome of SHM by repriming downstream of Ap sites on the leading strand thus preventing C>G transversions. These findings, therefore, further support mounting evidence that PrimPol's primary role in DNA damage tolerance is to reprime leading strand replication and not to perform TLS.

4.6. Why Doesn't the Pol α-Primase Complex Reprime Leading Strand Replication?

The emerging role for PrimPol in repriming leading strand replication begs the question; why doesn't the replicative Pol α-primase complex fulfil this role? In *E. coli*, DnaG, the replicative primase, efficiently reprimes replication ahead of replicase stalling DNA damage lesions, permitting bypass of the damage without dissociation of the replisome [13,14]. Likewise in yeast which lack PrimPol, leading strand repriming is presumably facilitated by Pol α-primase, suggesting that, at least in these organisms, the replicative primase has the capacity to also fulfil this role.

Whilst the answer to this question is not completely clear, PrimPol does have one advantage over Pol α-primase; it preferentially primes using dNTPs. This minimises the amount of RNA processing required on the leading strand. Although ribonucleotides are routinely incorporated during the initiation of each Okazaki fragment on the lagging strand and at replication origins on the leading strand, their persistent presence in DNA can lead to genomic instability [96]. Ribonucleotides incorporated during primer synthesis are routinely removed through Okazaki fragment maturation [97]. However, it is not clear how a DNA secondary structure or lesion requiring bypass upstream of the primer would affect this process.

Ribonucleotides incorporated by replicative Pols are removed by ribonucleotide excision repair (RER). Intriguingly, in RER deficient yeast leading strand ribonucleotides are removed through a topoisomerase I (Top1) mediated mechanism, which likely also removes a subset of ribonucleotides in RER proficient cells [97,98]. This mechanism of ribonucleotide removal, which does not appear to occur on the lagging strand, is susceptible to causing genome instability. This makes ribonucleotides present in the leading strand potentially more detrimental than those in the lagging strand. This is supported by observations that loss of RER and increased ribonucleotide incorporation by Pol ε, but not Pol α or Pol δ, is lethal [98].

Although RER deficient yeast are viable, loss of this pathway in mice results in embryonic lethality [99]. Thus, the greater pressure on higher eukaryotes to minimise the presence of ribonucleotides in their genomes may explain why PrimPol is employed for leading strand repriming using dNTPs in these organisms. However, this enzyme has been lost in some lower eukaryotes as an alternative repriming mechanism, possibly involving the replicative primase, appears to be available.

4.7. Why Is PrimPol Damage Tolerant In Vitro?

If PrimPol's primary role in vivo is to reprime DNA replication, why does the enzyme display TLS-like activity in vitro? Although it is possible that PrimPol's TLS-like activity is important in the cell, recent studies suggest that the enzyme's primase activity is more relevant for its in vivo role, as discussed above. This opens up the possibility that this TLS activity is a 'side effect' of being a primase and this is supported by a number of observations.

Recent studies of the RNA primase domains of human Pol α-primase provide insight into the unique way primases interact with their DNA template and primer [100,101]. The RNA primase associated with Pol α is a heterodimer composed of a small catalytic subunit, p49, and a large regulatory subunit, p58. These reports identify that the C-terminal domain of p58 binds to the DNA/RNA junction at the 5′-end of the RNA primer, whereas p49 binds and extends the 3′ end of the primer moving away from p58. The p49 subunit makes few contacts with the DNA/RNA, resulting in distributive activity. By only contacting the primer at the 5′ and 3′ ends, the primase is unable to sense modified nucleotides in the RNA strand, potentially explaining the propensity of primases to perform TLS-like extension [100,101]. The authors suggest that this binding mechanism is broadly applicable to most primases. In the context of PrimPol, the ZnF is likely functionally equivalent of p58. Indeed, both are flexibly tethered to the catalytic domain and required for template recognition during priming, although PrimPol's ZnF has only been shown to bind ssDNA [34,102]. Nevertheless, the ZnF domain may bind the ssDNA immediately upstream of the 5′ end of the primer.

The crystal structure of PrimPol's AEP domain potentially supports this model [35]. Here, only the templating base is held in the active site cleft, with the rest of the 5′ template strand directed out of the catalytic centre. Additionally, PrimPol lacks a thumb domain and makes few contacts with the primer strand. This potentially prevents the enzyme from sensing damaged bases in the template and allows them to be looped out. Furthermore, unlike TLS Pols, PrimPol does not possess an 'open' active-site cleft and is unable to accommodate bulky lesions such as CPDs and 6-4PPs [35]. This provides further evidence that PrimPol is not a 'true' TLS Pol, rather it loops out bulky-lesions during bypass, resulting in deletions.

The ability of primase-polymerases to perform TLS-like extension is well documented [24]. Some AEPs have co-opted this inherent catalytic versatility for use in other processes such as NHEJ, becoming specialised and in some instances, losing their ability to prime [24]. However, PrimPol's primase activity is critical for its role in vivo and thus it is possible that the TLS-like activities observed in vitro simply arise as a by-product of the structural features necessary for priming.

5. How Does PrimPol Get to Where It is Needed? The Recruitment of PrimPol to Stalled Replication Forks

The studies described above strongly indicate that PrimPol's main role in DNA replication is to reprime ahead of impediments on the leading strand. In order to fulfil this role, PrimPol must be efficiently recruited to ssDNA downstream of stalled replication forks. In this section, we will describe recent advances in our understanding of the interactions and mechanisms governing recruitment of PrimPol.

5.1. PrimPol Interacts with Single-Strand Binding Proteins

Replication fork stalling can cause uncoupling of leading and lagging strand synthesis, consequently generating ssDNA stretches on either strand due to continued unwinding by the replicative helicase [11]. The impact of this on the lagging stand is likely limited by the generation of new Okazaki fragments. However, in the absence of leading strand fork restart, extended uncoupling can produce stretches of ssDNA. In nuclear DNA replication, the resulting ssDNA is bound by RPA, which in turn can trigger the S phase checkpoint response [103].

Unlike TLS Pols, PrimPol does not interact with proliferating cell nuclear antigen (PCNA) [55]. However, it does interact with both the major nuclear and mitochondrial single-strand binding proteins (SSBs); RPA and mtSSB [28,55]. PrimPol's interaction with RPA is mediated by its C-terminal domain (CTD), which binds to the N-terminus of RPA70 (RPA70N), the largest subunit of the RPA heterotrimer [55]. The structural basis for PrimPol's interaction with RPA has recently been elucidated [104], identifying that PrimPol possesses two RPA binding motifs (RBMs) in its CTD (RBM-A and RBM-B), which both bind to the basic cleft of RPA70N, independently of each other. Interestingly, this cleft has previously been shown to interact with, and recruit, a number of different DNA damage response proteins, including RAD9, MRE11, ATRIP, and p53 [105].

Together, these studies indicate that PrimPol may also be recruited to stalled replication forks through its interaction with RPA; with mtSSB likely playing an analogous role in mitochondria.

5.2. RPA Recruits PrimPol to Stalled Replication Forks

Previously, it was identified that PrimPol's CTD is required for its function and co-localisation with RPA in vivo [28]. However, interpretation of these results is limited as removal of the whole CTD has been shown to reduce primase activity in vitro and may also abrogate interactions with other binding partners [34]. Structural studies of PrimPol-RPA complexes have enabled the in vivo analysis of point mutants that disrupt this interaction. These studies identified that PrimPol's RBM-A is the primary mediator of the RPA interaction in vivo, whilst RBM-B appears to play a secondary role. Furthermore, RBM-A mutants were unable to restore replication fork rates following UV-damage, in comparison to the wild-type or RBM-B mutant protein [104]. These findings revealed that PrimPol's interaction with RPA is required for its cellular role. Moreover, this study also showed that this interaction is responsible for the recruitment of PrimPol to chromatin, demonstrating that the enzyme is recruited to stalled replication forks by RPA [104]. Intriguingly, mutations of key residues in each RBM have been identified in cancer patient cell lines, adding further support that these motifs are important for PrimPol's function in vivo [104].

Aside, from identifying the mechanism by which PrimPol is recruited to stalled replication forks; these studies also add to the growing evidence supporting a role for PrimPol as a repriming enzyme. PrimPol's recruitment to RPA, and lack of interaction with PCNA, suggests it binds to ssDNA downstream of a stalled replicase on the leading strand, the ideal place to facilitate repriming following initial leading/lagging strand uncoupling to prevent excessive ssDNA generation. A recent report investigating the role of RAD51 recombinase (RAD51) in aiding replication across UV lesions supports this [106]. Here, RAD51 and MRE11 depletion was found to favour ssDNA accumulation at replication obstacles and subsequent PrimPol-dependent repriming. This also supports previous suggestions that excessive unwinding of DNA following stalling of the replicase is sufficient to promote ssDNA generation and repriming at replication impediments [12].

Further work is required to elucidate the exact mechanisms controlling PrimPol's recruitment by RPA to ssDNA. Interestingly, binding of MRE11 and RAD9 to RPA is enhanced upon RPA32C phosphorylation [107,108]. Thus, phosphorylation of RPA may act to signal recruitment of DNA damage response proteins, potentially including PrimPol [109].

6. Regulation of PrimPol during DNA Replication

Recent reports strongly indicate that PrimPol is recruited by RPA to the leading strand, following replicase stalling, in order to reprime replication and prevent genome instability. However, PrimPol is an error-prone enzyme and unscheduled or dysregulated activity could lead to mutagenesis [55]. In this section, we will discuss our current understanding of the mechanisms used to limit PrimPol's contribution to DNA synthesis during replication (Figure 3).

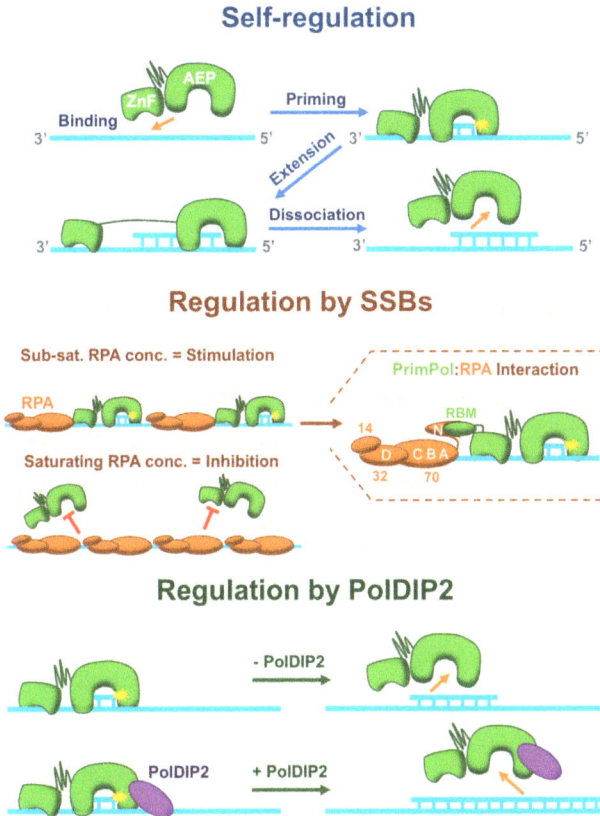

Figure 3. Regulation of PrimPol by its ZnF domain and interacting partners. **Top panel**: PrimPol is inherently self-regulatory due to the restraining effect of its ZnF domain. The AEP and ZnF domains of PrimPol form a hinge-like structure, connected by a flexible linker. Binding of PrimPol to ssDNA is mediated by the ZnF domain, which binds 3′ relative to the AEP domain on the template strand. Binding of the ZnF stabilises the AEP domain, permitting primer synthesis. The AEP then extends the primer, but is restricted by the maximum distance it can move away from the ZnF. The enzyme subsequently dissociates leaving behind a short primer. This mechanism limits the processivity of the PrimPol; **Middle panel**: PrimPol is regulated by single-strand binding proteins (SSBs). At sub-saturating concentrations of RPA, the protein acts to recruit PrimPol to the ssDNA template, consequently stimulating primer synthesis. *In vivo*, this interaction is primarily mediated by PrimPol's RBM-A, which binds to the basic cleft of RPA70N. At saturating RPA concentrations, when the ssDNA template is fully coated, PrimPol cannot gain access and primer synthesis is inhibited. This serves to limit where PrimPol can prime; **Bottom panel**: Polymerase-delta interacting protein 2 (PolDIP2 or PDIP38) enhances PrimPol's primer extension activity by binding the AEP domain and stabilising it on DNA.

6.1. Regulation of the Cellular Concentration of PrimPol

The simplest way to regulate the activity of a protein is by controlling its intracellular concentration. This is especially true for proteins that are only required to act in response to a specific stress, for example DNA damage response proteins. This strategy is utilised during the SOS response in *E. coli*. Here, ≈40 DNA damage response genes are upregulated in response to DNA damage [110].

In comparison to Prim1, PrimPol is expressed at very low levels in human U2OS cells (<500 protein copies per cell compared to ≈13,300) [111]. This is, however, similar to the expression level of TLS Pols, including η and κ. PrimPol mRNA expression peaks in G1-S phase, although the total protein levels remain roughly constant throughout the cell cycle [36]. Thus, the increased association of PrimPol with chromatin during the G1 and S phases of the cell cycle in unperturbed cells is a result of finer mechanisms controlling recruitment to DNA, rather than increased expression. This may also be the case with the increased recruitment of the enzyme to chromatin in response to DNA damage. Nevertheless, the low level of PrimPol expression, in comparison to the replicative primase, acts as the primary mechanism to restrict its contribution to 'normal' replication.

6.2. PrimPol Is Self-Regulating

The structural features afforded to PrimPol by virtue of being a primase also act as inherent regulatory mechanisms. As mentioned previously, PrimPol displays very low processivity. This distributive nature appears to be due to two key features. Firstly, the AEP catalytic domain has a much smaller 'footprint' than most polymerases, potentially explaining why the enzyme binds so poorly to DNA [35]. Secondly, the ZnF domain acts to negatively regulate PrimPol's processivity (Figure 3, top panel) [34].

It has been suggested that the p58 subunit of the replicative eukaryotic primase enforces a strict counting mechanism on the enzyme [112]. Here, the p58 and p49 subunits form a hinge-like structure. The enzyme binds to ssDNA in a 'closed' conformation, with p58 facilitating template recognition. The p49 subunit then initiates primer synthesis, moving away from p58 that binds the 5' end of the primer [101,112]. Thus, an inherent counting mechanism is conferred by the maximum distance p49 can elongate the primer strand away from p58. The ZnF domain of PrimPol is thought to act in a similar way [104]. In this scenario, the AEP domain and ZnF may form a hinge-like structure, connected by a flexible linker. The enzyme probably binds to DNA in a closed conformation assisted by the ZnF domain, which binds on the 3' side relative to the AEP domain on the template strand. The AEP domain can then synthesise and elongate the primer strand until further extension is restricted by the ZnF domain (Figure 3, top panel). It is also conceivable that the AEP and ZnF domains bind DNA in an open conformation, with the ZnF bound on the 5' side relative to the AEP on the template strand, extension would then be limited by inter-domain collisions. In the absence of the ZnF, PrimPol displays increased, but still poor, processivity due to the weak affinity of the AEP domain for the DNA template [34].

PrimPol is, therefore, self-regulating. The supervisory effect of the ZnF domain, which permits priming but limits elongation, coupled with the AEP's poor affinity for DNA, restricts the ability of PrimPol to partake in significant unregulated DNA synthesis during DNA replication.

6.3. Regulation by Single-Strand Binding Proteins

The ability of primases to bind and prime on ssDNA gives them the potential to facilitate unscheduled priming in vivo, wherever ssDNA is available. Despite limiting the synthesis of long DNA tracts, PrimPol's self-regulatory mechanisms do not restrict where it can prime. Dysregulated priming is potentially highly detrimental to the cell, as these primers could be extended by other Pols. To prevent this, PrimPol is also regulated by RPA and mtSSB (Figure 3, middle panel). Both of these SSBs stimulate the activity of their respective replicative Pols, δ and γ [113,114]. In contrast, both RPA and mtSSB severely restrict the polymerase activity of PrimPol [55]. Additionally, these SSBs can also inhibit primase activity, as is the case with Pol α-primase [55,115]. More recently, it was reported that RPA's effect on PrimPol's primase activity is highly concentration-dependent. In fact, sub-saturating concentrations of RPA dramatically stimulate primer synthesis but inhibition occurs as the concentration increases [104].

It is likely that both RPA and mtSSB act to prevent unscheduled priming events by blocking access to the DNA template. Thus, PrimPol requires a free ssDNA interface adjacent to the SSB in order to be recruited (Figure 3, middle panel). This recruitment likely acts to enhance PrimPol's poor affinity for DNA, providing a platform for primer synthesis.

RPA binds ssDNA with a defined polarity [116–119]. Initially, the DNA-binding domain A (DBD-A) and DBD-B oligonucleotide binding (OB) folds of RPA70 bind ssDNA in a tandem manner, forming an 8-nt binding complex. The interface in contact with DNA is then extended to 20–30 nts by the binding of DBD-C and DBD-D, which occurs in a defined $5'$-$3'$ direction on the template strand [120]. This would likely position the RPA70N domain, which recruits PrimPol, $5'$ relative to rest of the RPA molecule on the template strand (Figure 3, middle panel). This suggests that PrimPol binds ahead of RPA in vivo, with the ZnF contacting ssDNA adjacent to RPA and the AEP bound downstream.

The orientation of PrimPol's interaction with RPA may explain the inhibition observed in primer extension assays. By preferentially binding on the $5'$ side of RPA, PrimPol would not be able to access the primer stand at the $3'$ end of the template. Additionally, replicative Pols are thought to be able to easily displace RPA as they approach the protein from the $3'$ side, encountering the weakly bound DBD-D and DBD-C domains, before DBD-B and DBD-A [121]. This in turn shifts the equilibrium from the 20–30-nt RPA complex, to the more weakly bound 8-nt mode, thus permitting displacement. In contrast, if PrimPol binds to the $5'$ side of RPA, it would move away from the protein, preventing displacement in the same way. It is likely that this interaction also further enhances the regulation of PrimPol's processivity by 'holding' the ZnF domain and preventing continued extension by the AEP domain.

6.4. What Generates the ssDNA Interface Required for PrimPol Recruitment?

The requirement of a ssDNA interface downstream of RPA for efficient PrimPol recruitment begs the question: how is this free ssDNA interface generated in vivo? Although the answer to this question is currently unknown, one obvious solution would be through the action of the replicative helicase. Following stalling of the leading strand replicase, leading and lagging strand replication can become uncoupled. Here, the replisome progresses in the absence of DNA synthesis on the leading strand. Continued unwinding of duplex parental DNA by MCM generates ssDNA on the leading strand, which is bound by RPA. Consequently, an RPA/ssDNA interface for PrimPol binding could be generated directly behind the progressing MCM. Subsequent repriming by PrimPol would prevent extended leading/lagging strand uncoupling, allowing leading strand replication to resume at the progressing replisome. The short RPA-bound ssDNA gap left behind could then be filled by TLS or template switching mechanisms.

In support of this, it has recently been shown that the mitochondrial replicative helicase, Twinkle, can stimulate DNA synthesis by PrimPol, indicating that replicative helicases can potentially facilitate PrimPol activity in vivo [42]. It is interesting to note that many DNA primases interact with replicative helicases, with some even possessing their own helicase domains [112].

6.5. Regulation by PolDIP2

PolDIP2 was originally identified as a binding partner of the p50 subunit of Pol δ, in addition to PCNA [122]. More recently, PolDIP2 was shown to interact with Pols η, ζ, λ, and Rev1 [70,123]. In vitro, the protein stimulates the polymerase activity of Pol δ by increasing its affinity for PCNA, as well as enhancing TLS by Pols η and λ [70]. These observations have led to suggestions that PolDIP2 may play an important role in the switch between Pol δ and TLS polymerases during DNA replication [70,123].

PolDIP2 also significantly enhances the DNA binding and processivity of PrimPol's AEP domain (Figure 3, bottom panel) [41]. Additionally, PolDIP2 appears to be important for PrimPol's function in vivo, suggesting it may act as a way to positively regulate the enzyme's activity. Notably, however, this was not sufficient to relieve the negative effect of RPA or mtSSB on PrimPol's polymerase activity. It seems likely that PolDIP2 acts to assist PrimPol's AEP domain during primer extension after synthesis of the initial di-nucleotide, without necessarily allowing synthesis of long DNA tracts. Interestingly, PolDIP2 binds to PrimPol at a region in close proximity to motifs Ia and Ib, identified in the recent crystal structure of PrimPol [35,41]. These motifs harbour the majority of the residues responsible for mediating binding of the AEP domain to the DNA template. PolDIP2, therefore, potentially changes the conformation of this region to enhance PrimPol's affinity for the DNA template, resulting in increased DNA binding and processivity. Additionally, PolDIP2 may also serve as a hand-off mechanism to the replicative Pol, following primer synthesis by PrimPol (Figure 4).

Intriguingly, Pol δ has recently been implicated in extension of a small fraction of primers synthesised by Pol α-primase on the leading strand during DNA replication in yeast [124]. Given that yeast lack PrimPol, this small fraction of primers could in theory be products of repriming by Pol α-primase. This raises the fascinating possibility that Pol δ can serve to extend primers on the leading strand following a repriming event, before subsequent replacement by Pol ε. This could possibly be due to the stalling of Pol ε at the initial impediment, or alteration of the core replisome following leading/lagging strand uncoupling. In higher eukaryotes, these leading strand repriming events appear to be facilitated by PrimPol, not Pol α-primase. Thus, PolDIP2 may act as a hand-off mechanism from PrimPol to Pol δ, given the interaction with both proteins and ability of PolDIP2 to enhance the Pol δ/PCNA interaction. However, more work is required to investigate this possible mechanism (Figure 4).

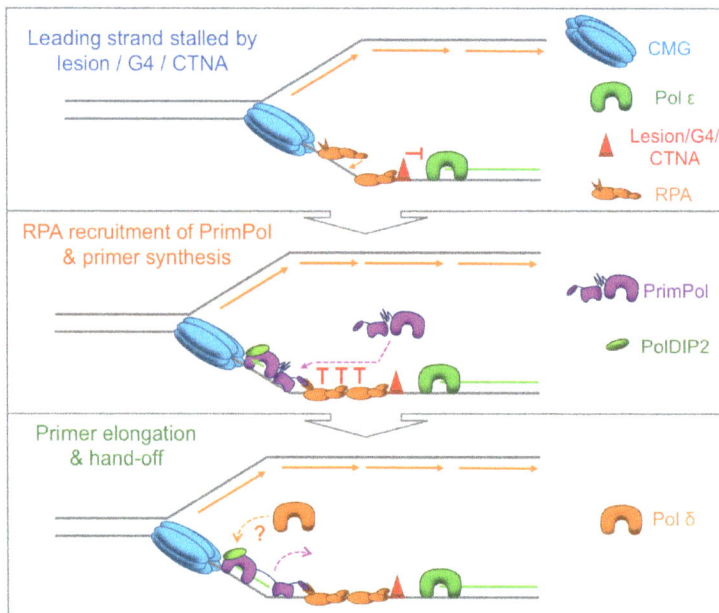

Figure 4. Role, recruitment, and regulation of PrimPol during DNA replication. **Top panel:** Pol ε is stalled on the leading strand by a lesion, secondary structure, or CTNA. Lagging strand replication continues, subsequently generating ssDNA on the leading strand. This ssDNA is bound by RPA as the CMG complex progresses; **Middle panel:** The generation of an RPA / ssDNA interface provides a platform for PrimPol recruitment. PrimPol requires a free ssDNA region adjacent to RPA and thus is recruited to the exposed ssDNA behind the CMG complex. This recruitment is facilitated by the interaction between PrimPol's RBMs and RPA70N. Following recruitment, PrimPol reprimes the leading strand; **Bottom panel:** PrimPol elongates its primer, assisted by PolDIP2, before further extension is restricted by its ZnF and RPA interaction. The primer is then handed-off to the replicative polymerase, possibly Pol δ, mediated by each protein's interaction with PolDIP2.

7. Conclusions and Perspectives

Nearly half a century ago, Rupp and Howard-Flanders identified the presence of ssDNA gaps left opposite UV photoproducts following DNA replication in nucleotide excision repair deficient *E. coli* [125]. A model was proposed which envisaged re-initiation of replication downstream of the damage on both leading and lagging strand templates; the first suggestion of repriming. The idea of leading strand re-initiation remained controversial until almost four decades later when origin-independent leading strand re-initiation was observed [126]. Follow-up studies confirmed that the replicative primase, DnaG, could reprime leading strand replication downstream of a lesion, whilst the replisome remained associated with the template [13,14]. Over recent years, evidence has accumulated to support leading strand repriming as a conserved mechanism for dealing with replisome-stalling impediments in eukaryotes [11,12,18] and recent studies have established that PrimPol's major role in eukaryotic organisms is to act as a primase that facilitates the bypass of a wide range of leading strand obstacles (Figure 4) [34,36,66–68,104,106].

Since the initial reports describing PrimPol only three years ago, studies from a number of laboratories have greatly increased our understanding of the role, recruitment, and regulation of the enzyme during DNA replication [26–28]. However, we are only just beginning to appreciate the novel roles that PrimPol plays in DNA replication and damage tolerance. The exact interplay between leading strand repriming by PrimPol and other DNA damage tolerance mechanisms, such as TLS, is still not yet clear. It is possible that DNA damage tolerance mechanisms work to complement repriming by filling in the resulting ssDNA gaps. Alternatively, repriming could occur when TLS at the replication fork fails, in order to prevent extended leading/lagging strand replication uncoupling. The redundancy between Pol α-primase and PrimPol in vivo is also an interesting avenue for future studies. The reason for the apparent requirement of PrimPol for leading strand repriming in higher eukaryotes, but not other organisms, is not yet completely clear. Although leading strand repriming is emerging as the primary role for PrimPol during DNA replication, the catalytic versatility of the enzyme may lend itself to disparate roles in other processes, such as transcription [43].

We now know that RPA serves to recruit PrimPol to stalled replication forks in the nucleus [104]. However, mtSSB has not yet been shown to play an analogous role in the mitochondria, although an interaction between these proteins in vivo has been reported [55]. Additionally, it is possible that post-translational modifications, as well as interactions with the replicative helicases, play a role in this process [42]. The necessity of appropriate recruitment and regulation of PrimPol in the cell is highlighted by the mutations of PrimPol's RBMs identified in cancer patient cell lines, which likely adversely affect recruitment of the enzyme [104]. The regulation of PrimPol appears to walk a fine line between preventing and causing genetic instability, as PrimPol is inherently error-prone and also been found to be over-expressed in some cancers, such as glioma [55,127]. Although we have highlighted some of the known mechanisms regulating PrimPol's activity here, it is likely that additional layers of regulation remain to be discovered.

The hypersensitivity to DNA damaging agents observed in absence of PrimPol legitimises the enzyme as a potential target for inhibition in combination with other DNA damage tolerance factors and DNA damaging chemotherapeutics [26,36,66]. Similarly, PrimPol homologues in trypanosomes have been identified as essential for survival and thus PrimPol-like proteins in other species may also be potential targets for anti-parasitic drugs [33]. Further studies will be important in determining the viability and usefulness of manipulating PrimPol in treating cancer and other diseases.

Acknowledgments: AJD's laboratory supported by grants from the Biotechnology and Biological Sciences Research Council (BBSRC: BB/H019723/1 and BB/M008800/1). TAG was supported by a University of Sussex PhD studentship. We thank members of our group, past and present, and collaborators who have made important contributions to studies cited in this review. Funding for open access charge: Research Councils UK (RCUK).

Data Statement: No new data were created during this study.

Conflicts of Interest: The authors declare no conflict of interest.

References

1. Johansson, E.; Dixon, N. Replicative DNA Polymerases. *Cold Spring Harb. Perspect. Biol.* **2013**, *5*, a012799. [CrossRef] [PubMed]
2. Zeman, M.K.; Cimprich, K.A. Causes and consequences of replication stress. *Nat. Cell Biol.* **2014**, *16*, 2–9. [CrossRef] [PubMed]
3. Ciccia, A.; Elledge, S.J. The DNA damage response: Making it Safe to play with knives. *Mol. Cell* **2010**, *40*, 179–204. [CrossRef] [PubMed]
4. Boyer, A.-S.; Grgurevic, S.; Cazaux, C.; Hoffmann, J.-S. The human specialized DNA polymerases and non-B DNA: Vital relationships to preserve genome integrity. *J. Mol. Biol.* **2013**, *425*, 4767–4781. [CrossRef] [PubMed]
5. Brambati, A.; Colosio, A.; Zardoni, L.; Galanti, L.; Liberi, G. Replication and transcription on a collision course: Eukaryotic regulation mechanisms and implications for DNA stability. *Front. Genet.* **2015**, *6*, 166. [CrossRef] [PubMed]
6. Kim, J.C.; Mirkin, S.M. The balancing act of DNA repeat expansions. *Curr. Opin. Genet. Dev.* **2013**, *23*, 280–288. [CrossRef] [PubMed]
7. Nick McElhinny, S.A.; Kumar, D.; Clark, A.B.; Watt, D.L.; Watts, B.E.; Lundström, E.-B.; Johansson, E.; Chabes, A.; Kunkel, T.A. Genome instability due to ribonucleotide incorporation into DNA. *Nat. Chem. Biol.* **2010**, *6*, 774–781. [CrossRef] [PubMed]
8. Bester, A.C.; Roniger, M.; Oren, Y.S.; Im, M.M.; Sarni, D.; Chaoat, M.; Bensimon, A.; Zamir, G.; Shewach, D.S.; Kerem, B. Nucleotide deficiency promotes genomic instability in early stages of cancer development. *Cell* **2011**, *145*, 435–446. [CrossRef] [PubMed]
9. Yeeles, J.T.P.; Poli, J.; Marians, K.J.; Pasero, P. Rescuing Stalled or Damaged Replication Forks. *Cold Spring Harb. Perspect. Biol.* **2013**, *5*, a012815. [CrossRef] [PubMed]
10. Svoboda, D.L.; Vos, J.M. Differential replication of a single, UV-induced lesion in the leading or lagging strand by a human cell extract: Fork uncoupling or gap formation. *Proc. Natl. Acad. Sci. USA* **1995**, *92*, 11975–11979. [CrossRef] [PubMed]
11. Lopes, M.; Foiani, M.; Sogo, J.M. Multiple mechanisms control chromosome integrity after replication fork uncoupling and restart at irreparable UV lesions. *Mol. Cell* **2006**, *21*, 15–27. [CrossRef] [PubMed]
12. Elvers, I.; Johansson, F.; Groth, P.; Erixon, K.; Helleday, T. UV stalled replication forks restart by re-priming in human fibroblasts. *Nucleic Acids Res.* **2011**, *39*, 7049–7057. [CrossRef] [PubMed]
13. Yeeles, J.T.P.; Marians, K.J. The *Escherichia coli* replisome is inherently DNA damage tolerant. *Science* **2011**, *334*, 235–238. [CrossRef] [PubMed]
14. Yeeles, J.T.P.; Marians, K.J. Dynamics of leading-strand lesion skipping by the replisome. *Mol. Cell* **2013**, *52*, 855–865. [CrossRef] [PubMed]
15. Goodman, M.F.; Woodgate, R. Translesion DNA Polymerases. *Cold Spring Harb. Perspect. Biol.* **2013**, *5*, a010363. [CrossRef] [PubMed]
16. Sale, J.E.; Lehmann, A.R.; Woodgate, R. Y-family DNA polymerases and their role in tolerance of cellular DNA damage. *Nat. Rev. Mol. Cell Biol.* **2012**, *13*, 141–152. [CrossRef] [PubMed]
17. Daigaku, Y.; Davies, A.A.; Ulrich, H.D. Ubiquitin-dependent DNA damage bypass is separable from genome replication. *Nature* **2010**, *465*, 951–955. [CrossRef] [PubMed]
18. Karras, G.I.; Jentsch, S. The RAD6 DNA damage tolerance pathway operates uncoupled from the replication fork and is functional beyond S phase. *Cell* **2010**, *141*, 255–267. [CrossRef] [PubMed]
19. Ulrich, H.D. Timing and spacing of ubiquitin-dependent DNA damage bypass. *FEBS Lett.* **2011**, *585*, 2861–2867. [CrossRef] [PubMed]
20. Edmunds, C.E.; Simpson, L.J.; Sale, J.E. PCNA ubiquitination and REV1 define temporally distinct mechanisms for controlling translesion synthesis in the avian cell line DT40. *Mol. Cell* **2008**, *30*, 519–529.
21. Ge, X.Q.; Jackson, D.A.; Blow, J.J. Dormant origins licensed by excess Mcm2–7 are required for human cells to survive replicative stress. *Genes Dev.* **2007**, *21*, 3331–3341. [CrossRef] [PubMed]
22. Branzei, D.; Szakal, B. DNA damage tolerance by recombination: Molecular pathways and DNA structures. *DNA Repair* **2016**, *44*, 68–75. [CrossRef] [PubMed]
23. Neelsen, K.J.; Lopes, M. Replication fork reversal in eukaryotes: From dead end to dynamic response. *Nat. Rev. Mol. Cell Biol.* **2015**, *16*, 207–220. [CrossRef] [PubMed]

24. Guilliam, T.A.; Keen, B.A.; Brissett, N.C.; Doherty, A.J. Primase-polymerases are a functionally diverse superfamily of replication and repair enzymes. *Nucleic Acids Res.* **2015**, *43*, 6651–6664. [CrossRef] [PubMed]

25. Jozwiakowski, S.K.; Borazjani Gholami, F.; Doherty, A.J. Archaeal replicative primases can perform translesion DNA synthesis. *Proc. Natl. Acad. Sci. USA* **2015**, *112*, E633–E638. [CrossRef] [PubMed]

26. Bianchi, J.; Rudd, S.G.; Jozwiakowski, S.K.; Bailey, L.J.; Soura, V.; Taylor, E.; Stevanovic, I.; Green, A.J.; Stracker, T.H.; Lindsay, H.D.; et al. PrimPol bypasses UV photoproducts during eukaryotic chromosomal DNA replication. *Mol. Cell* **2013**, *52*, 566–573. [CrossRef] [PubMed]

27. García-Gómez, S.; Reyes, A.; Martínez-Jiménez, M.I.; Chocrón, E.S.; Mourón, S.; Terrados, G.; Powell, C.; Salido, E.; Méndez, J.; Holt, I.J.; et al. PrimPol, an archaic primase/polymerase operating in human cells. *Mol. Cell* **2013**, *52*, 541–553. [CrossRef] [PubMed]

28. Wan, L.; Lou, J.; Xia, Y.; Su, B.; Liu, T.; Cui, J.; Sun, Y.; Lou, H.; Huang, J. hPrimpol1/CCDC111 is a human DNA primase-polymerase required for the maintenance of genome integrity. *EMBO Rep.* **2013**, *14*, 1104–1112. [CrossRef] [PubMed]

29. Aravind, L.; Leipe, D.D.; Koonin, E.V. Toprim—A conserved catalytic domain in type IA and II topoisomerases, DnaG-type primases, OLD family nucleases and RecR proteins. *Nucleic Acids Res.* **1998**, *26*, 4205–4213. [CrossRef] [PubMed]

30. Koonin, E.V.; Wolf, Y.I.; Kondrashov, A.S.; Aravind, L. Bacterial homologs of the small subunit of eukaryotic DNA primase. *J. Mol. Microbiol. Biotechnol.* **2000**, *2*, 509–512. [PubMed]

31. Della, M.; Palmbos, P.L.; Tseng, H.-M.; Tonkin, L.M.; Daley, J.M.; Topper, L.M.; Pitcher, R.S.; Tomkinson, A.E.; Wilson, T.E.; Doherty, A.J. Mycobacterial Ku and ligase proteins constitute a two-component NHEJ repair machine. *Science* **2004**, *306*, 683–685. [CrossRef] [PubMed]

32. Iyer, L.M.; Koonin, E.V.; Leipe, D.D.; Aravind, L. Origin and evolution of the archaeo-eukaryotic primase superfamily and related palm-domain proteins: Structural insights and new members. *Nucleic Acids Res.* **2005**, *33*, 3875–3896. [CrossRef] [PubMed]

33. Rudd, S.G.; Glover, L.; Jozwiakowski, S.K.; Horn, D.; Doherty, A.J. PPL2 translesion polymerase is essential for the completion of chromosomal DNA replication in the African trypanosome. *Mol. Cell* **2013**, *52*, 554–565. [CrossRef] [PubMed]

34. Keen, B.A.; Jozwiakowski, S.K.; Bailey, L.J.; Bianchi, J.; Doherty, A.J. Molecular dissection of the domain architecture and catalytic activities of human PrimPol. *Nucleic Acids Res.* **2014**, *42*, 5830–5845. [CrossRef] [PubMed]

35. Rechkoblit, O.; Gupta, Y.K.; Malik, R.; Rajashankar, K.R.; Johnson, R.E.; Prakash, L.; Prakash, S.; Aggarwal, A.K. Structure and mechanism of human PrimPol, a DNA polymerase with primase activity. *Sci. Adv.* **2016**, *2*, e1601317. [CrossRef] [PubMed]

36. Mourón, S.; Rodriguez-Acebes, S.; Martínez-Jiménez, M.I.; García-Gómez, S.; Chocrón, S.; Blanco, L.; Méndez, J. Repriming of DNA synthesis at stalled replication forks by human PrimPol. *Nat. Struct. Mol. Biol.* **2013**, *20*, 1383–1389. [CrossRef] [PubMed]

37. Frick, D.N.; Richardson, C.C. DNA Primases. *Annu. Rev. Biochem.* **2001**, *70*, 39–80. [CrossRef] [PubMed]

38. Biswas, N.; Weller, S.K. A Mutation in the C-terminal putative Zn2+ finger motif of UL52 severely affects the biochemical activities of the HSV-1 helicase-primase subcomplex. *J. Biol. Chem.* **1999**, *274*, 8068–8076. [CrossRef] [PubMed]

39. Chen, Y.; Carrington-Lawrence, S.D.; Bai, P.; Weller, S.K. Mutations in the putative Zinc-binding motif of UL52 demonstrate a complex interdependence between the UL5 and UL52 subunits of the human herpes simplex virus type 1 helicase/primase complex. *J. Virol.* **2005**, *79*, 9088–9096. [CrossRef] [PubMed]

40. Zafar, M.K.; Ketkar, A.; Lodeiro, M.F.; Cameron, C.E.; Eoff, R.L. Kinetic analysis of human PrimPol DNA polymerase activity reveals a generally error-prone enzyme capable of accurately bypassing 7,8-dihydro-8-oxo-2'-deoxyguanosine. *Biochemistry (Mosc.)* **2014**, *53*, 6584–6594. [CrossRef] [PubMed]

41. Guilliam, T.A.; Bailey, L.J.; Brissett, N.C.; Doherty, A.J. PolDIP2 interacts with human PrimPol and enhances its DNA polymerase activities. *Nucleic Acids Res.* **2016**, *44*, 3317–3329. [CrossRef] [PubMed]
42. Stojkovič, G.; Makarova, A.V.; Wanrooij, P.H.; Forslund, J.; Burgers, P.M.; Wanrooij, S. Oxidative DNA damage stalls the human mitochondrial replisome. *Sci. Rep.* **2016**, *6*, 28942. [CrossRef] [PubMed]
43. Martínez-Jiménez, M.I.; García-Gómez, S.; Bebenek, K.; Sastre-Moreno, G.; Calvo, P.A.; Díaz-Talavera, A.; Kunkel, T.A.; Blanco, L. Alternative solutions and new scenarios for translesion DNA synthesis by human PrimPol. *DNA Repair* **2015**, *29*, 127–138. [CrossRef] [PubMed]
44. Pitcher, R.S.; Brissett, N.C.; Picher, A.J.; Andrade, P.; Juarez, R.; Thompson, D.; Fox, G.C.; Blanco, L.; Doherty, A.J. Structure and function of a mycobacterial NHEJ DNA repair polymerase. *J. Mol. Biol.* **2007**, *366*, 391–405. [CrossRef] [PubMed]
45. Yakovleva, L.; Shuman, S. Nucleotide misincorporation, 3′-mismatch extension, and responses to abasic sites and DNA adducts by the polymerase component of cacterial DNA ligase D. *J. Biol. Chem.* **2006**, *281*, 25026–25040. [CrossRef] [PubMed]
46. Wang, T.S.-F.; Eichler, D.C.; Korn, D. Effect of manganese(2+) ions on the in vitro activity of human deoxyribonucleic acid polymerase .beta. *Biochemistry (Mosc.)* **1977**, *16*, 4927–4934. [CrossRef]
47. Pelletier, H.; Sawaya, M.R.; Wolfle, W.; Wilson, S.H.; Kraut, J. A Structural basis for metal ion mutagenicity and nucleotide selectivity in human DNA polymerase β. *Biochemistry (Mosc.)* **1996**, *35*, 12762–12777. [CrossRef] [PubMed]
48. Goodman, M.F.; Keener, S.; Guidotti, S.; Branscomb, E.W. On the enzymatic basis for mutagenesis by manganese. *J. Biol. Chem.* **1983**, *258*, 3469–3475. [PubMed]
49. El-Deiry, W.S.; Downey, K.M.; So, A.G. Molecular mechanisms of manganese mutagenesis. *Proc. Natl. Acad. Sci. USA* **1984**, *81*, 7378–7382. [CrossRef] [PubMed]
50. Vaisman, A.; Ling, H.; Woodgate, R.; Yang, W. Fidelity of Dpo4: Effect of metal ions, nucleotide selection and pyrophosphorolysis. *EMBO J.* **2005**, *24*, 2957–2967. [CrossRef] [PubMed]
51. Chabes, A.; Georgieva, B.; Domkin, V.; Zhao, X.; Rothstein, R.; Thelander, L. Survival of DNA damage in yeast directly depends on increased dNTP levels allowed by relaxed feedback inhibition of ribonucleotide reductase. *Cell* **2003**, *112*, 391–401. [CrossRef]
52. Shimizu, K.; Hashimoto, K.; Kirchner, J.M.; Nakai, W.; Nishikawa, H.; Resnick, M.A.; Sugino, A. Fidelity of DNA polymerase ε holoenzyme from budding yeast *Saccharomyces cerevisiae*. *J. Biol. Chem.* **2002**, *277*, 37422–37429. [CrossRef] [PubMed]
53. Minko, I.G.; Washington, M.T.; Kanuri, M.; Prakash, L.; Prakash, S.; Lloyd, R.S. Translesion synthesis past acrolein-derived DNA adduct, γ-hydroxypropanodeoxyguanosine, by yeast and human DNA polymerase η. *J. Biol. Chem.* **2003**, *278*, 784–790. [CrossRef] [PubMed]
54. Sabouri, N.; Viberg, J.; Goyal, D.K.; Johansson, E.; Chabes, A. Evidence for lesion bypass by yeast replicative DNA polymerases during DNA damage. *Nucleic Acids Res.* **2008**, *36*, 5660–5667. [CrossRef] [PubMed]
55. Guilliam, T.A.; Jozwiakowski, S.K.; Ehlinger, A.; Barnes, R.P.; Rudd, S.G.; Bailey, L.J.; Skehel, J.M.; Eckert, K.A.; Chazin, W.J.; Doherty, A.J. Human PrimPol is a highly error-prone polymerase regulated by single-stranded DNA binding proteins. *Nucleic Acids Res.* **2015**, *43*, 1056–1068. [CrossRef] [PubMed]
56. Ash, D.E.; Schramm, V.L. Determination of free and bound manganese(II) in hepatocytes from fed and fasted rats. *J. Biol. Chem.* **1982**, *257*, 9261–9264. [PubMed]
57. Markesbery, W.R.; Ehmann, W.D.; Alauddin, M.; Hossain, T.I.M. Brain trace element concentrations in aging. *Neurobiol. Aging* **1984**, *5*, 19–28. [CrossRef]
58. Versieck, J.; McCall, J.T. Trace elements in human body fluids and tissues. *CRC Crit. Rev. Clin. Lab. Sci.* **1985**, *22*, 97–184. [CrossRef] [PubMed]
59. Goldschmidt, V.; Didierjean, J.; Ehresmann, B.; Ehresmann, C.; Isel, C.; Marquet, R. Mg^{2+} dependency of HIV-1 reverse transcription, inhibition by nucleoside analogues and resistance. *Nucleic Acids Res.* **2006**, *34*, 42–52. [CrossRef] [PubMed]
60. Gee, J.B., II; Corbett, R.J.T.; Perlman, J.M.; Laptook, A.R. Hypermagnesemia does not increase brain intracellular magnesium in newborn swine. *Pediatr. Neurol.* **2001**, *25*, 304–308. [CrossRef]
61. Rampazzo, C.; Ferraro, P.; Pontarin, G.; Fabris, S.; Reichard, P.; Bianchi, V. Mitochondrial deoxyribonucleotides, pool sizes, synthesis, and regulation. *J. Biol. Chem.* **2004**, *279*, 17019–17026. [CrossRef] [PubMed]

62. Zhang, S.; Grosse, F. Accuracy of DNA primase. *J. Mol. Biol.* **1990**, *216*, 475–479. [CrossRef]

63. Sheaff, R.J.; Kuchta, R.D. Misincorporation of nucleotides by calf thymus DNA primase and elongation of primers containing multiple noncognate nucleotides by DNA polymerase alpha. *J. Biol. Chem.* **1994**, *269*, 19225–19231. [PubMed]

64. Cotterill, S.; Chui, G.; Lehman, I.R. DNA polymerase-primase from embryos of *Drosophila melanogaster*. DNA primase subunits. *J. Biol. Chem.* **1987**, *262*, 16105–16108. [PubMed]

65. Bailey, L.J.; Bianchi, J.; Hégarat, N.; Hochegger, H.; Doherty, A.J. PrimPol-deficient cells exhibit a pronounced G2 checkpoint response following UV damage. *Cell Cycle* **2016**, *15*, 908–918. [CrossRef] [PubMed]

66. Kobayashi, K.; Guilliam, T.A.; Tsuda, M.; Yamamoto, J.; Bailey, L.J.; Iwai, S.; Takeda, S.; Doherty, A.J.; Hirota, K. Repriming by PrimPol is critical for DNA replication restart downstream of lesions and chain-terminating nucleosides. *Cell Cycle* **2016**, *15*, 1997–2008. [CrossRef] [PubMed]

67. Schiavone, D.; Jozwiakowski, S.K.; Romanello, M.; Guilbaud, G.; Guilliam, T.A.; Bailey, L.J.; Sale, J.E.; Doherty, A.J. PrimPol is required for replicative tolerance of G quadruplexes in vertebrate cells. *Mol. Cell* **2016**, *61*, 161–169. [CrossRef] [PubMed]

68. Pilzecker, B.; Buoninfante, O.A.; Pritchard, C.; Blomberg, O.S.; Huijbers, I.J.; van den Berk, P.C.M.; Jacobs, H. PrimPol prevents APOBEC/AID family mediated DNA mutagenesis. *Nucleic Acids Res.* **2016**, *44*, 4734–4744. [CrossRef] [PubMed]

69. Haracska, L.; Yu, S.-L.; Johnson, R.E.; Prakash, L.; Prakash, S. Efficient and accurate replication in the presence of 7,8-dihydro-8-oxoguanine by DNA polymerase η. *Nat. Genet.* **2000**, *25*, 458–461. [PubMed]

70. Maga, G.; Crespan, E.; Markkanen, E.; Imhof, R.; Furrer, A.; Villani, G.; Hübscher, U.; Loon, B. van DNA polymerase δ-interacting protein 2 is a processivity factor for DNA polymerase λ during 8-oxo-7,8-dihydroguanine bypass. *Proc. Natl. Acad. Sci. USA* **2013**, *110*, 18850–18855. [CrossRef] [PubMed]

71. Zahn, K.E.; Wallace, S.S.; Doublié, S. DNA polymerases provide a canon of strategies for translesion synthesis past oxidatively generated lesions. *Curr. Opin. Struct. Biol.* **2011**, *21*, 358–369. [CrossRef] [PubMed]

72. Keen, B.A.; Bailey, L.J.; Jozwiakowski, S.K.; Doherty, A.J. Human PrimPol mutation associated with high myopia has a DNA replication defect. *Nucleic Acids Res.* **2014**, *42*, 12102–12111. [CrossRef] [PubMed]

73. Choi, J.-H.; Lindsey-Boltz, L.A.; Kemp, M.; Mason, A.C.; Wold, M.S.; Sancar, A. Reconstitution of RPA-covered single-stranded DNA-activated ATR-Chk1 signaling. *Proc. Natl. Acad. Sci. USA* **2010**, *107*, 13660–13665. [CrossRef] [PubMed]

74. Watson, J.D.; Crick, F.H. Molecular structure of nucleic acids; a structure for deoxyribose nucleic acid. *Nature* **1953**, *171*, 737–738. [CrossRef] [PubMed]

75. Bochman, M.L.; Paeschke, K.; Zakian, V.A. DNA secondary structures: Stability and function of G-quadruplex structures. *Nat. Rev. Genet.* **2012**, *13*, 770–780. [CrossRef] [PubMed]

76. Murat, P.; Balasubramanian, S. Existence and consequences of G-quadruplex structures in DNA. *Curr. Opin. Genet. Dev.* **2014**, *25*, 22–29. [CrossRef] [PubMed]

77. Maizels, N.; Gray, L.T. The G4 Genome. *PLoS Genet.* **2013**, *9*, e1003468. [CrossRef] [PubMed]

78. Cheung, I.; Schertzer, M.; Rose, A.; Lansdorp, P.M. Disruption of dog-1 in *Caenorhabditis elegans* triggers deletions upstream of guanine-rich DNA. *Nat. Genet.* **2002**, *31*, 405–409. [CrossRef] [PubMed]

79. Ribeyre, C.; Lopes, J.; Boulé, J.-B.; Piazza, A.; Guédin, A.; Zakian, V.A.; Mergny, J.-L.; Nicolas, A. The yeast Pif1 helicase prevents genomic instability caused by G-quadruplex-forming CEB1 sequences in vivo. *PLoS Genet.* **2009**, *5*, e1000475. [CrossRef] [PubMed]

80. Sarkies, P.; Reams, C.; Simpson, L.J.; Sale, J.E. Epigenetic instability due to defective replication of structured DNA. *Mol. Cell* **2010**, *40*, 703–713. [CrossRef] [PubMed]

81. León-Ortiz, A.M.; Svendsen, J.; Boulton, S.J. Metabolism of DNA secondary structures at the eukaryotic replication fork. *DNA Repair* **2014**, *19*, 152–162. [CrossRef] [PubMed]

82. Sarkies, P.; Murat, P.; Phillips, L.G.; Patel, K.J.; Balasubramanian, S.; Sale, J.E. FANCJ coordinates two pathways that maintain epigenetic stability at G-quadruplex DNA. *Nucleic Acids Res.* **2012**, *40*, 1485–1498. [CrossRef] [PubMed]

83. Schiavone, D.; Guilbaud, G.; Murat, P.; Papadopoulou, C.; Sarkies, P.; Prioleau, M.-N.; Balasubramanian, S.; Sale, J.E. Determinants of G quadruplex-induced epigenetic instability in REV1-deficient cells. *EMBO J.* **2014**, *33*, 2507–2520. [CrossRef] [PubMed]

84. Berdis, A.J. DNA Polymerases as therapeutic targets. *Biochemistry (Mosc.)* **2008**, *47*, 8253–8260. [CrossRef] [PubMed]

85. De Clercq, E.; Field, H.J. Antiviral prodrugs—The development of successful prodrug strategies for antiviral chemotherapy. *Br. J. Pharmacol.* **2006**, *147*, 1–11. [CrossRef] [PubMed]

86. Alexeyev, M.; Shokolenko, I.; Wilson, G.; LeDoux, S. The maintenance of mitochondrial DNA integrity—critical analysis and update. *Cold Spring Harb. Perspect. Biol.* **2013**, *5*, a012641.

87. Bianchi, J. Investigating the Role of a Novel Primase-Polymerase, PrimPol, in DNA Damage Tolerance in Vertebrate Cells. Ph.D. Thesis, University of Sussex, Brighton, UK, 2013.

88. Falkenberg, M.; Larsson, N.-G.; Gustafsson, C.M. DNA replication and transcription in mammalian mitochondria. *Annu. Rev. Biochem.* **2007**, *76*, 679–699. [CrossRef] [PubMed]

89. Loeb, L.A.; Monnat, R.J. DNA polymerases and human disease. *Nat. Rev. Genet.* **2008**, *9*, 594–604.

90. Wisnovsky, S.; Jean, S.R.; Kelley, S.O. Mitochondrial DNA repair and replication proteins revealed by targeted chemical probes. *Nat. Chem. Biol.* **2016**, *12*, 567–573. [CrossRef] [PubMed]

91. Singh, B.; Li, X.; Owens, K.M.; Vanniarajan, A.; Liang, P.; Singh, K.K. Human REV3 DNA Polymerase Zeta Localizes to Mitochondria and Protects the Mitochondrial Genome. *PLoS ONE* **2015**, *10*, e0140409.

92. Bharti, S.K.; Sommers, J.A.; Zhou, J.; Kaplan, D.L.; Spelbrink, J.N.; Mergny, J.-L.; Brosh, R.M. DNA Sequences Proximal to Human Mitochondrial DNA Deletion Breakpoints Prevalent in Human Disease Form G-quadruplexes, a Class of DNA Structures Inefficiently Unwound by the Mitochondrial Replicative Twinkle Helicase. *J. Biol. Chem.* **2014**, *289*, 29975–29993. [CrossRef] [PubMed]

93. Bailey, L.J.; Doherty, A.J. Mitochondrial DNA replication—A PrimPol perspective. *Biochem. Soc. Trans.*. (Under Revison).

94. Noia, J.M.D.; Neuberger, M.S. Molecular mechanisms of antibody somatic hypermutation. *Annu. Rev. Biochem.* **2007**, *76*, 1–22. [CrossRef] [PubMed]

95. Sale, J.E.; Batters, C.; Edmunds, C.E.; Phillips, L.G.; Simpson, L.J.; Szüts, D. Timing matters: Error-prone gap filling and translesion synthesis in immunoglobulin gene hypermutation. *Philos. Trans. R. Soc. Lond. B Biol. Sci.* **2009**, *364*, 595–603. [CrossRef] [PubMed]

96. Williams, J.S.; Lujan, S.A.; Kunkel, T.A. Processing ribonucleotides incorporated during eukaryotic DNA replication. *Nat. Rev. Mol. Cell Biol.* **2016**, *17*, 350–363. [CrossRef] [PubMed]

97. Williams, J.S.; Kunkel, T.A. Ribonucleotides in DNA: Origins, repair and consequences. *DNA Repair* **2014**, *19*, 27–37. [CrossRef] [PubMed]

98. Williams, J.S.; Clausen, A.R.; Lujan, S.A.; Marjavaara, L.; Clark, A.B.; Burgers, P.M.; Chabes, A.; Kunkel, T.A. Evidence that processing of ribonucleotides in DNA by topoisomerase 1 is leading-strand specific. *Nat. Struct. Mol. Biol.* **2015**, *22*, 291–297. [CrossRef] [PubMed]

99. Reijns, M.A.M.; Rabe, B.; Rigby, R.E.; Mill, P.; Astell, K.R.; Lettice, L.A.; Boyle, S.; Leitch, A.; Keighren, M.; Kilanowski, F.; et al. Enzymatic removal of ribonucleotides from DNA is essential for mammalian genome integrity and development. *Cell* **2012**, *149*, 1008–1022. [CrossRef] [PubMed]

100. Baranovskiy, A.G.; Babayeva, N.D.; Zhang, Y.; Gu, J.; Suwa, Y.; Pavlov, Y.I.; Tahirov, T.H. Mechanism of concerted RNA-DNA primer synthesis by the human primosome. *J. Biol. Chem.* **2016**, *291*, 10006–10020.

101. Baranovskiy, A.G.; Zhang, Y.; Suwa, Y.; Gu, J.; Babayeva, N.D.; Pavlov, Y.I.; Tahirov, T.H. Insight into the Human DNA Primase Interaction with Template-Primer. *J. Biol. Chem.* **2016**, *291*, 4793–4802.

102. Liu, L.; Huang, M. Essential role of the iron-sulfur cluster binding domain of the primase regulatory subunit Pri2 in DNA replication initiation. *Protein Cell* **2015**, *6*, 194–210. [CrossRef] [PubMed]

103. Zou, L.; Elledge, S.J. Sensing DNA Damage through ATRIP recognition of RPA-ssDNA complexes. *Science* **2003**, *300*, 1542–1548. [CrossRef] [PubMed]

104. Guilliam, T.A.; Brissett, N.C.; Ehlinger, A.; Keen, B.A.; Kolesar, P.; Taylor, E.; Bailey, L.J.; Lindsay, H.D.; Chazin, W.J.; Doherty, A.J. Molecular basis for PrimPol recruitment to replication forks by RPA. *Nat. Commun.*. (Under Review).

105. Xu, X.; Vaithiyalingam, S.; Glick, G.G.; Mordes, D.A.; Chazin, W.J.; Cortez, D. The basic cleft of RPA70N binds multiple checkpoint proteins, including RAD9, To Regulate ATR Signaling. *Mol. Cell. Biol.* **2008**, *28*, 7345–7353. [CrossRef] [PubMed]

106. Vallerga, M.B.; Mansilla, S.F.; Federico, M.B.; Bertolin, A.P.; Gottifredi, V. Rad51 recombinase prevents Mre11 nuclease-dependent degradation and excessive PrimPol-mediated elongation of nascent DNA after UV irradiation. *Proc. Natl. Acad. Sci. USA* **2015**, *112*, E6624–6633. [CrossRef] [PubMed]

107. Robison, J.G.; Elliott, J.; Dixon, K.; Oakley, G.G. Replication protein A and the Mre11·Rad50·Nbs1 complex co-localize and interact at sites of stalled replication forks. *J. Biol. Chem.* **2004**, *279*, 34802–34810.

108. Wu, X.; Shell, S.M.; Zou, Y. Interaction and colocalization of Rad9/Rad1/Hus1 checkpoint complex with replication protein A in human cells. *Oncogene* **2005**, *24*, 4728–4735. [CrossRef] [PubMed]

109. Oakley, G.G.; Patrick, S.M. Replication protein A: Directing traffic at the intersection of replication and repair. *Front. Biosci. J. Virtual Libr.* **2010**, *15*, 883–900. [CrossRef]

110. Michel, B. After 30 Years of Study, the Bacterial SOS Response Still Surprises Us. *PLoS Biol* **2005**, *3*, e255.

111. Beck, M.; Schmidt, A.; Malmstroem, J.; Claassen, M.; Ori, A.; Szymborska, A.; Herzog, F.; Rinner, O.; Ellenberg, J.; Aebersold, R. The quantitative proteome of a human cell line. *Mol. Syst. Biol.* **2011**, *7*, 549.

112. Kuchta, R.D.; Stengel, G. Mechanism and evolution of DNA primases. *Biochim. Biophys. Acta BBA Proteins Proteom.* **2010**, *1804*, 1180–1189. [CrossRef] [PubMed]

113. Tsurimoto, T.; Stillman, B. Multiple replication factors augment DNA synthesis by the two eukaryotic DNA polymerases, alpha and delta. *EMBO J.* **1989**, *8*, 3883–3889. [PubMed]

114. Oliveira, M.T.; Kaguni, L.S. Functional roles of the N- and C-terminal regions of the human mitochondrial single-stranded DNA-binding protein. *PLoS ONE* **2010**, *5*, e15379. [CrossRef] [PubMed]

115. Collins, K.L.; Kelly, T.J. Effects of T antigen and replication protein A on the initiation of DNA synthesis by DNA polymerase alpha-primase. *Mol. Cell. Biol.* **1991**, *11*, 2108–2115. [CrossRef] [PubMed]

116. Kolpashchikov, D.M.; Khodyreva, S.N.; Khlimankov, D.Y.; Wold, M.S.; Favre, A.; Lavrik, O.I. Polarity of human replication protein A binding to DNA. *Nucleic Acids Res.* **2001**, *29*, 373–379. [CrossRef] [PubMed]

117. De Laat, W.L.; Appeldoorn, E.; Sugasawa, K.; Weterings, E.; Jaspers, N.G.J.; Hoeijmakers, J.H.J. DNA-binding polarity of human replication protein A positions nucleases in nucleotide excision repair. *Genes Dev.* **1998**, *12*, 2598–2609. [CrossRef] [PubMed]

118. Iftode, C.; Borowiec, J.A. 5′ → 3′ Molecular Polarity of Human Replication Protein A (hRPA) Binding to Pseudo-Origin DNA Substrates. *Biochemistry (Mosc.)* **2000**, *39*, 11970–11981. [CrossRef]

119. Fan, J.; Pavletich, N.P. Structure and conformational change of a replication protein A heterotrimer bound to ssDNA. *Genes Dev.* **2012**, *26*, 2337–2347. [CrossRef] [PubMed]

120. Brosey, C.A.; Yan, C.; Tsutakawa, S.E.; Heller, W.T.; Rambo, R.P.; Tainer, J.A.; Ivanov, I.; Chazin, W.J. A new structural framework for integrating replication protein A into DNA processing machinery. *Nucleic Acids Res.* **2013**, *41*, 2313–2327. [CrossRef] [PubMed]

121. Iftode, C.; Daniely, Y.; Borowiec, J.A. Replication Protein A (RPA): The Eukaryotic SSB. *Crit. Rev. Biochem. Mol. Biol.* **1999**, *34*, 141–180. [CrossRef] [PubMed]

122. Liu, L.; Rodriguez-Belmonte, E.M.; Mazloum, N.; Xie, B.; Lee, M.Y. Identification of a novel protein, PDIP38, that interacts with the p50 subunit of DNA polymerase δ and proliferating cell nuclear antigen. *J. Biol. Chem.* **2003**, *278*, 10041–10047. [CrossRef] [PubMed]

123. Tissier, A.; Janel-Bintz, R.; Coulon, S.; Klaile, E.; Kannouche, P.; Fuchs, R.P.; Cordonnier, A.M. Crosstalk between replicative and translesional DNA polymerases: PDIP38 interacts directly with Polη. *DNA Repair* **2010**, *9*, 922–928. [CrossRef] [PubMed]

124. Daigaku, Y.; Keszthelyi, A.; Müller, C.A.; Miyabe, I.; Brooks, T.; Retkute, R.; Hubank, M.; Nieduszynski, C.A.; Carr, A.M. A global profile of replicative polymerase usage. *Nat. Struct. Mol. Biol.* **2015**, *22*, 192–198.

125. Rupp, W.D.; Howard-Flanders, P. Discontinuities in the DNA synthesized in an excision-defective strain of Escherichia coli following ultraviolet irradiation. *J. Mol. Biol.* **1968**, *31*, 291–304. [CrossRef]

126. Heller, R.C.; Marians, K.J. Replication fork reactivation downstream of a blocked nascent leading strand. *Nature* **2006**, *439*, 557–562. [CrossRef] [PubMed]

127. Yan, X.; Ma, L.; Yi, D.; Yoon, J.; Diercks, A.; Foltz, G.; Price, N.D.; Hood, L.E.; Tian, Q. A CD133-related gene expression signature identifies an aggressive glioblastoma subtype with excessive mutations. *Proc. Natl. Acad. Sci. USA* **2011**, *108*, 1591–1596. [CrossRef] [PubMed]

![genes logo](GCAT TACG GCAT *genes*)

MDPI

Review

Control of Genome Integrity by RFC Complexes; Conductors of PCNA Loading onto and Unloading from Chromatin during DNA Replication

Yasushi Shiomi * and Hideo Nishitani *

Graduate School of Life Science, University of Hyogo, Kamigori, Ako-gun, Hyogo 678-1297, Japan
* Correspondence: shiomi@sci.u-hyogo.ac.jp (Y.S.); hideon@sci.u-hyogo.ac.jp (H.N.)

Academic Editor: Eishi Noguchi
Received: 28 November 2016; Accepted: 21 January 2017; Published: 26 January 2017

Abstract: During cell division, genome integrity is maintained by faithful DNA replication during S phase, followed by accurate segregation in mitosis. Many DNA metabolic events linked with DNA replication are also regulated throughout the cell cycle. In eukaryotes, the DNA sliding clamp, proliferating cell nuclear antigen (PCNA), acts on chromatin as a processivity factor for DNA polymerases. Since its discovery, many other PCNA binding partners have been identified that function during DNA replication, repair, recombination, chromatin remodeling, cohesion, and proteolysis in cell-cycle progression. PCNA not only recruits the proteins involved in such events, but it also actively controls their function as chromatin assembles. Therefore, control of PCNA-loading onto chromatin is fundamental for various replication-coupled reactions. PCNA is loaded onto chromatin by PCNA-loading replication factor C (RFC) complexes. Both RFC1-RFC and Ctf18-RFC fundamentally function as PCNA loaders. On the other hand, after DNA synthesis, PCNA must be removed from chromatin by Elg1-RFC. Functional defects in RFC complexes lead to chromosomal abnormalities. In this review, we summarize the structural and functional relationships among RFC complexes, and describe how the regulation of PCNA loading/unloading by RFC complexes contributes to maintaining genome integrity.

Keywords: DNA replication; genome integrity; chromatin; PCNA; RFC complex; PCNA loader; PCNA unloader; RFC1; Ctf18; Elg1

1. Introduction

Genome integrity requires precise chromosome duplication. Duplication of genomic DNA occurs only once during S phase in the eukaryotic cell cycle [1]. Before replication is initiated, replication origins are licensed for replication by minichromosome maintenance (MCM) 2–7 complex loading onto origin recognition complex (ORC)-bound origins, assisted by Cdc6 and Cdt1 [2]. The next step is activation of the origins and the formation of replication forks. The active DNA helicase, CMG complex (comprising Cdc45, MCM2-7, and Sld5(go), Psf1(ichi), Psf2(ni), and Psf3(san) (GINS) complex), unwinds double-stranded DNA, and DNA polymerases are recruited for replication [3]. Accompanying these events are several other important processes, including repair; recombination; chromatin formation, modification, and remodeling; as well as the maintenance of epigenetic information and the prevention of re-replication during replication fork progression. Moreover, sister chromatid cohesion must occur, as their alignment is required for faithful chromosome segregation [4,5].

To carry out these various activities, the DNA replication fork requires many proteins that form a large complex, the replisome, to facilitate the efficient initiation and elongation of DNA synthesis and chromatin-associated events [6]. Among these proteins, in eukaryotes, the DNA sliding clamp proliferating cell nuclear antigen (PCNA) plays a fundamental role in coordinating multiple events on the DNA [7]. To perform all of its functions, both loading the PCNA onto DNA and removing it from DNA must be precisely regulated. To achieve this, PCNA uses the molecular PCNA ring-opening machinery, replication factor C (RFC) complex [8].

Here, we first describe PCNA and then focus on how PCNA loading and unloading are regulated while coupled to DNA replication. In particular, we highlight the role of the RFC complex as a PCNA loader or unloader that conducts replication-linked processes, and discuss how these functions are orchestrated to maintain genome integrity.

2. PCNA, the DNA Sliding Clamp in Eukaryotic Cells

2.1. Structure and Primary Function of PCNA

DNA replicative polymerases, particularly polymerase δ/ε, require additional factors to support DNA replication [9]. The DNA sliding clamp, PCNA, tethers DNA polymerases, strengthens the interactions of the polymerases with the template DNA, and enhances their processivity up to 1000-fold [10–12], which makes PCNA an essential processivity factor for DNA replication. Many PCNA-binding factors that are involved in replication-coupled processes have been identified [7].

PCNA is a ring-shaped homo trimer, in which the three subunits assemble in a head-to-tail manner [13,14] (Figure 1). PCNA is loaded onto the DNA in an orientation-dependent manner. The association between PCNA and DNA is stable as PCNA encircles DNA and can slide freely along the DNA due to the polarity repelling effects between the inner surface of the PCNA ring and the DNA. The front face of PCNA has amino acid polarity that interacts with DNA polymerases and numerous other DNA metabolic enzymes, most of which have a PCNA-interacting protein (PIP)-motif, to recruit and tether them correctly to the DNA [15] (Figure 2). Because the regions that interact with these enzymes often overlap, PCNA switches its binding partner depending on the circumstances of the replication fork progression [7]. In addition, PCNA couples the initiation of DNA replication to ubiquitin-mediated proteolysis [16]. Thus, as a platform, PCNA plays an important role in the replisome by accommodating multiple processes at the replication fork [6].

2.2. Post-Translational Modifications of PCNA

Various PCNA modifications also regulate the replisome depending on specific circumstances during DNA replication [18] (Figure 2). Following DNA damage, PCNA is monoubiquitinated at K164 in a Rad18-Rad6-dependent manner, which switches the affinity of PCNA from replicative polymerases to damage-tolerant translesion synthesis polymerases, such as polymerase η [19,20]. The translesion synthesis (TLS) polymerases can bypass DNA damage to continue replication, though this method of damage bypass is prone to error [21,22]. In contrast, polyubiquitination of the same site by Mms2-Ubc13 and Rad5 leads to the repair through template switching, which is essentially an error-free mechanism [19]. PCNA can also be SUMOylated (small ubiquitin-like modifier) at the same site as K164 and K127 in a Ubc9- and Siz1-dependent manner [18]. SUMOylation mediates the repression of unwanted homologous recombination through recruitment of the helicase Srs2, which is well characterized in yeast [23,24]. SUMOylated PCNA also exists in vertebrates. Acetylation of PCNA appears to have a role in enhancing the processivity of associated polymerases, and promotes the removal of chromatin-bound PCNA and its degradation during nucleotide excision repair [25,26]. A recent study revealed that K20 at the inner surface of the PCNA ring is acetylated by cohesion acetyltransferase Eco1 in response to DNA damage, induces alteration of the PCNA structure, and stimulates homologous recombination [27].

2.3. PCNA Requires Ring-Opening Factors to Regulate Its ON–OFF DNA Binding

Because PCNA performs many aspects of DNA replication-associated events when loaded onto the DNA, PCNA loading onto the DNA must be strictly regulated. Conversely, when PCNA completes its role, it must leave (or unload from) the DNA to suppress illegitimate enzymatic reactions. To bind and leave the DNA, PCNA must temporarily open its closed ring, which is achieved by PCNA ring-opening machinery, RFC complexes [8].

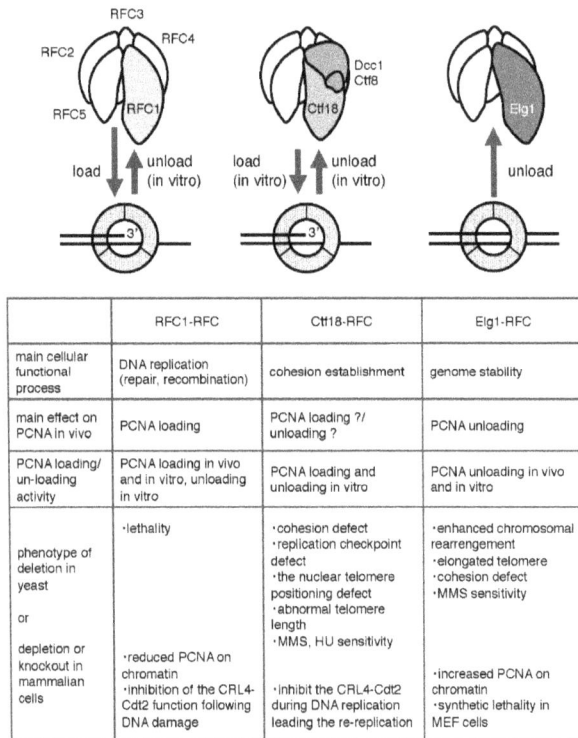

	RFC1-RFC	Ctf18-RFC	Elg1-RFC
main cellular functional process	DNA replication (repair, recombination)	cohesion establishment	genome stability
main effect on PCNA in vivo	PCNA loading	PCNA loading ?/ unloading ?	PCNA unloading
PCNA loading/ un-loading activity	PCNA loading in vivo and in vitro, unloading in vitro	PCNA loading and unloading in vitro	PCNA unloading in vivo and in vitro
phenotype of deletion in yeast or depletion or knockout in mammalian cells	·lethality ·reduced PCNA on chromatin ·inhibition of the CRL4-Cdt2 function following DNA damage	·cohesion defect ·replication checkpoint defect ·the nuclear telomere positioning defect ·abnormal telomere length ·MMS, HU sensitivity ·inhibit the CRL4-Cdt2 during DNA replication leading the re-replication	·enhanced chromosomal rearrengement ·elongated telomere ·cohesion defect ·MMS sensitivity ·increased PCNA on chromatin ·synthetic lethality in MEF cells

Figure 1. Summary of the functions of the three RFC complexes transacting on PCNA [17]. See text for details. The "?" marks in the table mean that the main effect of Ctf18-RFC (loading and unloading) on PCNA in vivo is not well understood.

3. Fundamental Features of RFC Complexes as PCNA Loaders/Unloaders

Eukaryotic cells have three RFC complexes that act on PCNA: RFC1-RFC, Ctf18-RFC, and Elg1-RFC, which essentially form hetero-pentameric complexes by sharing four small RFC subunits (RFC 2, 3, 4, and 5 [RFC2-5]), and each is distinguished by its largest subunit (i.e., RFC1, Ctf18, and Elg1, also called ATAD5 in human cells) [8] (Figure 1). Ctf18-RFC and Elg1-RFC are also called Ctf18-RFC-like complex (Ctf18-RLC) and Elg1-RFC-like complex (Elg1-RLC), respectively. All of these subunits, both large and small, belong to the AAA+ ATPase family [28,29]. The molecular morphologic similarity of these RFC complexes suggests that they all interact with PCNA and mediate the interactions between PCNA and DNA.

In addition to these complexes, eukaryotic cells have another RFC complex, Rad17-RFC (Rad24-RFC in *S. cerevisiae*), that acts to load the PCNA-like hetero-trimeric 9-1-1 complex (Rad9-Hus1-Rad1 in humans and Ddc1-Mec3-Rad17 in *S. cerevisiae*) at damaged DNA sites depending on checkpoint activation. We do not discuss this complex in this review and readers are referred to these excellent reviews [30,31].

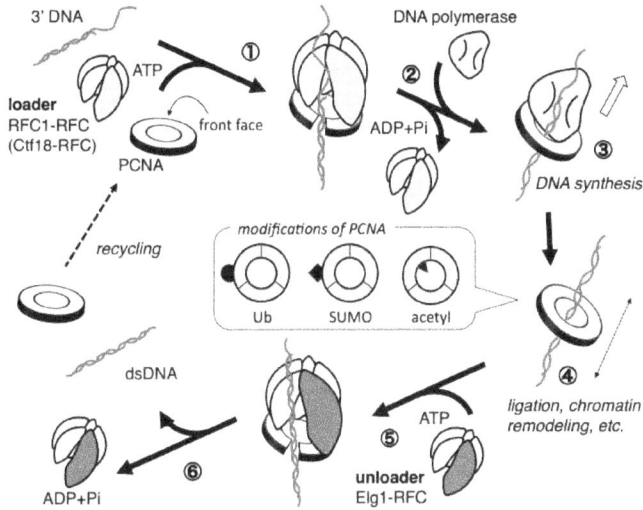

Figure 2. PCNA loading on and unloading from chromatin by RFC complexes during DNA synthesis. ① PCNA loader RFC1-RFC or Ctf18-RFC bind to PCNA and recognize the 3′ DNA template, and ATP binding triggers a conformational change of the RFC complex that allows for a tight interaction with PCNA and ring opening. ② ATP hydrolysis by the RFC loader complex is coupled with ring closure and the release of PCNA, finally encircling the DNA duplex. ③ DNA polymerases bind to chromatin-loaded PCNA, and DNA synthesis begins. ④ After the DNA synthesis is complete and DNA polymerase is released, PCNA recruits various enzymes for additional functions such as chromatin remodeling. PCNA slides along the double-stranded DNA to its functional sites. In ③ and ④, PCNA might be modified by mono-ubiquitin, poly-ubiquitin, SUMO (small ubiquitin-like modifier), or acetyl depending on the circumstances as illustrated in a dotted-line square (note that modification on single subunit of PCNA trimer is shown). ⑤ After the role of PCNA is completed, Elg1-RFC unloads PCNA from the double-stranded DNA in an ATP-dependent manner as a reverse reaction of PCNA loading. ⑥ During PCNA unloading, its modification might be removed so that it can be recycled. In this figure, nucleosomes and chromatin structures are omitted.

3.1. RFC1-RFC

3.1.1. Fundamental Features and Structure

A classic RFC complex, RFC1-RFC, comprises five subunits; the largest subunit is RFC1 and the four small subunits are RFC2-5. Common sequence motifs in these subunits are termed the RFC box, including the P-loop, a general Walker-type ATPase motif [32] (Figure 3). A yeast genetic study indicated that RFC1 is the only essential gene among the three RFC large subunits [33,34]. None of the other alternative RFC complexes is essential, alone or in combination [34]. RFC1 contains both N- and C-terminal extensions from the RFC box [35] (Figure 3). The C-terminals of the four small subunits and RFC1 are required to form the RFC1-RFC complex [36,37].

The biochemical activity of RFC1-RFC has been well analyzed, as described later, and structural analysis has provided details of the PCNA loading mechanism [38,39]. The crystal structure and electron microscopy images of the complex show that the five RFC1-RFC subunits are aligned in a circular shape with a gap between RFC1 and RFC5, making it well suited to interact with the PCNA ring [38,40,41] (Figures 1 and 2). Therefore, RFC1-RFC is generally regarded as the standard for studying other RFC complexes.

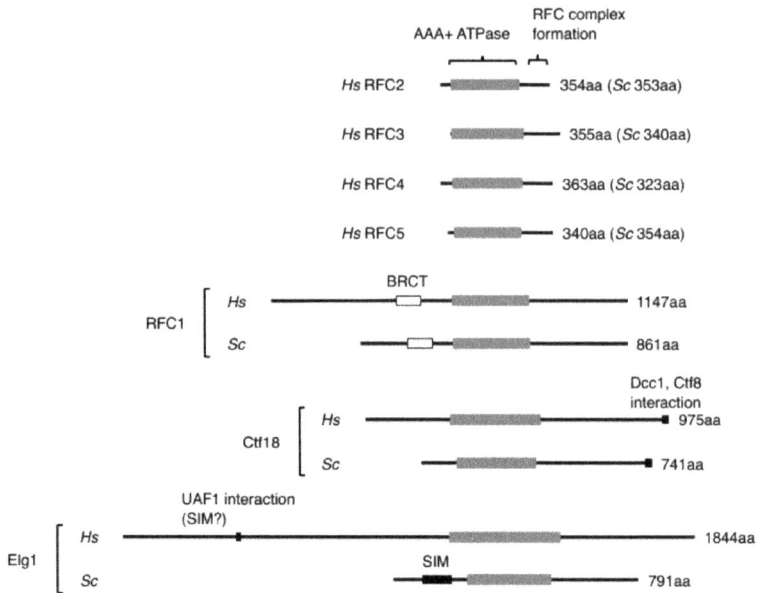

Figure 3. RFC subunit structures. The center of these subunits includes RFC boxes containing a P-loop, which is a general Walker-type ATPase motif. The C-terminal regions following the grey boxes representing the four small subunits (RFC2–5) are required for RFC complex formation. As for the largest RFC subunits, the domains required for complex formation are not well defined. The N-terminal of RFC1 contains a BRCT motif and the C-terminal of Ctf18 contains an interaction motif with Dcc1 and Ctf8. The N-terminal of Elg1 includes a SUMO (in *Sc*, called SIM) or UAF1 (in *Hs*) binding motif, which is potentially involved in PCNA binding. The UAF1 binding motif in *Hs* Elg1 likely has a role as a SIM, thus referred as "SIM?". *Hs*: Homo sapiens, *Sc*: Saccharomyces cerevisiae.

3.1.2. PCNA Loading/Unloading Activity of RFC1-RFC

The molecular role of RFC1-RFC for PCNA loading was first identified using purified complex from human HEK293 cells based on its requirement in the SV40 replication system in vitro [42,43]. Many biochemical analyses revealed that RFC1-RFC has multiple functions that allow for PCNA loading onto DNA at the 3′ primer/template junction in an ATP-dependent manner (Figure 2): PCNA binding, 3′ end of the primer DNA binding, and ATP binding trigger a conformational change of RFC1-RFC that allows it to bind tightly with PCNA and induce ring opening, and then ATP hydrolysis is associated with ring closure and release of the PCNA, which now encircles the DNA duplex. The binding partner of PCNA then switches from RFC1-RFC to DNA polymerase, and RFC1-RFC leaves the new DNA synthesizing complex, DNA polymerase/PCNA/DNA complex [6]. RFC1-RFC binds to a specific side of PCNA, the front face, and loads it in an orientation-dependent manner, so that the front face of PCNA that binds to its partner is oriented toward the elongating DNA (Figure 2). The ATP-driven PCNA-loading process by RFC1-RFC is discussed in detail in these excellent reviews [29,39,44–46].

In vitro experiments revealed that RFC1-RFC unloads PCNA from nicked or gapped circular DNA in an ATP-dependent manner as a reverse reaction of PCNA loading [47,48]. These observations led us to speculate that RFC1-RFC drives both PCNA loading and unloading during DNA replication. Whether unloading of PCNA occurs by RFC1-RFC in vivo, however, remains unclear. Interestingly, the subassembly complex of the four small subunits of RFC can also open PCNA and remove it from DNA, suggesting that all RFC complexes have the potential to unload PCNA [49].

3.2. Ctf18-RFC

3.2.1. Fundamental Features and Structure

Ctf18-RFC was the second PCNA-conducting RFC complex identified. It is a hetero-pentamer formed by the large subunit Ctf18 and two additional subunits, Dcc1 and Ctf8, binding with the small subunits RFC2-5 [50–53] (Figure 1). These additional subunits are unique to this RFC, interact with the C-terminal end of Ctf18, and are conceivably located outside of the circle created by the other five subunits [54,55] (Figure 1). Ctf18, Dcc1, and Ctf8 are conserved from yeast to humans [53,56,57].

Early genetic analysis in *S. cerevisiae* revealed mutations in Ctf18 in screens for genes important for preventing chromosome loss, and it was thus termed chromosome transmission fidelity (previously called Chl12; chromosome loss) [58–60]. The two additional subunits, Dcc1 (defect of chromosome cohesion) and Ctf8, were also identified as genes required for chromosome segregation [59]. The absence of Ctf18, Ctf8, or Dcc1 singly or in combination leads to precocious sister chromatid separation accompanied by pre-anaphase accumulation of cells that depends on the spindle assembly checkpoint [50,61]. Ctf18 is located at replication forks with Ctf4 and Eco1, which are required to establish cohesion, coupled with PCNA recruitment [62]. These findings indicate that Ctf18-RFC is primarily required for sister chromatid cohesion, and might be involved in the regulation of PCNA on chromatin.

3.2.2. PCNA Loading/Unloading Activity of Ctf18-RFC

Electron microscopy images of a recombinant pentameric Ctf18-RFC complex devoid of Dcc1 and Ctf8 subunits and designated here as Ctf18-RFC(5) are indistinguishable from RFC1-RFC, whose five subunits are aligned in a circle with a gap [54]. In addition, in the course of identifying human PCNA-interacting proteins by mass-spectrometric analysis, Ctf18 was identified together with RFC1 and four small subunits [63]. The results suggested that Ctf18-RFC also interacts with PCNA and functions as a PCNA loader. Actually, Ctf18-RFC loads PCNA onto nicked circular DNA or primed single-stranded DNA, which have a 3′ end, in vitro with both yeast and human recombinant proteins [52]. Consistent with this in vitro result, Ctf18 yeast mutants exhibit a reduced amount of chromatin-bound PCNA [62].

Interestingly, Ctf18-RFC(5) also effectively binds to and loads PCNA onto DNA, indicating that the two additional subunits are dispensable for PCNA-loading activity [54]. Thus, the Ctf18-RFC(5) supports DNA polymerase δ activity with PCNA on primed M13 single-stranded DNA in vitro, similar to human RFC1-RFC, demonstrating that Ctf18-RFC-loaded PCNA is functional. This Ctf18-RFC(5), however, cannot substitute for RFC1-RFC in the in vitro SV40 DNA replication system with a crude cell extract that includes the proteins required for replication, such as PCNA, RPA, and DNA polymerases, except RFC(s) [54]. Furthermore, in addition to loading activity, in vitro experiments with purified Ctf18-RFC showed unloading activity toward the primed DNA template [56]. The detailed activity of Ctf18-RFC for PCNA loading/unloading in the cells, however, is not fully understood.

3.3. Elg1-RFC

3.3.1. Fundamental Features and Structure

Elg1-RFC is the most recently identified RFC complex forming a hetero-pentamer by RFC2-5 and the large subunit Elg1 [64–66] (Figure 1). Elg1 has a much longer N terminus compared with the other large subunits, especially in humans (Figure 3). Elg1 was first identified in a series of genetic screens in yeast, in which mutants exhibited various defects leading to genomic instability (hence, its name—enhanced levels of genome instability) [64–72]. In mammals, the corresponding gene was isolated as ATAD5 (ATPase Family, AAA Domain Containing 5).

3.3.2. PCNA Unloading Activity of Elg1-RFC

Three independent groups demonstrated that Elg1-RFC functions as the major PCNA unloader during DNA replication, in both yeast and mammalian cells [73–75]. Suppression of Elg1 expression leads to an extreme accumulation of chromatin-bound PCNA and the corresponding PCNA foci are larger and more intense, indicating an extended lifespan of PCNA in replication factories. In contrast, overexpression of Elg1 results in a reduction of PCNA on chromatin (Figure 4A). Elg1 depletion also leads to an increase in the number of cells in S phase, indicating that abnormal levels of PCNA on chromatin affect cell-cycle progression [73,75].

By quantitative proteomic analysis of a yeast elg1 deletion strain, Kubota et al. initially observed a substantial accumulation of PCNA on the chromatin among other proteins [76]. They then used the auxin inducible degradation (AID)-Elg1 construct for timely depletion or induction of Elg1 in synchronized cell cultures [74,77]. Application of this system confirmed that the accumulation of PCNA and its SUMOylated forms on chromatin occurred in the course of the first cycle of DNA replication, and that lack of Elg1 resulted in a slight delay in S phase progression without checkpoint activation, similar to mammalian cells [74]. In the absence of Elg1, PCNA is not retained at specific sites on the chromatin, indicating that the Elg1-RFC unloads PCNA genome-wide, rather than only from specific chromosomal sites [78].

Partially purified Elg1-RFC from yeast or human cells unloads PCNA from chromatin isolated from an elg1 mutant in a yeast strain or from permeabilized nuclei from human cells in an ATP-dependent manner in vitro [74] (Figure 4B). These results strongly support that Elg1-RFC is a primary PCNA unloader. It remains unclear, however, whether Elg1-RFC is the only PCNA unloader during normal DNA replication, as Elg1 is not essential for cell division [34].

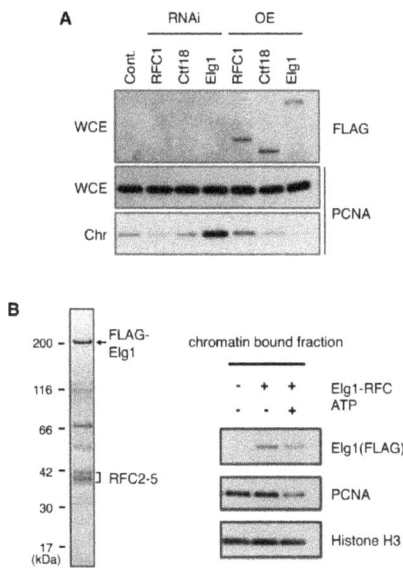

Figure 4. PCNA loading or unloading function of human RFC complexes. (**A**) Depletion by RNA interference (RNAi) or overexpression (OE) of the largest subunits of RFCs in human HEK293 cells. Whole cell extract (WCE) and chromatin-containing fractions (Chr) were prepared after centrifugation. The results demonstrated that RFC1-RFC and Elg1-RFC have a primary role in PCNA loading and unloading, respectively, in vivo. Depletion or overexpression of Ctf18 does not change the level of PCNA on chromatin in human cells; (**B**) PCNA unloading assay. Left panel: partially purified Elg1-RFC complex. HEK293T cells were co-transfected with FLAG-tagged Elg1 and RFC2-5, and complexes were purified with anti-FLAG antibody. Right panel: PCNA unloading assay. The purified Elg1-RFC was incubated with permeabilized cell nuclei containing PCNA-loaded chromatin in the presence or absence of ATP. The purified Elg1-RFC complexes unload PCNA from chromatin in an ATP-dependent manner.

4. The Three RFC Complexes Contribute to Genomic Integrity by Controlling PCNA Loading/Unloading

4.1. Roles during DNA Replication Progression

As described above, the three RFC complexes likely share the roles of PCNA loading and/or unloading in vivo. RFC1-RFC, and probably Ctf18-RFC, primarily function as PCNA loaders and Elg1-RFC primarily functions as an unloader. Indeed, as shown Figure 4A, our results clearly demonstrated that depletion or overexpression of RFC1 or Elg1 in human cells have opposite effects on the PCNA levels on chromatin. In a knockdown experiment, depletion of RFC1 led to decreased PCNA levels on chromatin, while depletion of Elg1 led to increased PCNA levels. An overexpression experiment produced completely opposite results.

Once DNA replication is initiated, PCNA must be loaded onto DNA, both on leading and lagging strands. As expected, PCNA is detected almost twice as often on the lagging strand than on the leading strand at the replication fork [79]. On chromatin, PCNA plays multiple roles; first, it clamps the polymerase for DNA synthesis, and then it recruits many of the enzymes required for the following chromosomal events. The fact that RFC1, but not Ctf18 or Elg1, is essential also reflects the importance of PCNA loading by RFC1-RFC and its potential unloading activity [34]. While it appears that Ctf18-RFC can load PCNA on chromatin, it cannot substitute for RFC1 deletion, probably because the PCNA-loading activity of Ctf18-RFC is weaker than that of RFC1-RFC in vitro, or Ctf18-RFC may load PCNA for specific purposes such as for establishing cohesion [54,62].

Elg1-RFC is the PCNA unloader during normal DNA replication. The absence of Elg1 leads to various types of chromosome instability, such as DNA damage sensitivity, replication defects, enhanced homologous recombination, gross chromosomal rearrangements, chromosome maintenance defects, elongated telomeres, and cohesion defects [64–72]. In mammals, defects in corresponding ATAD5 likewise cause genomic instability and predisposition to cancer in human and mouse cells [80]. Mouse embryonic fibroblasts derived from ATAD5 heterozygous mice are highly sensitive to DNA damaging agents, demonstrating high levels of aneuploidy and genomic instability in response to DNA damage [81]. In addition, altered levels of recruitment of the PCNA-interacting proteins on chromatin were observed [75] (see Section 5). All these abnormalities may be due to enhanced retention of PCNA on chromatin [82].

PCNA unloading must be coupled with the completion of the chromosome replication process, because not only delayed but also precocious unloading would cause abnormalities in DNA replication and its associated chromosomal events. Actually, defects of the Okazaki fragment ligase Cdc9 in yeast leads to PCNA accumulation on chromatin, similar to the accumulation caused by a lack of Elg1 [78]. Thus, PCNA unloading is at least dependent upon completion of the Okazaki fragment ligation during DNA replication in S phase. The unloading of PCNA may be also dependent on the completion of the nucleosome assembly, because the absence of its assembly due to inhibition of histone supply causes PCNA to accumulate on chromatin [83].

Is Elg1-RFC the only PCNA unloader? Even in the absence of Elg1, the PCNA retained on the chromatin is eventually removed in the M phase [75]; therefore, the PCNA unloading function could conceivably be performed by RFC1-RFC and/or Ctf18-RFC, as suggested by their biochemical and genetic analyses. One possible regulation mechanism is modification of the largest subunits of RFCs, which may switch on either the PCNA loading or unloading activity. It is also possible that PCNA eventually spontaneously dissociates from DNA without the help of any unloaders [49]. Of course, there may be other novel pathways that can remove PCNA from the chromatin, such as acetylation-mediated removal and degradation of PCNA [26].

4.2. Roles in Sister Chromatid Cohesion

Ctf18-RFC is required for establishing sister chromatid cohesion, which may involve PCNA loading that aids the function of cohesion establishment factor Eco1. Eco1 associates with PCNA and promotes cohesion by acetylating the cohesion subunit Smc3 during S phase [84–87]. The fact that loss of either Ctf18, Dcc1, or Ctf8 causes cohesion defects and that these molecules form a DNA polymerase ε binding module suggest that sufficient levels of PCNA on the leading strand must be supplied by Ctf18-RFC, which would support polymerase ε and Eco1 acetylation activity [55,88,89].

A previous report indicated that Elg1 also plays a role in sister chromatid cohesion [71]. A yeast strain with deletion of Elg1, *elg1Δ*, exhibits precocious sister chromatid separation like the *ctf18Δ* mutant. Although the frequency is lower than that of the Ctf18-deleted strain, the *elg1Δ* strain is synthetic lethal with the cohesion mutants *scc1* or *smc1*. It is probable that inefficient PCNA unloading also affects cohesion establishment. Eco1 is recruited by PCNA, and PCNA SUMOylation appears to counteract Eco1 activity [90]. It is therefore possible that the excess PCNA SUMOylation observed in an *elg1Δ* mutant on chromatin interferes with the function of Eco1 in establishing cohesion. Indeed, a yeast Elg1 mutation that leads to over-SUMOylated PCNA, also causes a cohesion defect [90].

4.3. Roles for Proteolysis to Prevent DNA Re-Replication

Regulation of PCNA loading and unloading also has an important role in the once-per-cell-cycle replication. Chromatin-loaded PCNA activates the ubiquitin ligase CRL4-Cdt2 to prevent re-replication in the same cell cycle [16,91]. Cdt1 is a factor that is required for licensing of replication origins in G1 phase [92]. Cdt1 has a PIP-degron composed of a PIP-box sequence and downstream basic amino acid(s) [93,94]. When PCNA is loaded on chromatin upon the initiation of S phase, Cdt1 associates through its PIP-box, exposes the PIP-degron to CRL4-Cdt2, and is ubiquitinated for degradation. Cyclin-dependent kinase (CDK) inhibitor p21 and histone H4K20 mono methyltransferase Set8, which are also involved in the regulation of origin licensing, have PIP-degrons, and are also degraded by the same mechanism [95–101]. These proteins begin to re-accumulate around the end of S phase or G2 phase, when all of the PCNA is unloaded. Therefore, timely degradation and accumulation of these proteins are important for correct regulation of DNA replication, which is likely ensured by correct PCNA loading and unloading in the cell cycle. Ctf18-RFC is involved in CRL4-Cdt2 recruitment to the site of PCNA foci to degrade Cdt1, because RNA interference treatment of Ctf18, but not other large subunits, leads to defects in CRL4-Cdt2 recruitment at the replication fork. In contrast, RFC1-RFC contributes to CRL4-Cdt2 activation not during S phase, but following UV damage [102].

4.4. Roles of RFCs in Other Events

Several lines of experiments demonstrated that RFC complexes are involved in DNA repair processes. For example, when cells are irradiated with UV or treated with DNA-damaging reagents, the DNA repair reaction occurs and PCNA accumulates at the DNA-damaged sites, even though the cells are not in S phase. Studies of nucleotide excision repair, base excision repair, and mismatch repair have all demonstrated indispensable roles of RFC1-RFC in a DNA repair reaction to load PCNA and DNA pol δ/ε repair synthesis [103–114]. Elg1 is also involved in the DNA damage response [80,115]. It is not known, however, whether Elg1-RFC unloads PCNA after repair synthesis of the excised DNA damage site. A method termed enrichment and sequencing of protein-associated nascent DNA (eSPAN) was developed to discriminate proteins enriched at either the nascent leading or lagging strands. This method revealed that in yeast cells, PCNA is unloaded from the lagging strands upon stalling replication fork with hydroxyurea, and this process is dependent on Elg1 [79]. Cells deficient in PCNA unloading (*elg1Δ*) exhibit increased spontaneous chromosome breaks; hence, Elg1 contributes to genome stability when the supply of nucleotides is limited or when the replication fork encounters obstacles that cause replication stress.

Several studies revealed other aspects of Ctf18-RFC function that may be distinct from its function in sister chromatid cohesion. In a genome-wide specific screen for mutants affecting replication initiation, Ctf18 was newly identified and shown to physically interact with ORC, Cdt1, and MCM proteins. Furthermore, depletion of Ctf18 reduces pre-RC formation during the M-to-G1 phase transition, prevents S phase entry, and retards S phase progression [116]. Ctf18 is also essential for activating the

DNA replication checkpoint upon the replication stress response [76,117,118]. Additionally, Dcc1 and Ctf8 are required for replication checkpoint activation, and for proper telomere length regulation and telomere intra-nuclear positioning [119]. Ctf18-RFC associates with DNA polymerase ε mediated by Dcc1 and Ctf8 at defective replication forks for activating the S phase checkpoint [88,89]. The Ctf18-RFC complex is also important for replication fork velocity and this effect seems to be linked to its major role in sister chromatid cohesion [120]. These findings represent new aspects of Ctf18-RFC's roles. It is not yet fully elucidated, however, how these roles are related to its PCNA loading and unloading activity.

5. RFC Complexes May Play Roles beyond PCNA Loading/Unloading

5.1. Extended Region of the Large Subunits of RFCs

All large subunits of RFCs, especially human RFCs, have extensions in both the N-terminal and C-terminal regions from the RFC boxes (Figure 3). The extended regions of the large subunit likely have multiple roles, such as modulating PCNA loading/unloading, coupling the PCNA on chromatin with other events, as well as a role completely separate from its PCNA loading/unloading activities.

The N-terminal extension of RFC1 is not essential for cell viability, nor is it required for in vitro clamp loading activity, but removal of this region results in DNA damage sensitivity in vivo, suggesting that it has additional roles outside of its primary function as a PCNA loader [37,121]. The N-terminal extension of RFC1 contains a region that shares homology with DNA ligases, known as the BRCA1 C-terminal (BRCT) domain, though it does not have ligase activity. The structure of the human BRCT domain in solution suggested a binding model between BRCT and 5'-phosphorylated double-stranded DNA [122].

Ctf18 has an extended region in the C-terminus. As mentioned, the C-terminal end of Ctf18 interacts with Dcc1 and Ctf8 and forms a DNA polymerase ε binding module that is conserved from yeast to human cells and is important for activating the DNA replication checkpoint [88,89].

Although the interacting domain on Elg1 was not mapped, a recent finding showed that the Drosophila KAT6 Enok acetyltransferase complex interacts with Elg1 and inhibits its unloading activity [123]. The N terminal region of yeast Elg1 has a SUMO-interacting motif (SIM) at the N-terminal, and Elg1-RFC preferentially interacts and unloads SUMOylated PCNA from chromatin [73]. The N-terminal domain of yeast Elg1 might make a crucial contribution to PCNA unloading, because this domain interacts with PCNA and is important for the in vivo function of Elg1. Indeed, cells expressing Elg1 lacking the N-terminal 215 amino acids exhibit increased methyl methanesulfonate sensitivity compared with wild-type cells, but less methyl methanesulfonate (MMS) than an *elg1Δ* mutant [124]. The SIM at the N-terminus of Elg1 interacts with its target SUMOylated PCNA (and also unmodified PCNA), which helps Elg1-RFC bind PCNA strongly through all five subunits and open the PCNA ring to release it from chromatin. Because PCNA must be unloaded from double-stranded DNA passing through the PCNA ring, the unloading steps might not be a simple reverse reaction of PCNA loading [78].

Human Elg1 has an extremely extended N-terminal region, whose full amino acid length is 2.3 times longer than that of yeast Elg1 (*Hs* 1844 aa vs. *Sc* 791 aa; Figure 3). Human Elg1 also has a SIM in the N-terminal region. In contrast to yeast Elg1, however, the motif interacts with a SUMO-like domain in the deubiquitination factor UAF1 [125]. Thus, human Elg1 regulates PCNA deubiquitination by recruiting the USP1-UAF1 complex to ubiquitinated PCNA on chromatin. The different reactions of N-terminal Elg1 between yeast and human cells may reflect the difference in the levels of modification of PCNA between the species. In both cases, the N-terminal region may help to detect modified PCNA on chromatin and to facilitate its unloading. This process may be coupled with removal of the modification, as the loading of modified PCNA at a new site may bring about an irregular reaction by modified PCNA on chromatin.

5.2. RFC Complexes Interact with and Regulate Proteins Other than PCNA

The RFC complexes reported so far have various functions through their interactions with other factors. RFC1 interacts with DNA ligase I and negatively regulates its activity [126]. RFC1 binds directly to Asf1, a histone deposition protein, and histone deacetylase 1, and may play a role in replication-coupled chromatin remodeling or replication fork progression [127,128]. Additionally, RFC1 is suggested to regulate transcription, as it interacts with several transcription factors [129–131]. We demonstrated that RFC1-RFC and Ctf18-RFC interact with polymerase η, but RFC1-RFC inhibits its activity and Ctf18-RFC stimulates its activity in vitro [132]. Ctf18 also interacts with DNA polymerase ε to stimulate DNA synthesis activity [88,89]. RFC1-RFC and Ctf18-RFC appear to interact with the E3 ubiquitin ligase CRL4-Cdt2. Especially, Ctf18-RFC plays a role to recruit CRL4-Cdt2 to PCNA foci during DNA replication [102]. Thus, RFC complexes have more roles beyond PCNA loading/unloading and fulfill multiple functions.

Elg1 depletion in human cells leads to changes in the chromatin-bound levels of chromatin proteins. Elg1-depleted cells have decreased levels of proteins, such as RanGEF RCC1, SMC3, HBO1, SNF2H, HP1α, and Rif1 [75]. Most of these proteins bind to chromatin and correlate with chromatin remodeling behind the replication fork. In contrast, the chromatin levels of factors more directly involved in DNA replication, such as both PCNA-binding proteins, polymerase δ, DNA ligase I, MSH2, and non-PCNA binding protein Mcm6, remain the same after Elg1 depletion. These findings suggest that DNA replication processes such as Okazaki fragment maturation can be fulfilled correctly even in the absence of Elg1, but PCNA remains on the chromatin behind the active replisomes. Such unremoved PCNA could inhibit the association of chromatin formation, modification, or modeling factors. It is also possible that the extended N-terminus of Elg1 contains unidentified domains that interact with and recruit factors for chromatin transactions. Defects in Elg1 would induce changes in chromosomal stability. Analysis of cells in which endogenous Elg1 is replaced with nested deletion constructs will be required to define the role of the N-terminal domain.

6. Conclusions and Perspective

PCNA loading and unloading must repeatedly occur to initiate DNA synthesis and after the completion of every Okazaki fragment, as well as at replication fork termination. Three PCNA-conducting RFC complexes share the role of ensuring appropriate PCNA loading or unloading. In addition, all three RFC complexes have additional functions other than PCNA loading/unloading. It is unclear, however, why all eukaryotic cells require three similar RFC complexes. Given that RFC1-RFC and Elg1-RFC have primary roles in PCNA loading and unloading, respectively, it remains to be clarified how strictly the labor of PCNA loading/unloading is divided and shared by the three RFC complexes. Many questions remain regarding the detailed functions of the three RFC complexes to maintain genome integrity.

Acknowledgments: This work was financially supported by Grants-in-Aid for Basic Scientific Research (C) (25430171) to Yasushi Shiomi; and by Grants-in-Aid for Basic Scientific Research (B) (26291025) and for Challenging Exploratory Research (26650064) from the Ministry of Education, Culture, Sports, Science, and Technology of Japan to Hideo Nishitani.

Conflicts of Interest: The authors declare no conflict of interest.

References

1. Nurse, P. Ordering S phase and M phase in the cell cycle. *Cell* **1994**, *79*, 547–550. [CrossRef]
2. Nishitani, H.; Taraviras, S.; Lygerou, Z.; Nishimoto, T. The human licensing factor for DNA replication Cdt1 accumulates in G1 and is destabilized after initiation of S-phase. *J. Biol. Chem.* **2001**, *276*, 44905–44911. [CrossRef] [PubMed]
3. Masai, H.; Matsumoto, S.; You, Z.; Yoshizawa-Sugata, N.; Oda, M. Eukaryotic chromosome DNA replication: Where, when, and how? *Annu. Rev. Biochem.* **2010**, *79*, 89–130. [CrossRef] [PubMed]
4. Nasmyth, K.; Haering, C.H. The structure and function of SMC and kleisin complexes. *Annu. Rev. Biochem.* **2005**, *74*, 595–648. [CrossRef] [PubMed]

5. Nasmyth, K.; Haering, C.H. Cohesin: Its roles and mechanisms. *Annu. Rev. Genet.* **2009**, *43*, 525–558.

6. Johnson, A.; O'Donnell, M. Cellular DNA replicases: Components and dynamics at the replication fork. *Annu. Rev. Biochem.* **2005**, *74*, 283–315. [CrossRef] [PubMed]

7. Moldovan, G.-L.; Pfander, B.; Jentsch, S. PCNA, the maestro of the replication fork. *Cell* **2007**, *129*, 665–679.

8. Kim, J.; MacNeill, S.A. Genome stability: A new member of the RFC family. *Curr. Biol. CB* **2003**, *13*, R873–R875. [CrossRef] [PubMed]

9. Waga, S.; Stillman, B. The DNA replication fork in eukaryotic cells. *Annu. Rev. Biochem.* **1998**, *67*, 721–751.

10. Bruck, I.; O'Donnell, M. The ring-type polymerase sliding clamp family. *Genome Biol.* **2001**, *2*, reviews3001.1–3001.3. [CrossRef] [PubMed]

11. Bravo, R.; Frank, R.; Blundell, P.A.; Macdonald-Bravo, H. Cyclin/PCNA is the auxiliary protein of DNA polymerase-delta. *Nature* **1987**, *326*, 515–517. [CrossRef] [PubMed]

12. Prelich, G.; Tan, C.K.; Kostura, M.; Mathews, M.; So, A.G.; Downey, K.M.; Stillman, B. Functional identity of proliferating cell nuclear antigen and a DNA polymerase-delta auxiliary protein. *Nature* **1987**, *326*, 517–520.

13. Krishna, T.S.; Kong, X.P.; Gary, S.; Burgers, P.M.; Kuriyan, J. Crystal structure of the eukaryotic DNA polymerase processivity factor PCNA. *Cell* **1994**, *79*, 1233–1243. [CrossRef]

14. Gulbis, J.M.; Kelman, Z.; Hurwitz, J.; O'Donnell, M.; Kuriyan, J. Structure of the C-terminal region of p21(WAF1/CIP1) complexed with human PCNA. *Cell* **1996**, *87*, 297–306. [CrossRef]

15. Tsurimoto, T. PCNA, a multifunctional ring on DNA. *Biochim. Biophys. Acta* **1998**, *1443*, 23–39. [CrossRef]

16. Havens, C.G.; Walter, J.C. Mechanism of CRL4(Cdt2), a PCNA-dependent E3 ubiquitin ligase. *Genes Dev.* **2011**, *25*, 1568–1582. [CrossRef] [PubMed]

17. Kubota, T.; Myung, K.; Donaldson, A.D. Is PCNA unloading the central function of the Elg1/ATAD5 replication factor C-like complex? *Cell Cycle Georget. Tex* **2013**, *12*, 2570–2579. [CrossRef] [PubMed]

18. Stelter, P.; Ulrich, H.D. Control of spontaneous and damage-induced mutagenesis by SUMO and ubiquitin conjugation. *Nature* **2003**, *425*, 188–191. [CrossRef] [PubMed]

19. Hoege, C.; Pfander, B.; Moldovan, G.-L.; Pyrowolakis, G.; Jentsch, S. RAD6-dependent DNA repair is linked to modification of PCNA by ubiquitin and SUMO. *Nature* **2002**, *419*, 135–141. [CrossRef] [PubMed]

20. Prakash, S.; Johnson, R.E.; Prakash, L. Eukaryotic translesion synthesis DNA polymerases: Specificity of structure and function. *Annu. Rev. Biochem.* **2005**, *74*, 317–353. [CrossRef] [PubMed]

21. Matsuda, T.; Bebenek, K.; Masutani, C.; Hanaoka, F.; Kunkel, T. A. Low fidelity DNA synthesis by human DNA polymerase-eta. *Nature* **2000**, *404*, 1011–1013. [PubMed]

22. Masutani, C.; Kusumoto, R.; Iwai, S.; Hanaoka, F. Mechanisms of accurate translesion synthesis by human DNA polymerase eta. *EMBO J.* **2000**, *19*, 3100–3109. [CrossRef] [PubMed]

23. Papouli, E.; Chen, S.; Davies, A.A.; Huttner, D.; Krejci, L.; Sung, P.; Ulrich, H.D. Crosstalk between SUMO and ubiquitin on PCNA is mediated by recruitment of the helicase Srs2p. *Mol. Cell* **2005**, *19*, 123–133.

24. Pfander, B.; Moldovan, G.-L.; Sacher, M.; Hoege, C.; Jentsch, S. SUMO-modified PCNA recruits Srs2 to prevent recombination during S phase. *Nature* **2005**, *436*, 428–433. [CrossRef] [PubMed]

25. Naryzhny, S.N.; Lee, H. The post-translational modifications of proliferating cell nuclear antigen: Acetylation, not phosphorylation, plays an important role in the regulation of its function. *J. Biol. Chem.* **2004**, *279*, 20194–20199. [CrossRef] [PubMed]

26. Cazzalini, O.; Sommatis, S.; Tillhon, M.; Dutto, I.; Bachi, A.; Rapp, A.; Nardo, T.; Scovassi, A.I.; Necchi, D.; Cardoso, M.C.; Stivala, L.A.; Prosperi, E. CBP and p300 acetylate PCNA to link its degradation with nucleotide excision repair synthesis. *Nucleic Acids Res.* **2014**, *42*, 8433–8448. [CrossRef] [PubMed]

27. Billon, P.; Li, J.; Lambert, J.-P.; Chen, Y.; Tremblay, V.; Brunzelle, J.S.; Gingras, A.-C.; Verreault, A.; Sugiyama, T.; Couture, J.-F.; et al. Acetylation of PCNA Sliding Surface by Eco1 Promotes Genome Stability through Homologous Recombination. *Mol. Cell* **2017**, *65*, 78–90. [CrossRef] [PubMed]

28. Ogura, T.; Wilkinson, A.J. AAA+ superfamily ATPases: Common structure–diverse function. *Genes Cells Devoted Mol. Cell. Mech.* **2001**, *6*, 575–597. [CrossRef]

29. Davey, M.J.; Jeruzalmi, D.; Kuriyan, J.; O'Donnell, M. Motors and switches: AAA+ machines within the replisome. *Nat. Rev. Mol. Cell Biol.* **2002**, *3*, 826–835. [CrossRef] [PubMed]

30. Sancar, A.; Lindsey-Boltz, L.A.; Unsal-Kaçmaz, K.; Linn, S. Molecular mechanisms of mammalian DNA repair and the DNA damage checkpoints. *Annu. Rev. Biochem.* **2004**, *73*, 39–85. [CrossRef] [PubMed]

31. Parrilla-Castellar, E.R.; Arlander, S.J.H.; Karnitz, L. Dial 9-1-1 for DNA damage: The Rad9-Hus1-Rad1 (9-1-1) clamp complex. *DNA Repair* **2004**, *3*, 1009–1014. [CrossRef] [PubMed]

32. Mossi, R.; Hübscher, U. Clamping down on clamps and clamp loaders—The eukaryotic replication factor C. *Eur. J. Biochem.* **1998**, *254*, 209–216. [PubMed]

33. Cullmann, G.; Fien, K.; Kobayashi, R.; Stillman, B. Characterization of the five replication factor C genes of *Saccharomyces cerevisiae*. *Mol. Cell. Biol.* **1995**, *15*, 4661–4671. [CrossRef] [PubMed]

34. Kim, J.; Robertson, K.; Mylonas, K.J.L.; Gray, F.C.; Charapitsa, I.; MacNeill, S.A. Contrasting effects of Elg1-RFC and Ctf18-RFC inactivation in the absence of fully functional RFC in fission yeast. *Nucleic Acids Res.* **2005**, *33*, 4078–4089. [CrossRef] [PubMed]

35. Bunz, F.; Kobayashi, R.; Stillman, B. cDNAs encoding the large subunit of human replication factor C. *Proc. Natl. Acad. Sci. USA* **1993**, *90*, 11014–11018. [CrossRef] [PubMed]

36. Uhlmann, F.; Cai, J.; Flores-Rozas, H.; Dean, F.B.; Finkelstein, J.; O'Donnell, M.; Hurwitz, J. In vitro reconstitution of human replication factor C from its five subunits. *Proc. Natl. Acad. Sci. USA* **1996**, *93*, 6521–6526. [CrossRef] [PubMed]

37. Uhlmann, F.; Cai, J.; Gibbs, E.; O'Donnell, M.; Hurwitz, J. Deletion analysis of the large subunit p140 in human replication factor C reveals regions required for complex formation and replication activities. *J. Biol. Chem.* **1997**, *272*, 10058–10064. [PubMed]

38. Bowman, G.D.; O'Donnell, M.; Kuriyan, J. Structural analysis of a eukaryotic sliding DNA clamp-clamp loader complex. *Nature* **2004**, *429*, 724–730. [CrossRef] [PubMed]

39. Yao, N.Y.; O'Donnell, M. The RFC clamp loader: Structure and function. *Subcell. Biochem.* **2012**, *62*, 259–279.

40. Shiomi, Y.; Usukura, J.; Masamura, Y.; Takeyasu, K.; Nakayama, Y.; Obuse, C.; Yoshikawa, H.; Tsurimoto, T. ATP-dependent structural change of the eukaryotic clamp-loader protein, replication factor C. *Proc. Natl. Acad. Sci. USA* **2000**, *97*, 14127–14132. [CrossRef] [PubMed]

41. O'Donnell, M.; Jeruzalmi, D.; Kuriyan, J. Clamp loader structure predicts the architecture of DNA polymerase III holoenzyme and RFC. *Curr. Biol.* **2001**, *11*, R935–R946. [CrossRef]

42. Tsurimoto, T.; Stillman, B. Functions of replication factor C and proliferating-cell nuclear antigen: functional similarity of DNA polymerase accessory proteins from human cells and bacteriophage T4. *Proc. Natl. Acad. Sci. USA* **1990**, *87*, 1023–1027. [CrossRef] [PubMed]

43. Tsurimoto, T.; Stillman, B. Purification of a cellular replication factor, RF-C, that is required for coordinated synthesis of leading and lagging strands during simian virus 40 DNA replication in vitro. *Mol. Cell. Biol.* **1989**, *9*, 609–619. [CrossRef] [PubMed]

44. Kelch, B.A.; Makino, D.L.; O'Donnell, M.; Kuriyan, J. Clamp loader ATPases and the evolution of DNA replication machinery. *BMC Biol.* **2012**. [CrossRef] [PubMed]

45. Hedglin, M.; Kumar, R.; Benkovic, S.J. Replication clamps and clamp loaders. *Cold Spring Harb. Perspect. Biol.* **2013**. [CrossRef] [PubMed]

46. Kelch, B.A. Review: The lord of the rings: Structure and mechanism of the sliding clamp loader. *Biopolymers* **2016**, *105*, 532–546. [CrossRef] [PubMed]

47. Yao, N.; Turner, J.; Kelman, Z.; Stukenberg, P.T.; Dean, F.; Shechter, D.; Pan, Z.Q.; Hurwitz, J.; O'Donnell, M. Clamp loading, unloading and intrinsic stability of the PCNA, beta and gp45 sliding clamps of human, *E. coli* and T4 replicases. *Genes Cells Devoted Mol. Cell. Mech.* **1996**, *1*, 101–113. [CrossRef]

48. Shibahara, K.; Stillman, B. Replication-dependent marking of DNA by PCNA facilitates CAF-1-coupled inheritance of chromatin. *Cell* **1999**, *96*, 575–585. [CrossRef]

49. Yao, N.Y.; Johnson, A.; Bowman, G.D.; Kuriyan, J.; O'Donnell, M. Mechanism of proliferating cell nuclear antigen clamp opening by replication factor C. *J. Biol. Chem.* **2006**, *281*, 17528–17539. [CrossRef] [PubMed]

50. Mayer, M.L.; Gygi, S.P.; Aebersold, R.; Hieter, P. Identification of RFC (Ctf18p, Ctf8p, Dcc1p): An alternative RFC complex required for sister chromatid cohesion in S. cerevisiae. *Mol. Cell* **2001**, *7*, 959–970. [CrossRef]

51. Naiki, T.; Kondo, T.; Nakada, D.; Matsumoto, K.; Sugimoto, K. Chl12 (Ctf18) forms a novel replication factor C-related complex and functions redundantly with Rad24 in the DNA replication checkpoint pathway. *Mol. Cell. Biol.* **2001**, *21*, 5838–5845. [CrossRef] [PubMed]

52. Bermudez, V.P.; Maniwa, Y.; Tappin, I.; Ozato, K.; Yokomori, K.; Hurwitz, J. The alternative Ctf18-Dcc1-Ctf8-replication factor C complex required for sister chromatid cohesion loads proliferating cell nuclear antigen onto DNA. *Proc. Natl. Acad. Sci. USA* **2003**, *100*, 10237–10242. [CrossRef] [PubMed]

53. Merkle, C.J.; Karnitz, L.M.; Henry-Sánchez, J.T.; Chen, J. Cloning and characterization of hCTF18, hCTF8, and hDCC1. Human homologs of a Saccharomyces cerevisiae complex involved in sister chromatid cohesion establishment. *J. Biol. Chem.* **2003**, *278*, 30051–30056. [CrossRef] [PubMed]

54. Shiomi, Y.; Shinozaki, A.; Sugimoto, K.; Usukura, J.; Obuse, C.; Tsurimoto, T. The reconstituted human Chl12-RFC complex functions as a second PCNA loader. *Genes Cells Devoted Mol. Cell. Mech.* **2004**, *9*, 279–290. [CrossRef] [PubMed]

55. Murakami, T.; Takano, R.; Takeo, S.; Taniguchi, R.; Ogawa, K.; Ohashi, E.; Tsurimoto, T. Stable interaction between the human proliferating cell nuclear antigen loader complex Ctf18-replication factor C (RFC) and DNA polymerase {epsilon} is mediated by the cohesion-specific subunits, Ctf18, Dcc1, and Ctf8. *J. Biol. Chem.* **2010**, *285*, 34608–34615. [CrossRef] [PubMed]

56. Bylund, G.O.; Burgers, P.M.J. Replication protein A-directed unloading of PCNA by the Ctf18 cohesion establishment complex. *Mol. Cell. Biol.* **2005**, *25*, 5445–5455. [CrossRef] [PubMed]

57. McLellan, J.; O'Neil, N.; Tarailo, S.; Stoepel, J.; Bryan, J.; Rose, A.; Hieter, P. Synthetic lethal genetic interactions that decrease somatic cell proliferation in Caenorhabditis elegans identify the alternative RFC CTF18 as a candidate cancer drug target. *Mol. Biol. Cell* **2009**, *20*, 5306–5313. [CrossRef] [PubMed]

58. Spencer, F.; Gerring, S.L.; Connelly, C.; Hieter, P. Mitotic chromosome transmission fidelity mutants in Saccharomyces cerevisiae. *Genetics* **1990**, *124*, 237–249. [PubMed]

59. Kouprina, N.; Tsouladze, A.; Koryabin, M.; Hieter, P.; Spencer, F.; Larionov, V. Identification and genetic mapping of CHL genes controlling mitotic chromosome transmission in yeast. *Yeast Chichester Engl.* **1993**, *9*, 11–19. [CrossRef] [PubMed]

60. Kouprina, N.; Kroll, E.; Kirillov, A.; Bannikov, V.; Zakharyev, V.; Larionov, V. CHL12, a gene essential for the fidelity of chromosome transmission in the yeast Saccharomyces cerevisiae. *Genetics* **1994**, *138*, 1067–1079.

61. Hanna, J.S.; Kroll, E.S.; Lundblad, V.; Spencer, F.A. Saccharomyces cerevisiae CTF18 and CTF4 are required for sister chromatid cohesion. *Mol. Cell. Biol.* **2001**, *21*, 3144–3158. [CrossRef] [PubMed]

62. Lengronne, A.; McIntyre, J.; Katou, Y.; Kanoh, Y.; Hopfner, K.-P.; Shirahige, K.; Uhlmann, F. Establishment of sister chromatid cohesion at the *S. cerevisiae* replication fork. *Mol. Cell* **2006**, *23*, 787–799. [CrossRef] [PubMed]

63. Ohta, S.; Shiomi, Y.; Sugimoto, K.; Obuse, C.; Tsurimoto, T. A proteomics approach to identify proliferating cell nuclear antigen (PCNA)-binding proteins in human cell lysates. Identification of the human CHL12/RFCs2-5 complex as a novel PCNA-binding protein. *J. Biol. Chem.* **2002**, *277*, 40362–40367. [CrossRef] [PubMed]

64. Ben-Aroya, S.; Koren, A.; Liefshitz, B.; Steinlauf, R.; Kupiec, M. ELG1, a yeast gene required for genome stability, forms a complex related to replication factor C. *Proc. Natl. Acad. Sci. USA* **2003**, *100*, 9906–9911. [CrossRef] [PubMed]

65. Bellaoui, M.; Chang, M.; Ou, J.; Xu, H.; Boone, C.; Brown, G.W. Elg1 forms an alternative RFC complex important for DNA replication and genome integrity. *EMBO J.* **2003**, *22*, 4304–4313. [CrossRef] [PubMed]

66. Kanellis, P.; Agyei, R.; Durocher, D. Elg1 forms an alternative PCNA-interacting RFC complex required to maintain genome stability. *Curr. Biol.* **2003**, *13*, 1583–1595. [CrossRef]

67. Huang, M.-E.; Rio, A.-G.; Nicolas, A.; Kolodner, R.D. A genomewide screen in Saccharomyces cerevisiae for genes that suppress the accumulation of mutations. *Proc. Natl. Acad. Sci. USA* **2003**, *100*, 11529–11534. [CrossRef] [PubMed]

68. Smolikov, S.; Mazor, Y.; Krauskopf, A. ELG1, a regulator of genome stability, has a role in telomere length regulation and in silencing. *Proc. Natl. Acad. Sci. USA* **2004**, *101*, 1656–1661. [CrossRef] [PubMed]

69. Smith, S.; Hwang, J.-Y.; Banerjee, S.; Majeed, A.; Gupta, A.; Myung, K. Mutator genes for suppression of gross chromosomal rearrangements identified by a genome-wide screening in Saccharomyces cerevisiae. *Proc. Natl. Acad. Sci. USA* **2004**, *101*, 9039–9044. [CrossRef] [PubMed]

70. Maradeo, M.E.; Skibbens, R.V. The Elg1-RFC clamp-loading complex performs a role in sister chromatid cohesion. *PLoS ONE* **2009**, *4*, e4707. [CrossRef] [PubMed]

71. Parnas, O.; Zipin-Roitman, A.; Mazor, Y.; Liefshitz, B.; Ben-Aroya, S.; Kupiec, M. The ELG1 clamp loader plays a role in sister chromatid cohesion. *PLoS ONE* **2009**, *4*, e5497. [CrossRef] [PubMed]

72. Shkedy, D.; Singh, N.; Shemesh, K.; Amir, A.; Geiger, T.; Liefshitz, B.; Harari, Y.; Kupiec, M. Regulation of Elg1 activity by phosphorylation. *Cell Cycle Georget. Tex* **2015**, *14*, 3689–3697. [CrossRef] [PubMed]

73. Lee, K.; Fu, H.; Aladjem, M.I.; Myung, K. ATAD5 regulates the lifespan of DNA replication factories by modulating PCNA level on the chromatin. *J. Cell Biol.* **2013**, *200*, 31–44. [CrossRef] [PubMed]

74. Kubota, T.; Nishimura, K.; Kanemaki, M.T.; Donaldson, A. D. The Elg1 replication factor C-like complex functions in PCNA unloading during DNA replication. *Mol. Cell* **2013**, *50*, 273–280. [CrossRef] [PubMed]

75. Shiomi, Y.; Nishitani, H. Alternative replication factor C protein, Elg1, maintains chromosome stability by regulating PCNA levels on chromatin. *Genes Cells* **2013**, *18*, 946–959. [CrossRef] [PubMed]

76. Kubota, T.; Hiraga, S.; Yamada, K.; Lamond, A.I.; Donaldson, A.D. Quantitative proteomic analysis of chromatin reveals that Ctf18 acts in the DNA replication checkpoint. *Mol. Cell. Proteomics* **2011**. [CrossRef] [PubMed]

77. Nishimura, K.; Fukagawa, T.; Takisawa, H.; Kakimoto, T.; Kanemaki, M. An auxin-based degron system for the rapid depletion of proteins in nonplant cells. *Nat. Methods* **2009**, *6*, 917–922. [CrossRef] [PubMed]

78. Kubota, T.; Katou, Y.; Nakato, R.; Shirahige, K.; Donaldson, A.D. Replication-Coupled PCNA Unloading by the Elg1 Complex Occurs Genome-wide and Requires Okazaki Fragment Ligation. *Cell Rep.* **2015**, *12*, 774–787. [CrossRef] [PubMed]

79. Yu, C.; Gan, H.; Han, J.; Zhou, Z.-X.; Jia, S.; Chabes, A.; Farrugia, G.; Ordog, T.; Zhang, Z. Strand-specific analysis shows protein binding at replication forks and PCNA unloading from lagging strands when forks stall. *Mol. Cell* **2014**, *56*, 551–563. [CrossRef] [PubMed]

80. Sikdar, N.; Banerjee, S.; Lee, K.; Wincovitch, S.; Pak, E.; Nakanishi, K.; Jasin, M.; Dutra, A.; Myung, K. DNA damage responses by human ELG1 in S phase are important to maintain genomic integrity. *Cell Cycle Georget. Tex* **2009**, *8*, 3199–3207. [CrossRef] [PubMed]

81. Bell, D.W.; Sikdar, N.; Lee, K.-Y.; Price, J.C.; Chatterjee, R.; Park, H.-D.; Fox, J.; Ishiai, M.; Rudd, M.L.; Pollock, L.M.; et al. Predisposition to cancer caused by genetic and functional defects of mammalian Atad5. *PLoS Genet.* **2011**, *7*, e1002245. [CrossRef] [PubMed]

82. Johnson, C.; Gali, V.K.; Takahashi, T.S.; Kubota, T. PCNA Retention on DNA into G2/M Phase Causes Genome Instability in Cells Lacking Elg1. *Cell Rep.* **2016**, *16*, 684–695. [CrossRef] [PubMed]

83. Mejlvang, J.; Feng, Y.; Alabert, C.; Neelsen, K.J.; Jasencakova, Z.; Zhao, X.; Lees, M.; Sandelin, A.; Pasero, P.; Lopes, M.; et al. New histone supply regulates replication fork speed and PCNA unloading. *J. Cell Biol.* **2014**, *204*, 29–43. [CrossRef] [PubMed]

84. Rowland, B.D.; Roig, M.B.; Nishino, T.; Kurze, A.; Uluocak, P.; Mishra, A.; Beckouët, F.; Underwood, P.; Metson, J.; Imre, R.; et al. Building sister chromatid cohesion: Smc3 acetylation counteracts an antiestablishment activity. *Mol. Cell* **2009**, *33*, 763–774. [CrossRef] [PubMed]

85. Zhang, J.; Shi, X.; Li, Y.; Kim, B.-J.; Jia, J.; Huang, Z.; Yang, T.; Fu, X.; Jung, S.Y.; Wang, Y.; et al. Acetylation of Smc3 by Eco1 is required for S phase sister chromatid cohesion in both human and yeast. *Mol. Cell* **2008**, *31*, 143–151. [CrossRef] [PubMed]

86. Unal, E.; Heidinger-Pauli, J.M.; Kim, W.; Guacci, V.; Onn, I.; Gygi, S.P.; Koshland, D.E. A molecular determinant for the establishment of sister chromatid cohesion. *Science* **2008**, *321*, 566–569.

87. Rolef Ben-Shahar, T.; Heeger, S.; Lehane, C.; East, P.; Flynn, H.; Skehel, M.; Uhlmann, F. Eco1-dependent cohesin acetylation during establishment of sister chromatid cohesion. *Science* **2008**, *321*, 563–566. [CrossRef] [PubMed]

88. García-Rodríguez, L.J.; De Piccoli, G.; Marchesi, V.; Jones, R.C.; Edmondson, R.D.; Labib, K. A conserved Polε binding module in Ctf18-RFC is required for S-phase checkpoint activation downstream of Mec1. *Nucleic Acids Res.* **2015**, *43*, 8830–8838. [CrossRef] [PubMed]

89. Okimoto, H.; Tanaka, S.; Araki, H.; Ohashi, E.; Tsurimoto, T. Conserved interaction of Ctf18-RFC with DNA polymerase ε is critical for maintenance of genome stability in *Saccharomyces cerevisiae*. *Genes Cells Devoted Mol. Cell. Mech.* **2016**, *21*, 482–491. [CrossRef] [PubMed]

90. Moldovan, G.-L.; Pfander, B.; Jentsch, S. PCNA controls establishment of sister chromatid cohesion during S phase. *Mol. Cell* **2006**, *23*, 723–732. [CrossRef] [PubMed]

91. Jin, J.; Arias, E.E.; Chen, J.; Harper, J.W.; Walter, J.C. A family of diverse Cul4-Ddb1-interacting proteins includes Cdt2, which is required for S phase destruction of the replication factor Cdt1. *Mol. Cell* **2006**, *23*, 709–721. [CrossRef] [PubMed]

92. Nishitani, H.; Lygerou, Z.; Nishimoto, T.; Nurse, P. The Cdt1 protein is required to license DNA for replication in fission yeast. *Nature* **2000**, *404*, 625–628. [PubMed]

93. Havens, C.G.; Walter, J.C. Docking of a specialized PIP Box onto chromatin-bound PCNA creates a degron for the ubiquitin ligase CRL4Cdt2. *Mol. Cell* **2009**, *35*, 93–104. [CrossRef] [PubMed]

94. Michishita, M.; Morimoto, A.; Ishii, T.; Komori, H.; Shiomi, Y.; Higuchi, Y.; Nishitani, H. Positively charged residues located downstream of PIP box, together with TD amino acids within PIP box, are important for CRL4(Cdt2) -mediated proteolysis. *Genes Cells Devoted Mol. Cell. Mech.* **2011**, *16*, 12–22. [CrossRef] [PubMed]

95. Abbas, T.; Sivaprasad, U.; Terai, K.; Amador, V.; Pagano, M.; Dutta, A. PCNA-dependent regulation of p21 ubiquitylation and degradation via the CRL4Cdt2 ubiquitin ligase complex. *Genes Dev.* **2008**, *22*, 2496–2506. [CrossRef] [PubMed]

96. Abbas, T.; Shibata, E.; Park, J.; Jha, S.; Karnani, N.; Dutta, A. CRL4(Cdt2) regulates cell proliferation and histone gene expression by targeting PR-Set7/Set8 for degradation. *Mol. Cell* **2010**, *40*, 9–21.

97. Centore, R.C.; Havens, C.G.; Manning, A.L.; Li, J.-M.; Flynn, R.L.; Tse, A.; Jin, J.; Dyson, N.J.; Walter, J.C.; Zou, L. CRL4(Cdt2)-mediated destruction of the histone methyltransferase Set8 prevents premature chromatin compaction in S phase. *Mol. Cell* **2010**, *40*, 22–33. [CrossRef] [PubMed]

98. Kim, Y.; Starostina, N.G.; Kipreos, E.T. The CRL4Cdt2 ubiquitin ligase targets the degradation of p21Cip1 to control replication licensing. *Genes Dev.* **2008**, *22*, 2507–2519. [CrossRef] [PubMed]

99. Nishitani, H.; Shiomi, Y.; Iida, H.; Michishita, M.; Takami, T.; Tsurimoto, T. CDK inhibitor p21 is degraded by a proliferating cell nuclear antigen-coupled Cul4-DDB1Cdt2 pathway during S phase and after UV irradiation. *J. Biol. Chem.* **2008**, *283*, 29045–29052. [CrossRef] [PubMed]

100. Oda, H.; Hübner, M.R.; Beck, D.B.; Vermeulen, M.; Hurwitz, J.; Spector, D.L.; Reinberg, D. Regulation of the histone H4 monomethylase PR-Set7 by CRL4(Cdt2)-mediated PCNA-dependent degradation during DNA damage. *Mol. Cell* **2010**, *40*, 364–376. [CrossRef] [PubMed]

101. Tardat, M.; Brustel, J.; Kirsh, O.; Lefevbre, C.; Callanan, M.; Sardet, C.; Julien, E. The histone H4 Lys 20 methyltransferase PR-Set7 regulates replication origins in mammalian cells. *Nat. Cell Biol.* **2010**, *12*, 1086–1093. [CrossRef] [PubMed]

102. Shiomi, Y.; Hayashi, A.; Ishii, T.; Shinmyozu, K.; Nakayama, J.; Sugasawa, K.; Nishitani, H. Two different replication factor C proteins, Ctf18 and RFC1, separately control PCNA-CRL4Cdt2-mediated Cdt1 proteolysis during S phase and following UV irradiation. *Mol. Cell. Biol.* **2012**, *32*, 2279–2288. [CrossRef] [PubMed]

103. Aboussekhra, A.; Biggerstaff, M.; Shivji, M.K.; Vilpo, J.A.; Moncollin, V.; Podust, V.N.; Protić, M.; Hübscher, U.; Egly, J.M.; Wood, R.D. Mammalian DNA nucleotide excision repair reconstituted with purified protein components. *Cell* **1995**, *80*, 859–868. [CrossRef]

104. Corrette-Bennett, S.E.; Borgeson, C.; Sommer, D.; Burgers, P.M.J.; Lahue, R.S. DNA polymerase delta, RFC and PCNA are required for repair synthesis of large looped heteroduplexes in Saccharomyces cerevisiae. *Nucleic Acids Res.* **2004**, *32*, 6268–6275. [CrossRef] [PubMed]

105. Hashiguchi, K.; Matsumoto, Y.; Yasui, A. Recruitment of DNA repair synthesis machinery to sites of DNA damage/repair in living human cells. *Nucleic Acids Res.* **2007**, *35*, 2913–2923. [CrossRef] [PubMed]

106. Holmes, A.M.; Haber, J.E. Double-strand break repair in yeast requires both leading and lagging strand DNA polymerases. *Cell* **1999**, *96*, 415–424. [CrossRef]

107. Kadyrov, F.A.; Dzantiev, L.; Constantin, N.; Modrich, P. Endonucleolytic function of MutLalpha in human mismatch repair. *Cell* **2006**, *126*, 297–308. [CrossRef] [PubMed]

108. Kolodner, R.D.; Marsischky, G.T. Eukaryotic DNA mismatch repair. *Curr. Opin. Genet. Dev.* **1999**, *9*, 89–96.

109. Matsumoto, Y. Molecular mechanism of PCNA-dependent base excision repair. *Prog. Nucleic Acid Res. Mol. Biol.* **2001**, *68*, 129–138. [PubMed]

110. Ogi, T.; Limsirichaikul, S.; Overmeer, R.M.; Volker, M.; Takenaka, K.; Cloney, R.; Nakazawa, Y.; Niimi, A.; Miki, Y.; Jaspers, N.G.; et al. Three DNA polymerases, recruited by different mechanisms, carry out NER repair synthesis in human cells. *Mol. Cell* **2010**, *37*, 714–727. [CrossRef] [PubMed]

111. Overmeer, R.M.; Gourdin, A.M.; Giglia-Mari, A.; Kool, H.; Houtsmuller, A.B.; Siegal, G.; Fousteri, M.I.; Mullenders, L.H.F.; Vermeulen, W. Replication factor C recruits DNA polymerase delta to sites of nucleotide excision repair but is not required for PCNA recruitment. *Mol. Cell. Biol.* **2010**, *30*, 4828–4839. [CrossRef] [PubMed]

112. Peng, Z.; Liao, Z.; Dziegielewska, B.; Matsumoto, Y.; Thomas, S.; Wan, Y.; Yang, A.; Tomkinson, A.E. Phosphorylation of serine 51 regulates the interaction of human DNA ligase I with replication factor C and its participation in DNA replication and repair. *J. Biol. Chem.* **2012**, *287*, 36711–36719. [CrossRef] [PubMed]

113. Pluciennik, A.; Dzantiev, L.; Iyer, R.R.; Constantin, N.; Kadyrov, F.A.; Modrich, P. PCNA function in the activation and strand direction of MutLα endonuclease in mismatch repair. *Proc. Natl. Acad. Sci. USA* **2010**, *107*, 16066–16071. [CrossRef] [PubMed]

114. Shivji, M.K.; Podust, V.N.; Hübscher, U.; Wood, R.D. Nucleotide excision repair DNA synthesis by DNA polymerase epsilon in the presence of PCNA, RFC, and RPA. *Biochemistry (Mosc.)* **1995**, *34*, 5011–5017.

115. Ogiwara, H.; Ui, A.; Enomoto, T.; Seki, M. Role of Elg1 protein in double strand break repair. *Nucleic Acids Res.* **2007**, *35*, 353–362. [CrossRef] [PubMed]

116. Ma, L.; Zhai, Y.; Feng, D.; Chan, T.; Lu, Y.; Fu, X.; Wang, J.; Chen, Y.; Li, J.; Xu, K.; et al. Identification of novel factors involved in or regulating initiation of DNA replication by a genome-wide phenotypic screen in Saccharomyces cerevisiae. *Cell Cycle Georget. Tex* **2010**, *9*, 4399–4410. [CrossRef] [PubMed]

117. Crabbé, L.; Thomas, A.; Pantesco, V.; De Vos, J.; Pasero, P.; Lengronne, A. Analysis of replication profiles reveals key role of RFC-Ctf18 in yeast replication stress response. *Nat. Struct. Mol. Biol.* **2010**, *17*, 1391–1397. [CrossRef] [PubMed]

118. Gellon, L.; Razidlo, D.F.; Gleeson, O.; Verra, L.; Schulz, D.; Lahue, R.S.; Freudenreich, C.H. New functions of Ctf18-RFC in preserving genome stability outside its role in sister chromatid cohesion. *PLoS Genet.* **2011**, *7*, e1001298. [CrossRef] [PubMed]

119. Hiraga, S.; Robertson, E.D.; Donaldson, A.D. The Ctf18 RFC-like complex positions yeast telomeres but does not specify their replication time. *EMBO J.* **2006**, *25*, 1505–1514. [CrossRef] [PubMed]

120. Terret, M.-E.; Sherwood, R.; Rahman, S.; Qin, J.; Jallepalli, P.V. Cohesin acetylation speeds the replication fork. *Nature* **2009**, *462*, 231–234. [CrossRef] [PubMed]

121. Gomes, X.V.; Gary, S.L.; Burgers, P.M. Overproduction in *Escherichia coli* and characterization of yeast replication factor C lacking the ligase homology domain. *J. Biol. Chem.* **2000**, *275*, 14541–14549.

122. Kobayashi, M.; Ab, E.; Bonvin, A.M.J.J.; Siegal, G. Structure of the DNA-bound BRCA1 C-terminal region from human replication factor C p140 and model of the protein-DNA complex. *J. Biol. Chem.* **2010**, *285*, 10087–10097. [CrossRef] [PubMed]

123. Huang, F.; Saraf, A.; Florens, L.; Kusch, T.; Swanson, S.K.; Szerszen, L.T.; Li, G.; Dutta, A.; Washburn, M.P.; Abmayr, S.M.; et al. The Enok acetyltransferase complex interacts with Elg1 and negatively regulates PCNA unloading to promote the G1/S transition. *Genes Dev.* **2016**, *30*, 1198–1210. [CrossRef] [PubMed]

124. Davidson, M.B.; Brown, G.W. The N- and C-termini of Elg1 contribute to the maintenance of genome stability. *DNA Repair* **2008**, *7*, 1221–1232. [CrossRef] [PubMed]

125. Lee, K.-Y.; Yang, K.; Cohn, M.A.; Sikdar, N.; D'Andrea, A.D.; Myung, K. Human ELG1 regulates the level of ubiquitinated proliferating cell nuclear antigen (PCNA) through Its interactions with PCNA and USP1. *J. Biol. Chem.* **2010**, *285*, 10362–10369. [CrossRef] [PubMed]

126. Levin, D.S.; Vijayakumar, S.; Liu, X.; Bermudez, V.P.; Hurwitz, J.; Tomkinson, A.E. A conserved interaction between the replicative clamp loader and DNA ligase in eukaryotes: Implications for Okazaki fragment joining. *J. Biol. Chem.* **2004**, *279*, 55196–55201. [CrossRef] [PubMed]

127. Franco, A.A.; Lam, W.M.; Burgers, P.M.; Kaufman, P.D. Histone deposition protein Asf1 maintains DNA replisome integrity and interacts with replication factor C. *Genes Dev.* **2005**, *19*, 1365–1375.

128. Anderson, L.A.; Perkins, N.D. The large subunit of replication factor C interacts with the histone deacetylase, HDAC1. *J. Biol. Chem.* **2002**, *277*, 29550–29554. [CrossRef] [PubMed]

129. Pennaneach, V.; Salles-Passador, I.; Munshi, A.; Brickner, H.; Regazzoni, K.; Dick, F.; Dyson, N.; Chen, T.T.; Wang, J.Y.; Fotedar, R.; et al. The large subunit of replication factor C promotes cell survival after DNA damage in an LxCxE motif- and Rb-dependent manner. *Mol. Cell* **2001**, *7*, 715–727. [CrossRef]

130. Hong, S.; Park, S.J.; Kong, H.J.; Shuman, J.D.; Cheong, J. Functional interaction of bZIP proteins and the large subunit of replication factor C in liver and adipose cells. *J. Biol. Chem.* **2001**, *276*, 28098–28105.

131. Anderson, L.A.; Perkins, N.D. Regulation of RelA (p65) function by the large subunit of replication factor C. *Mol. Cell. Biol.* **2003**, *23*, 721–732. [CrossRef] [PubMed]

132. Shiomi, Y.; Masutani, C.; Hanaoka, F.; Kimura, H.; Tsurimoto, T. A second proliferating cell nuclear antigen loader complex, Ctf18-replication factor C, stimulates DNA polymerase eta activity. *J. Biol. Chem.* **2007**, *282*, 20906–20914. [CrossRef] [PubMed]

![genes logo](GCAT TACG GCAT *genes*)

MDPI

Article

Two Archaeal RecJ Nucleases from *Methanocaldococcus jannaschii* Show Reverse Hydrolysis Polarity: Implication to Their Unique Function in Archaea

Gang-Shun Yi [1,†], Yang Song [2,†], Wei-Wei Wang [1], Jia-Nan Chen [1], Wei Deng [2], Weiguo Cao [3], Feng-Ping Wang [1,4], Xiang Xiao [1,4] and Xi-Peng Liu [1,4,*]

1   State Key Laboratory of Microbial Metabolism, School of Life Sciences and Biotechnology, Shanghai Jiao Tong University, 800 Dong-Chuan Road, Shanghai 200240, China; 13166228531@163.com (G.-S.Y.); www1037554814@sjtu.edu.cn (W.-W.W.); 14330101150369@sjtu.edu.cn (J.-N.C.); fengpingw@sjtu.edu.cn (F.-P.W.); xoxiang@sjtu.edu.cn (X.X.)
2   National Center for Protein Science, Chinese Academy of Sciences, Shanghai 201204, China; songyang@sibcb.ac.cn (Y.S.); dengwei@sibcb.ac.cn (W.D.)
3   Department of Genetics and Biochemistry, Clemson University, Clemson, SC 29634, USA; wgc@clemson.edu
4   State Key Laboratory of Ocean Engineering, School of Naval Architecture, Ocean and Civil Engineering, Shanghai Jiao Tong University, 800 Dong-Chuan Road, Shanghai 200240, China
*   Correspondence: xpliu@sjtu.edu.cn; Tel.: +86-21-3420-7205
†   These authors contributed equally to this work.

Academic Editor: Eishi Noguchi
Received: 24 June 2017; Accepted: 17 August 2017; Published: 24 August 2017

Abstract: Bacterial nuclease RecJ, which exists in almost all bacterial species, specifically degrades single-stranded (ss) DNA in the $5'$ to $3'$ direction. Some archaeal phyla, except Crenarchaea, also encode RecJ homologs. Compared with bacterial RecJ, archaeal RecJ exhibits a largely different amino acid sequence and domain organization. Archaeal RecJs from *Thermococcus kodakarensis* and *Pyrococcus furiosus* show $5' \rightarrow 3'$ exonuclease activity on ssDNA. Interestingly, more than one RecJ exists in some Euryarchaeota classes, such as Methanomicrobia, Methanococci, Methanomicrobia, Methanobacteria, and Archaeoglobi. Here we report the biochemical characterization of two RecJs from *Methanocaldococcus jannaschii*, the long RecJ1 (MJ0977) and short RecJ2 (MJ0831) to understand their enzymatic properties. RecJ1 is a $5' \rightarrow 3'$ exonuclease with a preference to ssDNA; however, RecJ2 is a $3' \rightarrow 5'$ exonuclease with a preference to ssRNA. The $5'$ terminal phosphate promotes RecJ1 activity, but the $3'$ terminal phosphate inhibits RecJ2 nuclease. Go-Ichi-Ni-San (GINS) complex does not interact with two RecJs and does not promote their nuclease activities. Finally, we discuss the diversity, function, and molecular evolution of RecJ in archaeal taxonomy. Our analyses provide insight into the function and evolution of conserved archaeal RecJ/eukaryotic Cdc45 protein.

Keywords: archaeal RecJ; Cdc45-MCM-GINS; nuclease; GINS; interaction

1. Introduction

Nucleases, including endonuclease and exonuclease, play important roles in DNA recombination and repair, degradation and recycling of DNA and RNA, and maturation of RNA and Okazaki fragments [1]. RecJ is a kind of nuclease involved in three DNA repair pathways: homologous recombination, mismatch repair (MMR), and base excision repair. RecJ nuclease belongs to the DHH phosphodiesterase superfamily with a conserved signature motif DHH. DHH motif is consisted of three successive conserved residues located at the corresponding N-terminal DHH domain, and the DHHA motif, located at the corresponding C-terminal domain, is a typical signature motif for classifying family of the DHH phosphodiesterase superfamily. Based on the sequence difference of the DHHA motif, the DHH phosphodiesterase superfamily can be split into DHHA1 and DHHA2 groups. The DHHA1 group has a typical DHHA1 motif of GGGHXXAAG, whereas the DHHA2 group lacks this typical motif or has a divergent or atypical motif. Based on the difference in biochemical properties and conserved motifs, DHH phosphodiesterases are classified into four families. Family 1 includes prokaryotic RecJ nuclease and eukaryotic Cdc45 protein [1–4], Family 2 is composed of various nanoRNases (Nrn), including NrnA [5] and NrnB [6], which specifically degrade short single-stranded (ss) RNA molecule [5–7]. Family 3 degrades the nucleotide derivatives, but not oligonucleotides, and includes eukaryotic Prune and PPX1 [8] and prokaryotic family II inorganic pyrophosphatase [9]. Family 4, HAN nuclease [10], is a fused protein containing an N-terminal domain and the C-terminal DHH phosphodiesterase domain, and is specific to archaea kingdom.

The family 1 DHH phosphodiesterase includes three subfamilies: bacterial RecJ, archaeal RecJ and eukaryotic Cdc45 protein. RecJ has a typical DHHA1 motif, but Cdc45 does not. Bacterial RecJ nuclease shows both single-stranded DNA (ssDNA)-specific $5' \rightarrow 3'$ exonuclease and deoxyribose phosphatase (dRPase) activities [3]. Its ssDNA-specific $5' \rightarrow 3'$ exonuclease is responsible for generating a long $3'$ ssDNA for strand invasion in homologous recombination [11], or a long ssDNA gap for DNA resynthesis by DNA polymerase in MMR [12]. The $5'$ dRPase of RecJ removes deoxyribose phosphate of the single-strand break generated by the cleavage of an abasic site by apurinic/apyrimidinic (AP) endonucleases in base excision repair [13]. Structurally, most bacterial RecJs, such as *Escherichia coli* RecJ, feature an N-terminal catalytic core, which consists of two domains of DHH and DHHA interconnected by a long helix, and a C-terminal oligonucleotide/oligosaccharide-binding (OB) domain that improves ssDNA-binding capability. Some bacterial RecJs, such as the RecJs of *Thermus thermophilus* and *Deinococcus radiodurans*, have an additional C-terminal domain [14,15]. The C-terminal domain IV of *D. radiodurans* RecJ (DrRecJ) can increase the $5' \rightarrow 3'$ nuclease activity by promoting ssDNA substrate binding and interacting with the HerA helicase [15].

Compared with the bacterial RecJ nuclease, little is known about archaeal RecJ nucleases. Research on archaeal RecJs mainly focused on their $5' \rightarrow 3'$ exonuclease activity on ssDNA [2,16], $3' \rightarrow 5'$ exonuclease activity on single-stranded RNA (ssRNA), and mismatched ribonucleotide of RNA/DNA hybrids [17]. The $3' \rightarrow 5'$ exonuclease on RNA possibly removes $3'$-mismatched ribonucleotides from the RNA primers in chromosomal DNA replication or is involved in the degradation of diverse ssRNAs [17]. The two potential *recj* genes from *Methanocaldococcus jannaschii* DSM 2661 can supply the capability of DNA recombination repair in an *recj*-deleted *E. coli* strain [16]. Unlike bacterial RecJs, archaeal RecJ nucleases only have two domains corresponding to the bacterial catalytic core domains of DHH and DHHA but lack the OB domain [14,17,18]. Moreover, archaeal RecJ proteins are longer by approximately 100 amino acid residues than the bacterial RecJ catalytic core domain [17]. This additional sequence forms a single domain, the minichromosome maintenance (MCM)-binding domain (MBD), in the topological structure of archaeal RecJ from *Thermococcus kodakarensis* [19], and occupies a location similar to the OB-fold domain of DrRecJ and *T. thermophilus* RecJ (TthRecJ) [14,20].

Despite the broad distribution of RecJ nuclease in bacteria and archaea, RecJ homolog does not exist in eukaryotes. Cdc45, an essential replication initiation protein whose site-mutations result in partial defect in DNA replication [21], shows low-sequence similarity to the conserved catalytic core of the RecJ nuclease subfamily; however, Cdc45 lacks most of the conserved motifs and residues that are essential for prokaryotic enzymatic activity [4,22]. Despite the lack of nuclease activity, Cdc45 retains ssDNA- and ssRNA-binding capability and functions as molecular wedge for DNA unwinding [22,23]. Recently, three groups of researchers reported their results on the structures of bacterial RecJs, archaeal RecJs, and human Cdc45 protein [19,20,24]. These proteins exhibited a similar overall topology, indicating their evolution from a common ancestor.

In addition to nuclease activity, archaeal RecJ also interacts with some subunits of DNA replisome, such as the Go-Ichi-Ni-San (GINS) complex, a central component in the archaeal DNA replication fork and replicative MCM helicase [2,25,26], to form a multi-subunit complex RecJ-MCM-GINS (RMG) [19,25]. Similar to archaeal RecJ, eukaryotic Cdc45 also interacts with MCM2–7 and GINS to form a complex Cdc45-MCM-GINS (CMG), which is believed to act as a DNA helicase during chromosome replication [27,28]. The crystal structure of human Cdc45 and cryo-electron microscopy (EM) structure of CMG provide not only a better understanding of the mechanism of subunit interaction in the CMG complex [24,29,30], but also clues regarding the subunit interaction in RMG [19].

BLAST with the *Pyrococcus furiosus* RecJ as a query sequence, it identified more than one RecJ gene in some archaea genomes, especially the methane-producing species. Previous works also found out the diversity of archaeal RecJ [16,31]. During the preparation of our manuscript, Ishino and coworker reported the biochemical characterization of two RecJs from *Thermoplasma acidophilum* [31]. TacRecJ1 is a ssDNA specific 5'exonuclease, and TacRecJ2 is a 3' exonuclease on both ssDNA and ssRNA. On the two RecJ nucleases from *M. jannaschii*, although they were primarily characterized [16], the protein preparations were largely impure, just the cell extract of an *E. coli* that was deleted the *recj* gene and supplied with one of two *M. jannaschii recj* genes [16]. To fully understand the enzymatic properties of two *M. jannaschii* RecJs, we recombinantly expressed, purified and biochemically characterized them in detail. Both RecJs are single-stranded DNA/RNA specific nucleases. RecJ1 (MJ0977) is a $5' \rightarrow 3'$ exonuclease with a weak preference to DNA, and RecJ2 (MJ0831) is a $3' \rightarrow 5'$ exonuclease with a preference to RNA. The terminal phosphate affected enzymatic activity differently. The 5' terminal phosphate promotes RecJ1 activity, but the 3' terminal phosphate inhibits RecJ2 nuclease. The GINS does not interact with either RecJ and thus does not promote their nuclease activities on ssDNA and ssRNA. Finally, the diversity, function in DNA repair, and molecular evolution of RecJ in archaeal taxonomy are discussed. Our results provide new clues to understand the functions of archaeal RecJ in nucleic acid metabolism and its evolution relationship with bacterial RecJ and eukaryotic Cdc45 protein.

2. Materials and Methods

2.1. Materials

KOD-plus DNA polymerase was purchased from Toyobo (Osaka, Japan). Nickel–nitrilotriacetic acid resin was purchased from Bio-Rad (Hercules, CA, USA). RNase A inhibitor was purchased from Takara (Shiga, Japan). Oligodeoxyribonucleotides and oligoribonucleotides (Table S1) were synthesized by Invitrogen (Carlsbad, CA, USA) and Takara (Shiga, Japan), respectively. The expression vectors of pDEST17 (Invitrogen) and pET28-sumo were used throughout this study. *E. coli* strain DH5α was used in the gene cloning and Rosetta 2(DE3)pLysS (Novagen) strain was used to express recombinant protein. All other chemicals and reagents were of analytic grade.

2.2. Preparation of Recombinant Proteins

Genes encoding for the archaeal RecJ nucleases (MJ0831 and MJ0977) and GINS (MJ0248) were amplified from *M. jannaschii* genomic DNA by polymerase chain reaction (PCR) using their respective primers (Table S1) and then inserted into pDEST17 or pET28-sumo, as described previously [17]. Amino acid substitutions were introduced into RecJs by PCR-mediated mutagenesis using KOD-plus DNA polymerase and the appropriate primers (Table S1). Nucleotide sequences were confirmed by DNA sequencing.

Recombinant plasmids were introduced into the Rosetta 2(DE3)pLysS strain of *E. coli* to express recombinant proteins. The expressions of recombinant proteins were induced by 0.5 mM isopropylthio-β-galactoside. The recombinant proteins were purified via immobilized Ni^{2+} affinity chromatography. The affinity purification was performed as follows: bacterial pellet was suspended in lysis buffer (20 mM Tris-HCl, pH 8.0; 0.3 M NaCl, 5 mM mercaptoethanol, 5 mM imidazole, 1 mM phenylmethylsulfonyl fluoride, and 10% glycerol) and then disrupted by sonication. After incubation for 30 min at 65 °C (not conducted for MjaGINS-sumo protein), cell extract was clarified by centrifugation at $12,000 \times g$ for 30 min. After loading the supernatant onto a column pre-equilibrated with lysis buffer, the resin was washed with >25 column volumes of lysis buffer containing 20 mM imidazole. Finally, bound proteins were eluted from the column using elution buffer (20 mM Tris-HCl, pH 8.0; 0.3 M NaCl, 5 mM mercaptoethanol, 200 mM imidazole, and 10% glycerol). After verifying the purity of eluate using 15% sodium dodecyl sulfate polyacrylamide gel electrophoresis (SDS-PAGE), samples were dialyzed against a storage buffer (20 mM Tris-HCl, pH 8.0; 0.1 M NaCl, and 50% glycerol) and stored in small aliquots at -20 °C.

2.3. Characterization of Methanocaldococcus jannaschii Enzymes

MJ0831 and MJ0977 were characterized in a standard reaction buffer consisting of 20 mM Tris-HCl (pH 7.5), 50 mM NaCl, 1 mM dithiothreitol (DTT), 2.0 mM $MnCl_2$, and 100 ng/μL BSA before optimization. Then, the pH value, ions strength, reaction temperature, and divalent ions were optimized on the basis of standard reaction buffer. Table S1 presents the oligoribonucleotides and oligodeoxyribonucleotides used in exonuclease activity assays. The dependence of activity on substrate structures on was characterized using ssDNA, double-stranded DNA (dsDNA), and dsDNA with $3'$ or $5'$ overhang. The effect of MjaGINS on two MjaRecJ nucleases was determined by assaying nuclease activity in the presence of increasing concentrations of MjaGINS. After incubation for a specified time at 50 °C, an equal volume of stopping buffer (90% formamide, 100 mM EDTA, and 0.2% SDS) was added to the reaction. Subsequently, the reactions were subjected to 15% 8 M urea-denatured polyacrylamide gel electrophoresis (PAGE). After electrophoresis, images of the gels were quantitated using FL5000 Fluorescent Scanner (FUJIFILM, Tokyo, Japan).

2.4. Determining the Interaction between MjaRecJs and MjaGINS

Two experiments were used to identify the possible interactions of the two MjaRecJs and MjaGINS. First, co-purification of RecJ and GINS was conducted to determine the interaction between RecJs and GINS. During co-purification, the purified MjaRecJ with a 6×His tag was mixed with the GINS, whose 6×His tag and sumo domain were removed by Uip protease, in a molecular ratio of 1:2. The mixtures were incubated at 37 °C for 30 min to form the possible complex. If GINS interacts with RecJ, It will be co-purified by the 6×His-RecJ. Second, RecJ and GINS were purified separately and then mixed in a molecular ratio of 1:2 to permit the formation of a possible complex. To check the existence of RecJ–GINS complex, Gel filtration chromatography was performed using a Hiload Superdex 200 column (GE Healthcare, Pittsburgh, PA, USA) pre-equilibrated with 20 mM HEPES (pH 7.0), 100 mM NaCl, 1 mM DTT, 0.1 mM EDTA, and 2% glycerol.

3. Results

3.1. Substrate Preferences of two MjaRecJs

Some archaeal species, such as *M. jannaschii* DSM 2661, contain more than one RecJ gene. Both RecJs of *M. jannaschii* have classical domain combinations, including the MBD domain specific to archaeal RecJ and Cdc45 protein. Two *M. jannaschii* RecJs, MjaRecJ1 (MJ0977), and MjaRecJ2 (MJ0831), have lower sequence similarity of approximately 30%, and show lower similarity to *T. kodakaraensis* RecJ (TkoRecJ), which is the only RecJ nuclease in *T. kodakaraensis* (Figure 1A). The two MjaRecJs have seven conserved motifs (I–VII), such as DHH (motif III) and DHHA1 (motif VII), which are common among many DHH phosphodiesterase families. TkoRecJ and MjaRecJs are different with regard to the conserved residues responsible for interacting with GINS (Figure 1A, red). A complete phylogenetic analysis of RecJs showed that RecJ1 and RecJ2 from some archaeal groups belong to two different branches. The RecJs from some archaea that contain a single *recj* gene such as TkoRecJ and *P. furiosus* RecJ (PfuRecJ), belong to the RecJ2 subfamily (Figure 1B). The bacterial RecJs form a distinct evolutionary branch that does not belong to any of archaeal RecJ groups.

To understand their enzymatic function, the two MjaRecJ proteins were recombinantly expressed, purified and biochemically characterized (Figure 1C). Activity assays confirmed that both MjaRecJs showed nuclease activity on ssDNA in opposite direction. *M. jannaschii* RecJ1 was probably a $5' \rightarrow 3'$ exonuclease (Figure 1D), and MjaRecJ2 was probably a $3' \rightarrow 5'$ exonuclease (Figure 1E). Their hydrolysis polarity was further confirmed using phosphothioate-modified substrates in next section. Changing the conserved motif the DHH to three alanines deprived the nuclease activity, indicating that DHH motif is essential for the nuclease activity (Figure 1D–E).

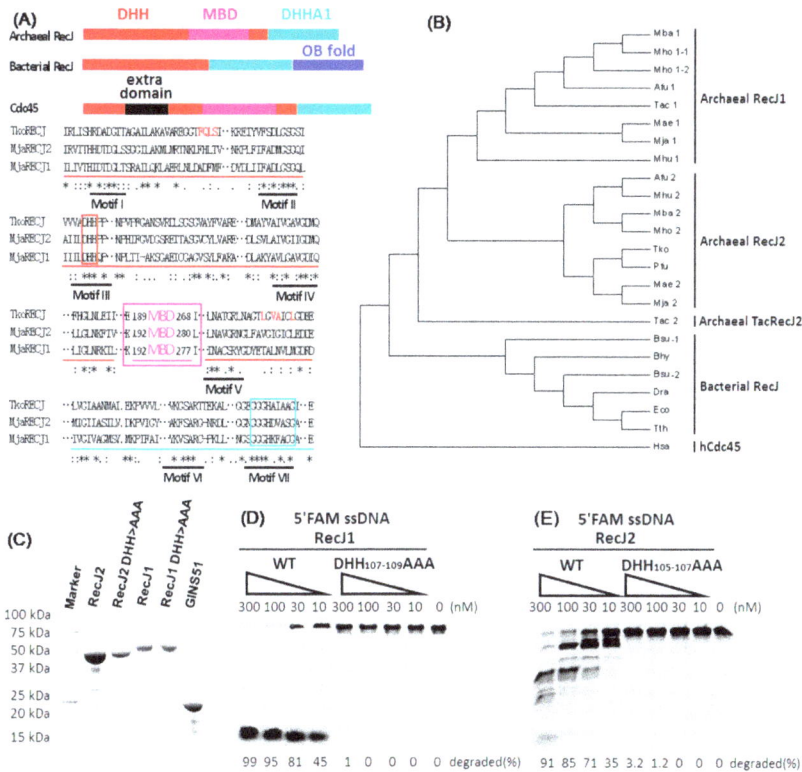

Figure 1. Two RecJs from *Methanocaldococcus jannaschii* demonstrate nuclease activity. (**A**) Multi-alignment of three archaeal RecJs. The domain combinations of archaeal and bacterial RecJs, and eukaryotic Cdc45 protein is compared on the top of panel A, and the names of each domain are indicated. For multi-alignment of RecJs, red, purple and cyan lines are used to represent the domains of DHH, MBD, DHHA, respectively. The conserved motifs are marked by black lines, and the motifs of DHH and DHHA1 are highlighted with red and cyan box, respectively. The middle domain, the MCM helicase Binding Domain (MBD) and its boundary are highlighted by a purple box with the indicated number of residues. The residues responsible for interaction with GINS51 subunit are shown in red. (**B**) The phylogenetic tree of RecJ homologs is built based on multi-alignment of these sequences. Archaeal RecJ1 and RecJ2 are classified based on their sequence similarity to *M. jannaschii* RecJ1 and RecJ2. RecJ homologs come from archaea of *Methanocaldococcus jannaschii* DSM 2661 (Mja), *Methanococcus aeolicus* Nankai-3 (Mae), *Methanospirillum hungatei* JF-1 (Mhu), *Methanosarcina barkeri* strain (str.) Fusaro (Mba), *Thermoplasma acidophilum* DSM 1728 (Tac), *Archaeoglobus fulgidus* DSM 4304 (Afu), *Methanomethylovorans hollandica* DSM 15,978 (Mho), *Pyrococcus furiosus* DSM 3638 (Pfu), and *Thermococcus kodakarensis* KOD1 (Tko); bacteria of *Deinococcus radiodurans* R1 (Dra), *Thermus thermophilus* HB8 (Tth), *Bacillus subtilis* str. 168 (Bsu), *Escherichia coli* K12 (Eco), and *Brachyspira hyodysenteriae* WA1 (Bhy); human Cdc45 (Hsa). (**C**) Expression and affinity purification of five *M. jannaschii* recombinant proteins. Increased amounts of wild-type (WT) or DHH motif mutated MjaRecJ1 (**D**) or MjaRecJ2 (**E**) were incubated with 200 nM 23 nt 5'FAM-labeled ssDNA substrates at 55 °C for 20 min in a standard reaction buffer. The degraded amount of substrate was quantified and listed at the bottom of the panel.

After positive detection of the nuclease activity, which is consistent with previous results [16], optimal reaction parameters with regard to pH, ion strength, divalent ions, and reaction temperature were determined for the two MjaRecJs (Figure 2). The RecJs displayed the highest activity at pH 8.0 (MjaRecJ1, Figure S1A) and 8.5 (MjaRecJ2, Figure S2A). Divalent ion manganese Mn^{2+} was the most effective metal cofactor (Figure S1B or Figure S2B), with the optimal concentration at 2.0 mM for MjaRecJ1 (Figure S1C) and 1.0 mM for MjaRecJ2 (Figure S2C), respectively. The two RecJs showed higher activity at lower concentrations of NaCl (Figure S1D or Figure S2D). Their optimal reaction temperatures differed. MjaRecJ1 and MjaRecJ2 showed the highest activities at 65 °C (Figure S1E) and 85 °C (Figure S2E), respectively. MjaRecJ2 is more thermostable than MjaRecJ1 (Figure S3); the result is consistent with those for the optimal reaction temperatures.

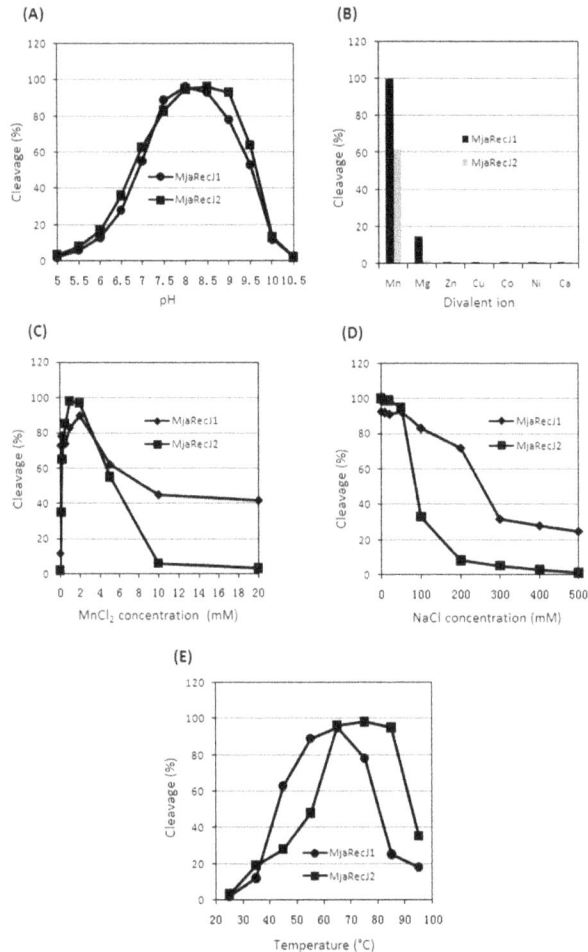

Figure 2. Optimization of ssDNA hydrolysis by MjaRecJs. pH value (**A**), divalent ions (**B**), concentration of divalent manganese ions (**C**), ion strength (**D**), and reaction temperature (**E**) were optimized for nuclease activities of two MjaRecJs (40 nM MjaRecJ1 or 50 nM MjaRecJ2) using a 23 nt single-stranded DNA (ssDNA) as substrate (200 nM). The degraded amount of substrate DNA was quantified and plotted vs. each value.

Since some nucleases hydrolyze both DNA and RNA, the (deoxy)ribose dependency of the two RecJs were characterized using ssDNA and ssRNA as substrates. MjaRecJs had a different (deoxy)ribose dependence as compared with bacterial RecJ, which only hydrolyzes ssDNA [3,14,15,20]. MjaRecJ1 could hydrolyze both ssDNA and ssRNA from the 5′ side (Figure 3A). MjaRecJ2 favored ssRNA hydrolysis with a clearly increased rate as compared with ssDNA substrate (Figure 3B). Therefore, the different (deoxy)ribose preferences of two MjaRecJs may suggest their different roles in nucleic acid metabolism in vivo.

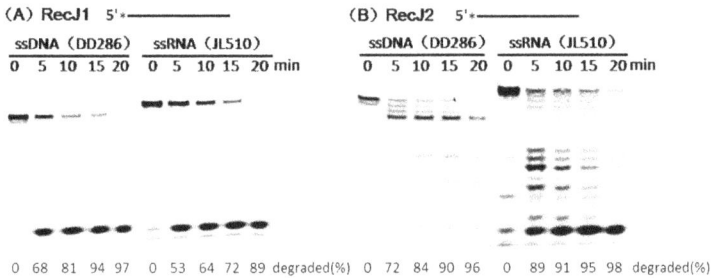

0 68 81 94 97   0 53 64 72 89 degraded(%)   0 72 84 90 96   0 89 91 95 98 degraded(%)

Figure 3. Preferences for (deoxy)ribose of two MjaRecJs. Reactions were performed at 55 °C with increasing time using 200 nM 23 nt 5′-FAM labeled ssDNA and ssRNA as substrates in their respective reaction buffer. 40 nM MjaRecJ1 and 50 nM MjaRecJ2 were used, respectively, in each reaction. The degraded amount of substrate was quantified at each time and listed at the bottom of the panel.

3.2. Hydrolysis Polarity of Two MjaRecJs

The fully phosphothioate-modified ssDNA and ssRNA were used as substrates to verify in detail the hydrolysis direction of two MjaRecJs. The existence of phosphothioate groups largely decreased the enzymatic hydrolysis rate and allowed capturing the image of each product during substrate degradation. For ssDNA degradation by MjaRecJ1 (Figure 4A), DNA ladders were generated from the 3′-FAM-labeled ssDNA, and only 1 nt products were generated from the 5′-FAM-labeled ssDNA. These results demonstrated that MjaRecJ1 degraded ssDNA from the 5′ end, and 5′-FAM group did not inhibit the hydrolysis of the first 5′ phosphodiester bond. For ssDNA degradation by MjaRecJ2 (Figure 4B), DNA ladders were generated from 5′-FAM-labeled ssDNA, and products did not appear for 3′-FAM-labeled ssDNA. These results confirmed that MjaRecJ2 degraded ssDNA in the 3′→5′ direction, and the 3′-FAM group strongly inhibited the hydrolysis of the 3′ first phosphodiester bonds. The degradation of fully-phosphothioate-modified ssDNA also showed that RecJ1 was more processive than RecJ2 (Figure 4A,B). For the fully phosphothioate-modified ssRNAs, MjaRecJ1 degraded ssRNA in the 5′→3′ direction (Figure 4C), and MjaRecJ2 degraded ssRNA in the 3′→5′ direction (Figure 4D). In summary, MjaRecJ1 was a 5′ exonuclease on both ssDNA and ssRNA, and MjaRecJ2 was a 3′ exonuclease on both ssDNA and ssRNA.

Using the partially phosphothioate-modified ssDNA as substrate, the two MjaRecJs also showed the same manner of degradation (Figure S4). The phosphothioate groups at the 5′ end clearly blocked the hydrolysis of ssDNA by MjaRecJ1 (Figure S4A; lanes 6, 10 and 12). When several phosphothioate groups exist at the 3′ end, they strongly blocked the degradation of ssDNA by MjaRecJ2 (Figure S4B; lanes 4, 6 and 12). The internal successive phosphothioate groups strongly hindered degradation before the modifications by MjaRecJs (Figure S4, lanes 4 and 10).

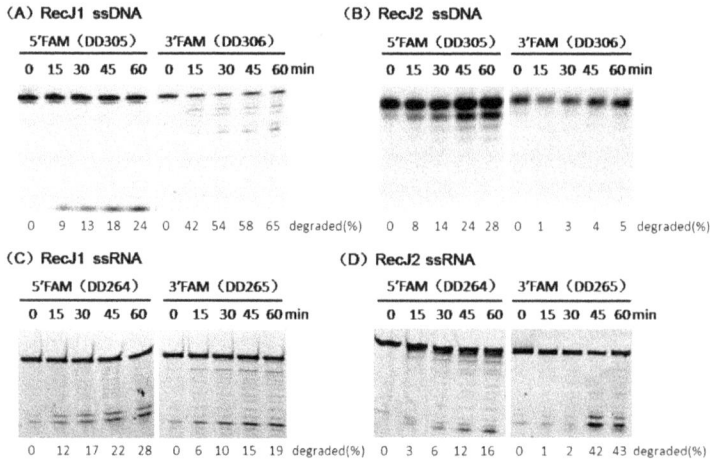

Figure 4. Hydrolysis polarities of two MjaRecJ nucleases. Two MjaRecJs (1.5 µM MjaRecJ1 or 10 µM MjaRecJ2) were incubated with fully phosphothioate-modified 200 nM 23 nt ssDNA or 17 nt ssRNA substrates in their respective reaction buffer at 55 °C with increasing time. SsDNA and ssRNA are labeled with fluorescence group fluorescein FAM at 5′ or 3′ end, respectively. The degraded amount of substrate was quantified at each time and listed at the bottom of the panel.

3.3. Opposite Effect of Terminal Phosphate Groups on MjaRecJs Activity

The terminal phosphate group generally affected exonuclease activity [32]. We characterized the effect of phosphate groups on the exonuclease activity of the two MjaRecJs (Figure 5). The 5′ phosphate group clearly promoted MjaRecJ1 activity on ssDNA (Figure 5A) but weakly affected the ssRNA substrate (Figure 5C). In contrast to MjaRecJ1, MjaRecJ2 showed a largely decreased activity on 3′-phosphorylated ssDNA and ssRNA (Figure 5B or Figure 5D). Furthermore, the 3′-phosphorylated ssRNA substrate displayed a more intensive inhibition than ssDNA.

Interestingly, 3′-phosphorylated ssDNA and ssRNA inhibited the 5′ exonuclease activity of MjaRecJ1 (Figure S5A, lane 4). Although MjaRecJ2 did not show clear 5′ exonuclease activity on ssDNA and ssRNA with a 5′-OH terminus (Figure S5B, lane 6), distinct 5′ exonuclease activity was observed on 5′-phosphorylated ssDNA and ssRNA (Figure S5B, lane 8). These results suggested that MjaRecJ2 may also function as a 5′ exonuclease especially on 5′-phosphorylated DNA.

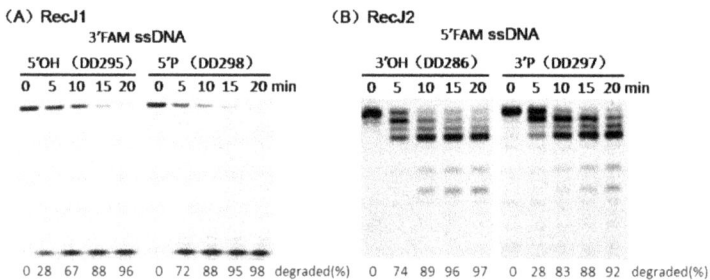

Figure 5. *Cont.*

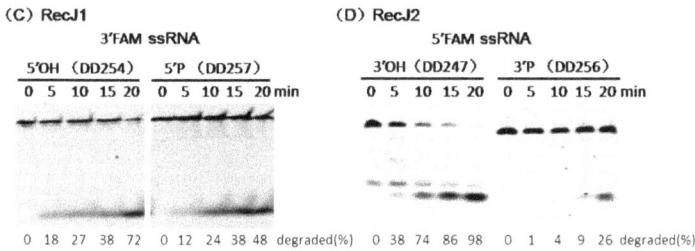

Figure 5. Effect of terminal phosphate group on *M. jannaschii* RecJ activity. Two MjaRecJs (40 nM RecJ1 or 50 nM RecJ2) were incubated with 200 nM 23 nt ssDNA or 12/16 nt ssRNA substrates in their respective reaction buffer at 55 °C with increasing time. The substrates are labeled with fluorescence group FAM at 5′ or 3′ end, and have 3′ or 5′ terminal phosphate groups. The degraded amount of substrate was quantified at each time and listed at the bottom of the panel.

3.4. Preferred Substrate Length of MjaRecJs

The two MjaRecJs had different preferences for substrate length. MjaRecJ1 could hydrolyze all length ssDNA (Figure 6A), and ssRNA ≥ 6nt (Figure 6B). MjaRecJ2 was preferable to shorter ssDNAs, which were degraded with extremely low efficiency when longer than 23 nt (Figure 6C). Only ssRNAs that are 12 or 16nt could be hydrolyzed, with a higher efficiency than ssDNA, by MjaRecJ2, and shorter ssRNAs (4 and 6 nt) are degraded from the 3′ end with very lower efficiency.

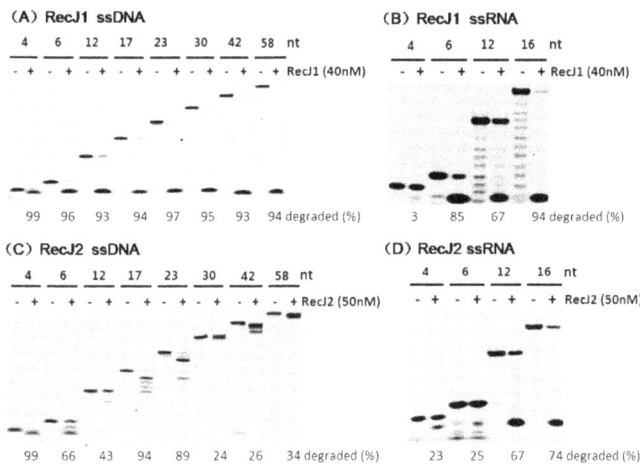

Figure 6. Substrate length preferences of MjaRecJs. Two MjaRecJs (40 nM MjaRecJ1 or 50 nM MjaRecJ2) were incubated with 200 nM 5′-FAM-labeled ssDNA or ssRNA with different lengths as substrates in their respective reaction buffer at 55 °C for 20 min. The degraded amount of substrate was quantified at each time and listed at the bottom of the panel.

3.5. Strand Preferences of MjaRecJs

Provided that the two MjaRecJs could efficiently hydrolyze ssDNA, double-stranded (ds) DNAs with different single-stranded structures were used to observe MjaRecJs activity on these molecules. Our results showed that ssDNA was the most favored substrate of two MjaRecJs (Figure 7). MjaRecJ1 hydrolyzed DNAs in the order of ssDNA > 5′ overhang > 5′ fork > 5′ blunt ≈ 5′ recess, and MjaRecJ2 followed the order of ssDNA > 3′ fork ≈ 3′ overhang >> 5′ blunt ≈ 5′ recess.

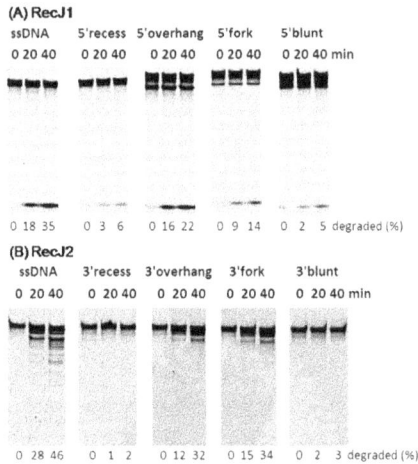

Figure 7. Selectivity of two MjaRecJs on DNA secondary structure. Two MjaRecJs (40 nM MjaRecJ1 or 50 nM MjaRecJ2) were incubated with 200 nM DNA substrates at 55 °C for 0, 20, and 40 min in their respective reaction buffer. DNA secondary structures are single-stranded, forked, overhanged, recessed, and blunt. The degraded amount of substrate DNA was quantified at each time and listed at the bottom of the panel.

3.6. No Interaction between MjaRecJs and MjaGINS

Considering that archaeal RecJ nuclease, TkoRecJ, forms a complex with the GINS [2,19,26], we characterized the possible interaction between MjaRecJs and MjaGINS. Surprisingly, both MjaRecJs did not form a complex with MjaGINS. However, the work on the RecJs from *Thermoplasma acidophilum* showed that TacRecJ2 forms a complex with GINS, but TacRecJ1 not [31]. Our result on MjaRecJ1 is consistent to that of TacRecJ1, but MjaRecJ2 showed the result contrary to TacRecJ2. The retention time of the mixtures of MjaRecJ1 and GINS were similar to that of any of the two proteins alone, indicating a lack of interaction between MjaRecJ1 and MjaGINS (Figure 8A). The mixtures of MjaRecJ2 and GINS did not generate a peak that moved faster than that of any of alone protein (Figure 8A), indicating that there was no interaction between MjaRecJ2 and MjaGINS. Another possibility is that MjaRecJ2 and GINS do interact but the complex takes a changed conformation and with the similar elution time to those of MjaRecJ2 or GINS. Surprisingly, MjaRecJ2 might exist in dimer based on its elution time (Figure 8A). It is possible that the dimer of MjaRecJ2 hinders its interaction with MjaGINS, for example, the dimer interface occupies the interaction surface for interacting with MjaGINS. Pulldown experiments using the mixtures of MjaRecJ and MjaGINS also confirmed that both MjaRecJs did not form a complex with MjaGINS (Figure 8B). However, the PfuRecJ forms a stable complex with PfuGINS (Figure S6A). Pulldown experiments using the induced *E. coli* cells co-expressing the MjaRecJ and MjaGINS further confirmed that no clear interaction existed between MjaRecJ and MjaGINS (Figure S6A). Since the tagged proteins were used in the pulldown experiment, the tag might have a potential negative effect on protein-protein interactions. Meanwhile, we also did a Microscale Thermophoresis (MST) experiment, an analysis technology for protein interaction. Our MST experiments also confirmed that the interaction does not exist between GINS and MajRecJs (Figure S6B). By checking the residues of MjaRecJs and TkoRecJ, we observed that the residues interacting with GINS have changed largely in MjaRecJ1, but they retained the most conservation in MjaRecJ2. *T. kodakaraensis* GINS51 promotes RecJ nuclease activity via forming a complex [2,19]. Consistent to the interaction results, MjaGINS51 had no promotion on the activities of two MjaRecJs (Figure S7), indirectly supporting the result that the two MjaRecJs do not interact with MjaGINS.

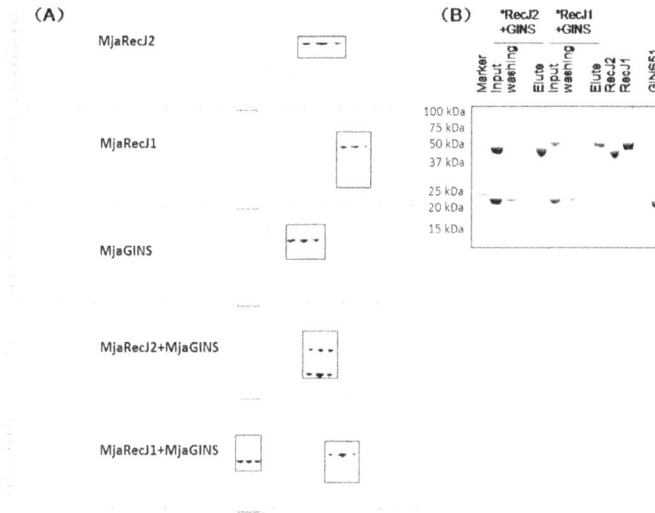

Figure 8. *Methanocaldococcus jannaschii* GINS does not interact with any of MjaRecJs. (**A**) Gel filtration was used to characterize the existence of complexes between MjaGINS51 and MjaRecJ1 or MjaRecJ2. Each peak was collected for protein identification by sodium dodecyl sulfate polyacrylamide gel electrophoresis (SDS-PAGE). (**B**) The pulldown experiments were used to characterize the interaction between *M. jannaschii* GINS and RecJ1 or RecJ2. 6×Histine-Tag RecJs were co-purified with the no His-Tag GINS using a Ni-NTA Resin. His-Tag MjaRecJ1 or MjaRecJ2 was mixed with MjaGINS (His-Tag free) in a molecular ratio of 1:2. After binding and washing, and resins were eluted with buffer containing 200 mM imidazole. Protein(s) in elutes were verified by 15% SDS-PAGE.

4. Discussion

4.1. Important Evolutionary Marker of Archaeal RecJ and Cdc45

Compared with bacterial RecJ, archaeal RecJ and hCdc45 possess a separate domain, namely the MBD domain (Figure 1A), that locates between the motifs IV and V of the DHH domain and participates in the interaction with MCM binding [13,19]. Although human Cdc45 and TkoGAN exhibit the similar structural fold [13,19], when compared with archaeal RecJ, hCdc45 has an additional sequence inserted between motifs III and IIIa of the DHH domain [17,19]. Perhaps, both mutations of the conserved residues and insertion of two additional domains caused Cdc45 to evolve into a protein that specifically binds ssDNA and prevents occasional slippage of the leading strand from the core channel of the CMG complex [33], or interacts with other DNA replication proteins, such as Sld3 [34]. It is also possible that a nuclease activity is at the heart of the ancestral replisome [35].

The similarity of crystal structures and conserved motifs may aid in the elucidation of the evolutionary origins of the RecJ/Cdc45 subfamily. It can be speculated that the ancestor of the RecJ/Cdc45 protein might originally evolve into bacterial and archaeal RecJ branches. Then, the bacterial RecJs had evolved into specific nucleases by adding the OB-fold domain. The archaeal RecJ branch, except for functioning as a nuclease in archaea, also had evolved into Cdc45 by inserting another domain between motifs III and IIIa. For archaea with two RecJs, the ancestor archaeal RecJ split into two groups: 3′ and 5′ exonucleases. The RecJ phylogenetic tree showed that TkoRecJ and PfuRecJ belongs to the archaeal RecJ2, but not RecJ1 the subfamily (Figure 1B). However, TkoRecJ and PfuRecJ have a nuclease activity similar to MjaRecJ1, but not to MjaRecJ2. Since TacRecJ2 interact with GINS, it is possible that the GINS-interaction characteristic makes PfuRecJ and TkoRecJ, which both interact with GINS, are more similar to archaeal RecJ2.

4.2. Hydrolysis Polarity of Archeal RecJs

The diversified hydrolysis polarity of archaeal RecJs might be universal in archaea. More than one archaea species possesses two or more *recJ* genes. We selected three other species, *Thermoplasma acidophilum*, *Archeoglobus fulgidus, and Methanococcus voltae*, to characterize the enzymatic activities of their RecJ homologs. For *T. acidophilum* we confirmed the 5′ exonuclease of RecJ1 and the 3′ exonuclease of RecJ2 [31]. The two RecJ1s from *A. fulgidus* and *M. voltae* were also 5′ exonuclease specific on ssDNA, while we could not identify any nuclease activity of their RecJ2s.

Both TacRecJ1 and TkoRecJ demonstrated 5′→3′ exonuclease activity only on ssDNA [2,31]. PfuRecJ, a homologue with higher sequence similarity to TkoGAN, can also hydrolyze ssDNA and ssRNA in the 5′→3′ and 3′→5′ direction [17], respectively. However, MjaRecJ1 hydrolyze both ssDNA and ssRNA in the 5′→3′ direction. To understand the hydrolysis polarity and its evolution in the DHH phosphodiesterase superfamily, the co-crystal structure of archaeal RecJ and ssDNA or ssRNA should be determined to characterize their catalytic mechanism.

4.3. Function of MjaRecJs in Archaeal DNA Replication and Repair

Both euryarchaeal and crenarchaeal GINS form a stable complex with archaeal Cdc45 homologs (RecJ in the former and RecJdbh in the latter) [2,19,25,36]. However, we did not confirm the interaction between MjaGINS and its two RecJs. Since the two MjaRecJs did not form a complex with MjaGINS, it suggests that the MjaRecJs do not participate in unwinding the chromosomal DNA during DNA replication. In future more experiments should be conducted to confirm whether an in vitro or in vivo interaction exists between MjaRecJs and GINS. On the other hand, the knockout of two *recJ* genes in archaea, which has genetic operation tools, should be done to confirm their functions in vivo based on the corresponding phenotypes of mutants.

Similar to the Eukaryotic CMG complex, archaea also have a complex RecJ-MCM-GINS (RMG) [25]. Updates to the function of RMG are still unknown. In Crenarchaea, RMG possibly functions as replicative DNA helicase [36]. The six-subunit complex of heterogenous GINS tetra-subunits and RecJdbh (namely, GC complex in a ratio of 2:2:2) in Crenarchaea *Sulfolobus* is specifically located in the replicative fork, indicating that the complex is essential for DNA replication [36]. However, the euryarchaeal *Haloferax volcanii* did not require the *recJ* gene for its normal growth [37]. Recent works on *T. kodakarensis* also demonstrated that GAN could be deleted with no discernable effects on viability and growth, indicating that it is not essential to the archaeal MCM replicative helicase [38].

Since MjaRecJ1 and MjaRecJ2 can complement the function of the deleted *recJ* gene during DNA recombination repair in *E. coli* [16], they also probably function in DNA repair processes, such as recombination repair, similar to that in bacterial RecJ [39]. Considering the existence of several different DNA resection pathways in prokaryotes [40,41], the two RecJs possibly undergo two-directional resection during the recombination repair of dsDNA break in *M. jannaschii*. TkoGAN might participate in primer removal during Okazaki fragment maturation cooperated with Fen1 and RNase HII. Failing in deleting both Fen1 and GAN genes suggested that both enzymes catalyze primer removal in vivo as a nuclease [38]. Similar to GAN, MjaRecJ1 might remove the RNA primer by its 5′-exonuclease on the flapped RNA section of Okazaki fragment. Since MjaRecJ2 has more pronounced 3′ exonuclease activity on ssRNA than on ssDNA; thus, it also may be responsible for degrading diverse abnormal ssRNAs (such as fragmental RNAs), as observed for the nanoRNase of DHH phosphodiesterase superfamily [5–7]. Therefore, more studies should be conducted to confirm the importance of *recJ* and *gins* genes in archaeal DNA replication and repair and to determine the functional diversity of archaeal RecJ and GINS homologs, especially in archaea with two RecJs and only one GINS51 subunit.

In summary, on the basis of identification of nuclease activity by Rajman & Lovett [16], we have further confirmed the reverse hydrolysis polarity of two MjaRecJs that are ideal models for investigating the molecular mechanism to determine the hydrolysis direction using structural and biochemical approaches. Meanwhile, the two MjaRecJs are also good models for studying the evolutionary pathway of archaeal RecJ and eukaryotic Cdc45 protein, and for elucidating the functions of RecJs in DNA replication and repair.

Supplementary Materials: The following are available online at www.mdpi.com/2073-4425/8/9/211/s1. Figure S1 Biochemical characterization of MjaRecJ1. Figure S2 Biochemical characterization of MjaRecJ2. Figure S3 Thermostabilities of MjaRecJs. Figure S4 Hydrolysis polarity of two MjaRecJs confirmed by special phosphothioate-modified substrates. Figure S5 Effect of terminal phosphate group on *M. jannaschii* RecJs activity. Figure S6 Interaction identification of MjaRecJs and MjaGINS. Figure S7 Effect of MjaGINS on MjaRecJs activity. Table S1 Oligonucleotides used in this research.

Acknowledgments: This work was supported by the National Natural Science Foundation of China (Grant Nos. 31371260, 41530967, J1210047), China Ocean Mineral Resources R&D Association (Grant No. DY125-22-04), and a research project of State Key Laboratory of Ocean Engineering. We are extremely grateful to the National Center for Protein Science Shanghai (Protein Expression and Purification system) for their instrument support and technical assistance.

Author Contributions: X.L. and G.Y. conceived and designed the experiments; G.Y., Y.S., W.W. and J.C. performed the experiments; all authors analyzed the data; X.L., W.D. F.W. and X.X. contributed reagents/materials/analysis tools; X.L. wrote the paper.

Conflicts of Interest: The authors declare that no conflicts of interest exist.

References

1. Yang, W. Nucleases: Diversity of structure, function and mechanism. *Q. Rev. Biophys.* **2011**, *44*, 1–93. [CrossRef] [PubMed]

2. Li, Z.; Pan, M.; Santangelo, T.J.; Chemnitz, W.; Yuan, W.; Edwards, J.L.; Hurwitz, J.; Reeve, J.N.; Kelman, Z. A novel DNA nuclease is stimulated by association with the GINS complex. *Nucleic Acids Res.* **2011**, *39*, 6114–6123. [CrossRef] [PubMed]

3. Han, E.S.; Cooper, D.L.; Persky, N.S.; Sutera, V.A.J.; Whitaker, R.D.; Montello, M.L.; Lovett, S.T. RecJ exonuclease: Substrates, products and interaction with SSB. *Nucleic Acids Res.* **2006**, *34*, 1084–1091. [CrossRef] [PubMed]

4. Sanchez-Pulido, L.; Ponting, C.P. Cdc45: The missing RecJ ortholog in eukaryotes? *Bioinformatics* **2011**, *27*, 1885–1888. [CrossRef] [PubMed]

5. Mechold, U.; Fang, G.; Ngo, S.; Ogryzko, V.; Danchin, A. YtqI from Bacillus subtilis has both oligoribonuclease and pAp-phosphatase activity. *Nucleic Acids Res.* **2007**, *35*, 4552–4561. [CrossRef] [PubMed]

6. Fang, M.; Zeisberg, W.M.; Condon, C.; Ogryzko, V.; Danchin, A.; Mechold, U. Degradation of nanoRNA is performed by multiple redundant RNases in Bacillus subtilis. *Nucleic Acids Res.* **2009**, *37*, 5114–5125. [CrossRef] [PubMed]

7. Wakamatsu, T.; Kim, K.; Uemura, Y.; Nakagawa, N.; Kuramitsu, S.; Masui, R. Role of RecJ-like protein with 5′-3′ exonuclease activity in oligo(deoxy)nucleotide degradation. *J. Biol. Chem.* **2011**, *286*, 2807–2816. [CrossRef] [PubMed]

8. Aravind, L.; Koonin, E.V. A novel family of predicted phosphoesterases includes Drosophila prune protein and bacterial RecJ exonuclease. *Trends Biochem. Sci.* **1998**, *23*, 17–19. [CrossRef]

9. Fabrichniy, I.P.; Lehtiö, L.; Tammenkoski, M.; Zyryanov, A.B.; Oksanen, E.; Baykov, A.A.; Lahti, R.; GolMBDan, A. A trimetal site and substrate distortion in a family II inorganic pyrophosphatase. *J. Biol. Chem.* **2007**, *282*, 1422–1431. [CrossRef] [PubMed]

10. Ishino, S.; Yamagami, T.; Kitamura, M.; Kodera, N.; Mori, T.; Sugiyama, S.; Ando, T.; Goda, N.; Tenno, T.; Hiroaki, H.; et al. Multiple interactions of the intrinsically disordered region between the helicase and nuclease domains of the archaeal Hef protein. *J. Biol. Chem.* **2014**, *289*, 21627–21639. [CrossRef] [PubMed]

11. Thoms, B.; Borchers, I.; Wackernagel, W. Effects of single-strand DNases ExoI, RecJ, ExoVII, and SbcCD on homologous recombination of recBCD+ strains of Escherichia coli and roles of SbcB15 and XonA2 ExoI mutant enzymes. *J. Bacteriol.* **2008**, *190*, 179–192. [CrossRef] [PubMed]

12. Burdett, V.; Baitinger, C.; Viswanathan, M.; Lovett, S.T.; Modrich, P. In vivo requirement for RecJ, ExoVII, ExoI, and ExoX in methyl-directed mismatch repair. *Proc. Natl. Acad. Sci. USA* **2001**, *98*, 6765–6770.

13. Dianov, G.; Sedgwick, B.; Daly, G.; Olsson, M.; Lovett, S.; Lindahl, T. Release of 5′-terminal deoxyribose-phosphate residues from incised abasic sites in DNA by the Escherichia coli RecJ protein. *Nucleic Acids Res.* **1994**, *22*, 993–998. [CrossRef] [PubMed]

14. Wakamatsu, T.; Kitamura, Y.; Kotera, Y.; Nakagawa, N.; Kuramitsu, S.; Masui, R. Structure of RecJ exonuclease defines its specificity for single-stranded DNA. *J. Biol. Chem.* **2010**, *285*, 9762–9769. [CrossRef] [PubMed]

15. Cheng, K.; Zhao, Y.; Chen, X.; Li, T.; Wang, L.; Xu, H.; Tian, B.; Hua, Y. A Novel C-Terminal Domain of RecJ is Critical for Interaction with HerA in Deinococcus radiodurans. *Front. Microbiol.* **2015**, *6*, 1302. [CrossRef] [PubMed]

16. Rajman, L.A.; Lovett, S.T. A thermostable single-strand DNase from Methanococcus jannaschii related to the RecJ recombination and repair exonuclease from Escherichia coli. *J. Bacteriol.* **2000**, *182*, 607–612. [CrossRef] [PubMed]

17. Yuan, H.; Liu, X.P.; Han, Z.; Allers, T.; Hou, J.L.; Liu, J.H. RecJ-like protein from Pyrococcus furiosus has 3′-5′ exonuclease activity on RNA: implication of its proofreading capacity on 3′-mismatched RNA primer in DNA replication. *Nucleic Acids Res.* **2013**, *41*, 5817–5826. [CrossRef] [PubMed]

18. Yamagata, A.; Kakuta, Y.; Masui, R.; Fukuyama, K. The crystal structure of exonuclease RecJ bound to Mn2+ ion suggests how its characteristic motifs are involved in exonuclease activity. *Proc. Natl. Acad. Sci. USA* **2002**, *99*, 5908–5912. [CrossRef] [PubMed]

19. Oyama, T.; Ishino, S.; Shirai, T.; Yamagami, T.; Nagata, M.; Ogino, H.; Kusunoki, M.; Ishino, Y. Atomic structure of an archaeal GAN suggests its dual roles as an exonuclease in DNA repair and a CMG component in DNA replication. *Nucleic Acids Res.* **2016**, *44*, 9505–9517. [CrossRef] [PubMed]

20. Cheng, K.; Xu, H.; Chen, X.; Wang, L.; Tian, B.; Zhao, Y.; Hua, Y. Structural basis for DNA 5′-end resection by RecJ. *eLife* **2016**, *5*, e14294. [CrossRef] [PubMed]

21. Fenwick, A.L.; Kliszczak, M.; Cooper, F.; Murray, J.; Sanchez-Pulido, L.; Twigg, S.R.; Goriely, A.; McGowan, S.J.; Miller, K.A.; Taylor, I.B.; et al. Mutations in CDC45, encoding an essential component of the pre-initiation complex, cause Meier-Gorlin syndrome and craniosynostosis. *Am. J. Hum. Genet.* **2016**, *99*, 125–138. [CrossRef] [PubMed]

22. Krastanova, I.; Sannino, V.; Amenitsch, H.; Gileadi, O.; Pisani, F.M.; Onesti, S. Structural and functional insights into the DNA replication factor Cdc45 reveal an evolutionary relationship to the DHH family of phosphoesterases. *J. Biol. Chem.* **2012**, *287*, 4121–4128. [CrossRef] [PubMed]

23. Szambowska, A.; Tessmer, I.; Kursula, P.; Usskilat, C.; Prus, P.; Pospiech, H.; Grosse, F. DNA binding properties of human Cdc45 suggest a function as molecular wedge for DNA unwinding. *Nucleic Acids Res.* **2014**, *42*, 2308–2319. [CrossRef] [PubMed]

24. Simon, A.C.; Sannino, V.; Costanzo, V.; Pellegrini, L. Structure of human Cdc45 and implications for CMG helicase function. *Nat. Commun.* **2016**, *7*, 11638. [CrossRef] [PubMed]

25. Marinsek, N.; Barry, E.R.; Makarova, K.S.; Dionne, I.; Koonin, E.V.; Bell, S.D. GINS, a central nexus in the archaeal DNA replication fork. *EMBO Rep.* **2006**, *7*, 539–545. [CrossRef] [PubMed]

26. Li, Z.; Santangelo, T.J.; Cuboňová, L.; Reeve, J.N.; Kelman, Z. Affinity purification of an archaeal DNA replication protein network. *mBio* **2010**, *1*, e00221-10. [CrossRef] [PubMed]

27. Gambus, A.; Jones, R.C.; Sanchez-Diaz, A.; Kanemaki, M.; van Deursen, F.; EMBDondson, R.D.; Labib, K. GINS maintains association of Cdc45 with MCM in replisome progression complexes at eukaryotic DNA replication forks. *Nat. Cell Biol.* **2006**, *8*, 358–366. [CrossRef] [PubMed]

28. Costa, A.; Ilves, I.; Tamberg, N.; Petojevic, T.; Nogales, E.; Botchan, M.R.; Berger, J.M. The structural basis for MCM2–7 helicase activation by GINS and Cdc45. *Nat. Struct. Mol. Biol.* **2011**, *18*, 471–477. [CrossRef] [PubMed]

29. Abid, A.F.; Renault, L.; Gannon, J.; Gahlon, H.L.; Kotecha, A.; Zhou, J.C.; Rueda, D.; Costa, A. Cryo-EM structures of the eukaryotic replicative helicase bound to a translocation substrate. *Nat. Commun.* **2016**, *7*, 10708. [CrossRef] [PubMed]

30. Yuan, Z.; Bai, L.; Sun, J.; Georgescu, R.; Liu, J.; O'Donnell, M.E.; Li, H. Structure of the eukaryotic replicative CMG helicase suggests a pumpjack motion for translocation. *Nat. Struct. Mol. Biol.* **2016**, *23*, 217–224. [CrossRef] [PubMed]

31. Ogino, H.; Ishino, S.; Kohda, D.; Ishino, Y. The RecJ2 protein in the thermophilic archaeon Thermoplasma acidophilum is a 3′-5′ exonuclease that associates with a DNA replication complex. *J. Biol. Chem.* **2017**, *292*, 7921–7931. [CrossRef] [PubMed]

32. Subramanian, K.; Rutvisuttinunt, W.; Scott, W.; Myers, R.S. The enzymatic basis of processivity in lambda exonuclease. *Nucleic Acids Res.* **2003**, *31*, 1585–1596. [CrossRef] [PubMed]

33. Petojevic, T.; Pesavento, J.J.; Costa, A.; Liang, J.; Wang, Z.; Berger, J.M.; Botchan, M.R. Cdc45 (cell division cycle protein 45) guards the gate of the Eukaryote Replisome helicase stabilizing leading strand engagement. *Proc. Natl. Acad. Sci. USA* **2015**, *112*, E249–E258. [CrossRef] [PubMed]

34. Bruck, I.; Kaplan, D.L. GINS and Sld3 compete with one another for Mcm2–7 and Cdc45 binding. *J. Biol. Chem.* **2011**, *286*, 14157–14167. [CrossRef] [PubMed]

35. Pellegrini, L. Structural insights into Cdc45 function: Was there a nuclease at the heart of the ancestral replisome? *Biophys. Chem.* **2016**, *225*, 10–14. [CrossRef] [PubMed]

36. Xu, Y.; Gristwood, T.; Hodgson, B.; Trinidad, J.C.; Albers, S.V.; Bell, S.D. Archaeal orthologs of Cdc45 and GINS form a stable complex that stimulates the helicase activity of MCM. *Proc. Natl. Acad. Sci. USA* **2016**, *113*, 13390–13395. [CrossRef] [PubMed]

37. Giroux, X.; MacNeill, S.A. Molecular Genetic Methods to Study DNA Replication Protein Function in Haloferax volcanii, A Model Archaeal Organism. *Methods Mol. Biol.* **2015**, *1300*, 187–218. [PubMed]

38. Burkhart, B.W.; Cubonova, L.; Heider, M.R.; Kelman, Z.; Reeve, J.N.; Santangelo, T.J. The GAN exonuclease, or the flap endonuclease Fen1 and RNase HII are necessary for viability of *Thermococcus kodakarensis*. *J. Bacteriol.* **2017**, *199*, e00141-17. [CrossRef] [PubMed]

39. Morimatsu, K.; Kowalczykowski, S.C. RecQ helicase and RecJ nuclease provide complementary functions to resect DNA for homologous recombination. *Proc. Natl. Acad. Sci. USA* **2014**, *111*, E5133–E5142. [CrossRef] [PubMed]

40. Wigley, D.B. Bacterial DNA repair: recent insights into the mechanism of RecBCD, AddAB and AdnAB. *Nat. Rev. Microbiol.* **2013**, *11*, 9–13. [CrossRef] [PubMed]

41. Hopkins, B.B.; Paull, T.T. The P. furiosus mre11/rad50 complex promotes 5′ strand resection at a DNA double-strand break. *Cell* **2008**, *135*, 250–260. [CrossRef] [PubMed]

GCAT
TACG
GCAT
genes

MDPI

Article

Error-Free Bypass of 7,8-dihydro-8-oxo-2'-deoxyguanosine by DNA Polymerase of *Pseudomonas aeruginosa* Phage PaP1

Shiling Gu [1,2], Qizhen Xue [2], Qin Liu [1], Mei Xiong [2], Wanneng Wang [1,*] and Huidong Zhang [2,*]

[1] College of Pharmacy and Bioengineering, Chongqing University of Technology, No. 69 Hongguang Street, Banan District, Chongqing 400054, China; shilinggu@foxmail.com (S.G.); liuqin@2014.cqut.edu.cn (Q.L.)
[2] Public Health Laboratory Sciences and Toxicology, West China School of Public Health, Sichuan University, No. 17 People's South Road, Chengdu 610041, China; xueqzhen@163.com (Q.X.); xhmei2016@126.com (M.X.)
* Correspondence: wannengw@cqut.cn (W.W.); huidong.zhang@foxmail.com (H.Z.)

Academic Editor: Eishi Noguchi
Received: 21 November 2016; Accepted: 30 December 2016; Published: 6 January 2017

Abstract: As one of the most common forms of oxidative DNA damage, 7,8-dihydro-8-oxo-2'-deoxyguanosine (8-oxoG) generally leads to G:C to T:A mutagenesis. To study DNA replication encountering 8-oxoG by the sole DNA polymerase (Gp90) of *Pseudomonas aeruginosa* phage PaP1, we performed steady-state and pre-steady-state kinetic analyses of nucleotide incorporation opposite 8-oxoG by Gp90 D234A that lacks exonuclease activities on ssDNA and dsDNA substrates. Gp90 D234A could bypass 8-oxoG in an error-free manner, preferentially incorporate dCTP opposite 8-oxoG, and yield similar misincorporation frequency to unmodified G. Gp90 D234A could extend beyond C:8-oxoG or A:8-oxoG base pairs with the same efficiency. dCTP incorporation opposite G and dCTP or dATP incorporation opposite 8-oxoG showed fast burst phases. The burst of incorporation efficiency ($k_{pol}/K_{d,dNTP}$) is decreased as dCTP:G > dCTP:8-oxoG > dATP:8-oxoG. The presence of 8-oxoG in DNA does not affect its binding to Gp90 D234A in a binary complex but it does affect it in a ternary complex with dNTP and Mg^{2+}, and dATP misincorporation opposite 8-oxoG further weakens the binding of Gp90 D234A to DNA. This study reveals Gp90 D234A can bypass 8-oxoG in an error-free manner, providing further understanding in DNA replication encountering oxidation lesion for *P.aeruginosa* phage PaP1.

Keywords: *P. aeruginosa* phage PaP1; DNA polymerase; 8-oxoG; steady-state kinetics; pre-steady-state kinetics; nucleotide incorporation

1. Introduction

Accurate synthesis of DNA is of great importance for genomic integrity in all forms of life. DNA replication is generally performed by DNA polymerases with high fidelity. Accurate DNA replication is under constant threat, which are formed within the genome [1]. DNA damage incurred by a multitude of factors constitutes an unavoidable challenge for the replication machinery [2]. Reactive oxygen species are a major source of DNA damage. One of the most common lesions induced by oxidative stress is 7,8-dihydro-8-oxo-2'-deoxyguanosine (8-oxoG) [3], which is representative of nucleoside damage and shows genotoxicity [4]. The deleterious effects of 8-oxoG on DNA replication can be attributed to its dual-coding potential that leads to G:C to T:A transversions [5].

Lesion tolerance is achieved, in part, by special translesion synthesis (TLS) polymerases, which are able to bypass lesions during DNA replication [6]. Compared with replicative DNA polymerases, TLS polymerases have relatively larger active sites, allowing them to accommodate mismatched base pairs and bulky DNA lesions at the cost of a lower fidelity [7,8]. Most DNA polymerase nucleotides incorporate opposite template 8-oxoG lesions with reduced efficiency and accuracy [9]. Error-free DNA synthesis involves 8-oxoG adopting an anti-conformation to the base pair with cytosine, whereas a mutagenic bypass involves 8-oxoG adopting a syn-conformation to the base pair with adenine [10].

Human DNA polymerase α, human DNA polymerase η, and *Bacillus stearothermophilus* DNA Polymerase I are capable of bypassing 8-oxoG in a mostly error-free manner [11]. The sole Y-family DNA polymerase Dpo4 in *Sulfolobus solfataricus*, can efficiently and reliably incorporate dCTP opposite 8-oxoG and extend from an 8-oxoG:C base pair with a mechanism similar to the bypass of undamaged DNA [12]. In crystal structures, Arg-332 in Dpo4 stabilizes the anti-conformation of the 8-oxoG template base, which results in increased efficiency for dCTP insertion and less favorable formation of a Hoogsteen pair between 8-oxoG and dATP [13].

Human DNA polymerase ι is error-prone in 8-oxoG bypass with low fidelity [14]. Human DNA polymerase β has similar efficiencies for dCTP and dATP insertion opposite 8-oxoG [15]. Human DNA polymerase κ bypasses 8-oxoG in an error-prone manner by mainly inserting dATP. Crystal structures of hPol κ ternary complex reveal nonproductive alignments of incoming nucleotides (dGTP or dATP) with 8-oxoG. The interactions between the N-clasp and finger domains of hPol κ stabilize the *syn* orientation of 8-oxoG that contributes to error-prone dATP incorporation. Mutation of Leu-508 into lysine at the little finger domain of hPol κ modulates the insertion of dCTP opposite 8-oxoG, leading to more accurate bypass [16].

Pseudomonas aeruginosa is difficult to treat because of drug resistance. PaP1, as a lytic phage of *P. aeruginosa*, is a potential alternative to treat *P. aeruginosa* infections. Recently, our group has identified that DNA polymerase (Gp90) in *P. aeruginosa* phage PaP1 is an A-family DNA polymerase containing ssDNA and dsDNA exonuclease activities [17]. As the sole DNA polymerase in PaP1 [18], Gp90 has a significant role in DNA synthesis in PaP1 propagation. Studies on DNA replication by Gp90 will contribute to our understanding of PaP1 propagation in hosts infected with *P. aeruginosa*. As one of the most common oxidation lesions, 8-oxoG may affect DNA replication by Gp90. We performed steady-state and pre-steady-state kinetic analyses of nucleotide incorporation opposite, or beyond, 8-oxoG using Gp90 to understand how PaP1 bypasses 8-oxoG.

During nucleotide incorporation, the nucleotide binding step, conformational change, and chemistry steps are three important elementary steps that directly determine the fidelity of 8-oxoG bypass [19]. Steady-state kinetic analysis of nucleotide incorporation can provide information on enzyme specificity and efficiency, but cannot determine these elementary steps. With the pre-steady-state kinetic method, these elementary steps are directly examined in the first turnover, eliminating the effect of the subsequent slow dissociation of polymerase from the DNA [20]. In this study, we performed steady-state and pre-steady-state kinetic analyses of dNTP incorporation opposite, or beyond, 8-oxoG and determined that Gp90 can ensure error-free bypass of 8-oxoG, indicating that PaP1 can tolerate oxidation lesions during its propagation.

2. Materials and Methods

2.1. Materials

Mutagenesis was performed using a QuikChange site-directed mutagenesis kit (Stratagene, La Jolla, CA, USA). Oligonucleotides were synthesized by Midland Certified Reagent Co (Midland, TX, USA). All unlabeled dNTPs and T4 polynucleotide kinases were obtained from Amersham Bio-sciences (Piscataway, NJ, USA). Ni-NTA mini-spin columns were purchased from GE Healthcare (Pittsburgh, PA, USA). [γ-^{32}P] ATP was obtained from PerkinElmer Life Sciences (Boston, MA, USA). Bio-spin columns were obtained from Bio-Rad (Hercules, CA, USA). Phage PaP1 was propagated and extracted, and its genomic DNA was extracted and purified as described previously [21,22]. Other commercially available reagents were of the highest quality.

2.2. Construction, Expression, and Purification of Gp90 Mutants

Glu-60, Asp-137 and Asp-234 were predicted as the potential exonuclease active residues of Gp90 by alignment of the sequences of Gp90, T7 DNA polymerase, and *Escherichia coli* polymerase I [17]. These residues in Gp90 were then mutated to alanine using a QuikChange site-directed mutagenesis kit. Three Gp90 mutants (Gp90 E60A, Gp90 D234A, and Gp90 E60A D137A D234A) were prepared. The primers used for mutagenesis were listed below:

E60A: sense, 5′-GTAGTAGCCGCCCACGGCGGTAACATTCTGGCGTTCTAC-3′;
    antisense, 5′-GCC GTGGGCGGCTACTACGACACAGTGATGACTGTAGCT-3′.
D137A: sense, 5′-ATTAACTTCGCCCTTATGTCGATGAAGCTTGTGGAAGATATG-3′;
    antisense, 5′-CGACATAAGGGCGAAGTTAATCATGTTGTGAGCCACTACGCG-3′.
D234A: sense, 5′-TGTATCTATGCCGTAAAGGCGAACACCGCTGTATGGCACTGG-3′;
    antisense, 5′-CGCCTTTACGGCATAGATACAGTAGTAAAGCATATCGGCTGC-3′.

The DNA sequences were confirmed by sequence analysis prior to bacterial expression. Wild-type and three mutants were expressed in *E. coli* A307 (DE3) cells, followed by purification through a HisTrap[TM] FF column (5 mL; GE Healthcare) as described previously [17].

2.3. Examination of Exonuclease Activities of Gp90 Mutants

The ssDNA or dsDNA exonuclease activities were determined by mixing 10 nM each of DNA polymerase with 20 nM ^{32}P-labeled 27-mer ssDNA or ^{32}P-labeled 27-mer/62-mer primer/template dsDNA, respectively, in a buffer containing 40 mM Tris-HCl (pH 7.5), 30 mM Mg^{2+} and 10 mM DTT at 37 °C. After 0.5, 1, 2, or 5 min, reactions were terminated with a quench solution containing 20 mM EDTA, 95% formamide (v/v), bromphenol blue, and xylene cyanol. The samples were then separated on a 20% polyacrylamide (w/v)/7 M urea gel. Products were visualized and quantified using a phosphorimaging screen and Quantity One[TM] software (Bio-Rad, Hercules, CA, USA) [23].

2.4. Primer Extension by Gp90 Mutants Using All Four dNTPs

A ^{32}P-labeled 27-mer primer, annealed to 62-mer template oligonucleotide, was extended in the presence of all four dNTPs in a buffer containing 40 mM Tris-HCl (pH 7.5), 30 mM Mg^{2+}, 10 mM DTT, and 50 mM potassium glutamate at 37 °C. The reactions were initiated by mixing 20 nM DNA substrates with 10 nM Gp90 or Gp90 mutants and 350 µM each of dNTP for 0.5, 1, 2, or 5 min. Reactions were terminated by a quench solution containing 20 mM EDTA, 95% formamide (v/v), bromphenol blue, and xylene cyanol. The samples were then separated on a 20% polyacrylamide (w/v)/7 M urea gel. Products were visualized and quantified using a phosphorimaging screen and Quantity One[TM] software. Primer extension beyond 8-oxoG were performed similarly by mixing 20 nM DNA substrates containing 8-oxoG with 10 nM Gp90 or Gp90 D234A and 350 µM each of dNTP and reacted for 0.5, 1, 2, or 5 min.

2.5. Steady-State Kinetics Analysis of Single-Base Incorporation and Next-Base Extension

Steady-state kinetic analysis of single-base incorporation and next-base extension by Gp90 D234A were performed using ^{32}P-labeled 27-mer/62-mer and ^{32}P-labeled 28-mer/62-mer dsDNA substrates, respectively (Table 1). The molar ratio of Gp90 D234A to DNA substrate was <0.10. The concentration of Gp90 D234A and reaction time were adjusted to control the extension of the primer <0.2 [24]. All reactions were performed in buffer containing 40 mM Tris-HCl (pH 7.5), 30 mM Mg^{2+}, 10 mM DTT, and 50 mM potassium glutamate. Reactions products were analyzed by gel electrophoresis, visualized using phosphorimaging and quantified by Quantity One[TM] software. Graphs of product formation rates versus dNTP concentrations were fit by nonlinear regression (hyperbolic fits) using GraphPad Prism Version 6.0 (San Diego, CA, USA) to determine k_{cat} and K_m values [25]. The misincorporation frequencies were calculated by dividing the misincorporation efficiency (k_{cat}/K_m) of incorrect dNTP by that of the correct dCTP.

Table 1. Oligodeoxynucleotieds used in this study.

27-mer	5′-GCTACAGAGTTATGGTGACGATACGTC-3′
28C-mer	5′-GCTACAGAGTTATGGTGACGATACGTCC-3′
28A-mer	5′-GCTACAGAGTTATGGTGACGATACGTCA-3′
30-mer	5′-TTTGCTACAGAGTTATGGTGACGATACGTC$_{dd}$-3′
62-mer	3′-CGATGTCTCAATACCACTGCTATGCAGG*CTATCTCGCCTAATGATATGATGTAATCTTAAGT-5′

G*: G or 8-oxoG.

2.6. Pre-Steady-State Kinetic Analysis

Rapid chemical quench experiments were performed using a model RQF-3 KinTek Quench Flow Apparatus (KinTek Corp, Austin, TX, USA) with 50 mM Tris-HCl (pH 7.5) buffer in the drive syringes [26]. Reactions were initiated by rapidly mixing 240 nM ^{32}P-labeled 27-mer/62-mer primer/template dsDNA substrate and 160 nM Gp90 D234A mixtures with an equal volume of 2 mM dNTP and 60 mM Mg^{2+} complex, incubated for a varied time (0.005–60 s) at 37 °C, and then quenched with 0.6 M EDTA. Substrate and product DNA were separated by electrophoresis on a 20% polyacrylamide (w/v)/7 M urea gel. The products were then visualized using phosphorimaging and quantified using Quantity One^{TM} software. The product and time were fit to Equation (1), corresponding dNTP incorporation in the first binding phase and the subsequent steady-state phase:

$$y = A\left(1 - e^{k_p\,t}\right) + k_{ss}\,t \tag{1}$$

where A is the amount of active complex formed in the first binding phase, nM; k_p is the dNTP incorporation rate in the first binding phase, s^{-1}; k_{ss} is the rate of steady-state dNTP incorporation, s^{-1}; and t is time, s^{-1}.

Nucleotide incorporation rates in the first binding phase could also be determined by rapidly mixing 400 nM Gp90 D234A and 200 nM DNA with an equal volume of different concentrations of dNTP and 60 mM Mg^{2+}, and incubated for a varied time. The product amount and time were fit to Equation (2), corresponding dNTP incorporation only in the first binding phase:

$$y = A\left(1 - e^{-k_{obs}\,t}\right) \tag{2}$$

where A is the amount of active complex formed in the first binding phase, nM; k_{obs} is the dNTP incorporation rate in the first binding phase (burst rate), s^{-1}; and t is time, s^{-1}.

Burst rates (k_{obs}) and concentrations of dNTP were fit to hyperbolic Equation (3) to obtain k_{pol} and $K_{d,dNTP}$ values:

$$k_{obs} = k_{pol}\,[dNTP]/([dNTP] + K_{d,dNTP}) \tag{3}$$

where k_{pol} is the maximal rate of dNTP incorporation, s^{-1}; and $K_{d,dNTP}$ is the equilibrium dissociation constant for dNTP in the burst phase, µM [27].

2.7. Biophysical Binding of Gp90 D234A to DNA Containing G or 8-oxoG

Surface plasmon resonance analysis was performed by using a Biacore-3000 instrument (Biacore, Uppsala, Sweden) [22]. An annealed DNA (600 RU) consisting of a primer (5′-Biotin-TTTGCTACAGAGTTATGGTGACGATACGTC$_{dd}$-3′) and a template (5′-TGAATTCTAAT GTAGTATAGTAATCCGCTCTATCGGACGTATCGTCACCATAACTCTGTAGC-3′) was coupled to a streptavidin (SA) chip. The C$_{dd}$ (double deoxycytosine) at the 3′-end of primercan stop DNA polymerization. A varying concentration of Gp90 or Gp90 D234A (20–600 nM) was flowed over the chip in a buffer containing 40 mM Tris-HCl (pH 7.5), 10 mM DTT, and 50 mM potassium glutamate at a flow rate of 10 μL/min at room temperature. In a control flow cell, biotin was used instead of the biotinylated DNA to compensate for background. The chip surface was regenerated by injection of 1 M NaCl solution at a flow rate of 100 μL/min. The binding signal was fitted to Equation (4) using a steady-state model provided by BIA evaluation 3.0.2 computational software (Biacore). The dissociation constants K_d were calculated using the steady-state average RU.

$$Y = B \times RU_{max}/(B + K_d) \tag{4}$$

where Y is the response signal corresponding to the binding, RU; B is the concentration of protein, nM; RU_{max} is the maximal binding amount, RU; and K_d is the dissociation constant, nM. All experiments were carried out three times, and standard errors were derived using Prism 6.0 software.

The similar binding assays were also performed in the presence of 30 mM $MgCl_2$ and 1 mM dCTP or dATP. In the presence of Mg^{2+} and dNTP, the active site of DNA polymerase would locate at the 3′-end of primer where dNTP is paired or mispaired opposite G or 8-oxoG in the template strand. The binding affinities of the polymerase to DNA were determined by the same methods as described above.

3. Results

3.1. Examination of Exonuclease and Polymerase Activities of Gp90 Mutants

Glu-60, Asp-137, and Asp-234 were predicted as the potential exonuclease active residues of Gp90 by alignment of the sequences of Gp90, T7 DNA polymerase and *E. coli* polymerase I [17]. SsDNA exonuclease activities of Gp90 and its three mutants were tested using a ^{32}P-labeled 27-mer ssDNA (Figure 1A). With increasing the reaction time, short ssDNA products were gradually produced by Gp90 and Gp90 E60A, demonstrating 3′–5′ ssDNA exonuclease activity. SsDNA exonuclease activity of Gp90 E60A was partially decreased compared with Gp90. Gp90 D234A and Gp90 E60A D137A D234A showed no obvious degraded ssDNA products, indicating that their ssDNA exonuclease activities have been eliminated.

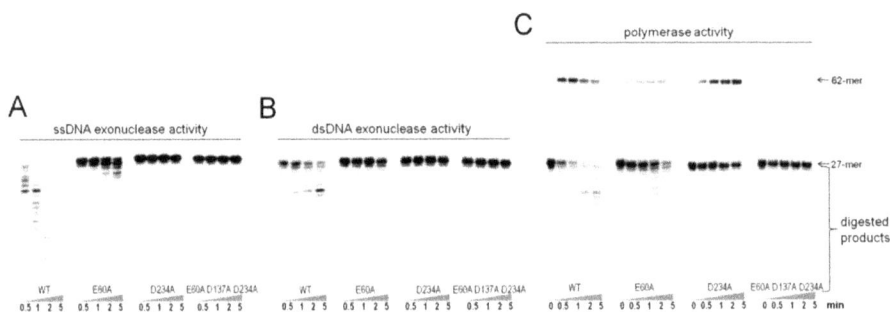

Figure 1. (**A**) ssDNA exonuclease assays. ssDNA exonuclease activities of Gp90 mutants were examined by mixing 10 nM each of DNA polymerase with 20 nM ^{32}P-labeled 27-mer ssDNA in a buffer containing 40 mM Tris-HCl (pH 7.5), 30 mM Mg^{2+}, and 10 mM DTT at 37 °C for 0.5, 1, 2, or 5 min. (**B**) dsDNA exonuclease assays. dsDNA exonuclease activities of Gp90 mutants were examined as described above, except for using ^{32}P-labeled 27-mer/62-mer dsDNA substrate. (**C**) Polymerase activity assays. Polymerase activities of Gp90 mutants were examined by mixing 10 nM polymerase with 20 nM ^{32}P-labeled 27-mer/62-mer dsDNA substrate and 350 μM each of dNTP in the same reaction buffer for 0.5, 1, 2, or 5 min. Representative data from multiple experiments are shown.

DsDNA exonuclease activities were examined using ^{32}P-labeled 27-mer/62-mer primer/template dsDNA substrate (Figure 1B). Compared with wild-type Gp90, Gp90 E60A exhibited partial dsDNA exonuclease activity. Short degraded products were gradually produced, demonstrating 3′–5′ dsDNA exonuclease activities. Gp90 D234A and Gp90 E60A D137A D234A were deficient in dsDNA exonuclease activity. Polymerase activities of Gp90 and its three mutants were also examined by full-length extension of a ^{32}P-labeled 27-mer/62-mer dsDNA substrate in the presence of all four dNTPs. The 27-mer primer was readily extended to 62-mer by Gp90 and Gp90 D234A (Figure 1C). Gp90 E60A showed partial polymerase activity and Gp90 E60A D137A D234A almost lost its polymerase activity. Therefore, mutation of Asp-234 to Ala can efficiently abolish exonuclease activity, but can almost retain the polymerase activity. Gp90 D234A was then used asexonuclease-deficient in this work.

3.2. Primer Extension beyond 8-oxoG by Gp90 D234A Using All Four dNTPs

Gp90 D234A readily extended 27-mer primer to 62-mer on unmodified template G (Figure 2). No intermediate product bands were observed, similar to previous results that Gp90 was a highly processive DNA polymerase [17]. The extension beyond 8-oxoG was partially inhibited, as evidenced by the presence of more unextended primer and less full-length extended products. Some 28-mer and 29-mer products were also observed for 8-oxoG. Therefore, the presence of 8-oxoG in the template partially inhibited DNA polymerization.

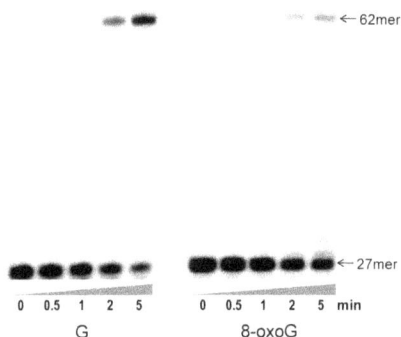

Figure 2. Extension of ^{32}P-labeled primer beyond 8-oxoG by Gp90 D234A in the presence of all four dNTPs. Extension assays were performed by mixing 10 nM Gp90 D234A, 20 nM ^{32}P-labeled 27-mer/62-mer dsDNA substrate, and 350 µM each of dNTP in a reaction buffer as described in Materials and Methods. Representative data from multiple experiments are shown.

3.3. Steady-State Kinetic Analysis of Single-Base Incorporation Opposite G or 8-oxoG by Gp90 D234A

The steady-state kinetic parameters (i.e., k_{cat} and K_m) for each single dNTP incorporation opposite G or 8-oxoG by Gp90 D234A were measured (Table 2). dCTP was preferentially incorporated opposite G and the misincorporation frequencies of other dNTPs were in the range of 10^{-4} to 10^{-5}. In detail, the k_{cat} values of all four dNTPs were similar, but the K_m values of three incorrect dNTPs were significantly increased compared with that of dCTP. For 8-oxoG, dCTP was still highly preferentially incorporated and the misincorporation frequencies of other dNTPs were in the range of 10^{-4} to 10^{-5}. Notably, all of incorporation efficiencies opposite 8-oxoG were significantly reduced compared with those opposite G. The efficiency of dCTP incorporation opposite 8-oxoG was reduced by 700-fold compared with that of dCTP opposite G because of the increased K_m value, but unchanged k_{cat} value. dNTP misincorporation opposite 8-oxoG generally resulted in higher K_m and lower k_{cat} values compared with that of dCTP incorporation opposite 8-oxoG.

Table 2. Steady-state kinetic analysis of single-base incorporation by Gp90 D234A.

Template Base	dNTP	$K_{m,dNTP}$ µM	k_{cat}, $\times 10^{-3}$min^{-1}	k_{cat}/K_m, µM^{-1}min^{-1}	Misincorporation Frequency
	C	$(1.3 \pm 0.1) \times 10^{-3}$	840 ± 10	650	
	A	12 ± 1	750 ± 50	0.06	9.3×10^{-5}
G	G	7.7 ± 0.5	800 ± 10	0.10	1.5×10^{-4}
	T	4.2 ± 0.5	430 ± 8	0.10	1.5×10^{-4}
	C	2.9 ± 0.1	850 ± 10	0.85	
8-oxoG	A	300 ± 36	190 ± 8	6.1×10^{-4}	7.2×10^{-4}
	G	73 ± 9	0.15 ± 0.01	2.1×10^{-6}	2.5×10^{-4}
	T	120 ± 9	0.25 ± 0.01	8.3×10^{-6}	1.0×10^{-5}

3.4. Steady-State Kinetic Analysis of Next-Base Extension beyond G or 8-oxoG by Gp90 D234A

Th steady-state kinetic parameters (i.e., k_{cat} and K_m) for the next-base extension beyond G or 8-oxoG by Gp90 D234A (Table 3) were measured. The C or A at the 3'-end of primer was paired or mispaired with template G or 8-oxoG, respectively (Table 1). dGTP was incorporated opposite the next template base C. The incorporation efficiency was approximately 10-fold higher in extension beyond C:G (primer:template) than A:G because of the higher k_{cat}, but unchanged K_m values. For template 8-oxoG, both C:8-oxoG and A:8-oxoG base pairs were similarly extended by Gp90 D234A, but the efficiencies were reduced 490-fold compared with the extension beyond the C:G base pair because of the increased K_m values. Gp90 D234A preferentially extended the C:G base pair rather than the A:G mispair, but extended beyond C:8-oxoG or A:8-oxoG base pairs with the same efficiency.

Table 3. Steady-state kinetic parameters for next-base extension by Gp90 D234A.

Template Base	Primer X	$K_{m,dGTP}$ µM	k_{cat}, $\times 10^{-2}$min^{-1}	k_{cat}/K_m, µM^{-1}min^{-1}	Efficiency relative to G:C
	C	0.05 ± 0.01	49 ± 1	9.8	1
G	A	0.04 ± 0.01	5 ± 0.2	1.3	8-fold less
	T	0.05 ± 0.01	1 ± 0.1	0.2	49-fold less
8-oxoG	C	20 ± 2	40 ± 1	2.0×10^{-2}	490-fold less
	A	20 ± 2	46 ± 1	2.3×10^{-2}	490-fold less

3.5. Pre-Steady-State Kinetic Analysis of Single dNTP Incorporation by Gp90 D234A

Generally, incorporation of a correct dNTP by most DNA polymerases shows a biphasic character (burst phase and linear steady-state phase) [23]. In the burst phase, dNTP is quickly incorporated opposite the template base during the first binding of polymerase to DNA (the first turnover); in the linear steady-state phase, polymerase is dissociated from DNA, then binds to DNA and incorporates dNTP, all of which are limited by the slow dissociation of polymerase from DNA. The presence of biphasic shapes indicates that dNTP incorporation is much faster than the subsequent dissociation of polymerase from DNA.

Excess molar concentration of DNA, compared with polymerase, was used to determine the rates of dNTP incorporation opposite G or 8-oxoG by Gp90 D234A in burst and linear steady-state phases [19]. Among four dNTPs, dCTP was preferentially incorporated opposite G and showed a fast burst phase, indicating that dCTP incorporation was faster than the dissociation of Gp90 D234A from DNA (Figure 3A). For dCTP, dATP, or dTTP incorporation opposite G, the product and reaction time exhibited a linear steady-state phase, without a fast burst phase. dCTP or dATP was preferentially incorporated opposite 8-oxoG and exhibited a fast burst phase (Figure 3B). dGTP or dTTP incorporations opposite 8-oxoG showed a linear steady-state phase. Therefore, dCTP or dATP incorporation opposite 8-oxoG was faster than the dissociation of Gp90 D234A from DNA and dTTP or dGTP incorporation opposite 8-oxoG.

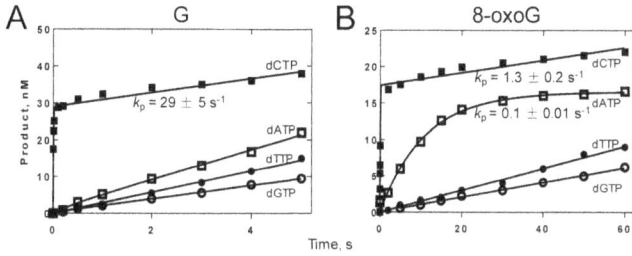

Figure 3. Pre-steady-state kinetic analysis of nucleotide incorporation by Gp90 D234A. Gp90 D234A (80 nM) was incubated with 100 nM ^{32}P-labeled 27-mer/62-mer primer/template containing G (**A**) or 8-oxoG (**B**), 1 mM each individual dNTP, and 30 mM Mg^{2+} in a RQF-3 KinTek quench flow apparatus as described in Materials and Methods. Representative data from multiple experiments are shown.

dCTP incorporation opposite G and dCTP or dATP incorporation opposite 8-oxoG exhibited fast burst phases. Moreover, the maximal rates of nucleotide incorporation (k_{pol}) and the apparent dissociation constants of dNTP from the Gp90-DNA-dNTP ternary complex ($K_{d,dNTP}$) were estimated by fitting the burst rates against dNTP concentrations to Equation (3) (Figure 4). k_{pol} of dCTP incorporation opposite G was 46 s^{-1} and $K_{d,dCTP}$ was 6 μM. For dCTP incorporation opposite 8-oxoG, k_{pol} was decreased by three-fold, $K_{d,dCTP}$ was increased by 20-fold, and total efficiency ($k_{pol}/K_{d,dCTP}$) was reduced by 59-fold compared with dCTP incorporation opposite G. The efficiency of dATP incorporation opposite 8-oxoG was further reduced by 188-fold compared with dCTP incorporation opposite 8-oxoG mainly because of the 115-fold reduction in k_{pol}.

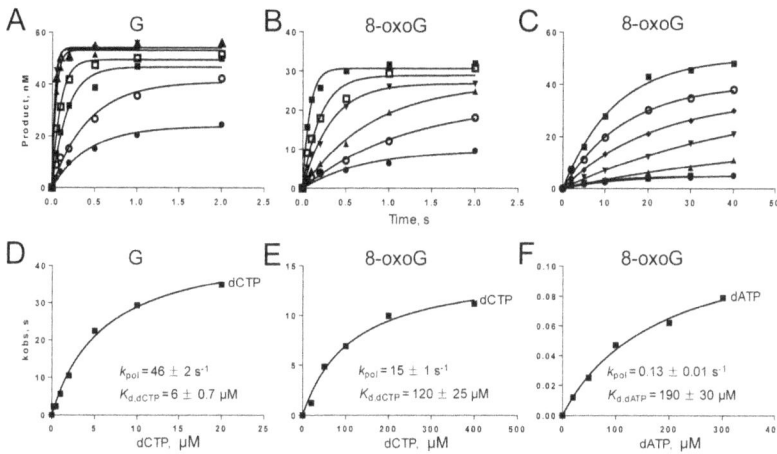

Figure 4. Pre-steady-state incorporation of a single dNTP opposite G or 8-oxoG by Gp90 D234A. Gp90 D234A (200 nM) incubated with 100 nM ^{32}P-labeled 27-mer/62-mer primer/template complexes was fast mixed with varying concentrations of dCTP to initiate reactions in a rapid quench-flow instrument. Plots of product concentrations versus time were fit to a single exponential equation to obtain k_{obs} at every dCTP concentration. Then, plots of burst rates (k_{obs}) versus dCTP concentrations were fit to a hyperbolic equation to obtain k_{pol} and K_d values. **A** and **D**, incorporation of dCTP opposite G. **B**, and **E**, incorporation of dCTP opposite 8-oxoG. **C** and **F**, incorporations of dATP opposite 8-oxoG. Representative data from multiple experiments are shown.

3.6. Binding of Gp90 D234A to the PrimerTemplate Containing 8-oxoG

The dissociation constants ($K_{d,DNA}$) between DNA and DNA polymerase were determined by surface plasmon resonance to determine whether 8-oxoG affects the binding affinity of Gp90 D234A to DNA [28]. DNA containing G or 8-oxoG (300 RU) was immobilized on the SA chip and different concentrations of polymerase were flowed over the chip to measure the binding of polymerase to DNA. $K_{d,DNA}$ was obtained by fitting the observed response signals against protein concentrations to Equation (4) using the steady-state model. In the absence of dNTP and Mg^{2+}, DNA polymerase randomly bound to DNA to form a binary complex. The binding of polymerase to DNA containing G or 8-oxoG exhibited a similar K_d of 108 and 116 nM, respectively (Figure 5), indicating that 8-oxoG does not affect the binding affinity of Gp90 D234A to DNA.

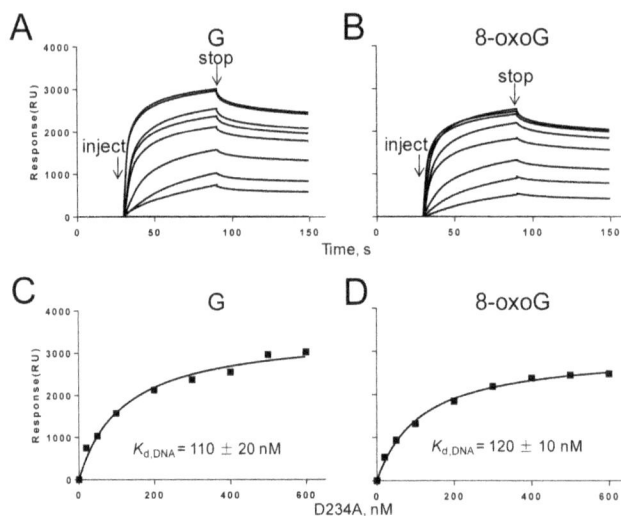

Figure 5. Biophysical binding of Gp90 D234A to DNA containing G or 8-oxoG in the absence of Mg^{2+} and dCTP. (**A,B**) Sensorgrams for binding of Gp90 D234A (20-600 nM) to DNA immobilized on the SA chip (300 RU) in buffer containing 40 mM Tris-HCl (pH 7.5), 10 mM DTT, and 50 mM potassium glutamate. (**C,D**) The binding affinities of Gp90 D234A to DNA were determined using the steady-state average response at each concentration of Gp90 D234A. The solid lines represent the theoretical curve calculated from the steady-state fit model (Biacore). Representative data from multiple experiments are shown.

In the presence of dNTP and Mg^{2+}, DNA polymerase, DNA, and dNTP formed a ternary complex, in which polymerase was preferentially positioned at the 3'-end of the primer strand. The $K_{d,DNA}$ values were significantly reduced compared with those without dNTP and Mg^{2+} (Figure 6). Therefore, the presence of dNTP and Mg^{2+} stabilized the binding of polymerase to DNA. Notably, $K_{d,DNA}$ of 8-oxoG complex (34 nM) was three-fold higher than that of the G complex (11 nM) in the presence of dCTP and Mg^{2+}, indicating that 8-oxoG in the template reduced the binding of Gp90 to DNA by three-fold. In the presence of dATP and Mg^{2+}, $K_{d,DNA}$ of 8-oxoG complex was further increased to 45 nM, showing that the presence of incorrect dATP further reduces the binding affinity of Gp90 D234A to DNA.

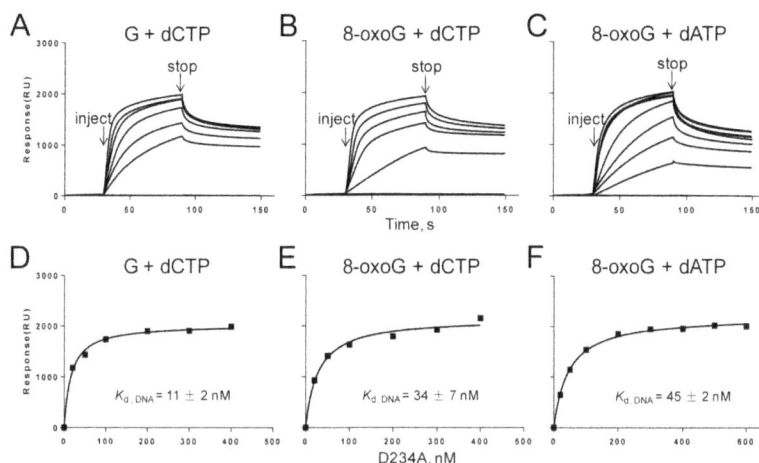

Figure 6. Biophysical binding of Gp90 D234A to DNA containing G or 8-oxoG in the presence of Mg^{2+} and dNTP. Scheme for measuring the interaction of Gp90 D234A with DNA that was prepared and immobilized onto the SA chip (300 RU). (**A–C**) Sensorgrams for the binding of Gp90 D234A (20–600 nM) to DNA immobilized on the SA chip (300 RU) in the presence of dCTP or dATP and Mg^{2+} in a buffer containing 40 mM Tris-HCl (pH 7.5), 10 mM DTT, and 50 mM potassium glutamate. (**D–F**) The binding affinities of Gp90 D234A to DNA were determined using the steady-state average response at each concentration of Gp90 D234A. The solid lines represent the theoretical curve calculated from the steady-state fit model (Biacore). Representative data from multiple experiments are shown.

4. Discussion

Gp90 is the sole DNA polymerase responsible for DNA replication in PaP1 [18]. The exonuclease activity of Gp90 should be eliminated to analyze 8-oxoG bypass by Gp90 kinetically. In our previous results, Glu-60, Asp-137, and Asp-234 were predicted as the exonuclease active residues of Gp90 [29]. Gp90 D234A, in which Asp-234 was replaced with Ala, can efficiently abolish the ssDNA and dsDNA exonuclease activities while maintaining its polymerase activity (Figure 1). Thus, Asp-234 should be a crucial residue for exonuclease activity. This residue corresponds to Asp-501 in *E. coli* DNA polymerase I and Asp-174 in T7 DNA polymerase based on sequence alignment analysis. Furthermore, Gp90 D234A was used as exonuclease-deficient DNA polymerase for kinetic analysis in this work.

Steady-state kinetic analysis dNTP incorporation opposite G or 8-oxoG provides information on the efficiency and accuracy of DNA replication [30]. Among four dNTPs, dCTP is preferentially incorporated opposite either G or 8-oxoG, yielding misincorporation frequencies of 10^{-4} to 10^{-5} for G and 8-oxoG (Table 2). Gp90 D234A can accurately bypass 8-oxoG without obvious dATP misincorporation. However, the incorporation efficiencies opposite 8-oxoG are significantly reduced. These results further confirm that 8-oxoG partially inhibits primer extension compared with G in the presence of four dNTPs.

Similar to Gp90 D234A, *S. solfataricus* DNA polymerase Dpo4 [12] and human DNA polymerase η [2] can also accurately bypass 8-oxoG without obvious dATP misincorporation, although the incorporation efficiencies are partially reduced. By contrast, T7 DNA polymerase [31], yeast DNA polymerase $η_{core}$ [24], human DNA polymerase ι [32] and human DNA polymerase β [15] lead to dCTP incorporation and obvious dATP misincorporation. Notably, human DNA polymerase κ [16] preferentially incorporates dATP opposite 8-oxoG and leads to G:C to T:A conversion.

Misincorporations opposite G inhibit next-base incorporation; whereas misincorporations opposite 8-oxoG do not affect next-base incorporation (Table 3). Gp90 D234A preferentially extends the

C:G base pair rather than the A:G mispair but shows the same priority in extension beyond C:8-oxoG or A:8-oxoG.

For most DNA polymerases, incorporation of a correct dNTP shows a biphasic character; whereas misincorporation shows only a linear steady-state phase. Unexpectedly, dCTP or dATP incorporation opposite 8-oxoG by Gp90 D234A shows a biphasic character (Figure 3), although dATP incorporation efficiency ($k_{pol}/K_{d,dATP}$) is lower than that of dCTP. Similarly, human Y-family DNA polymerase κ also shows a fast burst phase for dATP incorporation opposite 8-oxoG [16]. T7 DNA polymerase exhibits a similar burst rate for dCTP or dATP incorporation opposite 8-oxoG [31]. However, for most DNA polymerases, dATP misincorporation opposite 8-oxoG shows only a linear steady-state phase without a fast burst phase [9].

The physical binding of Gp90 D234A to DNA containing G or 8-oxoG was measured. In the absence of dNTP or Mg^{2+}, Gp90 D234A randomly binds to DNA to form a binary complex. The similar $K_{d,DNA}$ values (Figure 5) show that 8-oxoG in the template does not affect the binding of Gp90 D234A to DNA. In the presence of dNTP and Mg^{2+}, DNA polymerase is prone to bind at the 3′-end of the primer strand and form a ternary complex [22]. The lower $K_{d,DNA}$ values of ternary complexes compared with binary complexes indicate that the presence of dNTP and Mg^{2+} stabilizes the binding of polymerase to DNA for G and 8-oxoG. Notably, the $K_{d,DNA}$ value of 8-oxoG ternary complex was three-fold higher than that of G ternary complex in the presence of dCTP and Mg^{2+}, indicating that 8-oxoG in the template reduces the binding of Gp90 D234A to DNA (Figure 6). In the presence of dATP and Mg^{2+}, the misincorporation further weakens the binding affinity of Gp90 D234A to DNA.

Since the crystal structures of PaP1 DNA polymerase in complex with DNA containing 8-oxoG are not available, crystal structures of other DNA polymerases may provide insight in how Gp90 bypasses 8-oxoG. *S. solfataricus* DNA polymerase Dpo4 catalyzes 8-oxoG bypass efficiently and accurately. Crystal structures reveal the potential role of Arg332 in stabilizing the anti-conformation of 8-oxoG through the hydrogen bond or ion-dipole pair, which results in an increased enzymatic efficiency for dCTP insertion and a less favorable formation of a Hoogsteen pair between 8-oxoG and dATP [13]. DNA polymerase κ can bypass 8-oxoG in an error-prone manner by mainly inserting dATP. dATP:8-oxoG insertion events are two-fold more efficient than dCTP:G insertion events. Crystal structures of a complex of human Pol κ and DNA containing 8-oxoG show that the N-terminal extension of Pol κ stabilizes its little finger domain that surrounds the Hoogsteen base pair of 8-oxoG and incoming dATP, explaining the increase in efficiency for dATP incorporation opposite 8-oxoG [16].

5. Conclusions

In this work, we investigated the steady-state and pre-steady-state kinetics of nucleotide incorporation opposite G or 8-oxoG using Gp90 D234A, which has eliminated the ssDNA and dsDNA exonuclease activities. Among four dNTPs, dCTP was preferentially incorporated opposite G or 8-oxoG, exhibiting similar misincorporation frequencies of 10^{-4} to 10^{-5}. Misincorporation opposite G inhibits its subsequent extension, whereas misincorporation opposite 8-oxoG does not inhibit its subsequent extension. dCTP incorporation opposite G and dCTP or dATP incorporation opposite 8-oxoG show a fast burst phase, indicating that the incorporation step is faster than the subsequent dissociation of polymerase from DNA. The burst incorporation efficiency is decreased in the following order: dCTP:G > dCTP:8-oxoG > dATP:8-oxoG. 8-oxoG in the template does not affect the binding of Gp90 D234A to DNA in the binary complex, in contrast to that in the ternary complex in the presence of dCTP and Mg^{2+}. dATP misincorporation opposite 8-oxoG further weakens the binding affinity of Gp90 D234A to DNA. This study reveals that Gp90 D234A can ensure error-free bypass of 8-oxoG, providing an understanding of the DNA replication while encountering oxidation lesions for *P. aeruginosa* phage PaP1.

Acknowledgments: We thank Natural Science Foundation of China (NSFC Nos. 31370793 and 814220410) and the Youth 1000 Talent Plan to Huidong Zhang for supporting this work.

Author Contributions: Shiling Gu, Huidong Zhang and Wanneng Wang conceived and designed the experiments. Shiling Gu performed the experiments and analyzed the data. Qizhen Xue, Qin Liu and Mei Xiong contributed reagents and analysis. Shiling Gu and Huidong Zhang wrote the paper.

Conflicts of Interest: The authors declared no potential conflict of interest with respect to the research, authorship, and/or publication of this article.

References

1. Hogg, M.; Rudnicki, J.; Midkiff, J.; Reha-Krantz, L.; Doublie, S.; Wallace, S.S. Kinetics of mismatch formation opposite lesions by the replicative DNA polymerase from bacteriophage RB69. *Biochemistry* **2010**, *49*, 2317–2325. [CrossRef] [PubMed]

2. Patra, A.; Nagy, L.D.; Zhang, Q.; Su, Y.; Muller, L.; Guengerich, F.P.; Egli, M. Kinetics, structure, and mechanism of 8-oxo-7,8-dihydro-2′-deoxyguanosine bypass by human DNA polymerase eta. *J. Biol. Chem.* **2014**, *289*, 16867–16882. [CrossRef] [PubMed]

3. Taggart, D.J.; Fredrickson, S.W.; Gadkari, V.V.; Suo, Z. Mutagenic potential of 8-oxo-7,8-dihydro-2′-deoxyguanosine bypass catalyzed by human Y-family DNA polymerases. *Chem. Res. Toxicol.* **2014**, *27*, 931–940. [CrossRef] [PubMed]

4. Yin, Y.; Sasaki, S.; Taniguchi, Y. Effects of 8-halo-7-deaza-2′-deoxyguanosine triphosphate on DNA synthesis by DNA polymerases and cell proliferation. *Bioorg. Med. Chem.* **2016**, *24*, 3856–3861. [CrossRef] [PubMed]

5. Beckman, J.; Wang, M.; Blaha, G.; Wang, J.; Konigsberg, W.H. Substitution of Ala for Tyr567 in RB69 DNA polymerase allows dAMP to be inserted opposite 7,8-dihydro-8-oxoguanine. *Biochemistry* **2010**, *49*, 4116–4125. [CrossRef] [PubMed]

6. Schneider, S.; Schorr, S.; Carell, T. Crystal structure analysis of DNA lesion repair and tolerance mechanisms. *Curr. Opin. Struct. Biol.* **2009**, *19*, 87–95. [CrossRef] [PubMed]

7. Vooradi, V.; Romano, L.J. Effect of N-2-Acetylaminofluorene and 2-Aminofluorene Adducts on DNA Binding and Synthesis by Yeast DNA Polymerase η. *Biochemistry* **2009**, *48*, 4209–4216. [CrossRef] [PubMed]

8. Swanson, A.L.; Wang, J.; Wang, Y. In Vitro Replication Studies of Carboxymethylated DNA Lesions with *Saccharomyces cerevisiae* Polymerase η. *Biochemistry* **2011**, *50*, 7666–7673. [CrossRef] [PubMed]

9. Liu, B.; Xue, Q.; Tang, Y.; Cao, J.; Guengerich, F.P.; Zhang, H. Mechanisms of mutagenesis: DNA replication in the presence of DNA damage. *Mutat. Res. Rev. Mutat. Res.* **2016**, *768*, 53–67. [CrossRef] [PubMed]

10. Freudenthal, B.D.; Beard, W.A.; Wilson, S.H. DNA polymerase minor groove interactions modulate mutagenic bypass of a templating 8-oxoguanine lesion. *Nucleic Acids Res.* **2013**, *41*, 1848–1858. [CrossRef] [PubMed]

11. Patro, J.N.; Urban, M.; Kuchta, R.D. Interaction of Human DNA Polymerase α and DNA Polymerase I from *Bacillus stearothermophilus* with Hypoxanthine and 8-oxoguanine Nucleotides. *Biochemistry* **2009**, *48*, 8271–8278. [CrossRef] [PubMed]

12. Maxwell, B.A.; Suo, Z. Kinetic basis for the differing response to an oxidative lesion by a replicative and a lesion bypass DNA polymerase from *Sulfolobus solfataricus*. *Biochemistry* **2012**, *51*, 3485–3496. [CrossRef] [PubMed]

13. Eoff, R.L.; Irimia, A.; Angel, K.C.; Egli, M.; Guengerich, F.P. Hydrogen bonding of 7,8-dihydro-8-oxodeoxyguanosine with a charged residue in the little finger domain determines miscoding events in *Sulfolobus solfataricus* DNA polymerase Dpo4. *J. Biol. Chem.* **2007**, *282*, 19831–19843. [CrossRef] [PubMed]

14. Kim, J.; Song, I.; Jo, A.; Shin, J.H.; Cho, H.; Eoff, R.L.; Guengerich, F.P.; Choi, J.Y. Biochemical analysis of six genetic variants of error-prone human DNA polymerase iota involved in translesion DNA synthesis. *Chem. Res. Toxicol.* **2014**, *27*, 1837–1852. [CrossRef] [PubMed]

15. Brown, J.A.; Duym, W.W.; Fowler, J.D.; Suo, Z. Single-turnover kinetic analysis of the mutagenic potential of 8-oxo-7,8-dihydro-2′-deoxyguanosine during gap-filling synthesis catalyzed by human DNA polymerases lambda and beta. *J. Mol. Biol.* **2007**, *367*, 1258–1269. [CrossRef] [PubMed]

16. Irimia, A.; Eoff, R.L.; Guengerich, F.P.; Egli, M. Structural and functional elucidation of the mechanism promoting error-prone synthesis by human DNA polymerase kappa opposite the 7,8-dihydro-8-oxo-2′-deoxyguanosine adduct. *J. Biol. Chem.* **2009**, *284*, 22467–22480. [CrossRef] [PubMed]

17. Liu, B.; Gu, S.; Liang, N.; Xiong, M.; Xue, Q.; Lu, S.; Hu, F.; Zhang, H. Pseudomonas aeruginosa phage PaP1 DNA polymerase is an A-family DNA polymerase demonstrating ssDNA and dsDNA 3′–5′ exonuclease activity. *Virus Genes* **2016**, *52*, 538–551. [CrossRef] [PubMed]

18. Lu, S.; Le, S.; Tan, Y.; Zhu, J.; Li, M.; Rao, X.; Zou, L.; Li, S.; Wang, J.; Jin, X. Genomic and proteomic analyses of the terminally redundant genome of the *Pseudomonas aeruginosa* phage PaP1: Establishment of genus PaP1-like phages. *PLoS ONE* **2013**, *8*, e62933. [CrossRef] [PubMed]

19. Zhang, H.; Guengerich, F.P. Effect of N 2 -Guanyl Modifications on Early Steps in Catalysis of Polymerization by *Sulfolobus solfataricus* P2 DNA Polymerase Dpo4 T239W. *J. Mol. Biol.* **2010**, *395*, 1007–1018. [CrossRef] [PubMed]

20. Washington, M.T.; Wolfle, W.T.; Spratt, T.E.; Prakash, L.; Prakash, S. Yeast DNA polymerase eta makes functional contacts with the DNA minor groove only at the incoming nucleoside triphosphate. *Proc. Natl. Acad. Sci. USA* **2003**, *100*, 5113–5118. [CrossRef] [PubMed]

21. Lu, S.; Le, S.; Tan, Y.; Li, M.; Liu, C.; Zhang, K.; Huang, J.; Chen, H.; Rao, X.; Zhu, J. Unlocking the mystery of the hard-to-sequence phage genome: PaP1 methylome and bacterial immunity. *BMC Genom.* **2014**, *15*, 803. [CrossRef] [PubMed]

22. Zhang, H.; Yong, T.; Seung-Joo, L.; Wei, Z.; Jia, C.; Richardson, C.C. Helicase-DNA polymerase interaction is critical to initiate leading-strand DNA synthesis. *Proc. Natl. Acad. Sci. USA* **2011**, *108*, 9372–9377. [CrossRef] [PubMed]

23. Yang, J.; Rong, W.; Liu, B.; Xue, Q.; Zhong, M.; Hao, Z.; Zhang, H. Kinetic analysis of bypass of abasic site by the catalytic core of yeast DNA polymerase eta. *Mutat. Res.* **2015**, *779*, 134–143. [CrossRef] [PubMed]

24. Xue, Q.; Zhong, M.; Liu, B.; Tang, Y.; Wei, Z.; Guengerich, F.P.; Zhang, H. Kinetic analysis of bypass of 7,8-dihydro-8-oxo-2′-deoxyguanosine by the catalytic core of yeast DNA polymerase η. *Biochimie* **2016**, *121*, 161–169. [CrossRef] [PubMed]

25. Zhang, H.; Eoff, R.L.; Kozekov, I.D.; Rizzo, C.J.; Egli, M.; Guengerich, F.P. Versatility of Y-family *Sulfolobus solfataricus* DNA Polymerase Dpo4 in Translesion Synthesis Past Bulky N2-Alkylguanine Adducts. *J. Biol. Chem.* **2008**, *284*, 3563–3576. [CrossRef] [PubMed]

26. Zhang, H.; Eoff, R.L.; Kozekov, I.D.; Rizzo, C.J.; Egli, M.; Guengerich, F.P. Structure-Function Relationships in Miscoding by *Sulfolobus solfataricus* DNA Polymerase Dpo4: Guanine N2,N2-dimethyl substitution produces inactive and miscoding polymerase complexes. *J. Biol. Chem.* **2009**, *284*, 17687–17699. [CrossRef] [PubMed]

27. Zhang, H.; Bren, U.; Kozekov, I.D.; Rizzo, C.J.; Stec, D.F.; Guengerich, F.P. Steric and Electrostatic Effects at the C2 Atom Substituent Influence Replication and Miscoding of the DNA Deamination Product Deoxyxanthosine and Analogs by DNA Polymerases. *J. Mol. Biol.* **2009**, *392*, 251–269. [CrossRef] [PubMed]

28. Zhang, H.; Tang, Y.; Lee, S.-J.; Wei, Z.; Cao, J.; Richardson, C.C. Binding Affinities among DNA Helicase-Primase, DNA Polymerase, and Replication Intermediates in the Replisome of Bacteriophage T7. *J. Biol. Chem.* **2016**, *291*, 1472–1480. [CrossRef] [PubMed]

29. Liu, B.; Xue, Q.; Gu, S.; Wang, W.; Chen, J.; Li, Y.; Wang, C.; Zhang, H. Kinetic analysis of bypass of O(6)-methylguanine by the catalytic core of yeast DNA polymerase eta. *Arch. Biochem. Biophys.* **2016**, *596*, 99–107. [CrossRef] [PubMed]

30. Carlson, K.D.; Washington, M.T. Mechanism of efficient and accurate nucleotide incorporation opposite 7,8-dihydro-8-oxoguanine by *Saccharomyces cerevisiae* DNA polymerase eta. *Mol. Cell. Biol.* **2005**, *25*, 2169–2176. [CrossRef] [PubMed]

31. Katafuchi, A.; Nohmi, T. DNA polymerases involved in the incorporation of oxidized nucleotides into DNA: Their efficiency and template base preference. *Mutat. Res. Genet. Toxicol. Environ. Mutagen.* **2010**, *703*, 24–31. [CrossRef] [PubMed]

32. Taggart, D.J.; Dayeh, D.M.; Fredrickson, S.W.; Suo, Z. N-terminal domains of human DNA polymerase lambda promote primer realignment during translesion DNA synthesis. *DNA Repair* **2014**, *22*, 41–52. [CrossRef] [PubMed]

Erratum

Erratum: Gu, S. et al. Error-Free Bypass of 7,8-dihydro-8-oxo-2'-deoxyguanosine by DNA Polymerase of *Pseudomonas aeruginosa* Phage PaP1. *Genes* 2017, 8, 18

Shiling Gu [1,2], Qizhen Xue [2], Qin Liu [1], Mei Xiong [2], Wanneng Wang [1,*] and Huidong Zhang [2,*]

[1] College of Pharmacy and Bioengineering, Chongqing University of Technology,
     No. 69 Hongguang Street, Banan District, Chongqing 400054, China;
     shilinggu@foxmail.com (S.G.); liuqin@2014.cqut.edu.cn (Q.L.)
[2] Public Health Laboratory Sciences and Toxicology, West China School of Public Health,
     Sichuan University, No. 17 People's South Road, Chengdu 610041, China;
     xueqzhen@163.com (Q.X.); xhmei2016@126.com (M.X.)
*   Correspondence: huidong.zhang@foxmail.com (H.Z.); wannengw@cqut.cn (W.W.)

Received: 27 February 2017; Accepted: 27 February 2017; Published: 3 March 2017

The authors wish to make the following correction to their paper [1]. The title of the paper should be corrected to "Error-Free Bypass of 7,8-dihydro-8-oxo-2'-deoxyguanosine by DNA Polymerase of *Pseudomonas aeruginosa* Phage PaP1". Additionally, the affiliation 1 should be corrected to "College of Pharmacy and Bioengineering, Chongqing University of Technology, No. 69 Hongguang Street, Banan District, Chongqing 400054, China". The authors would like to apologize for any inconvenience caused. The change does not affect the scientific results. The manuscript will be updated and the original will remain online on the article webpage.

1. Gu, S.; Xue, Q.; Liu, Q.; Xiong, M.; Wang, W.; Zhang, H. Error-Free Bypass of 7,8-dihydro-8-oxo-2'-deoxyguanosine by DNA Polymerase of *Pseudomonas aeruginosa* Phage PaP1. *Genes* **2017**, *8*, 18. [CrossRef] [PubMed]

MDPI AG

St. Alban-Anlage 66

4052 Basel, Switzerland

Tel. +41 61 683 77 34

Fax +41 61 302 89 18

http://www.mdpi.com

Genes Editorial Office

E-mail: genes@mdpi.com

http://www.mdpi.com/journal/genes